Encyclopedia of Plant Physiology

New Series Volume 2 Part A

Editors

A. Pirson, Göttingen

M. H. Zimmermann, Harvard

Transport in Plants II

Part A Cells

Edited by

U. Lüttge and M. G. Pitman

Contributors

W. J. Cram J. Dainty G. P. Findlay T. K. Hodges
A. B. Hope D. H. Jennings U. Lüttge C. B. Osmond
M. G. Pitman R. J. Poole J. A. Raven F. A. Smith
N. A. Walker

With a Foreword by R. N. Robertson

Springer-Verlag Berlin Heidelberg New York 1976

With 97 Figures

ISBN-13: 978-3-642-66229-4 e-ISBN-13: 978-3-642-66227-0
DOI: 10.1007/978-3-642-66227-0

Library of Congress Cataloging in Publication Data. Main entry under title: Transport in plants. (Encyclopedia of plant physiology; v. Vol. 2, pt. A & B edited by U. Lüttge and M.G. Pitman. Bibliography: p. . Includes indexes. Contents: v. 1. Phloem transport. — v. 2. pt. A. Cells. pt. B. Tissues and organs. 1. Plant translocation. I. Canny, M.J. II. Zimmermann, Martin Huldrych, 1926– . III. Lüttge, Ulrich. QK871.T73 582'.041 75-20178

Softcover reprind of the hardcover 1st edition 1976

Foreword

As plant physiology increased steadily in the latter half of the 19th century, problems of absorption and transport of water and of mineral nutrients and problems of the passage of metabolites from one cell to another were investigated, especially in Germany. JUSTUS VON LIEBIG, who was born in Darmstadt in 1803, founded agricultural chemistry and developed the techniques of mineral nutrition in agriculture during the 70 years of his life. The discovery of plasmolysis by NÄGELI (1851), the investigation of permeability problems of artificial membranes by TRAUBE (1867) and the classical work on osmosis by PFEFFER (1877) laid the foundations for our understanding of soluble substances and osmosis in cell growth and cell mechanisms. Since living membranes were responsible for controlling both water movement and the substances in solution, "permeability" became a major topic for investigation and speculation. The problems then discussed under that heading included passive permeation by diffusion, Donnan equilibrium adjustments, active transport processes and antagonism between ions. In that era, when organelle isolation by differential centrifugation was unknown and the electron microscope had not been invented, the number of cell membranes, their thickness and their composition, were matters for conjecture. The nature of cell surface membranes was deduced with remarkable accuracy from the reactions of cells to substances in solution. In 1895, OVERTON, in U.S.A., published the hypothesis that membranes were probably lipid in nature because of the greater penetration by substances with higher fat solubility. However, in the early years of this century, RUHLAND in Germany, impressed by the way in which small molecules penetrated relatively much more quickly than slightly larger ones, suggested that the membranes must have sieve-like properties. The lipid and the sieve hypotheses rivalled each other until the 1930s when COLLANDER of Finland, with experimental evidence of his own, brought them together suggesting that the membranes were best described as a lipid-sieve.

Meanwhile, the obvious importance of the absorption of mineral nutrients led to investigations of the way in which they entered the cells. Thus, in the U.S.A., HOAGLAND established a flourishing school of mineral nutrition of plants and collected a great variety of talent in that school. Roots and single cells were the subjects of investigation. LUNDEGÅRDH, in Sweden, a distinguished autecologist with excellent experimental techniques, became increasingly interested in how the roots absorbed ions. Other investigators, trying to elucidate the fundamental physiology of cells, sought material which seemed to be suitable for the techniques then available. Thus STILES and KIDD, and later STEWARD, in England, and PETRIE in Australia, saw the value of slices of tissue. Others saw the advantages of the giant cells, so OSTERHOUT, in the U.S.A. began the work on *Chara* and on the marine alga *Valonia*. Some investigators realized the value of specialized cells such as the salt glands which were investigated by RUHLAND in Germany and ARISZ in Holland.

Techniques which could be used in the early part of this century were very limited compared with those available today; STRUGGER and SCHUMACHER in Germany and BROOKS in U.S.A. attempted deduction from experiments with the "vital" stains., i.e. those which stained the cells without killing them. HÖFLER, in Austria, compared rates of penetration by timing the deplasmolysis as the solutes, which had caused the plasmolysis, entered the cells. In those pre-isotope days, the uptakes of various inorganic ions and organic molecules were deduced from tedious chemical analysis, sometimes of the tissue, sometimes of the external solution. Ions with properties analagous to those of naturally occurring ions, e.g. Rb^+ for K^+ and Br^- for Cl^- were also used. When uptake of a single salt was being investigated, electrical conductivity of the external solution was a useful alternative.

The beginnings of the biophysical approach were apparent early in the century; electrophysiology had been used by OSTERHOUT, BLINKS, BROOKS, GELFAN and others in the U.S.A. and UMRATH in Austria was implanting electrodes in cells by 1930. The remarkable development of plant biophysics under MACAULEY in the University of Tasmania had also begun.

By 1930 some of the problems were clearly defined. At that time MILLER wrote in his textbook:

"Since the penetration of ions into cells ... apparently takes
place from a solution of low to a solution of high concentration
of these ions, the plant cell must do work in absorbing them ...
Just how such a process is brought about is not known ..."

During the next decade, various workers (STEWARD, LUNDEGÅRDH, ROBERTSON) investigated the relation between respiration and accumulation of ions (as active transport in plant cells was then termed) but no definitive hypothesis emerged. It was increasingly clear that some chemical combination was probably involved in the uptake and VAN DEN HONERT is credited with being the first person to state this clearly, in 1937, in relation to phosphate absorption by roots—the forerunner of the carrier hypothesis.

In addition to the limitations of techniques, progress before World War II was slow for other reasons. Individual workers had few research students or assistants associated with them and worked in isolation in their own countries. Transport by rail and ship was slow and opportunities for attendance at international meetings were limited. Some students did move from one country to another but travelling scholarships or fellowships were few in number. The result was that hypotheses tended to be developed in isolation and were not rapidly tested in other places. World War II decreased research and interrupted communication between scientists with these interests but, after the war, changes were rapid. In modern times in peaceful areas of the world, many countries foster exchanges of research workers with those from other countries, ideas are developed simultaneously and hypotheses are tested in many places. Frequent meetings and specialist conferences stimulate rapid development of ideas, exchanges of techniques and collaboration of all kinds.

I am pleased to have been asked to write the Foreword to this Volume of the Encyclopedia of Plant Physiology and, as a veteran with nearly 40 years' association with the subject, to comment on its history. My own experience in the field of active transport of electrolytes started just before World War II, and I have watched the developments which led to this volume in which the countries represented are Australia, Canada, Germany, the United Kingdom and the United States of America, and 19 of the 25 contributors have been associated with Cambridge or with Australia. About 50 years ago, an Australian, PETRIE, investigated the absorption of ions by carrot tissue slices and published his results in 1927. By that time PETRIE had become a research student at the University of Cambridge working with BRIGGS who thus became interested in the problem of ion absorption; so began the long collaboration, reflected in this volume, between Australian and Cambridge plant physiologists. Ten years later, I was the second Australian research student to work with BRIGGS in this field. Unlike PETRIE, who turned his attention to plant nutrition and growth, I continued research on ion absorption on my return to Australia. In Cambridge BRIGGS, with various students and collaborators, carried out further work on ion absorption, his own research and writing continuing since his retirement 14 years ago. In the post-war years a development of great importance was the establishment of the strong biophysical group under DAINTY, first in the University of Edinburgh and later in the University of East Anglia, and now at the University of Toronto. Cambridge and Edinburgh backgrounds came together when MACROBBIE moved to the Cambridge Botany School, developing further the interest in plant biophysics there. These converging chains of events account for much of the background to contributions in this Volume, showing the continuing interchange between Australia and the United Kingdom and the widening collaboration with other countries, especially Germany.

In a short Foreword it is not possible to mention all the names or survey all the historical events which, particularly since 1945, have led to the participation of the distinguished contributors in this volume. To give any idea of the work which has made the volume possible would make the Foreword as long as the rest of the book. It is worth repeating that so much of our increased understanding has been due to the development of new techniques. Tracers, for example, enabled us to measure both influx and efflux when previously only net flux could be measured (Part A, Chap. *5*). After USSING's definition of active transport (movement of an ion against its electrochemical potential gradient) was adopted, isotope techniques, coupled with the use of micro-electrodes to measure electrical potential differences (Part A, Chap. *4*) allowed us to characterize active transport for each ion. Electron microscopy, particularly with the technique for section cutting, defined where different organelles are and what membranes are associated with them (Part A, Chap. *1*). More recently it has been possible to measure intra-cellular concentrations by ion-selective electrodes and by electron microprobe. New techniques in biochemistry have resulted not only in better understanding of the reactions involved but also in better characterization of the solutes actually transported. Interpretation of biophysical measurements has been given a sound theoretical basis by the application of irreversible thermodynamics to active transport, particularly by KATCHALSKY.

It is about 20 years since the corresponding reviews were published in the volumes of the former series of the Encyclopedia. Appropriately, reflecting our increased knowledge, three Volumes of the present series deal with problems of solute uptake and transport. Volume 1 (edited by ZIMMERMANN and MILBURN) deals with long-distance transport in plants, this, Volume 2 (edited by LÜTTGE and PITMAN) is concerned with transport at the cellular level (Part A) and at the organ level (Part B), and Volume 3 (edited by HEBER and STOCKING) will deal with sub-cellular aspects of transport.

Volume 2 highlights the diversity of transport systems. At one end of the scale, simple unicellular organisms, normally submerged in an aqueous environment, have need for transport over short distances within the cell and across the cell surfaces. At the other end of the scale, complex multicellular plants, often very large, have roots in the soil where cellular transport is very different from that of cells in the leaves. Some plants contain specialized transport organs such as glands and some have transport problems between host and symbiont and between host and parasite. The whole evolutionary story of specialized transport mechanisms will be fascinating when it is understood.

I am sure that the Editors and Contributors to this volume hope it will be a stimulus to further satisfying hypotheses and to even deeper understanding; I believe it will be a success.

R.N. Robertson

Introduction

This volume (in two Parts A and B) is one of three volumes on transport in the *Encyclopedia of Plant Physiology, New Series*. Its subject matter bridges the topics of the other two volumes which deal with long-distance transport in the phloem (Vol. 1) and with transport processes within the cell (Vol. 3). Transport "in cells, tissues and organs" has biological characteristics different from the main themes of the other volumes. We felt justified in collecting material under the heading of cells, tissues and organs to point out generalizations in the work, even though in some cases processes are involved which are also topics in the other two volumes (e.g. energetics or membrane transport or loading of long distance transport pathways).

Transport in certain tissues shows properties different from transport into single cells, due to aspects of cell to cell organization in the tissue. Within such a tissue there often appears to be symplasmic continuity, and transport from cell to cell or across the tissue appears to be facilitated by interconnections of the cytoplasms of adjacent cells. Further, there may be structural adaptations of the apoplast that can give direction to transport across the tissue, or act as barriers to diffusion outside the symplast, as found for example in roots, glands and stomata. The properties of such systems and the possible control of transport in tissues and organs are treated in Part B of this volume. However, transport in tissues necessarily involves the movement of molecules across cell membranes at points of entry to the symplast and much has been learnt about these processes from study of membranes and transport in simple cells and parenchymatous tissues (storage tissues); this is the topic of Part A of this volume.

We have organized Part A into three sections. Section I aims to cover basic ideas used in experimentation and discussion of transport at the cellular level. These topics are membrane structure; biophysics of transport; measurement of electrical properties of membranes; and measurement of fluxes across plasmalemma and tonoplast.

In Section II experimental results are given for plants in which the transport studied is predominantly at the plasmalemma and tonoplast, or where symplasmic transport is negligible. Examples are taken from work on algae, fungi and certain higher plant storage tissues which contain mainly parenchyma cells.

The operation of many transport processes is linked to the cell's metabolism (Section III) by operation of specific carriers. Metabolism can be involved too in production of organic acid anions that contribute to maintenance of charge balance in the vacuoles, and in the excretion from the cells of H^+ or OH^- ions. It is remarkable how many transport processes seem to involve movement of protons into or out of the cell.

We thank the publishers for their support in this project and for attending to the tedious problem of proof production. One of the Series Editors, Professor

PIRSON, has given us valuable advice on organization of the material and shown continuing interest in the work, as well as understanding the problems facing us as Editors (such as acceptance by the Series Editors and the publishers of our request for extra space to include the amount of material we thought necessary!).

Many friends and colleagues have given encouragement and advice. To pick out individuals is not easy but special thanks are due to NOE HIGINBOTHAM, HEINZ KAUSS, ENID MACROBBIE and ULRICH ZIMMERMANN for their assistance at various stages. Colleagues in our departments in Darmstadt and Sydney have carried a heavy load of discussion and reading, particularly ANDRÉ LÄUCHLI, JOHN CRAM and ALAN WALKER. In Sydney, we are grateful to Mrs. S. TITT for secretarial assistance and in Darmstadt to GÜNTHER FINDENEGG, WOLFRAM ULLRICH and CORNELIA ULLRICH-EBERIUS for review and technical help, and to Frau T. KRIEGER for secretarial assistance.

Finally we thank the authors of this volume for the patience they have shown the Editors, who made an outrageous number of demands on authors, for reorganization and re-writing of their own chapters and for cross-reference and reviewing of related chapters. We enjoyed collaborating with the authors and have learnt much about ion transport in the process.

The organization of this volume developed from discussions at the Ion Transport Workshops at Liverpool and at Jülich. The Editors' job was made much easier by the friendliness and cooperation between members of this international group of workers in ion transport.

January 1976

ULRICH LÜTTGE
Darmstadt

MICHAEL PITMAN
Sydney

Instructions for the Reader

In using this book note that references to chapters in Part A are quoted as "Chap. *6*" whereas chapters in Part B are quoted as "Part B, Chap. *6*". Similarly parts of chapters are referred to as "*6*.4.2" if in Part A or as "Part B, *6*.4.2" if in Part B. In each case the italic number refers to the *Chapter* number and any roman numbers to the *part* within the chapter. The same conventions are used for Figures and Tables. "Fig. *5*.1" is Figure 1 in Chapter *5* in Part A; "Part B, Table *7*.2" refers to Table 2 in Chapter *7* of Part B. Chapters *3*, *4* and *5* in Part B have been subdivided, and in this Part (A) they are referred to as "Part B, Chap. *3.3*" and "Part B, *3.3*.4.1", respectively, depending on whether reference is made to the whole or part of the subchapter. The index covers both Parts A and B; italic numbers refer to pages in Part A and ordinary (roman) script to pages in Part B.

Contents of Part A

I. Theoretical and Biophysical Approaches

Survey of Part B

List of Contributors (Part A)

W.J. CRAM
School of Biological Sciences (A12),
University of Sydney, Sydney,
N.S.W. 2006/Australia

J. DAINTY
Department of Botany, University
of Toronto, Toronto M5S 1A1/Canada

G.P. FINDLAY
School of Biological Sciences,
The Flinders University of South
Australia, Bedford Park, S.A.
5042/Australia

T.K. HODGES
Department of Botany and Plant
Pathology, Purdue University,
West Lafayette, IN 47907/USA

A.B. HOPE
School of Biological Sciences,
The Flinders University of South
Australia, Bedford Park, South
Australia 5042/Australia

D.H. JENNINGS
Department of Botany, The University,
Liverpool L69 3BX/Great Britain

U. LÜTTGE
Botanisches Institut, Technische
Hochschule, Schnittspahnstr. 3—5,
6100 Darmstadt/Federal Republic of
Germany

C.B. OSMOND
Department of Environmental Biology,
Research School of Biological Sciences,
Australian National University,
Canberra City 2601/Australia

M.G. PITMAN
School of Biological Sciences (A 12),
University of Sydney,
Sydney, N.S.W. 2006/Australia

R.J. POOLE
Department of Biology,
McGill University, P.O. Box 6070,
Montreal H3C 3G1/Canada

J.A. RAVEN
Department of Biological Sciences,
University of Dundee, Dundee DD1 4HN/
Great Britain

Sir RUTHERFORD N. ROBERTSON
Director, Research School of
Biological Sciences,
Australian National University,
Canberra City, A.C.T. 2601/Australia

F.A. SMITH
Department of Botany, University of
Adelaide, Adelaide, S.A. 5001/Australia

N.A. WALKER
School of Biological Sciences (A 12),
University of Sydney, Sydney,
N.S.W. 2006/Australia

I. Theoretical and Biophysical Approaches

1. The Structure of Biological Membranes

N.A. WALKER

1. Introduction

When the previous edition of this handbook appeared, most if not all biologists accepted that the structure of biological membranes was probably the lipid-protein sandwich suggested by DANIELLI and DAVSON (1935) (Fig. *1*.1). This developed into the version shown in Fig. *1*.2, which became the universal model of membrane structure, repeated in text after text, challenged by some specialists but accepted by everyone else.

The current picture of membrane structure (Fig. *1*.3), first stated by VANDERKOOI (1972) and by SINGER (1972), was brought to general attention by SINGER and NICOLSON (1972). It retains one central feature of the older model, but differs in important ways. These differences are becoming more and more relevant to the working physiologist as explanation at the level of molecular structure becomes within his grasp.

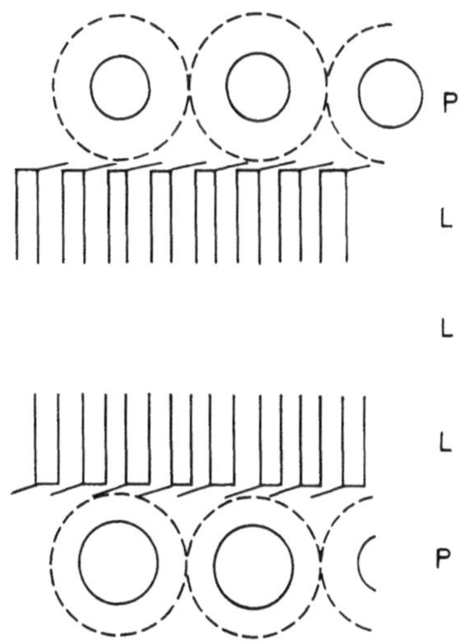

Fig. *1*.1. The membrane structure originally proposed by DANIELLI and DAVSON (1935), redrawn. *P* protein. *L* lipid

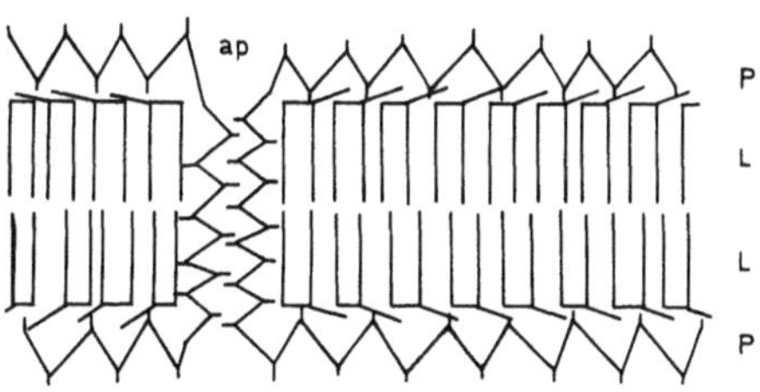

Fig. *1*.2. The membrane structure developed from that of Fig. *1*.1, and representative of a number of similar "sandwich" models. *P* protein, *L* lipid, *ap* aqueous pore

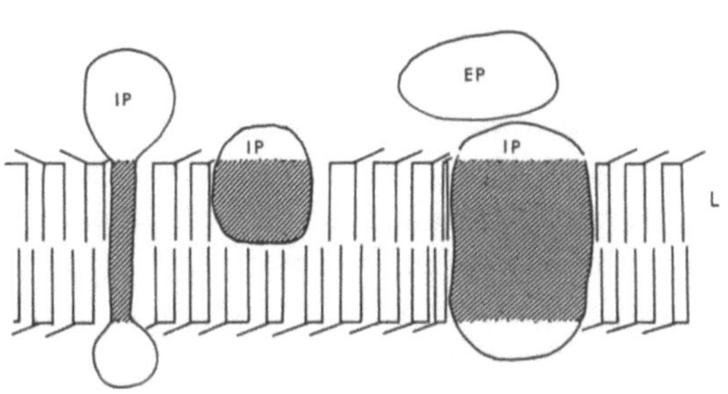

Fig. *1*.3. The current membrane model, based on the postulates of SINGER (1972) and of VANDERKOOI (1972, 1974). *EP* extrinsic protein, *IP* intrinsic protein, *L* lipid

This Chapter does not set out to be a comprehensive review of the field of membrane structure; but it is hoped to mention enough to show the importance of the subject to the physiological worker.

2. The Lipid-Protein Sandwich

2.1 The Danielli-Davson Model

The essential features of this model, as most people accepted it, were

(i) that it contained a continuous lipid layer which separated the two aqueous phases,

(ii) that the lipid layer was in fact a bilayer of amphiphilic lipid molecules such as lecithin and cholesterol,

(iii) that the layer was penetrated by aqueous pores, whose existence depended on protein which lined them (Fig. *1*.2), and

(iv) that the lipid bilayer itself was stabilized by a coating, on each face, of protein in the pleated sheet or β-configuration.

Since the studies of OVERTON it had been clear that lipophilic molecules generally penetrated living cells more rapidly than similar hydrophilic molecules; and that in a series of homologs the addition of —CH_2— increased permeating power while the addition of —OH, —NH_2 and ionized groups reduced it. COLLANDER's important work (e.g. 1954) on *Chara, Nitella* and *Nitellopsis* confirmed the connexion between oil/water partition and permeability and made it quantitative. It was, in essence, to accommodate this evidence that the DANIELLI-DAVSON model incorporated its lipids in a continuous film. That the film was generally supposed to be a bilayer followed first from the early study of GORTER and GRENDEL (1925), who showed that red cell ghosts contained enough lipids to form a close-packed monolayer of twice their own area. The bilayer is also the simplest way to imagine a film of amphiphilic lipid molecules to be arranged in an aqueous environment. Although DANIELLI and DAVSON originally envisaged that further lipid molecules might form a layer in the interior, no suitable, purely hydrophobic, molecules are known to be normal constituents of biological membranes. The classical work of FINEAN (1956), who investigated the structure of myelinated nerve by means of X-ray diffraction, showed that the bilayer model would fit the observed diffraction patterns. The electric capacitance of 10^{-2}F m^{-2} found for most biological membranes (see *4*.1.5) was also compatible with the bilayer model.

The observed, low, surface tensions of cell membranes caused HARVEY and DANIELLI (1938) to propose a coat of protein spread on each face of the lipid. This came to be thought of as a sheet of protein in the β-configuration (Fig. *1*.2). The precise way in which the protein sheet might be attached to the lipid bilayer was never convincingly described. One suggestion was that lipophilic side chains from the protein might enter the bilayer, passing between the polar headgroups. Perhaps because this seemed to be implausible, others suggested that each monolayer should be inverted, with the protein spread on the hydrophobic surfaces thus exposed. No evidence accumulated that either arrangement could exist.

2.2 Function and the Danielli-Davson Model

This model, as well as fitting adequately the X-ray diffraction patterns obtained from myelinated nerve fibres, provided convincingly for the passive permeability of the biological membrane. Neutral molecules could cross the membrane by solution in the lipids; their passage through the protein sheet was not usually described. The smaller hydrophilic molecules, like H_2O and CH_3OH, which had high permeabilities as COLLANDER showed, could be supposed to pass through the protein-lined aqueous pores; other pores of this kind might allow the passage of ions (K^+, Cl^-) whose passive permeation was demonstrable in many cells. Charges on the side groups of the proteins that lined the pores could make the pores selective for ions, so that a wide range of passive properties could be accommodated.

Specific, coupled transport was less well accounted for, the general opinion being that hydrophilic molecules might be made lipid-soluble by combination with a specific "carrier" molecule, the carrier-substrate combination diffusing across the lipid bilayer. Though such carriers are now known as naturally occurring offensive weapons (valinomycin, nigericin, etc. see Part B, *7.7.5*) this does not seem to be a good model for normal membrane transport.

2.3 The Unit Membrane

Using the term "unit membrane", J.D. ROBERTSON (1967) argued

(a) that all biological (cell and organelle) membranes had the same essential structure, giving rise to similar "tramline" staining patterns in transmission electron-micrographs at high resolution,

(b) that this common structure was that shown in Fig. *1.2*, with the additional idea that the two membrane faces were chemically different, and

(c) that nerve myelin was essentially membrane from the Schwann cell, so that it was a good model for all membranes in general.

Most workers now do seem to accept that there is some unity of structure among biological "tramline" membranes. We shall see however that there is considerable scope for variety within this unity.

The bilayer has won out, too, against competing suggestions that membrane lipids were arranged in spherical micelles, like soap micelles, which were then hexagonally packed into a sheet, stabilized by protein.

3. The Liquid Amphiphilic Layer

3.1 Introduction

The need to modify the DANIELLI-DAVSON model first clearly emerged in the sixties, as it became possible to measure the optical rotatory dispersion and the circular dichroism of the proteins in preparations of actual membranes (WALLACH and ZAHLER, 1966; LENARD and SINGER, 1966). These measurements showed that the protein of biological membranes contained almost none in the β-configuration; instead it was a mixture of α-helix and of random coil—a similar mixture to that found in the globular, water-soluble proteins. For such protein molecules, an obvious possibility was that they lay in the plane of the bilayer (WALLACH and ZAHLER, 1966). These molecules were already known to be removable from membrane preparations only by detergents and not by high ionic strength or low pH. This detergent extractability in fact defines "intrinsic" membrane proteins, whose role in membrane structure is discussed here.

There exist other proteins, associated with membrane preparations, which *are* extractable by acid or salt (e.g. cytochrome c, myelin basic protein). These "extrinsic" proteins are clearly bound to membranes largely by electrostatic interactions. Their effect upon membrane structure is not negligible, but is scarcely known, and is not further discussed. That they have great interest for membrane function is clear.

Thus it came to be thought that the intrinsic proteins of membranes were, like the lipids, amphiphilic, and that they were held into the membrane by the same factors that held the lipid molecules themselves.

The factors determining the stable structure of membranes have become better known and more clearly recognized: they are fundamentally (i) intermolecular attractive forces between water molecules and specific groups on other molecules (ia) electrostatic forces between dipolar water molecules and ionized groups, dipolar groups and induced dipoles (ib) hydrogen bonds between water molecules and such groups as —OH, =O, =NH or $—NH_2$, (ii) differences in the entropy of water molecules when they are in an aqueous environment or in one bordered by lipid. The nature of these forces and entropy differences emerges clearly in the nonelectrolyte section of the review by DIAMOND and WRIGHT (1969).

The structure of lowest energy that results from these forces and entropy differences depends of course upon the shapes of the molecules involved, and upon the proportions of lipids and water present in a given mixture. The same factors which cause soap and detergent (e.g. dodecyl sulphate) molecules to form spherical micelles, when their concentration is sufficiently high, cause membrane lipids to aggregate in ways that provide a larger volume-to-surface ratio—the bilayer is one such arrangement, though in general the spherical micelle is not.

3.2 The Liquid Lipid Bilayer

Considerable advances in X-ray diffraction techniques have made it possible to obtain high-resolution diffraction patterns from artificial lipid layers and from preparations of biological membranes other than the old standby, nerve myelin. After one or two false alarms, it became clear that for no preparation was there a repeat in the membrane plane such as would be characteristic of regularly packed lipid micelles; while in the direction normal to the membrane plane the electron-density profile was that characteristic of lipid bilayers oriented in the expected manner. The identification of the hydrocarbon chains and of the polar head-groups can be made with some certainty now that methods exist for providing an absolute electron-density scale. The methyl groups at the ends of the hydrocarbon tails produce a pronounced minimum of electron density, so that they are rather closely confined to a single plane. The rather weak intermolecular forces holding the two monolayers together in this plane make it very plausible that frozen tissue should often show fracture faces which include sections of this plane (BRANTON, 1969; PINTO DA SILVA and BRANTON, 1970).

In the plane of the bilayer, X-ray diffraction does show a repeat distance of great interest: at high temperatures it is at 0.46 nm, while at low temperatures it is at 0.42 nm (ENGELMAN, 1970; ESFAHANI et al., 1971). The latter is characteristic of the (hexagonal) packing of hydrocarbon chains in solid paraffins, while the former is characteristic of liquid paraffins.

DAVSON and DANIELLI (1943) had early suggested that at normal temperatures the lipids of the bilayer membrane would be in a liquid, rather than in a solid, state.

X-ray diffraction thus confirms the liquid-crystal bilayer arrangement of the lipids in biological membranes; and also indicates that the bilayer may show a phase transition, analogous to freezing, as the temperature is reduced.

This phase transition is detectable by other techniques as well. Differential scanning calorimetry of membrane preparations shows phase changes which are signalled by an input of latent heat: a high-temperature one, irreversible and identifiable with the denaturation of protein, and a low-temperature one, reversible and associated with the membrane lipids (STEIM et al., 1969; MELCHIOR et al., 1970).

The latter phase transition is also observable as a change in the mobility of "spin-labeled" amphiphilic molecules. These can be introduced into lipid bilayers and into cell membranes, and show an orientation which confirms the essential correctness of the lipid bilayer model (HUBBELL and MCCONNELL, 1969).

Frequently the "freezing" of membrane lipids takes place over a defined range of temperatures rather than at a single definite temperature. This is plausibly attributed to the freezing, at different temperatures, of different components of the lipid mixture of which the bilayer is composed.

The conclusion from all these studies is that most, or all, of the lipid of cell membranes is arranged in the form of a bilayer; and that the hydrocarbon tails are packed in the way characteristic of liquid hydrocarbons, at the temperatures at which the membrane normally functions. The importance of this for our understanding of membrane function will be obvious when we discuss the protein of the membrane.

3.3 Amphiphilic Proteins

The proteins of the membrane are, according to the postulate of VANDERKOOI and of SINGER and NICOLSON, globular, embedded in the bilayer, and held there by possessing a surface in part hydrocarbon and in part polar. Such protein molecules might emerge at one face of the bilayer, or at both (Fig. *1*.3). There is now much indirect evidence that this picture is essentially correct.

The freeze-fracture technique provides some of this evidence, and the pictures that result from it do indeed look compelling: even if interpreted with caution they do seem to show not only the smooth fracture faces already mentioned but bumps on these faces of about the size (8–9 nm) to be expected of protein molecules or oligomers (BRANTON, 1969; BRANTON and DEAMER, 1972). These bumps are not observed in freeze-fracture pictures of artificial lamellar lipid preparations, but are observed if the preparation contains intrinsic protein as well as lipid. The protein particles are often randomly distributed in the plane of the membrane, and this distribution can change rapidly. PINTO DA SILVA (1972) has shown that in the red cell membrane the particles are clustered at high and at low pH, and are randomly dispersed at two intermediate pH ranges, one of which is at the normal pH of plasma. The aggregation and dispersal both happen very rapidly, as would be expected if the protein particles are floating in a liquid bilayer.

Immunological techniques have been important in showing the random dispersal and the aggregation of membrane proteins, so that the interpretation of the freeze-fracture pictures is substantiated by a method which is specific for proteins. SINGER and NICOLSON (1972) show examples of this technique. The location of the face of the membrane at which the protein emerges can also be achieved by the use of immunological reagents, provided that the membrane preparation is one which allows their access to each face to be controlled. It is clear that for those membranes investigated all three classes of proteins occur, *viz.* accessible at face i, accessible at face o, accessible at both faces: this is shown also by experiments with proteolytic enzymes (STECK et al., 1971). Rhodopsin is an example of a protein accessible at one face of the membrane, while the glycoprotein of the erythrocyte is accessible at both faces. The amino acid sequence of this latter shows a central section of 23 non-ionic amino acids separating more polar sections (JACKSON et al., 1973). The central section would form an α-helix some 3.5 nm long, which could span the non-polar region of a bilayer.

The view that the lipids form the continuous phase in all biological membranes, with isolated proteins floating in the lipid, was the one put forward by SINGER and NICOLSON (1972). They argued that there was no evidence for long-range regularity in the arrangement of membrane proteins, so that the proteins could not form the continuous phase. However VANDERKOOI (1974) quotes several examples of membranes with such long-range order in the fresh (un-fixed) state, for example the acetylcholine receptor membrane of *Torpedo*. His own work shows that artificial lamellar preparations of 70% cytochrome oxidase and 30% lipid have a regular

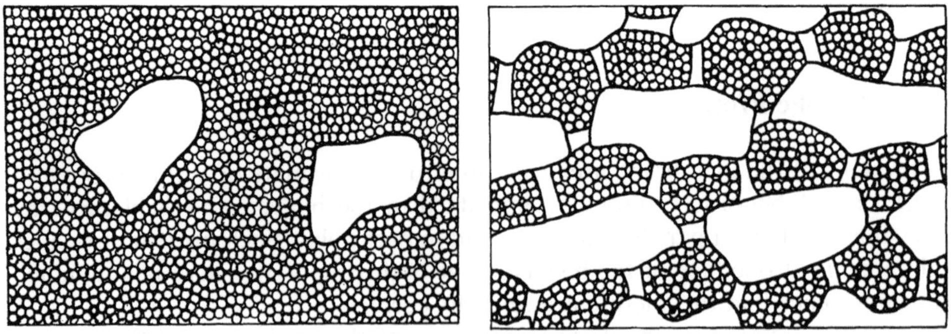

(a) (b)

Fig. *1*.4 a and b. Plan views of the range of structures represented by the sectional view of Fig. *1*.3 (a) Continuous lipid with intercalated (mobile) protein molecules or oligomers; (b) Continuous phase of regularly arranged protein with intercalated lipid. (b) is redrawn from VANDERKOOI (1974), and represents only one possible interpretation of his results. Small circles represent hydrocarbon chains of lipid molecules; clear areas represent protein particles arrangement of the protein, which does not exist if the proportion of lipid is higher. It seems then that we must admit that lamellar (membrane) systems may have structures in the range from (i) lipid bilayer, through (ii) lipid bilayer with isolated proteins units, to (iii) continuous protein bilayer with intercalated lipid (Fig. *1*.4).

It has become clear that membranes do not contain large amounts of "structural" protein; it is now accepted that the intrinsic proteins of biological membranes are largely functional.

3.4 Function and the Liquid Amphiphilic Layer

Since the arrangement of lipids in Fig. *1*.3 is essentially the same as that envisaged in the DANIELLI-DAVSON hypothesis, clearly the new model also explains the permeation of non-electrolytes which dissolve in lipid. Small, polar molecules which permeate more rapidly than this would suggest, and also ions, still need to be provided with special channels. One would expect these to be associated with protein, as DANIELLI suggested, although now one would expect the channel to be in the centre of a globular or cylindrical protein molecule that spanned the width of the bilayer: or in the centre of an oligomer of such molecules. Gramicidin A, a linear polypeptide antibiotic, (Part B, *7*.7.5) will form such channels, so that it is not really necessary to postulate very large protein molecules merely to explain the ionic conductivity of membranes. But a large, amphiphilic protein will be more likely to stay in the membrane in which it is put, while antibiotic molecules will be designed for smaller size and greater mobility from membrane to membrane.

Although Gramicidin A is too mobile to be a good model for permanent membrane structures, it has a property that introduces a topic of great functional interest. The conductivity conferred upon a membrane by this substance falls by two orders of magnitude when the membrane lipids freeze; the conductivity conferred by the mobile ion carrier nonactin falls by more than five orders of magnitude (SZABO et al., 1972). There is thus the possibility of experimental differentiation between processes which depend on movement among the lipid molecules (and which slow as the lipids freeze) and those which do not; this will be of increasing importance in elucidating molecular mechanisms in membrane function.

It is already known that the proton transport driven by light in membranes containing bacterial rhodopsin is not affected by the freezing of the membrane lipids (RACKER and HINKLE, 1974); further such studies are needed on isolated steps of the respiratory and photosynthetic pathways. The work of OVERATH et al., 1970) has shown that in *E. coli* the efflux of a glucose analog, thiomethylglucose, is slowed by membrane freezing. They and RAISON et al. (1971) have shown the effects of membrane freezing on complex reactions such as "respiration" and respiratory electron transport, demonstrating clearly the physiological importance of fluidity in membrane lipids.

SINGER and NICOLSON (1972) drew attention to the large energy barrier to the movement of a single-ended amphiphilic molecule from one half of the bilayer to the other, and the same must be true of double-ended ones. It is clear that in modeling the structures responsible for membrane functions only limited movements of protein molecules are to be thought of as probable. Yet these limited movements, together with changes in conformation, may transfer chemical groups from one protein molecule to another, open or close channels to the other side of the membrane and thus produce all the phenomena of membrane transport.

Good introductions to the literature are provided by SINGER and NICOLSON (1972), by OSEROFF et al. (1973), and by VANDERKOOI (1974). The physical techniques so briefly mentioned here, and their detailed results, are covered in the book by FOX and KEITH (1972). This Chapter aims merely to set up some general background for reading these reviews and the literature in general. The reader may also like to consult the colloquium proceedings edited by WALLACH and FISHER (1971): but should notice that here and in BRANTON and DEAMER (1972) the change of model from that of Fig. *1*.2 to that of Fig. *1*.3 is not yet complete.

References

BRANTON, D.: Membrane structure. Ann. Rev. Plant Physiol. **20**, 209–238 (1969).

BRANTON, D., DEAMER, D.W.: Membrane structure. Protoplasmatologia II E 1 (1972).

COLLANDER, R.: The permeability of *Nitella* cells to non-electrolytes. Physiol. Plantarum **7**, 420–445 (1954).

DANIELLI, J.F., DAVSON, H.A.: A contribution to the theory of the permeability of thin films. J. Cellular comp. Physiol. **5**, 495–508 (1935).

DAVSON, H.A., DANIELLI, J.F.: The permeability of natural membranes. Cambridge: Cambridge University Press 1943.

DIAMOND, J.M., WRIGHT, E.M.: Biological membranes: the physical basis of ion and non-electrolyte selectivity. Ann. Rev. Physiol. **31**, 581–646 (1969).

ENGELMAN, D.M.: X-ray diffraction studies of phase transitions in the membrane of *Mycoplasma laidlawii*. J. Mol. Biol. **47**, 115–117 (1970).

ESFAHANI, M., LIMBRICK, A.R., KNUTTON, S., OKA, T., WAKIL, S.J.: The molecular organization of lipids in the membrane of *Escherichia coli:* phase transitions. Proc. Natl. Acad. Sci. U.S. **68**, 3180–3184 (1971).

FINEAN, J.B.: Biochemical problems of lipids. London: Butterworths 1956.

FOX, C.F., KEITH, A.D.: Membrane molecular biology. Stamford, Conn: Sinauer Associates 1972.

GORTER, E., GRENDEL, F.: On bimolecular layers of lipoids on the chromocytes of the blood. J. Exptl. Med. **41**, 439–443 (1925).

HARVEY, E.N., DANIELLI, J.F.: Properties of the Cell Surface. Biol. Rev. **13**, 319–341 (1938).

HUBBELL, W.L., MCCONNELL, H.M.: Orientation and motion of amphiphilic spin labels in membranes. Proc. Natl. Acad. Sci. U.S. **64**, 20–27 (1969).

JACKSON, R.L., SEGREST, J.P., KAHANE, I., MARCHESI, V.T.: Studies on the major sialoglycoprotein of the human red cell membrane. Isolation and characterisation of tryptic glycopeptides. Biochemistry **12**, 3131–3188 (1973).

LENARD, J., SINGER, S.J.: Protein conformation in cell membrane preparations as studied by optical rotatory dispersion and circular dichroism. Proc. Natl. Acad. Sci. U.S. **56**, 1828–1835 (1966).

MELCHIOR, D.L., MOROWITZ, H.J., STURTEVANT, J.M., TSONG, T.Y.: Characterisation of the plasma membrane of *Mycoplasma laidlawii*. VII. Phase transitions of membrane lipids. Biochim. Biophys. Acta **219**, 114–122 (1970).

OSEROFF, A.R., ROBBINS, P.W., BURGER, M.M.: The cell surface membrane: biochemical aspects and biophysical probes. Ann. Rev. Biochem. **42**, 647–682 (1973).

OVERATH, P., SCHAIRER, H.U., STOFFEL, W.L.: Correlation of *in vivo* and *in vitro* phase transitions of membrane lipids in *Escherichia coli*. Proc. Natl. Acad. Sci. U.S. **67**, 606–612 (1970).

PINTO DA SILVA, P.: Translational mobility of the membrane intercalated particles of human erythrocyte ghosts. pH-dependent reversible aggregation. J. Cell Biol. **53**, 777–787 (1972).

PINTO DA SILVA, P., BRANTON, D.: Membrane splitting in freeze-etching. J. Cell Biol. **45**, 598–605 (1970).

RACKER, E., HINKLE, P.C.: Effect of temperature on the function of a proton pump. J. Membrane Biol. **17**, 181–188 (1974).

RAISON, J.K., LYONS, J.M., THOMSON, W.W.: The influence of membranes on the temperature-induced changes in the kinetics of some respiratory enzymes of mitochondria. Arch. Biochem. Biophys. **142**, 83–90 (1971).

ROBERTSON, J.D.: The organisation of cellular membranes. In: Molecular organisation and biological function (J.M. ALLEN, ed.), p. 65–106. New York: Harper and Row 1967.

SINGER, S.J.: A fluid lipid-globular protein mosaic model of membrane structure. Ann. N.Y. Acad. Sci. **195**, 16–23 (1972).

SINGER, S.J., NICOLSON, G.L.: The fluid mosaic model of the structure of cell membranes. Science **175**, 720–731 (1972).

STECK, T.L., FAIRBANKS, G., WALLACH, D.F.H.: Disposition of the major proteins in the isolated erythrocyte membrane. Proteolytic dissection. Biochemistry **10**, 2617–2624 (1971).

STEIM, J.M., TOURTELLOTTE, M.E., REINERT, J.C., MCELHANEY, R.N., RADER, R.L.: Calorimetric evidence for the liquidcrystalline state of lipids in a bio-membrane. Proc. Natl. Acad. Sci. U.S. **63**, 104–109 (1969).

SZABO, G., EISENMAN, G., MCLAUGHLIN, G.A., KRASNE, S.: Ionic probes of membrane structures. Ann. N.Y. Acad. Sci. **195**, 273–290 (1972).

VANDERKOOI, G.: Molecular architecture of biological membranes. Ann. N. Y. Acad. Sci. **195**, 6–15 (1972).

VANDERKOOI, G.: Organisation of proteins in membranes with special reference to the cytochrome oxidase system. Biochim. Biophys. Acta **344**, 307–345 (1974).

WALLACH, D.F.H., FISHER, H.: The dynamic structure of cell membranes. Berlin-Heidelberg-New York: Springer 1971.

WALLACH, D.F.H., ZAHLER, P.H.: Protein conformations in cellular membranes. Proc. Natl. Acad. Sci. U.S. **56**, 1552–1559 (1966).

2. Water Relations of Plant Cells

J. DAINTY

1. Introduction

There are many good treatments of water relations of plants, plant cells and plant tissues such as those by DAINTY (1963a), SLATYER (1967), BRIGGS (1967), WEATHERLEY (1970) and, most recently, the excellent general book by HOUSE (1974) covering both plant and animal cells and tissues. There are also many articles in the old Encyclopedia of Plant Physiology, and there is the collection of articles in the three volumes edited by KOZLOWSKI (1968a, b; 1972).

This present Chapter attempts to discuss what are felt to be the fundamental aspects of the water relations of plant cells. Its aim is to serve as a stimulus to workers in this field to fill in the glaring gaps in our knowledge of the basic parameters involved.

2. Water Potential

The theory of the "water relations" of plant cells and tissues is fairly well established now, although there are a few awkward points. It starts from the premise that the most suitable parameter for specifying the state of water in any system is the partial molar Gibbs free energy, or chemical potential, μ_w, of the water. The suitability of this quantity derives from its utility in quantitatively describing the flow of water from one system to another, for flow is best considered as being driven by differences of the chemical potential of water. Like all other expressions for energy, chemical potential is relative, i.e. only differences are meaningful. It is therefore necessary to measure the chemical potential of water relative to that in some convenient reference state. The reference state is chosen to be that of free, pure, water at the same temperature as the system under consideration and at atmospheric pressure (approximately one bar). It is also convenient, and it provides historical continuity, to express the chemical potential difference in units of energy per unit volume, or pressure. Taking these two conveniences together, the term water potential, Ψ, has been coined, defined by

$$\Psi = \frac{\mu_w - \mu_w^0}{\bar{V}_w} \tag{2.1}$$

as a practical measure of the free energy status of water in a given system.

The previous paragraph ignores a lot of history and lots of debate on nomenclature and other things (OERTLI, 1969; SPANNER, 1973), but it expresses the present general agreement to use water potential as a convenient measure of the free-energy state of water in the soil-plant-atmosphere system. $\bar{V}_w$, the partial molar volume of (liquid) water, arises naturally in the units conversion from the definition of chemical potential as $(\partial G/\partial n_w)_{P,T,n_j}$, the partial molar free energy (see SLATYER, 1967, for details).

Strictly speaking, this is as far as thermodynamics can take us in setting up an expression for the energy state of water. Ψ, or $\mu_w - \mu_w^0$, can be measured unambiguously by, for example, measuring the vapor pressure of water in equilibrium with the system (see BARRS, 1968, for details); in other words it is a thermodynamic observable. It is however necessary, for a proper understanding of the physics of water in plants, to split Ψ into what we believe are its components, even if there may be some doubts about the splitting process. It is common practice in textbooks of plant physiology to write Ψ as equal to the difference between hydrostatic pressure (P) and the osmotic pressure (π) and matric components (τ),

$$\Psi = P - \pi - \tau. \tag{2.2}$$

One can write, formally,

$$d\mu_w = \left(\frac{\partial \mu_w}{\partial P}\right)_{n_s, n_w} dP + \sum_j \left(\frac{\partial \mu_w}{\partial n_j}\right)_{P, n_{i \neq j}, n_w} dn_j + \left(\frac{\partial \mu_w}{\partial n_w}\right)_{P, n_s} dn_w \tag{2.3}$$

The first term in Eq. (2.3) expresses the dependence of the chemical potential on pressure. From general thermodynamic theory, $(\partial \mu_w/\partial P)_{n_s, n_w}$ is equal to $\bar{V}_w$. The second term expresses the contribution of the dissolved solutes to the chemical potential of the water. We know this as the osmotic component of the chemical potential and we can write

$$\left(\frac{\partial \mu_w}{\partial n_j}\right)_{P, n_{i \neq j}, n_w} dn_j = -\bar{V}_w \, d\pi_j \tag{2.4}$$

where π_j is the osmotic pressure arising from components j. The final term arises in systems like soil and cell walls where there is a large amount of "solid" matter in the system and the chemical potential will be a function of the water content of the system because of the large area of interface and the strong interactions between water and solid which go on at this interface and because of surface tension effects at air/water interfaces. This contribution to the chemical potential arises from the existence of a finely-divided matrix and is therefore called the matric contribution. We can define a matric potential, τ, analogously to an osmotic potential by writing

$$\left(\frac{\partial \mu_w}{\partial n_w}\right)_{P, n_s} dn_w = -\bar{V}_w \, d\tau \tag{2.5}$$

Thus Eq. (2.3) becomes

$$d\mu_w = \bar{V}_w (dP - \sum d\pi_j - d\tau)$$

or, putting $\sum d\pi_j = d\pi$

$$d\mu_w = \bar{V}_w (dP - d\pi - d\tau) \tag{2.6}$$

This Equation must be integrated between the water in the system and our free, pure, water reference state, making the reasonable assumption that $\bar{V}_w$ is sufficiently constant, to:

$$\mu_w - \mu_w^0 = \bar{V}_w (P - \pi - \tau)$$

or

$$\frac{\mu_w - \mu_w^0}{\bar{V}_w} = \Psi = P - \pi - \tau \tag{2.7}$$

This procedure shows how $\bar{V}_w$ arises naturally in the definition of Ψ. It is implied in the derivation that P is the hydrostatic pressure minus the atmospheric pressure, in bars.

As mentioned before there are some doubts about the validity of splitting the water potential into the above three, or any other, components. Some plant physiologists, e.g. BRIGGS (1967), question the reality of the matric potential term. Others say that the three terms cannot be additive, and so on. Some of the doubts arise from a realization, often not expressed, that free energy is made up of an enthalpic part, like pressure, and an entropic part such as is largely contributed by the osmotic potential (SPANNER, 1973). It can indeed be argued that in many circumstances the effect of the presence of a finely divided solid in a mixture of liquid and gas, such as soil, is to decrease the pressure in the liquid phase. This decrease of pressure can be calculated if the radii of curvature of the menisci of the air-water interfaces are known; the pressure in the liquid is given by $-2T/r$, where T is the surface tension and r the radius of curvature of the meniscus. If this pressure could be calculated or measured, then perhaps it could be included in the pressure term. Such effects arise from the strong force of attraction between the atoms (ions) at the surface of the solid and the adjacent water molecules. Forgetting about the possible existence of the air-water meniscus, the effect of these forces in the liquid is two-fold. In the first place the direct effect of the force on the water molecules orders them and puts them in a lower energy state than bulk free water. It is equivalent, as BRIGGS (1967) showed, to their being under a negative pressure; but the magnitude of this pressure cannot be computed easily. Both this force-field effect and the capillary effect (both arising from the same molecular interactions) are largely enthalpic in nature and could in principle be included in the P term in Ψ. There is however another effect of the solid-liquid interface. This interface is always electrically charged and is therefore the site of an electrical double layer. The charge on the solid surface is usually negative, hence cations will be concentrated in the double layer and the osmotic potential will be higher than in bulk solution. There is thus the possibility in sufficiently narrow pores of part of the matric potential being of the nature of the osmotic potential.

Another clear doubt about the splitting is—what constitutes a solute molecule? That is, what size does it have to be, to be part of the matrix and not contribute to the osmotic potential?

It seems clear that the splitting of the water potential into pressure, osmotic and matric components is an uneasy one. Strictly speaking it can only be used in an operational sense, i.e. if the components and the methods of measuring them are carefully specified in each system under consideration. This was made very clear by NOY-MEIR and GINZBURG (1967) in their theoretical analysis of plant-tissue water potentials. They showed that the integration of the basic Equation for $d\mu_w$ Eq. (2.3), to get Eq. (2.7) for the water potential can be carried out along several pathways. For instance the matric term can be obtained by carrying out the integration of $(\partial\mu_w/\partial n_w)_{P,n_s} dn_w$ with either $P=0$ or $P=$its final value, with either $n_s=0$ or the solute concentration at its final value. In principle a different value of τ would arise from these different mathematical procedures, which actually correspond to different experimental ways of measuring the matric potential. Clearly only if each component, P, π and τ, has no effect on the other two is the splitting of Ψ into P, π and τ unique. This is very unlikely; π and τ in particular are almost certainly interacting components as the earlier discussion of the matric potential indicated.

One might say that the splitting of Ψ into components is unnecessary because only the water potential is observable and important. Unfortunately this is not so. It is true that only the water potential is *thermodynamically* observable, but we are rarely dealing with thermodynamic equilibrium; flow processes, whether of water or metabolism, can depend on individual components such as hydrostatic or osmotic pressure, for instance. We cannot avoid therefore splitting up Ψ, but we must be aware of the dangers involved. (SPANNER, 1973, has suggested that a better split would be into enthalpic (or energetic) and entropic terms. However such a split would involve difficult temperature and calorimetric measurements which are certainly not in practical use at present.)

Fortunately it seems rare for all three components to be necessary. In the cell vacuole and in the xylem fluid, the only important components are P and π and the splitting is relatively unambiguous (the solute molecules are small). Soil and plant cell walls are systems in which the matric effect is usually dominating. In both these systems it seems best to define P as the *external* pressure. Since atmospheric pressure is the standard state, P becomes zero (apart from pressure contribution due to gravity) even though we know that part at least of the matric effect can be viewed as equivalent to a negative pressure in the water. The other contribution to Ψ in soils and cell walls will be an osmotic one arising from salts (usually) dissolved in the water. The osmotic pressure term has to be measured in a well-defined way so that it can be added to the matric potential unambiguously. Thus for soils and cell walls the two components necessary are π and τ. Protoplasm raises a difficulty. Normally it must be at the same turgor pressure P as the vacuole. But it does contain very large organelles, colloidal and membrane components which suggest an appreciable matric potential. And there is the above-mentioned difficulty of when does one count a "solute" molecule as contributing to the osmotic potential and when does one count it as part of the matrix? Just posing the question makes one realize again that the distinction can be quite artificial. I believe the only correct answer is that we do not know and it does not matter very much in considerations of water relations of the protoplasm. It is perhaps simplest to call the other-than-pressure component, the osmotic pressure.

The concept of water potential and its splitting into components can only refer, of course, to a single phase, e.g. cell wall, or protoplasm, or vacuole, or soil, etc. Therefore, as WEATHERLEY (1970) has forcibly pointed out, it does not seem to make much sense to talk about the water potential, and particularly its components, of a tissue such as a leaf. If all the phases of the tissue are in equilibrium then the water potential will be the same in each phase and it is thus correct to measure this water potential; it will have meaning. But it is dangerous to try and evaluate average values of the components P, π and τ.

Nevertheless the attempt, for leaves, has been made by some excellent investigators (GARDNER and EHLIG, 1965; WILSON, 1967a, b, c; NOY-MEIR and GINZBURG, 1967, 1969; BOYER, 1967; RAKHI, 1973). They all seem aware of the doubtfulness of their procedures, particularly the difficulties of ascribing meaning to the average values of π and τ. The best discussion of the difficulties involved is that of NOY-MEIR and GINZBURG (1967, 1969). The papers should be consulted for those who want an appreciation of the difficulties involved. Some of them will be referred to later in this Chapter.

3. The "Static" Water Relations of a Single Plant Cell

3.1 Components of Water Potential

Since this article is largely concerned with water relations at the single cell level and since such relations are fundamental to considerations of the water relations of the plant as a whole, it is appropriate to look at Ψ and its components for a single cell. At the outset a mature vacuolated plant cell will be assumed in equilibrium with its surroundings, of water potential Ψ^0. If the pressure inside the cell is P and the osmotic pressure of the vacuole (and, as argued earlier, the protoplasm) is π, then

$$\Psi = P - \pi = \Psi^0. \tag{2.8}$$

The matric potential of the vacuole is zero and that of the protoplasm can be taken to be zero for the purposes of the present formulation. If Ψ^0 is changed, then water will flow until a new equilibrium is reached at a new cell volume. We wish here to discuss how Ψ, P and π change with volume. (By cell volume is meant volume of vacuole plus protoplasm.)

Both P and π are functions of cell volume, V. For small changes in volume, dV, we can write

$$dP = \varepsilon\, dV/V \tag{2.9}$$

and

$$d\pi = -\pi\, dV/V. \tag{2.10}$$

Eq. (2.9) simply says that the cell wall has an elastic coefficient, ε, defined by the equation (note ε has the units of pressure). Eq. (2.10) assumes that the Boyle-Van't Hoff law, $\pi \cdot V = \text{constant}$, applies to the vacuole plus protoplasm, i.e. no correction is assumed for a non-osmotic volume. Combining the two Equations:

$$\begin{aligned} d\Psi &= dP - d\pi \\ &= (\varepsilon + \pi)\, dV/V. \end{aligned} \tag{2.11}$$

As we shall see later it is often convenient to think in terms of the cell capacity, C, defined, in analogy with an electrical capacitance, as the increment of volume (charge) per unit increment of potential. That is:

$$C = dV/d\Psi = V/(\varepsilon + \pi). \tag{2.12}$$

From Eq. (2.11) Ψ can, in principle, be derived as a function of volume provided that ε and π are known functions of volume. $\pi(V)$ presents little difficulty, but $\varepsilon(V)$ is a different matter. Alternatively from an experimental determination of $\Psi(V)$, $(\varepsilon + \pi)$ can be determined from the slope of a Ψ *versus* V graph.

It seems clear that the elastic modulus, ε, is an important parameter in cell water relations (DAINTY, 1972). Since it is often greater than π, it will control the way in which the water potential changes with cell volume. (As we shall see later it will also control the rate of swelling or shrinking of a plant cell.) This is particularly true in regions of moderate and high turgor pressure, P. (The maximum turgor pressure is of course equal to π at maximum volume, when $\Psi = 0$.)

One consequence of this is that plants, in principle, should not need to osmoregulate, providing that the changes in external water potential do not take the plant cell into regions of plasmolysis or zero turgor; for the pressure part handles most of the changes in Ψ, leaving π and therefore the solute, including ionic, concentration relatively constant. However this does not seem to be so (KAMIYA and KURODA, 1956; ZIMMERMANN and STEUDLE, 1971; KAUSS, 1973; BEN-AMOTZ and AVRON, 1973). Plant cells do osmoregulate; they do seem to prefer to be at a particular pressure and therefore must have pressure sensors to tell them to pump ions in or out or make or destroy some particular compound and thus get back to the desired pressure. (See also Chap. *11*.)

3.2 Measurement of ε for Giant Algal Cells

Quite a number of measurements, of various degrees of certainty, have now been made of ε, some of them as a function of volume. The most extensive, and reliable, investigations have been made using giant algal cells. As mentioned above, from the observed relationship between Ψ and V and, preferably, a knowledge of π, ε can be determined as a function of V using Eq. (*2*.11). (This is at present the only possible method for higher plant cells.) ε can also be obtained using Eq. (*2*.9), $P = \varepsilon dV/V$, from a direct determination of the turgor pressure/volume relationship. Such direct measurements have been made by KELLY et al. (1963), KAMIYA et al. (1963), VILLEGAS (1967), STEUDLE and ZIMMERMANN (1971, 1974), ZIMMERMANN and STEUDLE (1974a, 1974b), DAINTY et al. (1974), on giant algal cells of the Characeae and on *Valonia* species. Results from these measurements are collected together in Table *2*.1. It should be noted that ε is not usually defined by the above workers exactly by Eq. (*2*.9), but rather by a linear modification of it:

$$P = \frac{\varepsilon(V - V^*)}{V^*} \tag{2.13}$$

where V^* is some reference volume usually, though not always, either the volume

Table *2*.1. Values of ε for giant algal cells

Species	ε (bars)	Ref.
Nitella flexilis	70 for $P \leqq 1.5$ bar 300 for $P \sim 6$ bar	KAMIYA et al. (1963)
N. flexilis	60 for low P 200 for high P but $\varepsilon = f(V)$ as well	ZIMMERMANN and STEUDLE (1974b)
N. translucens	250 as $P \to 0$ 1,200 for $P = 8$ bar	KELLY et al. (1963)
Chara corallina	760 at high P	DAINTY et al. (1974)
Valonia ventricosa	180 for all P	VILLEGAS (1967)
V. utricularis	35 as $P \to 0$ 180 as $P \to 3$ bar	ZIMMERMANN and STEUDLE (1974a)

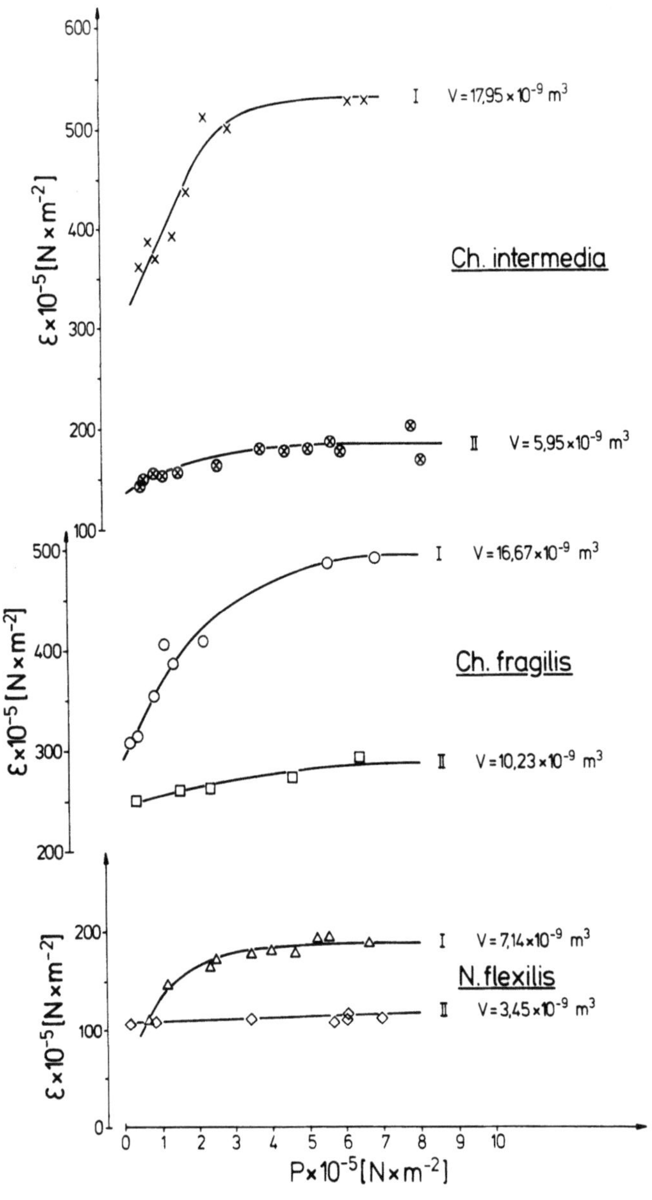

Fig. 2.1. Pressure dependence of the volumetric elastic modulus, ε, of *Chara* and *Nitella* internodal cells having different cell volume, *V*, (i.e. different cell lengths). Taken from ZIMMERMANN and STEUDLE (1975). (Note $N \times m^{-2} = Pa$)

at $P=0$ or the volume at maximum turgor. Inspection of the results on giant algal cells given in Table 2.1, shows that ε is quite high, some hundreds of bars, and that for Characean cells at least it is a strong function of turgor pressure P; it is largest at high turgor pressures and decreases quite markedly as the cells approach zero turgor (Fig. 2.1). The latest work of ZIMMERMANN and STEUDLE (1974b) also shows that the elastic behavior of Characean cells at a given turgor pressure, is a function of cell volume. They interpret this as implying the existence of two elastic moduli, one relating to the ends of the cells and the other to the general body of the cell.

3.3 Estimation of ε in Cells of Higher Plants

As indicated above, P cannot be directly measured in higher plant cells. ε must thus be estimated from experimental determinations of Ψ and π as a function of V for a tissue. A number of such estimations have been made. GARDNER and EHLIG (1965) and WILSON (1967a, b, c) measured Ψ as a function of relative water content R. (Relative water content is the water content of the tissue divided by the water content at full turgor). Both use psychrometric methods to measure Ψ on live tissue and π (or $\pi+\tau$) on killed tissue. WILSON (1967a, b, c) corrected his R values for the so-called bound water B in the tissues; in principle this should make his ε values more reliable. NOY-MEIR and GINZBURG (1967, 1969) measured Ψ on live tissue and π and τ on killed tissue as a function of water content of the tissue again by a kind of psychrometric method. They tried, unsuccessfully, to make a correction for "bound" water. The results of these three investigations, plus some others calculated from the literature by WILSON (1967a, b, c) are given in Table 2.2. In general the ε values are

Table 2.2. Values of ε derived from measurements of Ψ as a function of water content of leaves, by psychrometry

Species	ε (bars)	Ref.
Gossypium hirsutum	60 $P>2$ bar 15 $P<2$ bar	A
Helianthus annuus	47 $P>3.4$ bar 14 $P<3.4$ bar	A
H. annuus	7.3	B
Lotus corniculatus	60 $P>2$ bar 6.3 $P<2$ bar	A
Capsicum fruitescence	71 $P>2$ bar 4.4 $P<2$ bar	A
Brassica napus	9.5	B
Zea mays	15.8	B
Lycopersicum esculentum	21.5	quoted in B
Gossypium barbadense	29.6	quoted in B
Ligustrum lucidum	26.6	quoted in B
Pennisetum typhoides	33.4	quoted in B
Acacia aneura	84.3	quoted in B
Ceratonia siliqua	120–130 high P 34–45 low P	C
Platanus orientalis	100–105 high P 22–30 low P	C
Atriplex halimus	50 high P 16 low P	C

A is GARDNER and EHLIG (1965); B is WILSON (1967b); C is NOY-MEIR and GINZBURG (1969).
Where no turgor pressure is mentioned the authors have assumed that ε is independent of P.

much lower than those for the giant algal cells, but show the same rough trend of small values at low turgor pressures and higher values at high turgor pressures.

Another way of getting Ψ as a function of volume is by means of the so-called pressure-bomb technique. Here Ψ is measured as the gas pressure required to bring the xylem fluid, withdrawn after, for example, cutting a twig back until it just wets the cut end of the stem (Fig. 2.2). After such an equilibrium gas pressure is achieved, the gas pressure (P_g) can be increased. This "forces" water from the leaf cells, because increased gas pressure means increased Ψ, until a new equilibrium is reached. By repeating this process, all the time measuring the volume of water expressed, data are obtained giving gas pressure ($-\Psi$) as a function of volume expressed. Since the volume expressed is equal to the original volume of water in the cells minus the volume at a given value of Ψ, the data give the desired functional relationship between Ψ and V. Such data were first rather roughly obtained by SCHOLANDER et al. (1964) on some halophytes. More accurate data on a range of species has been obtained by TYREE and HAMMEL (1972). If the data are plotted as $1/P_g$ against volume expressed, V_e, a typical graph such as that shown in Fig. 2.3 results. Over a range of low gas pres-

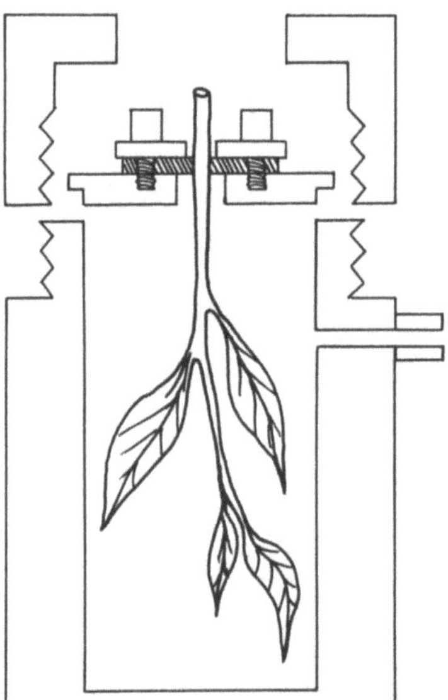

Fig. 2.2

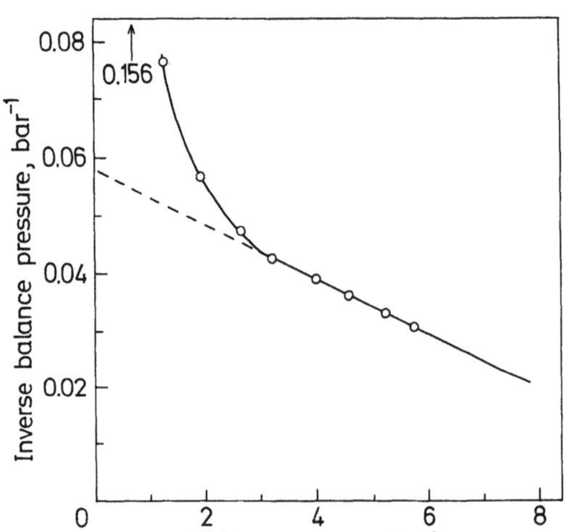

Fig. 2.3

Fig. 2.2. Diagram of a "pressure bomb" derived from the original version by SCHOLANDER et al. (1965). The vessel is made from thick metal to withstand high pressures. The screw-lid clamps the plate carrying the specimen onto the top of the vessel. The shoot is held into a hole made in a sheet of rubber which is then compressed by the bolts in the plate to seal the shoot. Compressed gas is fed into the vessel through the inlet tube and a measuring device used to record the pressure. A lens can be used to observe any solution flow from the cut end of the shoot

Fig. 2.3. Typical graph of inverse balance pressure against volume expressed obtained by pressure-bomb technique. Extrapolation of straight line portion to the ordinate gives the inverse of the osmotic pressure at $\Psi=0$; extrapolation of the straight line portion to the abscissa gives the volume of water in the symplasm; the intersection of the curved and straight portions is the point at which $P=0$. ε is determined from the curved portion. (Curve redrawn from TYREE et al., 1973) sures, P_g, there is a curvilinear relation between $1/P_g$ and V_e which at a certain pressure becomes linear and stays linear up to the usual limit of experimentally applied pressures. This linear portion is interpreted as occurring when the turgor pressure is zero; thus the only component of Ψ is π and we might expect a linear relation between $1/\pi$ and cell volume, and hence between $1/P_g$ and V_e. If this is correct then extrapolation of the straight line to the $V_e=0$ axis gives the initial osmotic pressure. Also extrapolation of the same straight line to the $1/P_g=0$ axis gives the volume of the cell water (by definition, this is the symplasm volume, see Part B, 2.1). All the necessary information is therefore in these curves to enable one to calculate from the curvilinear portions the dependence of average turgor pressure on cell volume and hence a value for ε. It should be realized that this approach measures the average ε for the tissue as a whole, not ε for a particular kind of cell.

TYREE and HAMMEL (1972) made such measurements on 7 woody species of South America (Chile). They found that the average turgor pressure, P, was related to cell volume by an Equation of the following form:

$$P=\varepsilon'\left(\frac{V-V_p}{V_p}\right)^n \tag{2.14}$$

where V_p is the volume at $P=0$ and n is a constant for a particular species. In their paper they give values for ε' (they call it ε) and n for the various species. n ranges from 1.8 to 3.7, thus confirming the general feeling that the elastic modulus strongly increases as the cell approaches maximum turgor. From the data given in the paper by TYREE and HAMMEL (1972) it can be calculated what the values of ε would be over the ranges of turgor pressure involved in their experiments. The results of these calculations are given in Table 2.3 and indicate that although the elastic modulus

Table 2.3. Values for ε calculated from data of TYREE and HAMMEL (1972)

Species	ε (bar)	
Pilgerodendron uvifera, twig I	20	$P \simeq 0$ bar
	115	$P \simeq 20$ bar
Pilgerodendron uvifera, twig II	8	$P \simeq 0$ bar
	125	$P \simeq 17$ bar
Podocarpus nudigenus	0	$P \simeq 0$ bar
	125	$P \simeq 10$ bar
Nothofagus betuloides	0	$P \simeq 0$ bar
	95	$P \simeq 12$ bar
Pernettya macronata	10	$P \simeq 0$ bar
	300	$P \simeq 23$ bar
Weinmannia trichosperma	8	$P \simeq 0$ bar
	100	$P \simeq 18$ bar
Abies concolor	8	$P \simeq 0$ bar
	100	$P \simeq 11$ bar
Picea glauca	2	$P \simeq 0$ bar
	75	$P \simeq 75$ bar

is of the order of 100 bars near full turgor it can become extremely small as the turgor pressure approaches zero. RICHARDS (1973) has shown that ε is a linear function of turgor pressure from a pressure-bomb analysis of sitka spruce *(Picea sitchensis)*.

One other estimate of ε is worth comment. HAMMEL (1967) has estimated, from elasticity data, that ε for hemlock *(Tsuga canadensis)* tracheids is about 2,300 bars. From this he deduces that when hemlock wood is subjected to sub-zero temperatures, the conversion of only 3% of the water in the tracheids to ice will change the negative pressure normally present in tracheids to a positive one. He claims that this very high ε for conifer tracheids is therefore an essential part of the mechanism for suppressing gaseous embolisms in conducting tissue during freezing.

Summarizing the available information on this important parameter, it can be said that ε is in general strongly dependent on the turgor pressure. In the range of P from 0 to about 2 bars ε can be very low; at full turgor ε varies between 50 and several hundred bars depending on the type of cell and the species. Due to the dependence of ε on P (and hence on cell volume relative to that at zero water potential), measurements of ε must be treated with caution. To date, exact measurements of ε have been made only with giant algal cells where P and ε can be measured directly and simultaneously.

4. Transport of Water across Cell Membranes

The transport of matter is properly dealt with by the theory of irreversible thermodynamics. We are concerned largely with the transport of water where the barriers to transport may be permeable to solute as well as to water molecules. The appropriate theory of irreversible thermodynamics has been adequately described in a number of places (DAINTY, 1963a; SLATYER, 1967; HOUSE, 1974) and need only be outlined here.

Basically the theory of irreversible thermodynamics predicts, from a consideration of the production of entropy during an irreversible process, that the flux of a component is a function of *all* the forces acting on *all* the mobile components of the system. It is then convenient to assume that, when the rates of flow are not too large, the relationship between fluxes and forces is a linear one of the form:

$$J_i = \sum L_{ij} X_j \tag{2.15}$$

where J_i is the flux of component i, X_j is the force on component j and L_{ij} is the so-called phenomenological or Onsager coefficient relating the flow of i to the force on j. A further part of the theory is the recognition that the L_{ij}'s are not all independent. In fact according to the Onsager reciprocity relation:

$$L_{ij} = L_{ji}. \tag{2.16}$$

The theory also tells us what to take for the J's and X's, for the product of J and X must give the rate of increase of entropy. If J_i is in mol m^{-2} s^{-1} then X_i must be the gradient or difference in chemical potential; if J is a volume flux in m^3 m^{-2} s^{-1} then X is a difference in pressure, and so on.

On the basis of the above simple ideas it is not difficult to formulate transport processes in the correct way (KEDEM and KATCHALSKY, 1958; DAINTY, 1963a). For the case of two aqueous solutions of a non-electrolyte solute separated by a membrane permeable to both the solute and water the Equations are:

$$J_v = L_p(\Delta P - \sigma \Delta \pi), \tag{2.17}$$

$$J_s = \omega \Delta \pi + J_v(1-\sigma)\ \bar{c}_s, \tag{2.18}$$

where J_v is the volume, essentially the water, flux in $m\ s^{-1}$, ΔP and $\Delta \pi$ are the differences in hydrostatic and osmotic pressures between the two solutions in bars, L_p is the so-called hydraulic conductivity in $m\ s^{-1}\ bar^{-1}$, σ is the reflection coefficient to be discussed below, J_s is the solute flux in $mol\ m^{-2}\ s^{-1}$, ω is a solute permeability coefficient at zero volume flow and $\bar{c}_s$ is a mean solute concentration between the two solutions. Note that with two permeating components there are *three* parameters determining the relations between flows and forces; these three parameters are derived from the four (2×2) L_{ij}'s, with $L_{12} = L_{21}$ by the reciprocity relation. L_p and ω are permeability coefficients of a kind. The reflection coefficient σ is a measure of a combination of the relative permeabilities of water and solute and of the frictional interaction between them as they cross the membrane (KEDEM and KATCHALSKY, 1961; DAINTY and GINZBURG, 1963). As can be seen from Eq. (2.17) it is the ratio of the apparent to the real osmotic pressures, i.e. $\Delta P / \Delta \pi$ when $J_v = 0$.

Eqs. (2.17) and (2.18) are derived from the basic equations of irreversible thermodynamics in which the force on water is the gradient, or difference, of the chemical potential of water and the force on solute is the gradient, or difference of the chemical potential of the solute. If the membrane involved is semi-permeable, then the flux of water is simply proportional to the force on the water, i.e. we can write in our nomenclature:

$$J_v = L_p\ \Delta \Psi. \tag{2.19}$$

The effect of departure from semi-permeability is largely to modify the $\Delta \pi$ component of $\Delta \Psi$. Here again it is necessary to distinguish between the various component potentials that go to make up Ψ. Taking into account matric potentials, as we must when dealing with flows in soil and cell walls, we can reasonably generalize Eq. (2.17) to

$$J_v = L_p(\Delta P - \sigma \Delta \pi - \Delta \tau). \tag{2.20}$$

Clearly if we have both non-permeating, suffix imp, and permeating, suffix p, solutes we can generalize (with only a slight loss of thermodynamic correctness) further to

$$J_v = L_p(\Delta P - \Delta \pi_{imp} - \Delta \tau - \sum \sigma_p\ \Delta \pi_p). \tag{2.21}$$

L_p and σ are the important parameters determining water flow. We shall discuss later what sparse knowledge we have about their values.

4.1 The "Dynamic" Water Relations of a Single Plant Cell

We will consider the same kind of idealized mature, vacuolated, single plant cell as we did earlier in this Chapter. Suppose the cell of area A and volume V is initially in equilibrium with its surroundings which are at a water potential of Ψ_1^0. Then, if the membrane is semi-permeable:

$$\Psi_1 = P_1 - \pi_1 = \Psi_1^0. \tag{2.22}$$

If Ψ_1^0 is increased by $\Delta\Psi^0$ to Ψ_2^0, water enters the cell and swelling takes place at a rate given by:

$$J_v = \frac{1}{A}\frac{dV}{dt} = L_p[\Psi_2^0 - P + \pi]. \tag{2.23}$$

If $\Delta\Psi^0$ and therefore ΔP, $\Delta\pi$ and ΔV are small, A and L_p can be treated as constant, and we can write:

$$P = \varepsilon\frac{(V - V_1)}{V_1} + P_1 \tag{2.24}$$

and

$$\pi = \frac{\pi_1 V_1}{V} = \pi_1\left[1 - \frac{(V - V_1)}{V_1}\right]. \tag{2.25}$$

Thus Eq. (2.23) becomes

$$\frac{1}{A}\frac{dV}{dt} = L_p\left[\Delta\Psi^0 - \frac{(\varepsilon + \pi_1)}{V_1}(V - V_1)\right] \tag{2.26}$$

which on integration becomes

$$V = V_1 + (V_2 - V_1)[1 - \exp(-t/t_c)] \tag{2.27}$$

where

$$t_c = \frac{V_1}{AL_p(\varepsilon + \pi_1)}.$$

Since only small changes in Ψ^0, etc. have been considered the suffixes 1 can be dropped in the Equation for t_c and we can write for the swelling, or shrinking, time constant:

$$t_c = \frac{V}{AL_p(\varepsilon + \pi)} \tag{2.28}$$

where the parameter values are all evaluated at some particular value of, say, V.

The time course of swelling or shrinking of a plant cell will thus be exponential, with the time constant given in Eq. (2.28), for small changes in the external water potential. For larger changes the time course will deviate from the exponential, for t_c will no longer be a constant mainly because of the change of ε with, say, turgor pressure. It is interesting to work out rough values for t_c at this point. For a turgid

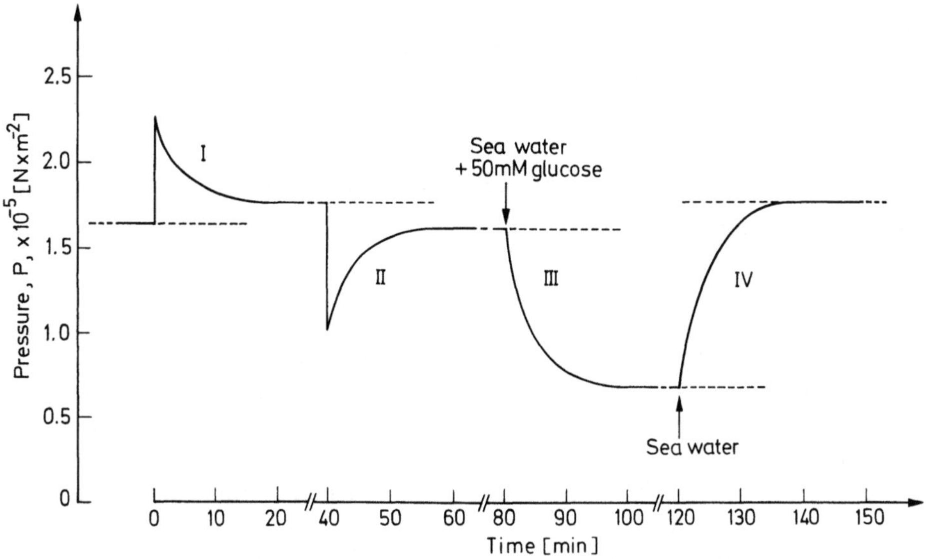

Fig. 2.4. Pressure changes with time in a *Valonia* cell measured with the apparatus shown in Fig. 2.6. The stationary pressures were altered with the aid of the micrometer screw (curves *I* and *II*) or by changing the water potential of the sea water (curves *III* and *IV*). From the time constant of swelling or shrinking L_p can be calculated using an Equation similar to Eq. (2.28). (Redrawn from STEUDLE and ZIMMERMANN, 1971)

Nitella cell with $V/A = 150\ \mu m$, $L_p = 10^{-7}\ m\ s^{-1}\ bar^{-1}$ and $\varepsilon + \pi = 600$ bar, $t_c = 2.5$ s. For a turgid higher plant cell with $V/A = 10\ \mu m$, $L_p = 10^{-9}\ m\ s^{-1}\ bar^{-1}$ and $\varepsilon + \pi = 100$ bar, $t_c = 100$ s. For a somewhat flaccid higher plant cell with $V/A = 10\ \mu m$, $L_p = 10^{-10}\ m\ s^{-1}\ bar^{-1}$ and $\varepsilon + \pi = 10$ bar, $t_c = 10^4\ s \hat{=} 3h$. We shall see later whether the values of L_p used here are reasonable or not. Experimental measurements of change in P in giant algal cells also show the time course of the approach to equilibrium (Fig. 2.4). In this example t_c was about 200 s.

A useful way of looking at the movement of water into and out of a plant cell is *via* the electrical analogy introduced earlier. That is we can think of the cell as having a capacity C being charged, with volume, through a resistor, R, the driving force being the water potential difference between inside and outside. We can then write:

$$\text{Current} = \frac{dV}{dt} = \frac{\Psi^0 - \Psi}{R} = \frac{\Psi^0 - V/C}{R}. \tag{2.29}$$

As is well known, the time constant for this charging of a capacity is given by:

$$t_c = RC. \tag{2.30}$$

We worked out earlier, Eq. (2.12), that

$$C = \frac{dV}{d\Psi} = \frac{V}{(\varepsilon + \pi)} \tag{2.12}$$

thus from Eq. (2.28) we can write:

$$R = 1/AL_p. \tag{2.31}$$

The units of C are $m^3\ bar^{-1}$ and those of R are $s\ bar\ m^{-3}$. This way of thinking about the uptake and loss of water by cells is extremely useful; its utility is well exemplified in the model of COWAN (1972) for the whole plant water relations.

4.2 The Hydraulic Conductivity and Reflection Coefficient for Giant Algal Cells

It is appropriate to discuss L_p and σ for giant algal cells at this stage, immediately after outlining the theory of water movements in single plant cells. For the theory is only applicable to single, isolated, cells and it is only for certain giant algal cells that we have reliable values of L_p and σ; we have none for higher plant cells!

For a typical giant algal cell, surrounded by an aqueous medium, L_p and σ are defined by Eqs. (2.17) or (2.21) (without the $\Delta\tau$). The now classical way of measuring L_p for the long, cylindrical, Characean cells is by the method of transcellular osmosis. This was developed by KAMIYA and TAZAWA (1956) and used by DAINTY and HOPE (1959), DAINTY and GINZBURG (1964a), DAINTY et al (1974) and members of the school of KAMIYA and TAZAWA to measure L_p for several Characean species. A diagram of the experimental arrangement for measuring L_p by the method of transcellular osmosis is given in Fig. 2.5. Essentially transcellular water flow is induced by suddenly changing the fluid in the open chamber from water to a solution of a non-permeating solute of suitable osmotic pressure, π. The initial rate of water flow $\dot{V}$ from the water side, through the cell, to the solution side can easily be shown to be given by:

$$\dot{V}=\frac{A_1 A_2}{A_1+A_2} L_p \pi. \tag{2.32}$$

A_1 and A_2 are the cell areas exposed to the fluids in the two chambers. The initial rate of flow must be measured for after a time $\dot{V}$ decreases because the internal solutes in the vacuole become polarized by the flow. The values of L_p obtained by this technique, by the authors quoted above, for *Nitella flexilis*, *N. translucens* and *Chara corallina* (formerly *australis*) are of the order of $10^{-7}\ m\ s^{-1}\ bar^{-1}$, which is quite a high value for a water permeability.

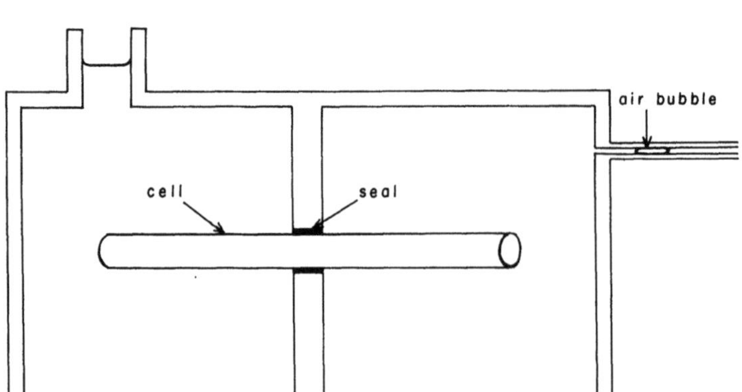

Fig. 2.5. Diagrammatic representation of the transcellular osmosis apparatus

However the first experiments by KAMIYA and TAZAWA (1956) disclosed that the situation was more complex than indicated by the simple Eq. (2.32) above. This equation says that the rate of transcellular flow, $\dot{V}$, would be unaffected, if the cell were placed asymmetrically such that $A_1 \neq A_2$, whether the solution were on the short or the long side of the cell. KAMIYA and TAZAWA (1956) demonstrated with *N. flexilis* that this was not true and explained the difference between the two rates of flow, to the short end *vs.* to the long end, by saying that L_p for water entering the cell, endosmosis L_{pen}, was different from, indeed greater than, L_p for water leaving the cell, exosmosis L_{pex}. The rate of flow when $L_{pen} \neq L_{pex}$ is given by:

$$\dot{V} = \frac{L_{pen}\, L_{pex}\, A_1\, A_2}{L_{pen}\, A_2 + L_{pex}\, A_1} \cdot \pi. \tag{2.33}$$

This Equation applies when the water goes into end 2 and out at end 1. It can easily be seen that if the experiment is performed with the cell reversed, i.e. into end 1 and out of end 2, a different value of $\dot{V}$ is obtained.

DAINTY and HOPE (1959) and DAINTY and GINZBURG (1964a) found similar, though not such marked, polarities with *C. corallina* and *N. translucens*. At one time DAINTY (1963b) thought that the difference between L_{pen} and L_{pex} was more apparent than real. He thought that L_{pex} was apparently less than L_{pen} because the outflowing water at the exosmosis side would "sweep away" the osmoticum and so reduce the driving force at the exosmosis side. Although this effect is certainly real and must be taken into account, DAINTY and GINZBURG (1964a) proved experimentally that this was not the whole explanation. They suggested that L_{pex} was probably less than L_{pen} because the lower water potential at the exosmosis side affected the membrane permeability perhaps by a measure of dehydration. TAZAWA and KAMIYA (1966) also carefully investigated the relevance of the "sweeping away" effect and concluded that, irrespective of it, there is a true polarity of L_p.

A new experimental and theoretical analysis of this polarity in water permeability or apparent rectification of water movement across Characean cell membranes has been carried out by the Osaka School (KIYOSAWA and TAZAWA, 1973; TAZAWA and KIYOSAWA, 1973). They have shown that it is possible by measuring transcellular osmosis and turgor pressure simultaneously to calculate L_{pen} and L_{pex} separately for each half of the cell. They have also corrected their values for the L_p of the cell wall, which shows no polarity. For both *N. flexilis* and *C. corallina* (which they call *C. australis*) they find, expressing their results in our units (m s^{-1} bar^{-1}):

$$L_{pen}^{-1} = 4.2 \cdot 10^6 - 1.1 \cdot 10^5\, \pi$$
$$L_{pex}^{-1} = 4.2 \cdot 10^6 + 2.9 \cdot 10^5\, \pi$$

where π is the external solution osmotic pressure used to induce transcellular osmosis. The first terms in these two Equations implicitly contain a term dependent on the internal osmotic pressure of the cell (see TAZAWA and KAMIYA, 1966). The surprising feature of this result is the dependence of L_{pen} on the osmotic pressure of the solution which only the exosmosis side actually sees. KIYOSAWA and TAZAWA (1973) explain this by pointing out that the flow of water induced by π will change the cytoplasmic osmotic pressure because the tonoplast is expected to have a finite

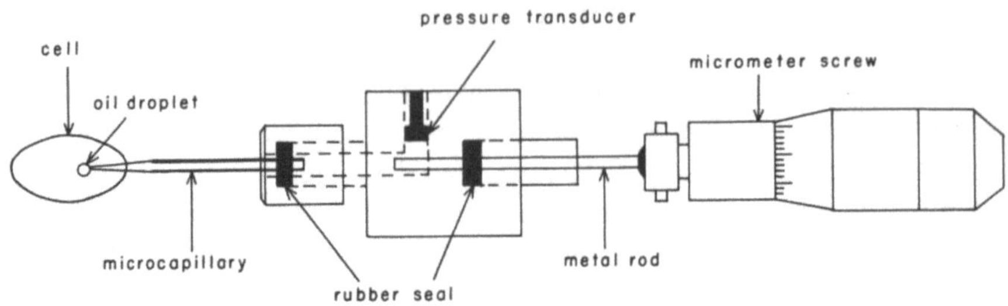

Fig. 2.6. Apparatus for measuring L_p and ε for single cells by changing the volume (and pressure) by a given amount and then measuring the change of pressure with time. (Redrawn from ZIMMERMANN and STEUDLE, 1974a)

resistance to water flow. The change in cytoplasmic osmotic pressure is dependent on water flow rate and hence on π and, since the cytoplasmic osmotic pressure may affect the membrane(s), an effect of π on L_{pen} is reasonable.

Another way of measuring L_p for single plant cells is to measure the rate of swelling or shrinking and from the time constant given in Eq. (2.28) to calculate L_p; of course $(\varepsilon+\pi)$ must be known. An early measurement of L_p by this technique was carried out by KELLY et al. (1963) on *N. translucens;* they obtained a value similar to that obtained by transcellular osmosis, *viz.* about 10^{-7} m s^{-1} bar^{-1}.

A new method of measuring turgor pressures, invented by ZIMMERMANN et al. (1969), has enabled ZIMMERMANN and his colleague STEUDLE to make very precise measurements of L_p (and of σ and ε) for giant algal cells. The apparatus is shown diagrammatically in Fig. 2.6. Essentially a microcapillary filled with an oil is inserted into the cell. The oil will immediately come to the same hydrostatic pressure as the cell contents and this pressure is recorded by a pressure transducer (as an electrical signal). The volume of the cell, and hence its pressure, can be instantaneously changed by a turn of screw which pushes the oil slightly forward or backward, thus changing the meniscus between oil and water at the tip of the microcapillary. This pressure change means a change in cell Ψ and water then flows in or out of the cell to restore Ψ to its pre-change value. The consequent pressure changes are recorded by the pressure transducer, and the volume changes are of course known from the turn of the screw. With the aid of a little mathematics, based on the theory already outlined, the values of ε and L_p are obtained quite straight-forwardly and accurately.

The values of ε obtained by this technique have already been given in Table 2.1. The value of L_p for *Nitella flexilis* obtained by STEUDLE and ZIMMERMANN (1974) was similar to that obtained by KAMIYA and TAZAWA (1956) and KIYOSAWA and TAZAWA (1973), that is about $2.5\cdot10^{-7}$ m s^{-1} bar^{-1}. STEUDLE and ZIMMERMANN (1974) also found some degree of polarity, i.e. L_{pex} was less than L_{pen} whether the flows were induced by changes in external osmotic pressure or in internal hydrostatic pressure. They also found that as the external osmotic pressure was increased L_{pex} decreased and so did L_{pen}, though much less strongly. But their most surprising observation was the marked dependence of L_p on turgor pressure, P. With *N. flexilis*, for $P>2$ bar, L_p was constant; for $P<2$ bar L_p increased to about $6\cdot10^{-7}$ m s^{-1} bar^{-1}. Since they showed that the cell wall itself had an L_p of about $6\cdot10^{-7}$ m s^{-1} bar^{-1}, the actual increase of the membrane L_p must have been very much greater than the factor of 2.5 which the uncorrected results indicate.

For *Valonia utricularis* ZIMMERMANN and STEUDLE (1974a) carried out a similar investigation. The normal turgor pressure of this species is about 1.5 bar. Above 0.5 bar L_p was about $9 \cdot 10^{-9}$ m s^{-1} bar^{-1} and independent of pressure. Below 0.5 bar L_p rapidly increased as P decreased, increases by factors between 3 and 10 being observed and these uncorrected for the L_p of the cell wall!

These interesting results throw considerable doubt on the relevance of the classical way of measuring L_p using plasmolyzed protoplasts (STADELMANN, 1963).

One other measurement of L_p on *Valonia ventricosa* should be given. This was obtained by GUTKNECHT (1968) by a fairly conventional method. L_p was $1.85 \pm 0.27 \cdot 10^{-9}$ m s^{-1} bar^{-1}; thus *V. ventricosa* is about 5 times less permeable to water than is *V. utricularis* and about 100 times less permeable than the internodal cells of the Characeae. It is interesting to note that *V. ventricosa* is the only plant cell on which a proper measurement has been made of the permeability, P_d, to labeled water (GUTKNECHT, 1968). It was found that $P_d = L_p RT/\bar{V}_w$ thus proving that in this species, molecules of water moved through the membrane(s) one by one and not by a bulk flow through narrow pores (DAINTY, 1963a).

Table 2.4. Reflection coefficients for some solutes of some giant algal cells

Solute	*N. flexilis*	*N. translucens*	*C. corallina*	*V. utricularis*
sucrose	0.97			1
glucose	0.96			0.95
glycerol	0.80			0.81
acetamide	0.91			0.79
urea	0.91	1	1	0.76
formamide	0.79	1	1	
ethylene glycol	0.94	1	1	
isopropanol	0.35	0.27		
n-propanol	0.17	0.16	0.22	
ethanol	0.34	0.29	0.27	
methanol	0.31	0.25	0.30	

Data on *N. flexilis* are from STEUDLE and ZIMMERMANN (1974). Data on *N. translucens* and *C. corallina* are from DAINTY and GINZBURG (1964b). Data on *V. utricularis* are from ZIMMERMANN and STEUDLE (1970).

Only three somewhat comprehensive sets of determinations of reflection coefficients seem to have been made: by DAINTY and GINZBURG (1964b) on *N. translucens* and *C. corallina*, by ZIMMERMANN and STEUDLE (1970) on *V. utricularis* and by STEUDLE and ZIMMERMANN (1974) on *N. flexilis*. The measurements are essentially based upon the recognition that the osmotic part of the driving force on water for a permeating solute is $\sigma\pi$, so that the ratio of flows produced by equal osmotic pressures of permeating and non-permeating solutes is $\sigma/1$ or, alternatively, the ratio of the pressures needed to produce zero flows is $\sigma/1$. DAINTY and GINZBURG (1964b) used the method of transcellular osmosis, whereas the German workers used their pressure transducer technique (Fig. 2.6). The results obtained are given in Table 2.4.

4.3 The "Dynamic" Water Relations of Plant Tissues and L_p for Higher Plant Cells

Since at the present time it is not technically possible to study single higher plant cells and measure ε, L_p, σ, etc. for them, we must try and deduce these parameters from appropriate experiments on plant tissue. Of course experiments of this kind are of interest, in their own right, for understanding the physiology of plant-water relations irrespective of whether ε, L_p, etc. can be extracted from such experiments. The "static" experiments which lead to an estimate of ε for higher plant cells, particularly leaf cells, have already been described. Here, the main concern will be with what can be deduced from rates of swelling or shrinking of plant tissues and, to some extent, from labeled water exchange studies.

A typical experiment is to take a disk of, say, storage tissue, which has been equilibrated for a longish time in a solution of a certain water potential, and transfer it to a solution of a different water potential. The disk will swell or shrink as water enters or leaves the cells and we would like to be able to describe the kinetics of this process. The events are quite complex. At the instant of change of the external solution concentration (assuming a non-permeating solute) the cells know nothing of the change, for first the new solution has to penetrate the unstirred layer (DAINTY, 1963a) about 50 µm thick surrounding the disk; this only takes a few seconds and is not normally of much importance in these kinds of experiments. Then the solutes of the osmoticum have to diffuse through the extracellular space of the tissue and therefore each cell, depending on its position, experiences a new external osmotic pressure which changes with time as the diffusion proceeds. Clearly, on this picture, the rates of swelling and shrinking will depend on *both* extracellular solute diffusion *and* cell membrane permeation. Either process may be rate-limiting in principle. For instance KOHN and DAINTY (1966), who studied the swelling and shrinking of disks of *Beta vulgaris* and *Helianthus tuberosus*, concluded that extracellular solute diffusion was the rate-controlling process; not only was the half-time for swelling or shrinking linearly dependent on the *square* of the disk thickness, but also actual measurements of the diffusion coefficient for sucrose (the osmoticum) in the extracellular space agreed fairly well with that deduced from the rates of swelling or shrinking. Thus if KOHN and DAINTY (1966) are correct in both their picture of the process and their analysis, no information at all about L_p can be obtained from such experiments.

However, in a series of papers, GLINKA and REINHOLD (1962, 1964, 1971) seem to have demonstrated that there is some degree of rate control by the membranes in the swelling and shrinking of their tissues. For instance in their 1964 paper they used disks of carrot storage tissue and showed that high CO_2 concentrations slowed down, by about 40%, the rates of swelling and shrinking in mannitol solutions. There was still some rate control by solute diffusion, for the use of raffinose instead of mannitol as the osmoticum also slowed down the rates; but even with raffinose, CO_2 slowed down the rate still further.

More recently STUART (1973) has shown that the rate of shrinking of potato tuber tissue in mannitol solutions is decreased by KCN, DNP, ammonium carbonate and oligomycin and these decreases are wholly or partly prevented by ATP and CTP. These effects, as also those of GLINKA and REINHOLD (1964), can of course just as well be on ε as on L_p. Because it is clear that there is still at least partial rate control by solute diffusion, it is hazardous to try and calculate a value for $L_p(\varepsilon+\pi)$ which is the appropriate combination of parameters governing rate control by membrane permeation.

But is the picture used so far correct? It is one of independent cells in a tissue in which the only movement of water is from the cells into the extracellular space,i.e. no movement of water from cell to cell is contemplated. An alternative picture, put forward by PHILIP (1958a, b, c), is as follows. A change in concentration of the external osmoticum is first sensed by cells at the surface of the tissue and a change of the water potential of these cells begins. This change is then sensed by the next cells in and so on. The movement of water is envisaged as taking place entirely from cell to cell, none in the extracellular space. PHILIP (1958b) made a simple one-dimensional model of this process and showed that the propagation of volume (or water potential, or turgor pressure, or osmotic pressure, for they are all linearly related in his simple model) is related to time and distance by a diffusion equation with an apparent diffusion coefficient given by:

$$D = \frac{\alpha l}{2} \cdot L_p(\varepsilon + \pi) \tag{2.34}$$

where α is a shape factor, approximately equal to one, and l is the linear dimension of a cell. Thus on this special model of rate control by membrane permeation, swelling and shrinking obey diffusion kinetics and, if it is correct, $L_p(\varepsilon + \pi)$ can be determined. It should be noted that PHILIP (1958b) ignored the likely presence of plasmodesmata between adjoining cells.

The most extensive application of this theory is in the work of KLEPPER and her associates on the diurnal swelling and shrinking, in diameter, of stems of cotton (*Gossypium hirsutum*). The paper by MOLZ and KLEPPER (1972) seems to be the most pertinent. It has been proved experimentally that the volume changes are confined to the phloem and associated tissues external to the xylem; these are about 1.2 mm thick. The value of D in Eq. (2.34) which best fitted their results was $8.0 \cdot 10^{-11}$ m^2 s^{-1}. In another paper KLEPPER et al. (1973) showed that D is strongly temperature-dependent, particularly in the 20–30° C range, which of course supports a membrane permeation concept. Such a value of D indicates, if theory is correct, a value for $L_p(\varepsilon + \pi)$ of about 10^{-5} m s^{-1}, i.e. if $(\varepsilon + \pi) = 100$ bars, L_p is about 10^{-7} m s^{-1} bar^{-1}.

Which of these two pictures of the kinetics of swelling and shrinking is correct? The answer is tied up with the question of whether the apoplastic or the symplasmic pathway for water is the more important in plant tissues (TYREE, 1970; WEATHERLEY, 1970; BOYER, 1974). What evidence there is suggests that the main pathway, i.e. the pathway of lower resistance, is the apoplastic or extracellular pathway. If this is so then the first picture is more correct than the second. In a recent paper, MOLZ and IKENBERRY (1974) have erected a theory in which both pictures are taken into consideration, i.e. the propagation of water potential, say, in a tissue occurs both *via* the cell walls and from cell to cell in the PHILIP (1958b) sense. Undoubtedly this is a more realistic picture although much more complicated. It seems clear that good values of L_p for higher plant cells are going to be quite difficult to obtain from the kinetics of tissue swelling and shrinking; certainly no reliable values are at present available.

Some people have studied the exchange of labeled water between a plant tissue and its bathing medium. This is a possible way of getting L_p providing that the exchange is rate-controlled by membrane permeation and providing it can be then assumed that the permeability coefficient, P_d, so calculated is related to L_p by the formula (DAINTY, 1963 a):

$$P_d = L_p RT/\bar{V}_w. \tag{2.35}$$

Again there are two extreme pictures of the exchange process: either the cells exchange with the extracellular space independently of each other, or the labeled water moves from cell to cell throughout the tissue. The experimentally-derived time constants would have different interpretations in the two cases.

WOOLLEY (1965) investigated the rate of exchange of labeled water in 3-day-old maize roots. He judged that the exchange followed cylindrical diffusion kinetics. The half-times at 25° C and at 3° C were 26 s and 480 s respectively; the half-time for killed roots was 7 s and for roots in the presence of such inhibitors as CCCP and DNP was about 150 s. Taken together these results suggest a strong degree of control by membrane permeation. On PHILIP's (1958 b) picture, P_d for, say, a cortical cell would be about $5 \cdot 10^{-5}$ m s^{-1}, i.e. L_p from Eq. (2.35) would be about $3 \cdot 10^{-8}$ m s^{-1} bar^{-1}. On the other picture, cells exchanging independently with the extracellular space, P_d is about $1.4 \cdot 10^{-7}$ m s^{-1} and L_p about 10^{-10} m s^{-1} bar^{-1}. I feel that the latter is likely to be closer to the true L_p.

In a series of papers, JARVIS and HOUSE (1967, 1969) and HOUSE and JARVIS (1968) have also investigated the rate of exchange of labeled water in young maize roots. They too found evidence from studies on live and dead roots, from temperature studies and the effects of inhibitors, that the exchange was partly controlled by membrane permeation. They rejected PHILIP's (1958 b) model and calculated a P_d value of $4.5 \cdot 10^{-8}$ m s^{-1} which would give the very low L_p value of $3 \cdot 10^{-11}$ m s^{-1} bar^{-1}. Similar studies have been carried out by GLINKA and REINHOLD (1972) on carrot tissue. One can calculate from their data a P_d value of about 10^{-8} m s^{-1} and a L_p value of less than 10^{-11} m s^{-1} bar^{-1}!

I confess to a feeling that there is an air of unreality about these L_p values from isotopic exchange. We have not yet got the right models.

5. Conclusion

I have only considered certain aspects of the water relations of plant cells and, of necessity, tissues. They are the ones which, naturally, I feel are the most important at the present time. It seems that the theory of the water relations of a single plant cell is adequate and well understood. In principle it is well understood for a tissue too, but the complexity at present is such that theory is not too useful, although MOLZ and IKENBERRY (1974) have taken an important step in the right direction. What should be evident from this article is the great lack of relevant experimental data on the important parameters involved, which I hope might prove a stimulation to remedy this defect.

References

BARRS, H.D.: Determination of water deficits in plant tissues. In: Water deficits and plant growth (T.T. KOZLOWSKI, ed.), vol. I, p. 235–368. New York-London: Academic Press 1968.

BEN-AMOTZ, A., AVRON, M.: The role of glycerol in the osmotic regulation of the halophilic alga *Dunaliella parva*. Plant Physiol. **51**, 875–878 (1973).

BOYER, J.S.: Matric potentials of leaves. Plant Physiol. **42**, 213–217 (1967).

BOYER, J.S.: Water transport in plants: Mechanism of apparent changes in resistance during absorption. Planta **117**, 187–207 (1974).

BRIGGS, G.E.: Movement of water in plants. Oxford: Blackwell 1967.

COWAN, I.R.: Oscillations in stomatal conductance and plant functioning associated with stomatal conductance: Observations and a model. Planta **106**, 185–219 (1972).

DAINTY, J.: Water relations of plant cells. Advan. Bot. Research I, 279–326 (1963a).

DAINTY, J.: The polar permeability of plant cell membranes to water. Protoplasma **57**, 220–228 (1963b).

DAINTY, J.: Plant cell-water relations: The elasticity of the cell wall. Proc. Roy. Soc. Edinburgh A **70**, 89–93 (1972).

DAINTY, J., GINZBURG, B.Z.: Irreversible thermodynamics and frictional models of membrane processes, with particular reference to the cell membrane. J. Theoret. Biol. **5**, 256–265 (1963).

DAINTY, J., GINZBURG, B.Z.: The measurement of hydraulic conductivity (osmotic permeability to water) of internodal characean cells by means of transcellular osmosis. Biochim. Biophys. Acta **79**, 102–111 (1964a).

DAINTY, J., GINZBURG, B.Z.: The reflection coefficient of plant cell membranes for certain solutes. Biochim. Biophys. Acta **79**, 129–137 (1964b).

DAINTY, J., HOPE, A.B.: The water permeability of cells of *Chara australis* R.Br.. Australian J. Biol. Sci. **12**, 136–145 (1959).

DAINTY, J., VINTERS, H., TYREE, M.T.: A study of transcellular osmosis and the kinetics of swelling and shrinking in cells of *Chara corallina*. In: Membrane transport in plants (U. ZIMMERMANN, J. DAINTY, eds.), p. 59–63. Berlin-Heidelberg-New York: Springer 1974.

GARDNER, W.R., EHLIG, C.F.: Physical aspects of the internal water relations of plant leaves. Plant Physiol. **40**, 705–710 (1965).

GLINKA, Z., REINHOLD, L.: Rapid changes in permeability of cell membranes to water brought about by carbon dioxide and oxygen. Plant Physiol. **37**, 481–486 (1962).

GLINKA, Z., REINHOLD, L.: Reversible changes in the hydraulic permeability of plant cell membranes. Plant Physiol. **39**, 1043–1050 (1964).

GLINKA, Z., REINHOLD, L.: Abscisic acid raises the permeability of plant cells to water. Plant Physiol. **48**, 103–105 (1971).

GLINKA, Z., REINHOLD, L.: Induced changes in permeability of plant cell membranes to water. Plant Physiol. **49**, 602–606 (1972).

GUTKNECHT, J.: Permeability of *Valonia* to water and solutes: apparent absence of aqueous membrane pores. Biochim. Biophys. Acta **163**, 20–29 (1968).

HAMMEL, H.T.: Freezing of xylem sap without cavitation. Plant Physiol. **42**, 55–66 (1967).

HOUSE, C.R.: Water transport in cells and tissues. London: Edward Arnold 1974.

HOUSE, C.R., JARVIS, P.: Effect of temperature on the radial exchange of labelled water in maize roots. J. Exptl. Bot. **19**, 31–40 (1968).

JARVIS, P., HOUSE, C.R.: The radial exchange of labelled water in maize roots. J. Exptl. Bot. **18**, 695–706 (1967).

JARVIS, P., HOUSE, C.R.: The radial exchange of labelled water in isolated steles of maize roots. J. Exptl. Bot. **20**, 507–515 (1969).

KAMIYA, N., KURODA, K.: Artificial modification of the osmotic pressure of the plant cell. Protoplasma **46**, 423–436 (1956).

KAMIYA, N., TAZAWA, M.: Studies of water permeability of a single plant cell by means of transcellular osmosis. Protoplasma **46**, 394–422 (1956).

KAMIYA, N., TAZAWA, M., TAKATA, T.: The relation of turgor pressure to cell volume in *Nitella* with special reference to mechanical properties of the cell wall. Protoplasma **57**, 501–521 (1963).

KAUSS, H.: Turnover of galactosylglycerol and osmotic balance in *Ochromonas*. Plant Physiol. **52**, 613–615 (1973).

KEDEM, O., KATCHALSKY, A.: Thermodynamic analysis of the permeability of biological membranes to non-electrolytes. Biochim. Biophys. Acta **27**, 229–246 (1958).

KEDEM, O., KATCHALSKY, A.: A physical interpretation of the phenomenological coefficients of membrane permeability. J. Gen. Physiol. **45**, 143–179 (1961).

KELLY, R.B., KOHN, P.G., DAINTY, J.: Water relations of *Nitella translucens*. Trans. Bot. Soc. Edinburgh **39**, 373–391 (1963).

KIYOSAWA, K., TAZAWA, M.: Rectification characteristics of *Nitella* membranes in respect to water permeability. Protoplasma **78**, 203–214 (1973).

KLEPPER, B., MOLZ, F.J., PETERSON, C.M.: Temperature effects on radial propagation of water potential in cotton stem bark. Plant Physiol. **52**, 565–568 (1973).

KOHN, P.G., DAINTY, J.: The measurement of permeability to water in disks of storage tissue. J. Exptl. Bot. **17**, 809–821 (1966).

KOZLOWSKI, T.T., ed.: Water deficits and plant growth. 3 vols. New York-London: Academic Press 1968a, b, 1972.

MOLZ, F.J., IKENBERRY, E.: Water transport through plant cells and cell walls: Theoretical development. Soil Sci. Soc. Amer. Proc. **38**, 699–704 (1974).

MOLZ, F.J., KLEPPER, B.: Radial propagation of water potential in stems. Agron. J. **65**, 469–473 (1972).

NOY-MEIR, I., GINZBURG, B.Z.: An analysis of the water potential isotherm in plant tissue. I. The theory. Australian. J. Biol. Sci. **20**, 695–721 (1967).

NOY-MEIR, I., GINZBURG, B.Z.: An analysis of the water potential isotherm in plant tissue. II. Comparative studies on leaves of different types. Australian J. Biol. Sci. **22**, 35–52 (1969).

OERTLI, J.J.: Terminology of plant-water energy relations. Z. Pflanzenphysiol. **61**, 264–265 (1969).

PHILIP, J.R.: The osmotic cell, solute diffusibility, and the plant water economy. Plant Physiol. **33**, 264–271 (1958a).

PHILIP, J.R.: Propagation of turgor and other properties through cell aggregations. Plant Physiol. **33**, 271–274 (1958b).

PHILIP, J.R.: Osmosis and diffusion in tissue: Half-times and internal gradients. Plant Physiol. **33**, 275–278 (1958c).

RAKHI, M.: An apparatus for investigation of components of the water potential of leaves. Fiziol. Rastenii **20**, 215–221 (1973).

RICHARDS, G.P.: Some aspects of the water relations of Sitka spruce. Ph.D. Thesis, University of Aberdeen, Scotland (1973).

SCHOLANDER, P.F., HAMMEL, H.T., HEMMINGSEN, E.A., BRADSTREET, E.D.: Hydrostatic pressure and osmotic potential in leaves of mangroves and some other plants. Proc. Natl. Acad. Sci. U.S. **52**, 119–125 (1964).

SLATYER, R.O.: Plant-water relationships. New York-London: Academic Press 1967.

SPANNER, D.C.: The components of the water potential in plants and soils. J. Exptl. Bot. **24**, 816–819 (1973).

STADELMANN, E.: Vergleich und Umrechnung von Permeabilitätskonstanten für Wasser. Protoplasma **57**, 660–678 (1963).

STEUDLE, E., ZIMMERMANN, U.: Hydraulische Leitfähigkeit von *Valonia utricularis*. Z. Naturforsch. **26b**, 1302–1311 (1971).

STEUDLE, E., ZIMMERMANN, U.: Determination of the hydraulic conductivity and of reflection coefficients in *Nitella flexilis* by means of direct cell-turgor pressure measurements. Biochim. Biophys. Acta **332**, 399–412 (1974).

STUART, D.M.: Reduction of water permeability in potato tuber slices by cyanide, ammonia, 2,4-dinitrophenol, and oligomycin and its reversal by adenosine 5′-triphosphate and cytidine 5′-triphosphate. Plant Physiol. **51**, 485–488 (1973).

TAZAWA, M., KAMIYA, N.: Water permeability of a Characean internodal cell with special reference to its polarity. Australian J. Biol. Sci. **19**, 399–419 (1966).

TAZAWA, M., KIYOSAWA, K.: Analysis of transcellular water movement in *Nitella*: A new procedure to determine the inward and outward water permeabilities of membranes. Protoplasma **78**, 349–364 (1973).

TYREE, M.T.: The Symplast Concept: A general theory of symplastic transport according to the thermodynamics of irreversible processes. J. Theoret. Biol. **26**, 181–214 (1970).

TYREE, M.T., DAINTY, J., BENIS, M.: The water relations of hemlock (*Tsuga canadensis*). I. Some equilibrium water relations as measured by the pressure-bomb technique. Canad. J. Botany **51**, 1471–1480 (1973).

TYREE, M.T., HAMMEL, H.T.: The measurement of turgor pressure and the water relations of plants by the pressure-bomb technique. J. Exptl. Bot. **24**, 267–282 (1972).

VILLEGAS, L.: Changes in volume and turgor pressure in *Valonia* cells. Biochim. Biophys. Acta **136**, 590–593 (1967).

WEATHERLEY, P.E.: Some aspects of water relations. Advan. Bot. Research **3**, 171–206 (1970).

WILSON, J.W.: The components of leaf water potential. I. Osmotic and matric potentials. Australian J. Biol. Sci. **20**, 329–347 (1967a).

WILSON, J.W.: The components of leaf water potential. II. Pressure potential and water potential. Australian J. Biol. Sci. **20**, 349–357 (1967b).

WILSON, J.W.: The components of leaf water potential. III. Effects of tissue characteristics and relative water content on water potential. Australian J. Biol. Sci. **20**, 359–367 (1967c).

WOOLLEY, J.T.: Radial exchange of labelled water in intact maize roots. Plant Physiol. **40**, 711–717 (1965).

ZIMMERMANN, U., RÄDE, H., STEUDLE, E.: Kontinuierliche Druckmessung in Pflanzenzellen. Naturwiss. **56**, 634 (1969).

ZIMMERMANN, U., STEUDLE, E.: Bestimmung von Reflexionkoeffizienten an der Membran der Alge *Valonia utricularis*. Z. Naturforsch. **25b**, 500–504 (1970).

ZIMMERMANN, U., STEUDLE, E.: Effects of potassium concentration and osmotic pressure of sea water on the cell-turgor pressure of *Chaetomorpha linum*. Marine Biol. **11**, 132–149 (1971).

ZIMMERMANN, U., STEUDLE, E.: The pressure-dependence of the hydraulic conductivity, the membrane resistance and membrane potential during turgor pressure regulation in *Valonia utricularis*. J. Membrane Biol. **16**, 331–352 (1974a).

ZIMMERMANN, U., STEUDLE, E.: Hydraulic conductivity and volumetric elastic modulus in giant algal cells: Pressure and volume-dependence. In: Membrane transport in plants (U. ZIMMERMANN, J. DAINTY, eds.), p. 64–71. Berlin-Heidelberg-New York: Springer 1974b.

ZIMMERMANN, U., STEUDLE, E.: The hydraulic conductivity and volumetric elastic modulus of cells and isolated cell walls of *Nitella* and *Chara* spp.: Pressure and volume effects. Australian J. Plant Physiol. **2**, 1–13 (1975).

3. Membrane Transport: Theoretical Background

N. A. Walker

1. Introduction

It is the aim of this Chapter to set out and to classify the equations that seem most useful to those who work on transport across biological membranes. So that they may be the more confidently used—or rejected—the equations will be accompanied by details of the assumptions upon which they are based. The intention is to be of service; the licence is claimed, which is allowed to some servants, to pass occasional moral judgements.

The equations to be discussed all refer to the model, shown in Fig. 3.1, in which a membrane M, through which material particles may move, forms the only connexion between the aqueous phases a and b. These phases are of uniform composition, hence the stirring arrangements of Fig. 3.1; they differ in composition and in such intensive properties as pressure and electric potential, but not in temperature or gravitational potential[1].

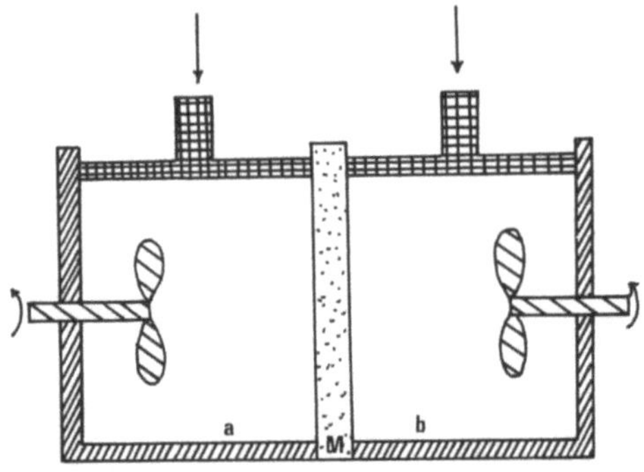

Fig. 3.1. Diagrammatic representation of the system to which the equations in this Chapter apply. The membrane M separates two aqueous phases a and b, within which there are no differences of composition. a and b have the same temperature and solvent (water) but have different hydrostatic pressures, compositions and electric potentials

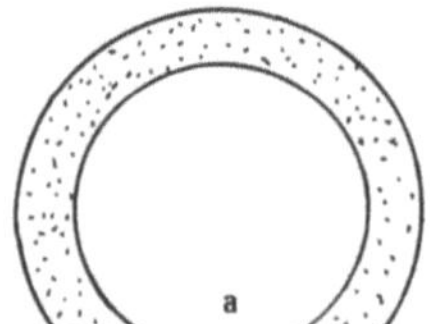

Fig. 3.2. The same equations may be applied to a simple cell and membrane if diffusion or stirring keeps the composition of the phases a and b constant in the face of transport through M. The membrane M is uniform, and thin

[1] In many practical stituations the unstirred layers on each side of the membrane must be considered, as discussed by Dainty (1963). Non-isothermal systems are sometimes considered, for example those which exhibit thermo-osmosis (Spanner, 1954).

The same equations will also apply to the more biological model shown in Fig. *3.2*. Real biological membranes are believed to be microscopic mosaics (Chap. *1*), with different tiny areas of different transport properties: real biological systems often consist of two or more membranes in series. Whether a given equation will be applicable to such a case must be investigated, not taken for granted. Only equilibrium equations are automatically valid for such cases.

The Equations we will be interested in provide us with relations between the activities of chemical components, their rates of passage between a and b, and properties of the system such as temperature, potential difference between a and b, and so on.

2. Origins of Equations

If we consider first that various mechanisms exist by which a material particle may be transferred between a and b, we can initially classify equations as (A) mechanism-independent, or (B) mechanism-dependent. Those of class A are found to be deduced from the laws of thermodynamics, while those of B are obtained by kinetic arguments based on the properties of the assumed model.

Classical thermodynamics can provide us with equations that apply to equilibrium states, which become inequalities when equilibrium does not exist. Non-equilibrium thermodynamics can provide rate equations of a rather general kind, which are in class A; but most rate equations are in class B. The reader is urged to pursue this question in the book by KATCHALSKY and CURRAN (1965).

3. Equilibrium Equations (Class A)

3.1 Solvent Transport

Equilibrium, for a component of the system of Figs. *3.1* and *3.2*, with respect to transport between a and b, is the condition in which the change in free energy (ΔG) resulting from such transport is zero. For a component without electrostatic charge, this change in free energy is measured by the difference in chemical potential, which must be zero at equilibrium. For small n:

$$\Delta G = n(\mu_a - \mu_b) = 0, \tag{3.1}$$

where ΔG is the change in free energy of the system produced by the transport of n moles of the component *from* b *to* a.

Expanding the chemical potential difference we obtain:

$$(\mu_a^0 - \mu_b^0) + \bar{V}(P_a - P_b) + RT \ln (a_a/a_b) = 0. \tag{3.2}$$

If we are considering the solvent, μ^0 is constant, so:

$$\bar{V}(P_a - P_b) + RT \ln (a_a/a_b) = 0.$$

The activity a can be written as the product of the mole fraction x and the activity coefficient γ, which latter will often be near unity in value:

$$\bar{V}(P_a - P_b) + RT \ln (\gamma_a x_a/\gamma_b x_b) = 0. \tag{3.3}$$

Re-arranging gives:

$$(P_a - P_b) + (RT/\bar{V}) \ln \gamma_a x_a - (RT/\bar{V}) \ln \gamma_b x_b = 0. \tag{3.4}$$

This is more familiar as:

$$(P_a - P_b) + (\pi_a - \pi_b) = 0. \tag{3.5}$$

In the field of water relations of plants, P_a and π_a are two of the three components of the water potential, Ψ; see Chap. 2, Eq. (2.2).

In dilute solutions x (for the solvent) is near 1.0, so that Eq. (3.4) can be reduced to:

$$(P_a - P_b) + RT(c_a - c_b) = 0, \tag{3.6}$$

where c is the *solute* concentration. This derivation of the van't Hoff law is dealt with very lucidly by THAIN (1967).

3.2 Solute Transport

3.2.1 Non-Electrolyte Solutes

Eq. (3.2) applies equally to non-electrolyte solutes, and it can be reduced to Eq. (3.3) as before, if a and b are aqueous solutions. For solutes however the two terms in this equation are unlikely to be equal in magnitude except by accident, since the values of $\bar{V}$ and T usually encountered mean that enormous pressure differences would be needed to balance even small concentration ratios[2]. So Eq. (3.3), as applied to solutes is usually equivalent to:

$$a_a - a_b = 0, \tag{3.7}$$

or neglecting activity coefficients and neglecting second-order effects:

$$c_a - c_b = 0. \tag{3.7a}$$

[2] If the concentration ratio is only 2/1, and if $\bar{V}$ is $10^{-4}\,m^3\,mol^{-1}$, the pressure difference at equilibrium would be 16.9 MPa, or 167 atm. Thus in biological systems pressure differences do not have significant effects upon concentration equilibria.

3.2.2 Electrolyte Solutes

The proper counterpart of Eq. (3.1) for use with ionized components is obtained by re-writing it with the electrochemical potentials:

$$\bar{\mu}_a - \bar{\mu}_b = 0, \tag{3.8}$$

which can be expanded to read:

$$(\mu_a^0 - \mu_b^0) + \bar{V}(P_a - P_b) + RT \ln(\gamma_a x_a / \gamma_b x_b) + zF(\psi_a - \psi_b) = 0. \tag{3.9}$$

Again the first term vanishes, for our systems, and the second is usually insignificant, so that Eq. (3.9) reduces to the Nernst equation:

$$RT \ln(a_a/a_b) + zF\psi_{ab} = 0. \tag{3.10}$$

This is often simplified by assuming that the activity ratio is the same as the concentration ratio, giving:

$$\psi_{ab} = -(RT/zF) \ln(c_a/c_b). \tag{3.10a}$$

This neglect of activity coefficients will often not be justified: so the writing of the Nernst equation in the form with concentrations should not be encouraged. Eq. (3.10) represents the equilibrium for the *isolated* transport of a single ion; if this transport is enzymically coupled to another transport or chemical reaction, the set of reactions so formed will have a quite different equilibrium condition.

3.3 Coupled Sets of Reactions

The simplest set of coupled reactions consists of two transport reactions tightly coupled together. Such a set is variously called a secondary active transport system, or an exchange-diffusion or co-transport system, or an antiporter or symporter (MITCHELL, 1967). If we take Y and Z to be the chemical species transported between a and b, we can represent their transport reactions by the usual chemical equations:

$$Y_a \rightleftharpoons Y_b,$$

and

$$Z_a \rightleftharpoons Z_b.$$

When these are summed with stoichiometric coefficients n_Y and n_Z we get the overall coupled reaction:

$$n_Y Y_a + n_Z Z_a \rightleftharpoons n_Y Y_b + n_Z Z_b.$$

The equilibrium for this coupled set will be given by:

$$n_Y(\bar{\mu}_{Y,a} - \bar{\mu}_{Y,b}) + n_Z(\bar{\mu}_{Z,a} - \bar{\mu}_{Z,b}) = 0. \tag{3.11}$$

Neglecting pressure terms this becomes:

$$n_Y RT \ln(a_{Y,a}/a_{Y,b}) + n_Z RT \ln(a_{Z,a}/a_{Z,b}) + (n_Y z_Y + n_Z z_Z) F\psi_{ab} = 0. \quad (3.12)$$

In the special case, often postulated, of an electrically neutral exchange, $n_Y z_Y$ equals $-n_Z z_Z$, and the equilibrium is not affected by ψ_{ab}, the condition being:

$$n_Y \ln(a_{Y,a}/a_{Y,b}) + n_Z \ln(a_{Z,a}/a_{Z,b}) = 0. \quad (3.13)$$

Symport reactions are represented by positive values of n_Y and n_Z; for antiport reactions n_Y and n_Z have opposite signs.

An exactly analogous treatment can be given of the case in which a chemical (scalar) reaction is supposed to be coupled to a transport (vector) reaction—the classic examples being the Na^+/K^+ ATPase of the animal cell membrane and the H^+ ATPase of specialized energy-transducing membranes. If the chemical reaction has a free-energy change of ΔG_{ch}, the equilibrium will be at:

$$n_{ch} \Delta G_{ch} + n_Y(\bar{\mu}_{Y,a} - \bar{\mu}_{Y,b}) = 0, \quad (3.14)$$

where the difference in electrochemical potential for the transported particle, Y, can be simplified as before. This case covers those systems known as primary active transport systems, with the addition where necessary of terms referring to another transported particle Z.

The reading of MITCHELL (1970) is recommended, which deals with coupling between transport and chemical reactions in an instructive way.

3.4 Using Equilibrium Equations

When mechanisms are unknown, the equilibrium relations for each major component of a biological system are the first tools to use in the search for a rational description of the observed phenomena. Surprisingly often, a biological system will be at, or can be brought to, an approximate equilibrium for the transport of one or more of its components. Apart from the well-known case of water (in most cells and tissues), some examples are:

Ca^{2+}, K^+, Cl^- etc.	in cell walls of e.g. water plants
K^+, Cl^-	in fibre of striated muscle
K^+	in some plant cells
Na^+	in giant axon (at peak of action potential)
$2H^+$ ATPase	postulated in *Chara* cell membrane
K^+	in mitochondria etc. treated with sufficient valinomycin
K^+/H^+ exchange	in mitochondria etc. treated with sufficient nigericin
H^+	in mitochondria etc. treated with sufficient dinitrophenol, etc.

The use of Eqs. (3.5a) and (3.7a) is straightforward, while the question of the activity coefficients is dealt with in the text-book by THAIN already cited, and in standard works on physical chemistry. If one is to be consistent in the application of S.I. units, Eq. (*3.5*a) is in pascal, while the units of RT are joule mole^{-1} and those of c are mole metre^{-3}. Similarly *3.10*a is in volts, and F in coulomb mole^{-1}, with RT in joule mole^{-1} as before. In the case of a coupled set of reactions, it is necessary only to decide the values of n, the stoichiometric coefficients, and in particular their signs. Thus for the $2H^+$ ATPase of *Chara*, the overall reaction postulated is:

$$ATP + H_2O + 2H_c^+ = ADP + H_3PO_4 + 2H_o^+.$$

This is the sum of partial reactions:

(i) $ATP + H_2O = ADP + PO_4$,
(ii) $2H_c^+ = 2H_o^+$.

If each moves from left to right, the respective molar values of ΔG are $\Delta G_{(i)} = \Delta G_{ATP} = -53$ kJ mol^{-1} (calculated from chemical data) and

$$\Delta G_{(ii)} = 2(\bar{\mu}_{H^+,o} - \bar{\mu}_{H^+,c})$$
$$= 2[\ln 10 \cdot RT(pH_c - pH_o) - F\psi_{co}].$$

When the outside pH is 6, that inside is about 7.7 (WALKER and SMITH, 1975) while ψ_{co} is about -180 mV.

Thus $\Delta G_{(i)}$ is about $+54$ kJ mol^{-1}, so that

$$\Delta G = \Delta G_{(i)} + \Delta G_{(ii)} = 1 \text{ kJ mol}^{-1}.$$

WALKER and SMITH (1975) used such a calculation to show that ψ_{co} in *Chara* is close to the equilibrium potential for a two-proton ATPase reaction, and that it could arise from such a mechanism. The term in $pH_c - pH_o$, whose origin is in the activity ratio term in $\bar{\mu}$, was small enough to be neglected in the calculations of SLAYMAN et al. (1973), for *Neurospora*.

3.5 Equilibrium as the Limit of the Possible

When equilibrium does not occur, equilibrium equations form the limits in inequalities based on the equations already discussed. They derive from the second law of thermodynamics, in the form that states that an isolated system will change spontaneously only under the condition that $\Delta G \leqq 0$.

So when a transport can be shown to have occurred, or better still to be occurring in the laboratory, for which $\Delta G \leqq 0$, it can be directly inferred that the species and the membrane do not form an isolated system, but interact energetically with some other component.

As an example, consider an ionic species found in a cell, whose interior it has reached from the environment. The condition $\Delta G \leqq 0$ for inward transport leads to $\bar{\mu}_c \leqq \bar{\mu}_o$. If we then find this condition disobeyed, as we do for Cl^- in the vacuole of *Chara*, where $\bar{\mu}_v - \bar{\mu}_o = 28$ kJ mol^{-1}, we are forced to conclude that this ion, the membrane and the compartments do not constitute an isolated system; an exchange of energy between the ion and some other component must be postulated. Deductions of this kind are of great value in the early stages of an investigation, and may even be resorted to later when other methods fail. For example, WALKER and SMITH (1975) calculated values of ΔG for the postulated coupled set:

$$Cl_o^- + H_o^+ \rightleftharpoons Cl_c^- + H_c^+,$$

and found positive values under some conditions.

4. Rate Equations (Class A: Mechanism – Independent)

The rate equations of class A, which are derived from the laws of non-equilibrium thermodynamics, contain no assumptions about transport mechanisms, but always do contain the restriction that the system is near to equilibrium, and the assumption (for mathematical convenience) that the flows are linear functions of the forces. Within these limits some rate equations have been developed which are of considerable importance, since they provide relations of the correct formal structure, containing the correct number of valid parameters.

4.1 Solvent and Non-Electrolyte Solute Transport

These are considered together because the formalism of non-equilibrium thermodynamics allows us to deal with any degree of interaction between transported species: classical thermodynamics was able to provide, as we have just seen, equations for the equilibrium of the transport of individual particles, or, at the other extreme, of tightly coupled sets of reactions. If M in Fig. 3.1 represents a single, uniform membrane (note these further assumptions) then non-equilibrium thermodynamics shows that for water and a single solute:

$$J_w = L_{ww}(\mu_{w,a} - \mu_{w,b}) + L_{ws}(\mu_{s,a} - \mu_{s,b}), \quad (3.15)$$

$$J_s = L_{sw}(\mu_{w,a} - \mu_{w,b}) + L_{ss}(\mu_{s,a} - \mu_{s,b}). \quad (3.16)$$

Since the Onsager relations identify L_{sw} as equal to L_{ws}, these equations define three parameters L which relate the two flows J to the two driving forces $\Delta\mu$. The parameters L are independent of the values of J and μ, but may e.g. be concentration-dependent. These equations are more commonly used in their re-arranged form:

$$J_v = L_p(P_a - P_b) + \sigma L_p(\pi_a - \pi_b), \quad (3.17)$$

$$J_D = \sigma L_p(P_a - P_b) + L_D(\pi_a - \pi_b), \quad (3.18)$$

where in the process of arranging that one flow, J_v, and both forces, ΔP and $\Delta\pi$, are the familiar and easily-measured ones, one has to replace the solute flow by J_D, the relative velocity of solute to solvent. J_s can be recovered as:

$$J_s = \omega_s(\pi_a - \pi_b) + J_v(1-\sigma)\,\bar{c}_s. \quad (3.19)$$

These equations, derived first by KEDEM and KATCHALSKY (1958), are discussed in the texts already referred to, and in HOUSE (1974). For more than one solute, Eq. (3.17) becomes:

$$J_v = L_p(P_a - P_b) + \sum_i \sigma_i(c_{i,a} - c_{i,b})\, RTL_p \tag{3.20}$$

while Eqs. (3.18) and (3.19) are replaced by sets of equations.

The three independent parameters defined by Eqs. (3.17) and (3.18) are L_p, L_D and σ: of these L_p, the hydraulic conductivity, and σ, the reflection coefficient, are of some importance to biologists. The former measures the permeability of the membrane to pressure-driven volume flow, while the latter measures the "osmotic effectiveness" of the solute for the particular membrane.

Both L_p and σ can be predicted by an argument based on an assumed mechanism, so that their values can be used as a test for the existence of that mechanism. If it is assumed that water is transported across a membrane by diffusion alone, and not by bulk flow, then:

$$L_p = P_w \bar{V}_w / RT, \tag{3.21}$$

where P_w is measured by a tracer experiment and L_p by a volume flow experiment. If this equation does not hold, and the measurement is not seriously affected by errors arising from unstirred layers, it can be inferred that bulk flow channels do exist in the membrane. If a membrane is assumed to have bulk flow channels into which solute molecules may enter, it can be shown (DAINTY and GINZBURG, 1963) that:

$$\sigma_s = 1 - P_s \bar{V}_s / RTL_p - A_{sf}/A_{wf}, \tag{3.22}$$

where the subscripts s refer to the solute in question, and the final term represents the result of frictional interaction in the channels between solvent, solute and membrane. The term in P_s represents the reduction in osmotic *volume* flow caused by the movement of solute volume in the opposite direction to that of solvent volume (refer again to Eq. (3.17)). A measurement of σ, P_s and L_p can therefore be used to evaluate the term A_{sf}/A_{wf}, which is non-zero if the solute can enter bulk flow channels. It is instructive to examine the papers of DAINTY and GINZBURG (1964), which are well discussed in HOUSE (1974).

Too little attention is commonly paid to the problems of applying these equations to biological membranes, which are not after all believed to be uniform in their properties but to be mosaiclike. The reader should consult HOUSE (1974) and then the papers of KEDEM and KATCHALSKY (1963). The important principle to grasp is that in a mosaic membrane there may be unobservable (but real) circulating flows of solvent or solute.

The application to systems in which the membrane is really two membranes separated by a compartment is still more fraught with danger, as DAINTY and GINZBURG (1964) pointed out. Some analysis relevant to this application is to be found in the references just quoted. Some of the surprising properties of two-membrane systems are derived and expounded by PATLAK et al. (1963). Here as elsewhere there is no substitute for the analysis of each case. The properties of a two-membrane system will depend for example on the constraints applied to the inner compartment (constant volume, constant pressure, constant content, etc.). They will be different also if there is any departure from the steady state.

No equations are quoted for the case of non-electrolyte flows coupled to chemical reactions—they could be obtained by reduction of the equations about to be discussed.

4.2 Solvent and Electrolyte Solute Transport

Two treatments are available for this case, both of which are instructive, though they have not been widely applied in practice. HOSHIKO and LINDLEY (1967) considered the case in which a and b contain a single electrolyte solute, and M contains a system which couples the transport of one of the ions of the electrolyte to a chemical reaction (one ion is actively transported). They arrived at the following set of practical equations, quoted in a form due to HOUSE (1974):

$$J_v = L_p \Delta P + \sigma L_p \Delta \pi - (P_E L_p/\kappa) I + V J_{ch}, \tag{3.23a}$$

$$J_s = P_s \Delta c_s + (1-\sigma) \bar{c}_s J_v + (\tau/F) I + U J_{ch}, \tag{3.23b}$$

$$I = (RT\kappa\tau/F\bar{c}_s) \Delta c_s + \kappa E_{ab} - P_E J_v + \varepsilon J_{ch}, \tag{3.23c}$$

$$J_{ch} = (RTkU/\bar{c}_s) \Delta c_s + (k\varepsilon/\kappa) I + (KV/L_p) J_v + k A_{ch}. \tag{3.23d}$$

These show how the flows depend on the forces (and, in this particular form, on other flows) by way of the ten parameters, as listed:

Flows	J_v	volume flux
	J_s	salt flux
	I	electric current density
	J_{ch}	rate of chemical reaction
Forces	ΔP	hydrostatic pressure difference
	Δc_s	salt concentration difference
	E_{ab}	difference in electric potential between electrodes reversible for the ion not subject to coupled transport ($E_{ab} = \psi_{ab} - \psi_j$)
	A_{ch}	affinity of the chemical reaction
Coefficients	L_p	hydraulic conductivity
	σ	reflection coefficient
	P_E	electroosmotic pressure coefficient
	P_s	salt permeability coefficient
	ε	electrogenicity coefficient
	U	salt-pump coefficient
	V	volume-pump coefficient
	k	chemical reaction rate coefficient
	τ	transfer number of the ion subject to coupled transport
	κ	electric conductance.

These equations, though complex, are worth study. They introduce the ideas that the unequal transport of cation and anion will produce the flow of an electric current, and that the operation of the pump may produce an electric current directly ($\varepsilon \neq 0$), and that the electric PD (which contributes to E_{ab}) will affect ion transport. Interactions between solvent and solute are expressed through the familiar coefficient σ, but since only one solute is assumed present, solute-solute interactions are not dealt with. These interactions are explicitly allowed for in a further set of equations for two salts and solvent, for which the original publication should be consulted.

The Hoshiko-Lindley equations also introduce the idea of electroosmosis and of its converse, streaming potentials. If in Eq. (3.23c), we set the chemical reaction rate J_{ch} at zero, and for simplicity also set c_a equal to c_b, then:

$$E_a - E_b = (P_E/\kappa) J_v , \tag{3.24}$$

so that a PD across the membrane can be set up by a volume flow. It is only necessary to imagine that $E_a - E_b$ is a stiff parameter (see 3.5.2) and J_v is pliant to see that a PD could set up a volume flow.

Note that these equations may be applicable to a mosaic membrane, but will not be applicable to a two-membrane system.

4.2.1 Flux Ratios

So far we have dealt only with net fluxes of components such as are in principle chemically determinable. The existence of isotopes which are physically distinguishable but nearly identical in chemical properties has made it possible to measure fluxes of particles "which reached phase y by time t_2 having been in phase x at time t_1"—so-called tracer fluxes. Though the first successful development and use of an equation in these fluxes was by Ussing (1949), we will quote first the more general (class A) form derived for a multicomponent system with a uniform membrane by Kedem and Essig (1965). They treated each particle (ion) as a separate entity, so that the electric current flow no longer appears explicitly; the processes which are represented explicitly are the flows of particles and of the chemical reaction (only one is considered) linked to transport. They obtain:

$$RT \ln (\phi_{j,ab}/\phi_{j,ba}) = (\mathcal{R}^*/\mathcal{R}) \left[(\bar{\mu}_{j,a} - \bar{\mu}_{j,b}) - \int_0^\delta \sum r_{jk} J_k \, dx - \int_0^\delta r_{jch} J_{ch} \, dx \right]. \tag{3.25}$$

The first term on the right-hand side expresses the effect of the electrochemical potential difference for j on its flux-ratio, the parameter ($\mathcal{R}^*/\mathcal{R}$), whose value is always positive, represents the effect of interactions between different particles of j as they cross the membrane. When they move quite independently, $\mathcal{R}^*/\mathcal{R} = 1$. The summation on the right-hand side represents the effects of all the other transported components on the flux-ratio of j—frictional drag or secondary active transports are represented by appropriate non-zero values of the r_{jk} coefficients. The remaining term in Eq. (3.25) represents the effect of the chemical reaction on the flux-ratio of i: the primary active transport of i would be represented by a non-zero value of r_{jch}.

USSING derived a less general form of this equation, by integrating the diffusion equation (it is thus literally a class B equation):

$$RT\ln(\phi_{j,a}/\phi_{j,b}) = (\bar{\mu}_{j,a} - \bar{\mu}_{j,b}) + (G_x/G_w)\ln(x_{w,a}\phi_{w,ab}/x_{w,b}\phi_{w,ba}) \tag{3.26}$$

where the passage of j across the membrane is by independent diffusion of single particles, and the second term on the right-hand side expresses the frictional drag of water on j, in channels through the membrane. There was no explicit term for active transport, the violation of Eq. (3.26) being taken as evidence for such a mechanism. Since USSING's successful application of Eq. (3.26) to transport by frog skin, the Equation has become rather popular, although its uses have not always been entirely proper. A common example is the use of Eq. (3.26) shorn of its drag term, as a test of active transport—rather than as a test for the absence of independent diffusion. It cannot be assumed that the "shorn" version of Eq. (3.26), namely:

$$RT\ln(\phi_{j,a}/\phi_{j,b}) = (\bar{\mu}_{j,a} - \bar{\mu}_{j,b}), \tag{3.26a}$$

will hold for any membrane. Thus MEARES and USSING (1959a and b) showed that for ion-exchange-resin membranes, Eq. (3.26a) was a rather bad fit to the data for Na^+ and Cl^- fluxes, and that a large correction for drag had to be made. For biological membranes Eq. (3.26a) has been shown not to fit the flux of K^+ in the giant axon (HODGKIN and KEYNES, 1955) and in the Characean cell (WALKER and HOPE, 1969). In each of these cases a parameter equivalent to $(\mathscr{R}^*/\mathscr{R})$ was found to have a value of about 2.5.

5. Rate Equations (Class B: Mechanism-Dependent)

5.1 Electric Potential Differences

So far the electric potential difference has appeared as a driving force affecting the transport of charged particles, though its presence in Eq. (3.25) for example is signalled only by the bar over the $\bar{\mu}_i$. Often we are interested in *predicting* the value of the steady PD, and it is appropriate to consider first briefly the mechanism by which electric PDs arise in membrane systems.

The presence of a non-zero electric charge density in an element of volume produces a change in the local electric field, given by Poisson's equation:

$$\rho\varepsilon_0/\varepsilon = \nabla^2\psi. \tag{3.27}$$

For the particular case of the parallel-plate condenser, which is applicable to most membrane problems, the integration of this equation gives:

$$\psi_{ab} = -q/C. \tag{3.28}$$

So that a potential difference ψ between a and b (Fig. 3.1) is caused by the displacement of charge q, from a to b, of magnitude given by Eq. (3.28). It is usual to assume that the actual charges whose displacement sets up PDs across biological membranes are ionic: though it seems extremely probable that in thylakoid, crista and procaryote membranes a net transport of electrons is achieved by specific systems, and that this is a major factor in determining ψ_{io} in these special cases.

The charges being ionic, the displacement of charge q is equivalent to the displacement of $q/\bar{z}F$ mol of substance, where $\bar{z}$ is a mean valence (weighted by transport number). There will be changes of concentration, if the volumes of a and b are finite, whenever q changes[3].

We should notice that the time differential of Eq. (3.28) is:

$$I_c/C = d\psi_{ab}/dt, \tag{3.28a}$$

where I_c is the displacement current which must flow in the external circuit whenever the PD ψ_{ab} across a capacitance is changing. If there are currents I_j which flow in parallel with the capacitance, as in a biological membrane in which the ionic conductances are in parallel with the capacitance, then the total current I is given by:

$$I = \sum_j I_j + C\, d\psi/dt. \tag{3.28b}$$

5.2 Stiff and Pliant Parameters

Before considering Equations that predict the value of the steady PD, it is desirable to draw attention to a matter that is usually implicit in their derivation or application. It seems useful to distinguish between those parameters of a system which are *stiff* i.e. whose values do not readily change, and those which are *pliant*, i.e. whose values are readily perturbed; the terms will be relative, not absolute, and will be meaningless unless the kind of perturbation is stated. If we consider a turgid plant cell bathed in a solution of an impermeant solute, and think of the solute concentration being perturbed, it is clear that the volume of the protoplast is relatively stiff, while the internal hydrostatic pressure is pliant to such a perturbation. A protoplast, with no cell wall, would in the same situation have a pliant volume and a stiff internal pressure, as would a plasmolysed cell. Consider a cell of volume $10^{-15}\,m^3$, of surface area $6 \cdot 10^{-10}\,m^2$, and let the interior (a) contain an impermeant anion whose concentration is $100\,mol\,m^{-3}$ (mM); the exterior (b) has an infinite volume and contains KCl at a concentration of $1\,mol\,m^{-3}$ (mM). The specific capacitance of the membrane is $10^{-2}\,F\,m^{-2}$: we will first think of it as permeable to K^+ alone. If the perturbation we consider is the transport of a small quantity of K^+ across M, say Q_K, then:

$$\frac{d\psi_{ab}}{dQ_K} = zF/C = F/C \tag{3.29}$$
$$= 1.6 \cdot 10^{16}\,V\,mol^{-1} \quad \text{in our example.}$$

---

[3] Consider a cubic cell of edge 10 µm, full of an aqueous solution and with a surface of capacitance $10^{-2}\,F\,m^{-2}$. Its capacitance C is then 6 pF, and the setting up of a PD of 200 mV will require the displacement of charge to be 1.2 pC. This is equivalent to an internal concentration change of $0.12\,mol\,m^{-3}$ (mM) of a univalent ion.

The Nernst Eq. (3.10) shows that we should consider also the perturbation of its second term by a change in Q_K:

$$\frac{d}{dQ_K}\frac{RT}{F}\ln([K]_b/[K]_a)=\frac{RT}{F}\left\{\frac{1}{[K]_b}\cdot\frac{d[K]_b}{dQ_K}-\frac{1}{[K]_a}\cdot\frac{d[K]_a}{dQ_K}\right\},$$
$$=-RT/F[K]_a V_a \quad \text{in our example,} \tag{3.30}$$
$$=-2.5\cdot 10^{14}\ \text{V mol}^{-1}.$$

The two terms in the Nernst equation then have different stiffness, and movement of K^+ will adjust ψ_{ab} much more rapidly than it will adjust the concentration ratio term, until equilibrium is reached. In such a system it is often indeed assumed that ψ_{ab} is completely pliant, and free to take up any value required by the concentration ratio term. Our sum indicates however, as MITCHELL (1966) showed in his classic paper, that objects somewhat smaller than cells, mitochondria for example, could have a rather stiff ψ and pliant concentration term. The reader may care to do the sum for an object of area $4\cdot 10^{-11}$ m^2 and of volume 10^{-18} m^3.

5.3.1 Equations for Steady Flow or PD

If one assumes that the PD and the ion flows are steady in time, and that the only source of electric current across the membrane is the transport of ions, then the PD will have that value which makes:

$$I=\sum_j I_j=0, \tag{3.31}$$

where the I_j are currents due to the transport of the individual species j, each I_j being a function of ψ_{ab}, and the sum being taken over all contributing (transportable) species. This relation is derived from Eq. (3.28b). The problem then is to find the form of the function $I_j(\psi_{ab})$, which involves (i) assuming a transport mechanism and (ii) performing an integration so that the function I_j contains only externally observable quantities such as concentrations and potentials in a and b.

If all the transported species have the same valence, z, it is possible to show that the following equation describes the PD between solutions separated by a membrane in which ions diffuse without energetic interactions:

$$\psi_{ab}=-(RT/zF)\ln\left(\sum_j P_j a_{j,a}\Big/\sum_j P_j a_{j,b}\right), \tag{3.32}$$

where the summation is over all the transported species, and where the parameters P_j are permeability coefficients, and are independent of changes in external solutions etc. This Equation is well discussed by SANDBLOM and EISENMAN (1967), who show that it can be expected to hold for a mosaic membrane. The permeability coefficients appear only as ratios in this equation, as can be clearly seen if we divide by say P_K:

$$\psi_{ab}=-(RT/F)\ln\left\{\frac{[K^+]_a+\sum\alpha_j a_{j,a}}{[K^+]_b+\sum\alpha_j a_{j,b}}\right\}, \tag{3.33a}$$

so that P_K can never be determined from measurement of ψ_{ab}. By symmetry, no single P_j can be found. The permeability ratios α_j can be found, however; in the simple case of a uniform membrane whose surfaces are in equilibrium with the solutions:

$$\alpha_j = P_j/P_K = b_j u_j / b_K u_K, \tag{3.33b}$$

where b are partition coefficients and u are mobilities. See the discussion by JOHNSON et al. (1954, p. 572) on this point. The significance of partition coefficients in ionic selectivity is well reviewed by DIAMOND and WRIGHT (1969).

In the case of transported ions of any valence, GOLDMAN (1943) showed that for a uniform membrane in which the electric field is constant, the current I_j carried by each ion has the form:

$$I_j = -z_j^2 P_j (F^2 \psi_{cd}/RT) \left\{ \frac{a_{j,c} - a_{j,d} \exp(-F\psi_{cd}/RT)}{1 - \exp(-F\psi_{cd}/RT)} \right\}. \tag{3.34}$$

In this equation the phase labels a and b have been replaced by labels c and d, which refer to the (inaccessible) membrane faces in equilibrium with a and b.

In two common cases of transported ions of mixed valence, this relation for I_j has been combined with Eq. (3.31) to give relations between ψ and ion concentrations. If we write S_c^+ to mean the sum, over all species of valence $+1$, of $P_j a_c$ it will allow us to write the Eqs. as:

(i) for $z_j = \pm 1$,

$$\psi_{cd} = (RT/F) \ln (S_d^+ + S_c^-)/(S_c^+ + S_d^-), \tag{3.35}$$

(ii) for $z_j = +1$ or $+2$,

$$\psi_{cd} = (RT/F) \ln \left\{ \frac{(S_d^+ - S_c^+) + 4(S_d^+ + 2S_d^{2+})(S_c^+ + 2S_c^{2+})^{\frac{1}{2}} + (S_d^+ - S_c^+)}{2(S_c^+ + 2S_c^{2+})} \right\}. \tag{3.36}$$

Early derivations of these relations (e.g. HODGKIN and KATZ, 1949) tended to ignore problem of potential differences at the membrane interfaces a–c and d–b. Writing Eqs. (3.35) and (3.36) as equations in ψ_{ab} and $S_a^{\pm}$ and $S_b^{\pm}$ is equivalent to assuming the membrane surface PDs to be zero, as JOHNSON et al. (*loc. cit.*) showed. Compared then with Eq. (3.32), Eq. (3.35) has lost generality in two ways: a linear gradient $\Delta\psi$ is assumed, and the problem of surface potentials has become unresolved. A more general, mathematical discussion of the range of applicability of Eqs. (3.32) and (3.35) is provided by SANDBLOM and EISENMANN (1967), which the reader should consult. They showed that Eq. (3.35) might be expected to hold, with constant values of α_j, for a membrane with independent diffusion only, if the total ionic concentrations in a and b were equal. This deals with both the objections to Eq. (3.35) mentioned above. It is not likely to be true of most plant cells, however.

As an alternative to equations based on the Goldman assumption, the Planck-Teorell equations are based on the assumption that at every point in the membrane there is zero net charge. The reader is referred to PITMAN (1969) for an example of the use of these equations, and to HOPE and WALKER (1961).

Note that the net flux of an ion, J_j, in mol $m^{-2} s^{-1}$, is given by $J_j = I_j/z_j F$, so that Eq. (3.34) yields the net flux. It yields also the tracer fluxes, ϕ_{cd} and ϕ_{dc}, if $a_{j,d}$ and $a_{j,c}$ respectively are set to zero.

As all uses of equations of this kind are more or less empirical, argument about the assumptions generally gives place to considerations of fit to the data (HOPE and WALKER, 1961) or of convenient form. Eqs. (3.35) and (3.36) should only be used as operational definitions of the permeability ratios α_j, which cannot be assumed to be constants until they are shown to be so.

5.3.2 PDs in Coupled Systems

BRIGGS (1962) pointed out that a more general assumption than that the independent ion currents summed to zero (Eq. (3.31)), was that:

$$I = \sum_j I_j + I_{ch}, \tag{3.37}$$

where I_{ch} is the current produced by a set of coupled reactions (active transport system). For simplicity he took I_{ch} to be constant, depending on metabolism but not on ψ or on ion concentrations. His equation did not give ψ explicitly, but allowed it to be calculated. Using the Goldman assumption, he showed that:

$$\begin{aligned}[\mathrm{K}]_a + \alpha[\mathrm{Na}]_a + \beta[\mathrm{Cl}]_b = &-(RT/F)(I_{ch}/P_K \psi_{ba})(1 - \exp(F\psi_{ba}/RT)) \\ &+([\mathrm{K}]_b + \alpha[\mathrm{Na}]_b + \beta[\mathrm{Cl}]_a)\exp(F\psi_{ba}/RT).\end{aligned} \tag{3.38}$$

This can be re-arranged to give:

$$\begin{aligned}\psi_{ba} = &(RT/F) \\ &\cdot \ln\left\{\frac{P_K[\mathrm{K}]_a + P_{Na}[\mathrm{Na}]_a + P_{Cl}[\mathrm{Cl}]_b + (I_{ch}RT/F\psi_{ba})(1 - \exp(F\psi_{ba}/RT))}{P_K[\mathrm{K}]_b + P_{Na}[\mathrm{Na}]_b + P_{Cl}[\mathrm{Cl}]_a}\right\},\end{aligned} \tag{3.39}$$

where again the problem of the membrane surfaces is ignored. Eq. (3.39) can be simplified by assuming that the term in (1 − exp) is small; the Equation then is very similar to that of MORETON (1969). (This paper is an instructive example of casting old equations in new and more useful forms: its reading is recommended.)

It would be a more natural assumption nowadays that I_{ch} would be a function of ψ_{ba} and of the ionic concentrations, as implied by the HOSHIKO-LINDLEY Eqs. (3.23a to d). No coupled system is well enough known however to set up equations for the functional dependence of I_{ch} on ψ. The way forward would seem to be to determine this function empirically, relying on a formalism such as that of HOSHIKO and LINDLEY (1967).

5.4.1 Equations for Conductance

Differentiation of Eq. (3.34) with respect to ψ_{ab} yields equations for the electric con- conductance due to ion j; and the membrane conductance is given by:

$$g_M = \sum_j g_j. \tag{3.40}$$

If an ion is at equilibrium, its partial conductance g_j can be got (without the constant-field assumption) as:

$$g_j = (\mathscr{R}^*/\mathscr{R})\, F^2 \phi_j / RT. \tag{3.41}$$

This equation, due in its original form to HODGKIN and KEYNES (1955), is useful as not being invalidated by self-interaction, which is allowed for in the factor $(\mathscr{R}^*/\mathscr{R})$. Equations based on the Goldman assumption are not valid if self-interaction occurs (as it does for K^+ in *Chara* and *Nitella*—WALKER and HOPE, 1969).

5.4.2 Conductance in Coupled Systems

A case of some interest is that in which a membrane contains a mechanism coupling a chemical reaction to an ion transport, e.g. that of H^+. If this reaction is not too far from equilibrium, as seems possible for *Neurospora* and more likely for *Chara* cell membranes (SLAYMAN et al., 1973; WALKER and SMITH, 1975) then the formalism of irreversible thermodynamics may guide us to write:

$$I_{ch} = nzF\,J_{ch}, \tag{3.42}$$

where J_{ch} is the reaction rate of the reaction

$$n\mathrm{H_a^+} + \mathrm{ATP} + \mathrm{H_2O} = n\mathrm{H_b^+} + \mathrm{ADP} + \mathrm{H_3PO_4},$$

and where near equilibrium we can assume (cf. 3.3) linearity of flow with force:

$$J_{ch} = L_{ch}[n(\bar{\mu}_{H^+,a} - \bar{\mu}_{H^+,b}) + \Delta G_{ATP}]. \tag{3.43}$$

The membrane conductance due to this mechanism is:

$$g_{ch} = \partial I_{ch}/\partial \psi_{ab}, \quad \text{whence:}$$
$$= n^2 F^2 L_{ch}. \tag{3.44}$$

Assumptions about mechanism would be necessary to find an expression for L_{ch}, but biological or physiological isolation of the transport ATPase should allow the separate estimation of its magnitude. The results of SPANSWICK (1972, 1974) already suggest that L_{ch} may be a function of [ATP]. Again, it seems that empirical measurement of the components of conductance will be more profitable than the construction of equations based on simple assumed mechanisms. What is known of the kinetics of the mitochondrial ATPase (MITCHELL and MOYLE, 1971) makes it unlikely that simple assumptions will represent the system at all well.

References

BRIGGS, G.E.: Membrane potential differences in *Chara australis*. Proc. Roy. Soc. (London), Ser. B **156**, 573–577 (1962)

DAINTY, J.: The polar permeability of plant cell membranes to water. Protoplasma **57**, 220–228 (1963).

DAINTY, J., GINZBURG, B.Z.: Irreversible thermodynamics and frictional models of membrane processes, with particular reference to the cell membrane. J. Theoret. Biol. **5**, 256–265 (1963).

DAINTY, J., GINZBURG, B.Z.: The reflexion coefficient of plant cell membranes for certain solutes. Biochim. Biophys. Acta **79**, 129–137 (1964).

DIAMOND, J.M., WRIGHT, E.M.: Biological membranes: The physical basis of ion and non-electrolyte selectivity. Ann. Rev. Physiol. **31**, 581–646 (1969).

GOLDMAN, D.E.: Potential, impedance and rectification in membranes. J. Gen. Physiol. **27**, 37–60 (1943).

HODGKIN, A.L., KATZ, B.: The effect of sodium ions on the electrical activity of the giant axon of the squid. J. Physiol. (London) **108**, 37–77 (1949).

HODGKIN, A.L., KEYNES, R.D.: The potassium permeability of a giant nerve fibre. J. Physiol. (London) **128**, 61–68 (1955).

HOPE, A.B., WALKER, N.A.: Ionic relations of cells of *Chara australis* IV. Membrane potential differences and resistances. Australian J. Biol. Sci. **14**, 26–44 (1961).

HOSHIKO, T., LINDLEY, B.D.: Phenomenological description of active transport of salt and water. J. Gen. Physiol. **50**, 729–758 (1967).

HOUSE, C.R.: Water transport in cells and tissues. London: Edward Arnold 1974.

JOHNSON, F.H., EYRING, H., POLISSAR, M.J.: The kinetic basis of molecular biology. New York: Wiley and Sons 1954.

KATCHALSKY, A., CURRAN, P.D.: Non-equilibrium thermodynamics in biophysics. Cambridge, Mass.: Harvard University Press 1965.

KEDEM, O., ESSIG, A.: Isotope flows and flux ratios in biological membranes. J. Gen. Physiol. **48**, 1047–1070 (1965).

KEDEM, O., KATCHALSKY, A.: Thermodynamic analysis of the permeability of biological membranes to non-electrolytes. Biochim. Biophys. Acta **27**, 229–246 (1958).

KEDEM, O., KATCHALSKY, A.: Permeability of composite membranes. 1, 2 and 3. Trans. Faraday Soc. **59**, 1918–1930; **59**, 1931–1940; **59**, 1941–1953 (1963).

MEARES, P., USSING, H.H.: The fluxes of sodium and chloride ions across a cation-exchange resin membrane. 1. Trans. Faraday Soc. **55**, 142–155 (1959a).

MEARES, P., USSING, H.H.: The fluxes of sodium and chloride ions across a cation-exchange resin membrane. 2. Trans. Faraday Soc. **55**, 244–254 (1959b).

MITCHELL, P.: Chemiosmotic coupling in oxidative and photosynthetic phosphorylation. Biol. Rev. Cambridge Phil. Soc. **41**, 445–502 (1966).

MITCHELL, P.: Translocations through natural membranes. Advan. Enzymol. **29**, 33–87 (1967).

MITCHELL, P.: Reversible coupling between transport and chemical reactions. In: Membranes and ion transport (E.E. BITTAR, ed.), vol. I, p. 192–356. New York: Wiley and Sons 1970.

MITCHELL, P., MOYLE, J.: Activation and inhibition of mitochondrial adenosinetriphosphatase by various anions. J. Bioenerg. **2**, 1–11 (1971).

MORETON, R.B.: An investigation of the electrogenic sodium pump in snail neurones, using the constant-field theory. J. Exptl. Biol. **51**, 181–201 (1969).

PATLAK, C.S., GOLDSTEIN, C.A., HOFFMAN, J.F.: The flow of solute and solvent across a two-membrane system. J. Theoret. Biol. **5**, 426–442 (1963).

PITMAN, M.G.: Simulation of Cl^- uptake by low-salt barley roots as a test of models of salt uptake. Plant Physiol. **44**, 1417–1427 (1969).

SANDBLOM, J.P., EISENMAN, G.: Membrane potentials at zero current. The significance of a constant ionic permeability ratio. Biophys. J. **7**, 217–242 (1967).

SLAYMAN, C.L., LONG, W.S., LU, C.Y.-H.: Electrogenic H^+ transport in *Neurospora*. J. Membrane Biol. **14**, 305–338 (1973).

SPANNER, D.C.: The active transport of water under temperature gradients. Symp. Soc. Exptl. Biol. **8**, 76–93 (1954).

SPANSWICK, R.M.: Evidence for an electrogenic pump in *Nitella translucens*. I. Biochim. Biophys. Acta **288**, 73–89 (1972).

SPANSWICK, R.M.: Evidence for an electrogenic pump in *Nitella translucens*. II. Biochim. Biophys. Acta **332**, 387–398 (1974).

THAIN, J.F.: Principles of osmotic phenomena. London: Roy. Inst. Chem. 1967.

USSING, H.H.: The distinction by means of tracers between active transport and diffusion. Acta Physiol. Scand. **19**, 43–56 (1949).

WALKER, N.A., HOPE, A.B.: Membrane fluxes in electric conductance in characean cells. Australian J. Biol. Sci. **22**, 1179–1195 (1969).

WALKER, N.A., SMITH, F.A.: Intracellular pH in *Chara corallina* measured by DMO distribution. Plant Sci. Letters **4**, 125–132 (1975).

4. Electrical Properties of Plant Cells: Methods and Findings

G.P. FINDLAY and A.B. HOPE

1. Methods

1.1 The Potential Difference

The starting point in any study of the electrophysiological properties of cells is the measurement of the electrical potential difference between two phases or between two points in one phase. Changes in the potential difference resulting from an applied electric current may then be used to calculate resistance and capacitance.

1.1.1 The Potential Difference between Aqueous Phases

In biological systems it is a potential difference between two aqueous phases that is usually measured, but it is necessary to eliminate as far as possible, or at least take into account, potential differences that are present at metal-liquid or liquid-liquid junctions between the measuring electrodes and the aqueous phases. The usual procedure is to place a KCl solution between the liquid phase of the measuring electrode and the aqueous phase, so introducing two diffusion potentials instead of one. But the sum of these is considerably smaller than the original because the transport numbers of K^+ and Cl^- are almost identical, and fairly independent of concentration.

Calomel half-cells are often used to measure the potential difference (PD) as shown in the following diagram:

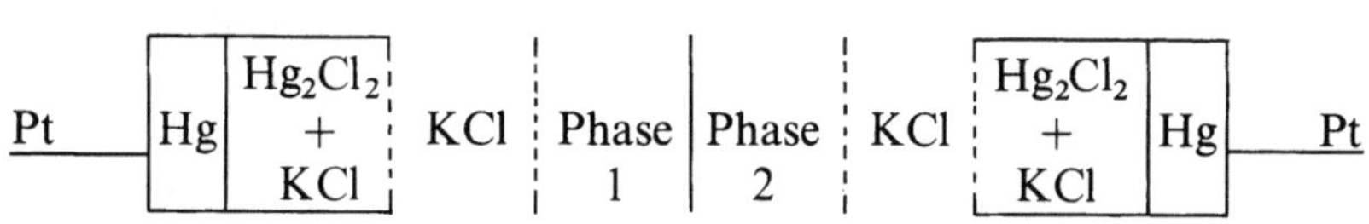

The PD, ψ, appearing between the platinum terminals is given by:

$$\psi = \psi_{2\text{-}1} + \psi_{\mathrm{KCl/Phase\ 2}} - \psi_{\mathrm{KCl/Phase\ 1}} \tag{4.1}$$

as the other junction PD's $\psi_{\mathrm{Hg/Hg_2Cl_2}}$ and $\psi_{\mathrm{Hg_2Cl_2/KCl}}$ cancel out between the phases. Provided the junction PDs between KCl and phases 1 and 2 are small then $\psi \simeq \psi_{2\text{-}1}$. The interposed KCl solution is referred to as a salt bridge, and experimentally may have a concentration in the range 0.1 to 3.0 M KCl, and is often contained in a gel of 1–3% agar.

1.1.2 Microelectrodes

The electrical properties of cells are largely determined by membranes bounding the cells and phases within cells (HOPE, 1971). Consequently most electrical investigations involve the measurement of PD between the interior of a cell and the outside milieu. A major experimental problem particularly with cells of higher plants is the connection of a KCl salt-bridge from the electrode to the interior of the cell, the volume of which may be as small as $2 \cdot 10^{-15}$ m^3. The connection must be made through the cellulose wall and the cell membranes, without disturbing the cell. The usual practice is to make this connection with a micro-salt-bridge. Usually this is referred to as a microelectrode, although strictly speaking the terminology is incorrect as the electrode itself is not miniaturized.

The microelectrode is a pyrex glass tube, drawn out at one end to a fine point of diameter ranging from 0.5 µm to 10 µm, and filled with 3 M KCl. With suitable mechanical and visual control the microelectrode can be inserted into a cell. In electrical terms the KCl-filled microelectrode is simply an insulated conductor connected to the appropriate phase in the cell. The connection to the solution bathing the cell is made by placing there a salt-bridge, usually consisting of 3 M KCl in 2% agar to minimize diffusion of KCl into the bathing solution and of diameter considerably greater than that of the microelectrode. The potential difference between the electrodes connecting micro-salt-bridge and reference salt bridge is then a measure of the potential difference between the inside of the cell and the outside. The PD between two phases within the cell can be measured between microelectrodes inserted into each phase. Calomel half-cells are most commonly used as electrodes for electrical measurements in plant cells. The Ag/AgCl electrode is also used. For ease of use a plastic tube containing 3 M KCl in 2% agar forms a flexible connection between the electrode and the microelectrode.

Manufacture. Microelectrodes can be made either by hand or with a machine. The method for construction by hand is as follows. The middle 0.5 to 1 cm of a piece of pyrex glass tubing, length 10–15 cm and diameter 1–4 mm, is heated to red heat in a flame. The tube is then removed from the flame and the ends quickly pulled apart until the two parts of the tube separate. With experience and some skill it is possible to make microelectrodes with tip diameters down to 5 µm. A somewhat more reliable method consists of drawing out one end of a glass tube to a capillary of diameter 20–50 µm, and forming a hook at this end. A weight attached to the hook will provide the force to draw out the tube to the required diameter when a small flame or heated filament is brought close to the capillary.

In microelectrode-pulling machines the microelectrodes are usually drawn out in two stages. A length of glass tube is usually held vertically by a chuck at each end. The chuck at one end is fixed, and at the other is attached to a sliding iron arm that is surrounded by an electromagnet. An electrically-heated filament surrounds a short middle section of the tube, and as the glass is heated the weight of the arm draws out the tube. At a pre-determined stage an extra force is applied by switching on electric current to the electromagnet. In some machines the direction of movement is horizontal, and the two forces pulling the microelectrode are separately controlled by changing the current through the electromagnet at an appropriate time.

There are several variables which determine the shape and tip diameter of a microelectrode, but temperature is the most important variable. The higher the filament temperature, the longer will be the microelectrode. A detailed account of the manufacture of microelectrodes is given by FRANK and BECKER (1964).

Filling. There are various methods for filling microelectrodes with solutions of electrolyte. If the tip diameter is greater than 5 μm the simplest procedure is to fill part of the shank of the microelectrode from a syringe with a square-cut end and then apply pressure, either from the syringe with its end pressed against the tapered part of the microelectrode, or by removing the syringe and applying air pressure above the solution in the shank. Another method is to fill the microelectrode as close to the tip as possible, and then slide a thin glass rod, drawn out to a fine point, down into the tapered part of the microelectrode, and by moving it back and forward gradually eliminate air bubbles until solute flows by capillary action to the tip.

The latter method can also be used to fill microelectrodes with tip diameters down to about 1 μm but there are several other methods commonly used when tip diameters are in the range 0.5–2 μm. These methods, which have the virtue that a number of microelectrodes can be filled at the one time, involve boiling microelectrodes in 3 M KCl, either at atmospheric pressure, or under reduced pressure. A gentler technique, which causes less breakage of very fine tips is first to boil the microelectrodes in methyl alcohol under reduced pressure, and then transfer them to 3 M KCl solution, where over a period of 1 to 2 days the methyl alcohol is replaced by KCl by diffusion.

Tip Diameter and Electrical Resistance. Microelectrodes for vacuolar insertions into cells of charophyte plants should have a tip diameter in the range from 2–10 μm, to ensure gushing of vacuolar sap into the microelectrode (see *4*.1.2.1). For cytoplasmic insertions a tip diameter of 1–2 μm is usually satisfactory. The electrical resistance of a microelectrode filled with 3 M KCl ranges from about 100 KΩ for a tip diameter about 10 μm to about 1 MΩ for a tip diameter of 1 μm. The tip diameter of a microelectrode can be measured with the light microscope for diameter 1 μm or greater. Often, however, small particles unseen under the microscope block the tip, and a measurement of the electrical resistance of the microelectrode will give a value higher than normal.

Microelectrodes for insertion into cells of higher plants should have tip diameters less than 1 μm, a value at the limit of resolution of the light microscope. Although the tip diameter can be estimated from the electrical resistance of the microelectrodes, such estimates are not always reliable as apparently similar microelectrodes can have widely differing resistances.

ROBINSON and SCOTT (1973) have devised a simple and reliable method for estimating tip diameter from the rate at which microelectrodes fill with electrolyte. The barrel of a microelectrode held in the air is filled with a syringe to within 1 mm of the tip. The electrolyte then moves towards the tip by capillary action, and the remaining air is expelled through the tip. The internal diameter d_0 of the tip is then estimated from the rate at which the meniscus of the KCl solution moves towards the tip, and is given by:

$$d_0 = d_1 \, (4 v_1 \eta / 3 \tau \tan \theta)^{1/3} \qquad (4.2)$$

where v_1 is the speed of the meniscus at a point where the internal diameter is d_1, η is the viscosity of air, 2θ is the taper angle, and τ is the surface tension for the electrolyte. This method is a sensitive indicator of microelectrode tip diameter since $v_1 \propto d_0^3$.

1.1.3 Micromanipulators

The insertion of a microelectrode into a plant cell requires adequate visual control under the microscope, and reasonably precise mechanical control of the movement. Although there are many types of micromanipulators (KOPAC, 1964), those with a screw movement in each of the three dimensions have proved satisfactory in electrophysiological studies of plant cells. The main requirement for a micromanipulator is that the microelectrode can be advanced smoothly in a straight line, without rotation or vibration, up to and into the cell. A disadvantage of this type of micromanipulator is that the forward velocity of the microelectrode is necessarily small because of the gearing-down of the angular velocity of the screw. An increased velocity can be imparted to the microelectrode, often ensuring a better insertion, if the micromanipulator or the apparatus holding the cell is gently tapped with the hand after the microelectrode tip has been pushed up against the cell wall. Micromanipulators which move microelectrodes with velocities of up to 10^4 µm s^{-1} have been described by ELLIS (1962) and LASSEN and STEN-KNUDSEN (1968). In these micromanipulators the movement is produced by the bending of piezoelectrical material when an electric PD is applied to it. These types of micromanipulators have more obvious application for electrical measurements in animal cells, particularly red blood cells, but could also be used to insert microelectrodes into the small cells of higher plants.

1.1.4 Electrometers

In general the PD between a microelectrode and the reference salt-bridge will require amplification before it can be measured and suitably displayed, because the power available from the biological source is insufficient to drive the measuring devices. For instance, in a cell of *Chara corallina* the PD between the vacuole and the external solution is about -150 mV, and the total resistance is about $2 \cdot 10^4\ \Omega$. If (for example) the electrometer is not to change the PD by more than 0.1% or 0.15 mV, the current flowing in the measuring circuit must not exceed $7.5 \cdot 10^{-8}$ A. Thus the permissible power dissipation in the cell and the microelectrode of total resistance 5 MΩ is $(7.5 \cdot 10^{-8})^2 \cdot 5 \cdot 10^6$, about $3 \cdot 10^{-8}$ W. By comparison, the power input required for a potentiometric recorder is of the order of 10^{-3} W. In the small cells of higher plants where the total resistance could be 200 MΩ or more, the permissible power dissipation would be as low as 10^{-14} W. Thus a power amplification of between 10^5 and 10^{11} is required for the measurements of PD's in plant cells. Furthermore, the input resistance to the amplifier, being in series with the combined resistance of the cell and micropipette, needs to be 1,000 times greater than this combined resistance in order that the measured PD is within 0.1% of the actual PD.

High-performance operational amplifiers with a field-effect transistor input and integrated circuits provide a suitable first stage for an electrometer that meets the requirements described above. Such amplifiers have an imput impedance of $10^{11}\ \Omega$ or greater, and are considerably more stable than high-input-impedance thermionic electrometer valves. The output from the electrometer is usually displayed on a chart recorder, or a cathode ray oscilloscope.

1.2 What an Inserted Microelectrode Actually Measures

1.2.1 Charophyte Cells

The insertion of a microelectrode into a plant cell can often be observed under the microscope, and the location of the tip determined visually. WALKER (1955) described micro-electrode experiments, under such visual control, with *Nitella sp.*; see Fig. *4*.1. He found that as the microelectrode was inserted into the cell, the turgor pressure in the cell forced vacuolar sap into the tip, thus breaking any cytoplasm surrounding the tip, and ensuring that the electrolyte in the microelectrode made contact with the vacuolar sap. After this gush of sap into the microelectrode, the tip was almost immediately blocked by particles, some of which were crystalline, suspended in the sap.

WALKER also found that the cytoplasm gradually welled up along a microelectrode which had been inserted into the vacuole and eventually flowed over the tip. The potential measured by the microelectrode changed by only about 10 mV when this occurred, but later, when a callose deposition had also built up along the microelectrode, and eventually covered the tip, the magnitude of the potential decreased by about another 50 mV. WALKER concluded that the potential of the cytoplasmic phase was about 10 mV more negative than the vacuolar phase, and that the larger change in potential resulted from a sealing process, in which a new plasmalemma was formed over the microelectrode, thus effectively excluding it from the cell (cf. UMRATH, 1932). WALKER's experiments have provided a basis for modern electrophysiological studies of plant cells. FINDLAY (1959, 1961) measured the action potential in cells of *Nitella sp.* either with a microelectrode inserted to a depth of 20 to 50 µm, and subsequently covered with cytoplasm, or inserted completely across the cell, with tip in the cytoplasm opposite. He concluded, by comparison with measurements from microelectrodes in the vacuole, that the action potential occurred almost completely at the plasmalemma, and that practically no transient in PD took place at the tonoplast. However, FINDLAY and HOPE (1964a) inserted thickwalled microelectrodes, with external diameter about 2 µm and internal diameter less than 0.5 µm, into cells of *Chara corallina* to a depth of 20 to 50 µm, without a gush of sap into the tip and the usual accompanying stoppage of cytoplasmic streaming. Under these conditions the tip of the microelectrode remained in the cytoplasm, and steady potentials could be recorded for several minutes before a sudden sealing of the tip, and consequent loss of potential, occurred. It was found that the cytoplasmic potential was consistently more negative, by up to 20 mV, than the vacuolar potential. During the action potential, there were changes in potential difference at both plasmalemma and tonoplast. Further observations have clearly demonstrated that consistent measurements of cytoplasmic potential are only obtained when the microelectrode is inserted 20–50 µm into the cell and directly into the cytoplasm, without gushing and stoppage of cytoplasmic streaming. Often in *Chara corallina* a microelectrode initially inserted into the vacuole but whose tip subsequently becomes covered with cytoplasm, records a fluctuating potential somewhere between the cytoplasmic and vacuolar levels. An action potential recorded under these conditions is close to that appropriate to the vacuolar phase.

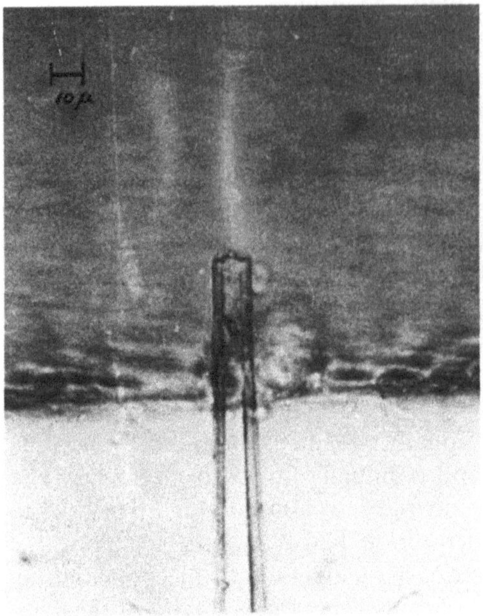

(a)

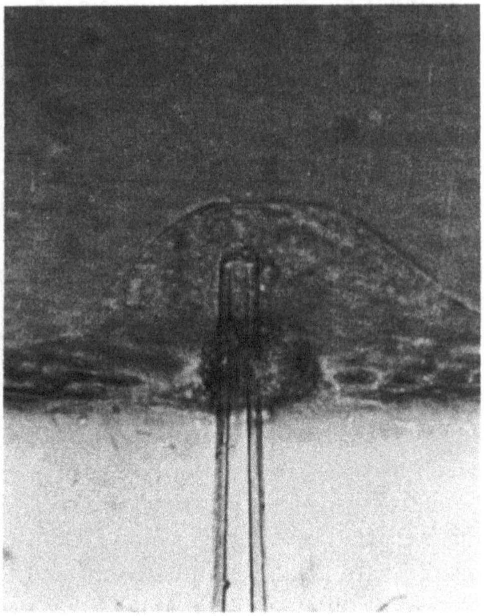

(b)

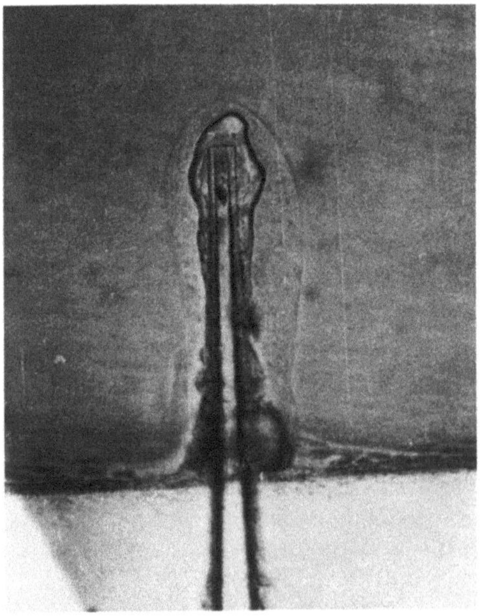

(c)

Fig. *4*.1 a–c. Photographs of a micro-electrode tip inserted into a *Nitella* cell. This tip is 10 µm in diameter, rather larger than those generally used. (a) Tip in vacuole immediately after insertion. (b) Tip covered with cytoplasm, 3 min later. (c) Tip covered with thick seal. (From WALKER, 1955)

1.2.2 Higher Plant Cells

From the experimental point of view, charophyte cells have the distinct advantage of large size. Cells of *Chara corallina* used for electrophysiological studies typically have volumes of 10 to 20 mm^3. Cells of higher plants, such as *Pisum, Avena, Hordeum* and *Elodea* have volumes of the order of 10^{-5} mm^3 and the volume contained in the final 20 μm length of microelectrode of 8° taper is then approximately 1% of the cell volume. Thus, unless the tip of the microelectrode is rapidly plugged by material in the cell when it is inserted, there will be a decrease in cell volume sufficient to alter considerably the turgor pressure of the cell. There is a number of reports of difficulties in obtaining steady measurements of PD in cells of higher plants (POOLE, 1966). One of the reasons for instability in the recorded potential may be that the tip is in the vacuole, and there is a loss of sap into the microelectrode. Another perhaps more obvious reason for instability in the PD is spasmodic electrical leakage in the region of insertion. Various authors, e.g. ETHERTON and HIGINBOTHAM (1960), GREENHAM (1966), POOLE (1966), HIGINBOTHAM et al. (1970) have published results of both vacuolar and cytoplasmic measurements, with the microelectrode tip located visually. However, the observations of FINDLAY (1962), FINDLAY and HOPE (1964a) and POOLE (1966) suggest that even with the tip of the microelectrode apparently in the vacuole, a true vacuolar potential may not be recorded because the tip may still be covered with cytoplasm or membrane material.

An interesting new approach was made by LÜTTGE and ZIRKE (1974) who inserted microelectrodes into cells of centrifuged leaves of the moss *Mnium cuspidatum*. These authors felt that their estimates of ψ_{co}, based on visual location of the tip of the microelectrode in the transparent cytoplasm, were reliable. However, two sets of values were obtained for ψ_{vc}, depending on conditions of centrifugation and time of cultivation of the moss in the laboratory, so that estimates of the tonoplast potential difference remained ambiguous.

1.3 Measurement of Membrane Resistance

The electrical resistance of a membrane can be estimated from the steady change in PD produced by the passage of an electric current through the membrane. Experimentally, the simplest procedure is to apply a current from a high impedance source. In practice this is obtained by placing a large resistance, about 10^9 Ω, in series with the membrane resistance in the current circuit. If it is assumed that the cell membrane does not have an inductive component, any changes in PD appearing when the current flows will be due to the resistive and capacitative elements of the membrane.

The most usual measurement made is that of the resistance of the membrane at its resting PD, given by $R=(\partial\psi/\partial I)_{I\to 0}$; but if ψ as a function of I over a range of values of I is also determined, the slope resistance, $\partial\psi/\partial I$, or the chord resistance, $\Delta\psi/\Delta I$ can also be calculated. These quantities may be related to the resistance of unit area of the membrane (units Ω m^2) provided the current density is uniform (see *4*.1.3.1 below). Sometimes the membrane conductance is more useful than resistance, for example in discussing ion fluxes, and is the reciprocal of resistance; the units, if referred to unit membrane area, are S m^{-2}.

1.3.1 Giant Algal Cells

The usual procedure for measuring resistance of membranes in a giant algal cell is the insertion of one microelectrode into the cell to pass current, and one or two others to measure the potential of the vacuole and the cytoplasm. The cells are sufficiently large to allow the insertion of three microelectrodes without appreciably affecting the electrical properties of the cells. However, the time between insertion of the microelectrodes and the measurement of resistance should be considered, because SPANSWICK (1970) has shown that for cells of *Nitella translucens* the measured vacuolar resistance increased by 50% during the first 5 h after the careful insertion of the microelectrodes. This result does not appear to have been found in other charophyte cells.

The geometry of cells also has to be taken into account in any estimates of membrane resistance. When current is injected into a long cylindrical cell of *Chara* or *Nitella*, the current density at the membrane will not be uniform because the cell behaves like a short, leaky coaxial cable (COLE and HODGKIN, 1939; WILLIAMS et al., 1964). If a pulse of constant current I is injected into a cell of length $2\,l$ at its midpoint, then the voltage response $\delta\psi_x$ at any distant point x from the current electrode is given by:

$$\delta\psi_x/I=(r_M/2\lambda)\cosh\{(l\text{-}x)/\lambda\}/\sinh(l/\lambda) \qquad (4.3)$$

where r_M is the resistance times unit length of the cell membrane, and λ is the characteristic length of the cell. λ is a measure of the attenuation of the voltage response along the cell and is given by $\lambda=[r_M/(r_i+r_o)]^{\frac{1}{2}}$ where r_i is essentially the resistance per unit length of the vacuolar sap, and r_o the resistance per unit length of the external solution. The Equation for the voltage response, given above, is not particularly convenient to use, but approximations are possible. WALKER (1960) used the relation $r_M=[R-R_L/12]\,A$ where $R=\delta\psi/I$ and R_L is the longitudinal resistance of the cylinder of cell sap. This approximation is good for $R_L<10\,r_M/A$, the situation that applied in the *Nitella* cells that WALKER examined. HOGG et al. (1968a) have devised a simple method for estimating the membrane resistance in charophyte cells. They noted first that Eq. (*4*.3) for the voltage response may be written as $R'_M=k\,R_M$, where R_M is the membrane resistance, uncorrected for the characteristic length, R'_M the corrected value, and k is a factor. An examination of k as a function of X $(=x/\lambda)$ and L $(=l/x)$ then shows that when $x=0.42\,l$, $k=1$ and is almost independent of λ over a wide range. Thus the membrane resistance can be measured simply by inserting the current-carrying microelectrode at the midpoint of the cell, and recording the change in PD at distance $0.42\,l$ from the current microelectrode, whence $R_M=R'_M$.

In marine algae the measurement of membrane resistance is more straight-forward. In the cells of such genera as *Griffithsia* (FINDLAY et al., 1969), *Chaetomorpha* (FINDLAY et al., 1971) and *Valonia* (GUTKNECHT, 1967) which have either spherical or oval cells, the current density is fairly uniform over the area of the cell, provided that the tip of the current microelectrode is not too close to the plasmalemma (PICKARD, 1971). The cylindrical cells of *Acetabularia* (SADDLER, 1970, GRADMANN, 1970) have values of λ about 7.5 cm, and thus $R_M=R'_M$.

1.3.2 Higher Plant Cells

There are considerable problems in measuring the membrane PD in higher plant cells with one inserted microelectrode; these problems are compounded when a second microelectrode is inserted to pass electric current and thus measure membrane resistance. HIGINBOTHAM et al. (1964) measured the resistance between the vacuole and outside of cells of *Avena* coleoptiles, whilst GREENHAM (1966) measured the resistance of the plasmalemma and tonoplast in root hair cells of cucumber, oats and maize. However, in neither of these reports are the actual membrane PD's given, and it is thus not possible to be sure that the electrical properties of the cells were unaffected by the insertion of two microelectrodes.

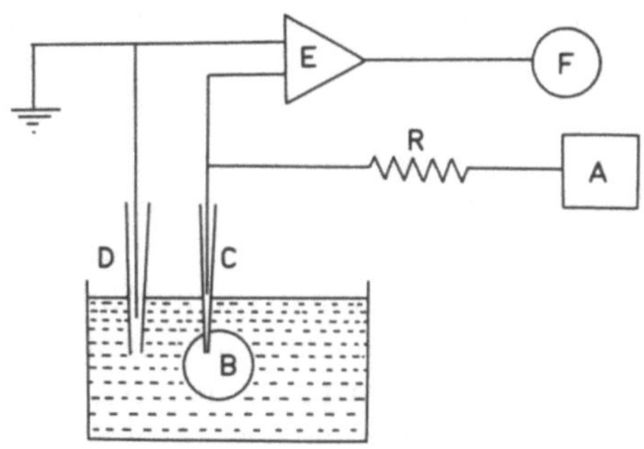

Fig. *4*.2. A diagram of the circuit to measure resistance in higher plant cells. *A* waveform generator; *R* 10^9 Ω resistance; *B* cell; *C* and *D* intracellular and extracellular electrodes respectively; *E* electrometer; *F* oscilloscope. (From ANDERSON et al., 1974)

ANDERSON et al. (1974) appear to have overcome the problems associated with inserting two microelectrodes into a cell by using a single intracellular microelectrode to measure membrane resistance, a technique similar to that described by BRENNECKE and LINDEMANN (1971). A block diagram of the measuring circuit used by ANDERSON et al. is shown in Fig. *4*.2.

The microelectrode resistance R_E (~ 5 MΩ) is in parallel with stray capacitance and this part of the circuit with a time constant of about 10 µs is in series with the parallel resistance, R_M, and capacitance C, of the membrane, with a time constant of about 1 ms. Pulses of current of constant amplitude, and frequency 10 kHz, are produced by generator A in series with a resistance of 10^9 Ω (much greater than $R_M + R_E$). When these pass through the inserted microelectrode and across the cell membrane, a practically steady PD given by $R_M I/2$ appears across the membrane resistance whilst the change in PD across R_E follows, with some attenuation, the current wave-form. In the measuring circuit shown, the oscilloscope displays the resting PD across the membrane when no current flows. When current pulses are applied, the periodic changes in PD of magnitude $R_E I$ across the resistance of the microelectrode are superimposed on the steady displacement of the membrane PD. The precautions necessary in using this technique, the most important being the prevention of electrode polarization caused by the flow of the electric current, are discussed by ANDERSON et al.

Another problem in measuring the resistance of higher plant cells, particularly those within a tissue, is the presence of plasmodesmata through which there are cytoplasmic connections from cell to cell. An appreciable quantity of electric current flowing from an intracellular microelectrode to the outside may pass from cell to cell through the plasmodesmata, rather than directly through the cell membrane. Various aspects of this electric coupling between cells are dealt with later in this Chapter, and in Part B, Chap. *2*.

1.4 The Control of Membrane Potential Difference by Voltage Clamping

The flow of electric current, I, across a cell membrane may be divided into a capacitative current, arising from the charging or discharging of the membrane capacitance, and an ionic current which depends on a flow of ions across the membrane. The current I is given by $I = C_M \frac{\partial \psi}{\partial t} + I_i$, provided the membrane capacity behaves like a perfect condenser. If the membrane PD ψ is kept constant, then $\partial\psi/\partial t = 0$, and $I = I_i$. Thus if the membrane PD is changed and maintained constant at a new level, the membrane current after an initial transient will be wholly ionic. However, such a control of the membrane potential requires that a source of potential difference of zero impedance be applied across the membrane.

The technique required in achieving such control was developed by MARMONT (1949) and COLE (1949) and later used extensively by HODGKIN et al. (1952) in their detailed studies of the action potential in the squid giant axon. In this technique, known as voltage clamping, the cell membrane becomes an element in a negative feedback control circuit. In the circuit the membrane PD recorded by a microelectrode inserted into the cell is compared with a control PD by means of a differential amplifier. The output of this amplifier supplies current to a second inserted microelectrode. If the membrane and control PD's are equal, the output PD is zero but any change appearing in the input PD is amplified, and the resultant output causes a flow of electric current through the second microelectrode and across the cell membrane. This current flow changes the membrane PD in a direction opposite to the initial change in input PD, and returns it to the control value, so that the input PD remains approximately zero. Thus the membrane PD follows the control PD. The higher the gain of the amplifier the more closely will the membrane PD follow the control PD. There are practical limits however. If the gain is too high the feedback circuit may become unstable and oscillations in PD occur in it.

In the experiments of HODGKIN et al. the membrane PD was measured by an electrode inserted at the ligatured end of a segment of axon. The current electrode was an axial wire also inserted at the end of the axon. In recent improvements in the voltage clamp technique (MOORE and COLE, 1963) the axial current electrode is retained, the membrane PD is measured by a KCl-filled microelectrode inserted through the membrane of the axon, and greater control of the PD over a well-defined length of axon has been achieved.

FINDLAY (1961, 1962) used the voltage-clamp technique for a study of the action potential in *Nitella*. In these experiments KCl-filled microelectrodes were used both to measure vacuolar PD and to pass current. FINDLAY followed the basic procedures used in most voltage-clamp analyses of membrane electrical behavior. The membrane potential difference was changed stepwise, and the current required to hold the membrane PD at the new level was measured. However, the high-resistance KCl-filled microelectrode effectively limited the gain in the feedback loop, and there was often inadequate control of the membrane potential. The problem was overcome to some extent by increasing the gain of the feedback amplifier, although often instabilities and consequent oscillations occurred in the feedback loop, but the current microelectrode eventually limited the magnitude of the current, particularly for outward current flow from the tip of the microelectrode.

KISHIMOTO (1961) and WILLIAMS and BRADLEY (1968) were able to achieve reasonable control of the vacuolar PD by voltage-clamping only a portion of a *Nitella* cell. This method also has the advantage that over the length of cell examined, which is small by comparison with the characteristic length, the current flow through the membrane will be reasonably uniform.

FINDLAY and HOPE (1964a, b) and FINDLAY (1964) improved the earlier voltage-clamp methods for *Chara* and *Nitella* by using a metal electrode, inserted through the end wall of the cell, positioned along the longitudinal axis, and extending for most of the length of the cell. Problems both of resistance of the current electrode and the non-uniform current flow are effectively eliminated in this arrangement, a diagram of which is shown in Fig. 4.3.

FINDLAY and HOPE also achieved a significant advance by clamping the PD across the plasmalemma alone, as well as the PD across plasmalemma and tonoplast in series, as in earlier experiments.

More recently KISHIMOTO (1972) has described a voltage-clamping method similar to that of FINDLAY (1964) but in which the metal current electrode only extended a short distance into the cell. It was only over this distance that the PD was controlled, and then only between vacuole and outside.

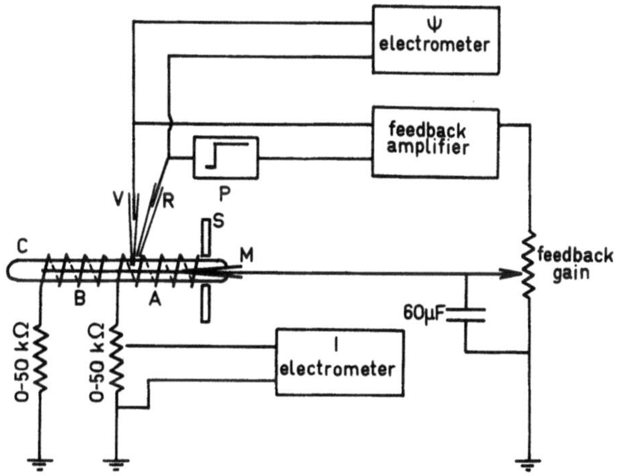

Fig. 4.3. The experimental arrangement for voltage-clamping the membrane PD in *Chara corallina*, showing the following parts: *C Chara* cell; *V* microelectrode in the vacuole or the cytoplasm; *R* external reference electrode; *S* stock to hold the cell in place, *P* electric current pulse generator; *M* intracellular metal current electrode; *A*, *B* separate sections of the external current electrode. The current flowing through a known area of cell is measured in the *A* section. (Modified from FINDLAY, 1964)

1.5 Electrical Capacitance of Membranes

In plant cells, the PD between the vacuole and outside changes to a new level with an approximately exponential time course for a step-wise change in current flow. This response of the vacuolar PD indicates that the cell membranes contain a capacitative, as well as a resistive component. The capacitative component of cell membranes is also apparent in the response of the membrane PD to a flow of sinusoidal alternating current. In order to be able to estimate the membrane capacitance, however, an equivalent electrical circuit for the cell must be postulated.

1.5.1 Charophyte Cells

HODGKIN and RUSHTON (1946) showed that the axon of a nerve cell behaved electrically as a coaxial transmission line of infinite length. WILLIAMS et al. (1964) have shown that charophyte cells, in particular *Nitella translucens,* behave more like short transmission lines terminated by an infinite impedance. The electrical equivalent circuit is shown in Fig. 4.4. WILLIAMS et al. have estimated the membrane capacitance and resistance in *Nitella translucens* by both direct current and alternating current methods. They treated the plasmalemma and tonoplast as one composite membrane. In the direct current method the exponential response of the membrane PD to a square current pulse yields a time constant $\tau_M = R_M C_M$ where R_M is the membrane resistance, and C_M the membrane capacitance. In the simple charging of a capacitance through a resistance, τ_M is given by the time for the change in PD to reach 63% of its final value. For an infinitely long cable, HODGKIN and RUSHTON (1946) have shown that τ_M is the time for the membrane PD to reach 84% of its final value. For a short cable τ_M lies somewhere between these values. Although it is possible to calculate the appropriate value, WILLIAMS et al. did not do so, but merely gave the expected range of value of C_M for the range of τ_M from 63% to 84%. HOGG et al. (1969), by solving the time-dependent equation for the short cable, were able to calculate the appropriate time constant.

In the alternating current method the input current, sinusoidal in waveform and of frequency 25 Hz, flowed from a microelectrode inserted at the mid-point of the cell, and the response in membrane PD was measured at two points at different distances from the current electrode. The resulting sinusoidal membrane PDs at these two points differ in amplitude and phase, and the membrane capacitance can be calculated from the appropriate equations. The details are omitted here as the equations are not particularly simple, and the calculations tedious.

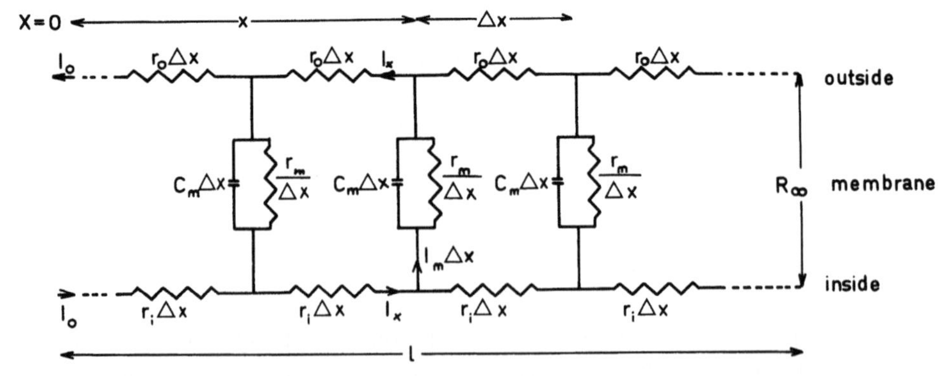

Fig. 4.4. The equivalent electrical circuit for the half-length of a *Nitella* cell. The symbols are described in the text. (From WILLIAMS et al., 1964)

The capacitance of membranes of charophyte cells can also be measured by using the same type of axial metal current electrode as in voltage clamp experiments. The cable-like properties of the cell are eliminated by this procedure, and the membrane then behaves as a simple parallel resistance and capacitance circuit. When sinusoidal current flows across the cell membrane, the response of the membrane PD is also sinusoidal, but differing in phase. The amplitude I of the current waveform, in terms of the amplitude ψ of the PD waveform is:

$$I/\psi A=\{(1/R_{\mathrm{M}}^{2})+\omega^{2}C_{\mathrm{M}}^{2}\}^{\frac{1}{2}} \tag{4.4}$$

and the phase difference δ is given by:

$$\tan\delta=\omega R_{\mathrm{M}}C_{\mathrm{M}} \tag{4.5}$$

where R_{M} is the membrane resistance, C_{M} is the membrane capacitance, ω the angular frequency of the sinusoidal signal and A the area of the membrane. These two equations can be solved to give:

$$R_{\mathrm{M}}=(A\psi)/(I\cos\delta) \tag{4.6}$$

and

$$C_{\mathrm{M}}=(I\sin\delta)/(\omega A\psi) \tag{4.7}$$

H.G.L. Coster and J.R. Smith (personal communication) have used this procedure to make measurements of the capacitance of *Chara* membranes as a function of the frequency of the applied AC current. A sinusoidal current is passed through the cell membrane by the internal current electrode to an external current electrode. The PD across the membrane (plasmalemma and tonoplast in series) is measured by an inserted microelectrode and an external microelectrode close to the cell surface, both connected to a differential electrometer.

A digital computer controls the frequency of the current generator, and the current and PD signals are recorded by the computer's analog-to-digital converter 40 times per cycle of the signal. These signals are averaged over many periods. This allows a small signal to be used which causes only a small change in the membrane PD. The averaged signal can be stored in the computer memory.

The phase and amplitude of the current and PD signals are calculated from the digitized data in the computer memory by fitting a sine wave to the data by the method of least squares. The membrane capacitance and conductance are calculated from the relative phases and amplitudes, and can be displayed on a storage oscilloscope. The computer is programmed so that the measurements are repeated for a number of preselected frequencies. In this way the frequency dependence of the membrane capacitance and conductance can be obtained. The great advantage of this system over ordinary "bridge" systems is that it operates over a wide frequency range, and down to very low frequencies, less than 1 Hz.

1.5.2 Higher Plant Cells

The capacitance of membranes of higher plant cells is probably most simply estimated from the response of the membrane PD to a square pulse of current flowing across it. If the cell is within a tissue, there are uncertainties about the path of extracellular current flow, and consequently the external resistivity r_2. However, if it is assumed that r_2 is small in relation to r_1, then the characteristic length λ is given by $\lambda=(r_M/r_i)^{\frac{1}{2}}$. For a cylindrical cell of diameter 20 μm, internal resistance of 0.8 Ω m, and length 100 μm, $\lambda=250$ μm and the cell will probably behave electrically like a short coaxial cable terminated by an infinite impedance. Then as the characteristic length is considerably greater than the length of the cell, τ_M $(=R_M C_M)$ will be given by the time the PD response takes to get to 63% of its final value, and the membrane capacitance can be calculated.

2. Electrical Properties of Cells—A Perspective

2.1 Potential Difference

2.1.1 The Membrane Potential Difference

All living plant cells probably exhibit a transmembrane PD in which the cytoplasm is electrically negative to the external medium. A less well-characterized PD exists across the tonoplast. A ready explanation of the transmembrane PDs found in plant cells is that they are diffusion potentials: ionic concentration differences exist across the plasmalemma and tonoplast and these membranes take the place of the boundary layers in the classical situation for diffusion potentials between simple aqueous solutions of salts.

The concentration gradients across the cell membranes are the result, very likely, of two quite different factors: (i) the presence of indiffusible ions on cytoplasmic molecules or of mobile, but membrane-restricted ions and (ii), a modification of the population of the counterions in this phase by the processes of active transport (BRIGGS et al., 1961). The tendencies for the ions to diffuse down their concentration gradients, a tendency related to the permeability of the intervening membrane to these ions, gives rise to small charge separations (see *4*.2.3) and hence transmembrane PDs (see also *3*.5.1).

Nevertheless, there are now many instances where the explanation in terms of a diffusion potential is inadequate. It appears that a specific type of membrane translocator may also tend to separate electric charge across the membrane, contributing to a resultant PD that is often much greater than the magnitude of any possible diffusion PD (see *3*.4.2 and *3*.5.3.2). The translocator is then said to be electrogenic. Several important transport systems have been discovered because they are electrogenic (*6*.6.6; *7*.2.1.4; *12*.3.2; Part B *3*.*3*.2).

In the absence of electrogenesis, a membrane PD may often be described to a good approximation by an equation that includes the concentrations of potassium and sodium on each side of the membrane, plus permeability terms. A good example is the plasmalemma PD in the marine alga *Griffithsia pulvinata* (FINDLAY et al.,

1969) in which (as in Eq. *3*.32):

$$\psi_{co} = (RT/F) \ln \frac{P_K[K^+]_o + P_{Na}[Na^+]_o}{P_K[K^+]_c + P_{Na}[Na^+]_c} \tag{4.8}$$

These quantities have already been defined in *3*.5.3.1, where it is shown that the meaning of ion permeability depends on the particular model used. However, approximately constant values for the ratio P_{Na}/P_K, *viz.* 0.003–0.01, were obtained for *Griffithsia* spp. The validity of Eq. (*4*.8) is illustrated in Fig. *4*.5. Studies such as this show simply that the plasmalemma is much more permeable to K^+ than to Na^+ and, since terms including Cl^- were not necessary to describe the PD, that the plasmalemma is also probably cation-selective. Further indications about membrane structure have been obtained from electrical measurements, as seen below.

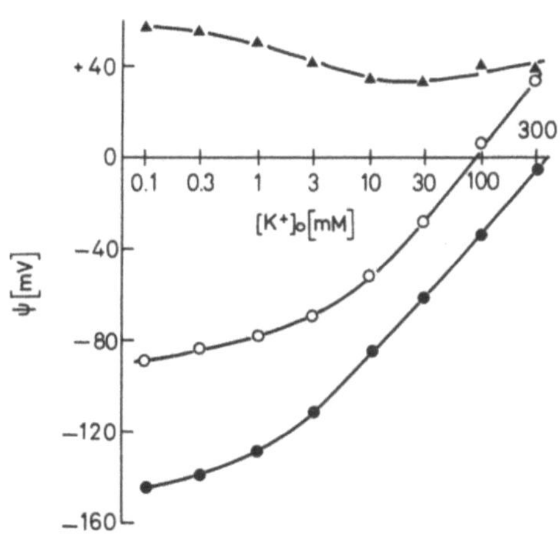

Fig. *4*.5. Potential differences as functions of the external potassium concentration $[K^+]_o$, in cells of *Griffithsia pulvinata*. Open circles: ψ_{vo}; closed circles, ψ_{co}; triangles, ψ_{vc}. The curve for ψ_{co} was obtained from the relationship described in the text, with $\alpha = P_{Na}/P_K = 0.002$, and $[K^+]_c + \alpha[Na^+]_c = 350$ mM. The other curves were fitted without a theoretical basis. (From FINDLAY et al., 1969)

Part of the "transmembrane" PD may occur across the cell wall, which is usually physically inseparable from the plasmalemma. Provided the ion-exchange system of the cell wall is in thermodynamic equilibrium with the external medium, conclusions about membrane permeability ratios of like ions made by regarding all the intracellular PD as across the plasmalemma are not invalid.

The resting PDs for some carefully studied systems are compiled in Tables in Chap. *6* (Tables *6*.3 and *6*.4) and Part B. Chap. *3.3* (Table *3*.8). From a consideration of equations for PD it is obvious that comparisons between systems are not easy without a knowledge of the composition of the external media, and without allowing for possible electrogenesis. Some of the less common conclusions from studies of the membrane PD are:

(a) In certain marine genera, the plasmalemma is very chloride-permeable relative to cations and so also is the tonoplast, since the vacuolar PD is close to the Nernst or equilibrium PD for this ion species (GUTKNECHT, 1966).

(b) The PD across the tonoplast of *Chaetomorpha* exists in two well-separated states. Either ψ_{vc} is about +75 or +46 mV. The latter value could reflect a reduced chloride permeability at the tonoplast under circumstances as yet ill-defined (FINDLAY et al.,1971). ZIMMERMANN and STEUDLE (1970) found a dependence of the PD (ψ_{vo}) in *Valonia utricularis* such that turgor pressure when raised from the normal 100 kPa (~1 atm) to 300 kPa caused ψ_{vo} to change from about +2 mV to −35 mV, accompanied by a temporary decrease in cell conductance. These authors interpreted the effects in terms of changes in an electrogenic potassium pump at the tonoplast, already postulated by GUTKNECHT (1968) to be pressure-sensitive.

(c) The discrimination of the plasmalemma in favor of K^+ is absent in freshly-cut low-salt barley root segments, where $P_{Na}/P_K \simeq 0.8$ (PITMAN et al., 1970); and in newly-fertilized *Fucus* eggs this ratio decreases from about 0.1 to 0.001 in 12–14 h (BENTRUP, 1970).

2.1.2 Extracellular PDs

Extracellular PDs around roots and coleoptiles reflect the fact that current is flowing in the external medium with the tissue as a return circuit. The EMF determining the current has not been identified but is presumably related to electrogenic ion-transporting systems. Investigators have implicated the currents in a feedback loop containing hormonal influences and those of ion permeability, membrane potential and osmotic potential (SCOTT, 1957, 1962), the feedback loop generating oscillations of *c.* 300 s period in PD and growth rate. NEWMAN (1963) found a correlation between extracellular PDs in *Avena* coleoptiles and the translocation of auxin from tip to base. NEWMAN and BRIGGS (1962) described phytochrome-mediated changes in PD in the same tissue. Nearer the cellular level are the experiments of JAFFE (1968) and BENTRUP (1968), dealing with the electric fields around developing eggs of *Fucus* and their influence on development. JAFFE et al. (1974) have described the ion fluxes that constitute the extracellular current and a detailed picture of the bioelectric field correlates of growth and development is beginning to emerge.

2.2 Conductance

2.2.1 Membrane Conductance

The conductance per unit area of plant cell membranes is usually in the range 0.5–20 S m^{-2}. This low value reflects the limited ability of the cytoplasmic membranes to conduct electric current, presumably by means of the ions K^+, Na^+, Cl^-, H^+, etc. The conductance of membranes lies between that of a 5 nm film of oil (~10^{-4} S m^{-2}) and that of the same thickness of good-quality distilled water (~10^4 S m^{-2}). "Black-film" membranes made from phospholipids also have a conductance as low as 10^{-4} S m^{-2}, so cytoplasmic membranes do *not* resemble the artificial bilayers in respect of conductance. Thus cytoplasmic membranes probably have specific conducting areas, not composed of hydrophobic lipid or phospholipid molecules. Unfortunately the area involved in ion conduction is not calculable from the observed conductance since the latter depends as well on the ion mobility and concentration within the conductive areas. Almost no-one has been brave enough to speculate on the values these quantities may have.

An electrogenic pump may contribute to the total membrane conductance. In a simple model, the electrogenic pump and passive conducting areas could be regarded as being composed of special protein modules located in the lipid matrix (see *1*.3.3; *4*.2.7). The parallel conductances of these areas would add together:

$$g_M = \partial J_{pump}/\partial\psi_M + \partial J_{passive}/\partial\psi_M.$$

Since the rate of working of an electrogenic pump will depend on the PD through which the ions are transported, $\partial J_{pump}/\partial\psi_M$ is not zero (see also *3*.5.4.2).

Membranes also show rectification (HOPE and WALKER, 1961; FUJITA, 1962) a phenomenon generally expected of an asymmetrical ion distribution. Current in one direction is less, for a given change in PD, than in the reverse direction, due to changes in the ion concentration in the membrane pores or conducting areas. For example, when the ions move from the solution more dilute in the ions to which the membrane is permeable, the conductance is smaller.

In examining rectification in *Chara corallina*, COSTER (1965) discovered "punch-through" in which the membrane conductance increases non-destructively and reversibly when the plasmalemma is hyperpolarized to the region of −300 mV. The implications of this for membrane models have been noted elsewhere (COSTER, 1973b) and below.

2.2.2 Membrane Conductance and Permeability

The theory given in Chap. *3* suggests that estimates of ion permeabilities may be made from measurements of PD and conductance. The estimates depend on assumptions as to the species of ions involved in the diffusion potential, and furthermore are invalid in the presence of electrogenic effects. In the plasmalemma of *Griffithsia*, which is apparently free of the latter complication and in which the PD is a diffusion potential determined by K^+ and Na^+, P_K and P_{Na} may be estimated as $3 \cdot 10^{-7}$ and 10^{-9} m s^{-1} respectively. Often, permeability estimated in this way does not equal that calculated from the size of passive ionic flux. This discrepancy has been noted repeatedly (HOPE and WALKER, 1961; MACROBBIE, 1962; WILLIAMS et al., 1964; WALKER and HOPE, 1969; FINDLAY et al., 1970) but the current, fashionable explanations implicating either proton conduction or an electrogenic pump do not suffice in *Griffithsia*. A degree of self-interaction between unidirectional fluxes still seems possible.

2.2.3 Intercellular Conductance

Intercellular connections between cells are provided by plasmodesmata so that the tissue may need to be treated as a syncytium for some considerations. Electrical measurements suggest that movement between certain cells is relatively easy for ions and possibly small molecules (see Part B, Chap. *2* for details). In three such studies (Table *4*.1) the connections were shown to have a conductance a great deal higher in value than the cell membranes, but less by 60–300 times that expected of the observed areas on the assumption that they contain fluid of the same conductivity as the cytoplasm.

Table 4.1. The electrical conductance between cells. $g_{intercellular}$ are estimates using the apparent area of contact. $g_{plasmod.}$ are estimates using the total cross-sectional area of the plasmodesmata. $g'_{plasmod.}$ are calculated on the assumption that all the cross-sectional area of plasmodesmata is filled with electrolyte solution of the same conductivity as the cell contents

Cell	$g_{intercell.}$ ($S\,m^{-2}$)	$g_{plasmod.}$ ($S\,m^{-2}$)	$g'_{plasmod.}$ ($S\,m^{-2}$)	Ref.
Chara braunii	60	—	—	SIBAOKA (1966)
Nitella translucens	6	1,000	33,000	SPANSWICK and CONSTERTON (1967)
Elodea canadensis	200	22,000	10^6	SPANSWICK (1972a)

2.3 Membrane Capacitance

The cell membranes of numerous cell types act as capacitors, with values 0.01–0.03 $F\,m^{-2}$ (CURTIS and COLE, 1938; GOLDSMITH et al., 1972). In simple terms, which may not much longer be acceptable (see below) the membranes are predominantly insulating sheets: if the effective thickness is about 4 nm and the dielectric constant 5, a capacitance of 0.01 $F\,m^{-2}$ is expected. New techniques (COSTER, 1973a) have recently allowed a new approach to the phenomenon of capacitance, with implications, again, for our concept of the plasma membrane. If the new model is correct, the capacitance should be a function of membrane potential. This would render more complex the estimation of membrane capacitance from the "time constant" for attainment of a steady PD following a current injection (cf. WILLIAMS et al., 1964, and others).

The charge stored on 1 m^2 of a plasmalemma with 150 mV across it is 1.5 mC or 15 nmol. This is usually not a significant proportion of the total charge within the cytoplasm which is essentially still neutral. In a sphere of diameter 1 μm containing 0.1 M electrolyte, the charge stored on the membrane capacitor is still only about 0.1% of the total available inside the sphere (see also 3.5.2).

2.4 The Action Potential

2.4.1 Charophyte Cells

Early Work. The cells of many charophyte plants such as *Chara, Nitella* and *Nitellopsis* exhibit characteristic transient changes in membrane PD and conductance when the PD is depolarized to a threshold level (Fig. 4.6). These transient changes are true action potentials similar in some respects to those in nerve cells. Early experimental observations by HÖRMANN (1898) and EWART (1903) showed that in charophytes the cessation of cytoplasmic streaming when a suitable stimulus was applied to an internodal cell was closely related to the occurrence of an action potential. OSTERHOUT (1931, 1934, 1936, 1943) and UMRATH (1930, 1932, 1953, 1934, 1953, 1954) examined many aspects of excitatory behavior in a number of charophyte species. UMRATH used inserted microelectrodes to measure the potential of the vacuole, and was probably the first person to use such microelectrodes to study the electrical properties of plant cells. In 1939, COLE and CURTIS measured the change in membrane conductance in *Nitella* during an action potential. They used external electrodes and an electrical bridge method and found that the conductance increased at the peak of the action potential by up to 200 times the resting conductance. UMRATH (1940) found a similar change in *N. mucronata*, but a much smaller change in *N. opaca*. COLE and CURTIS also found that the membrane capacitance changed by about 15% during the action potential. HARVEY (1942a) showed that in cells of *N. fragilis*, action potentials could be initiated by exposing the cells to intense flashes of UV light.

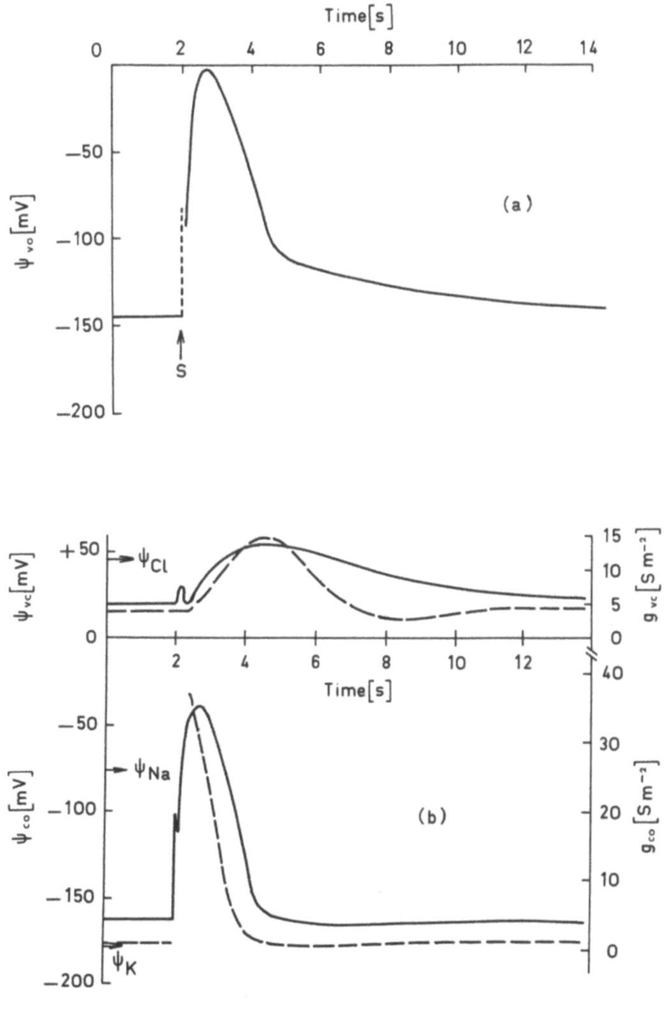

Fig. *4*.6. (a) The vacuolar action potential in a cell of *Chara corallina*. A depolarizing electrical stimulus, S, of about 100 ms in duration was applied at t=2. (b) Action potentials recorded simultaneously across the plasmalemma and tonoplast in the same cell of *Chara corallina*. The dashed lines are the approximate time courses of conductance during the action potential. ψ_K, ψ_{Na} and ψ_{Cl} are the calculated electrochemical equilibrium potentials for the plasmalemma. (From HOPE and WALKER, 1975)

Recent Work. From the late 1950's there has been an increasing interest in the action potential largely brought about by the great advances made by HODGKIN and HUXLEY (1952) in establishing the ionic basis of the action potential in nerve cells. Improvements in technique, particularly the use of fine-tipped intracellular microelectrodes (WALKER, 1955) have made possible detailed studies of the action potential in charophyte cells, and of course, more wide-ranging investigations of the electrophysiology of plant cells generally.

Various investigations of the action potential have been made, by ODA (1956, 1961, 1962), GAFFEY and MULLINS (1958), KISHIMOTO (1959, 1961, 1964, 1965, 1966a, 1966b, 1968b, 1972), FINDLAY (1959, 1961, 1962, 1964, 1970), HOPE (1961), SPYROPOULOS et al. (1961) MULLINS (1962), FUJITA (1962), FINDLAY and HOPE (1964a, 1964b), HOPE and FINDLAY (1964), HAAPANEN and SKOGLUND (1967), WILLIAMS and BRADLEY (1968), COSTER et al. (1968), KOPPENHÖFER (1972a, 1972b) and STRUNK (1971).

Initiation of the Action Potential. Various types of stimuli may initiate an action potential in a charophyte cell. The basic requirement for the production of an action potential appears to be a depolarization of the plasmalemma PD above a threshold level. KISHIMOTO (1968a) made quantitative measurements of the response of *Chara australis* internodes to mechanical stimulation. The sudden application of a force to a region of a cell for 1 s depolarized the vacuolar PD. The magnitude of the depolarization increased with increasing stimulus. Following the cessation of the stimulus, the PD recovered with an approximately exponential time course, with half-time of 5 s. When the depolarization exceeded the threshold, an action potential occurred. A mechanical stimulus as short as 1 ms could cause an action potential. KISHIMOTO also found that the responses to a number of subthreshold stimuli could summate, and an action potential was produced when the PD went above threshold. However, the magnitude of the response of the vacuolar PD also depended on the rate of increase of the stimulating force. Presumably, a slowly-increasing force allowed a dissipation of the force within the cell by the loss of water across the cell membranes. The response to any mechanical stimuli always occurred in the region where the stimulus was applied.

Action potentials can also be initiated by sudden thermal stimuli (HILL, 1935; HARVEY, 1942b). A sudden change of temperature of about $-15°$ C or $+25°$ C is necessary to produce an action potential. GAFFEY (1972) has shown that irradiation of cells of *Nitella gracilis* with α-particles initiated action potentials when the dose rate, applied to 13% of the length of a cell, was about 1 to $2 \cdot 10^5$ rad s^{-1}. The threshold for excitation depended on the radiation previously absorbed by the cell. Of particular interest is GAFFEY's observation that irradiation of a cell with α-particles hyperpolarized the membrane PD, and that the action potential did not occur until the radiation ceased. This behavior is similar to "anode-break" excitation (KATZ, 1937) which occurs in some cells following the cessation of a pulse of hyperpolarizing current across the membrane.

A pulse of electric current provides a convenient, reproducible (and probably the most direct) stimulus to initiate an action potential in both animal and plant cells. FUJITA and MIZUGUCHI (1955) have shown that in *Nitella* the electric stimulus necessary to change the membrane PD to the threshold level obeys the classical strength-duration relationship (KATZ, 1937). The threshold stimulating voltage V, is given by $V=(\alpha/t)+\beta$, where t is the duration of the stimulus, and α, and β are constants, β called the rheobase, and α/β the chronaxie. There is some deviation from this relationship for short pulses. This relationship arises because the passage of a minimum quantity of electric charge, given by the product of stimulus current intensity and time, is required to discharge the membrane capacitance and depolarize the membrane PD to threshold level. Similar relationships have been shown to hold in *Nitella* sp. (FINDLAY, 1959) and *N. gracilis* (GAFFEY, 1972). FUJITA and MIZUGUCHI also showed that as a result of a subthreshold "conditioning" stimulus, the threshold to a subsequent stimulus is reduced.

The Ionic Basis of the Action Potential. GAFFEY and MULLINS (1958) in an important paper have outlined the ionic basis of the action potential as measured in the vacuole of *Chara globularis*. Although this is a corticated species, it appears that the interpretation of their results as applied to the main internodal cell is largely unaffected by the presence of the surrounding layer of cortical cells. GAFFEY and MULLINS determined the electrochemical equilibrium potentials for Na^+, K^+ and Cl^- in the vacuole from the Nernst equation, and showed that while the resting PD was -181 mV, $\psi_{Na}=-155$, $\psi_K=-184$ and $\psi_{Cl}=+202$. The ion furthest from equilibrium was Cl^- and thus likely by analog with Na^+ ions in nerve cells (HODGKIN and HUXLEY, 1952) to play a role in the action potential. Furthermore, the effluxes of K^+ and Cl^- from the cell were increased during periods when action potentials were produced, but there was no change in the Na^+ efflux. The effluxes of K^+ and Cl^- had values up to 100 µmol m^{-2} $impulse^{-1}$. GAFFEY and MULLINS concluded that the action potential in *C. globularis* is produced by a transient increase in the permeability of the plasmalemma of the internodal cell to Cl^- following the depolarization of the membrane PD above the threshold. The PD then moves towards ψ_{Cl} at $+202$ mV, as P_{Cl} increases, but in so doing, moves away from ψ_K at -184 mV, causing an increase in K^+ efflux. When P_{Cl} begins to decrease, the extra K^+ efflux returns the membrane PD to the resting level.

HOPE (1961) and FINDLAY (1961, 1962) proposed an alternative explanation of the action potential from consideration of the effects of varying the concentration of Ca^{2+} in the external bathing solution. HOPE found that cells of *Chara australis* were not excitable when calcium ions were absent from the external solution. He also found that the peak of the action potential measured in the vacuole was a linear function of log $[Ca^{2+}]_o$ such that $\psi_{peak}=\text{constant}+29\log_{10}[Ca^{2+}]_o$. HOPE also found that changes in $[Cl^-]_o$ in the range 0–10 mM, although changing the resting PD somewhat, had little effect on the peak of the action potential. He concluded that during the action potential the plasmalemma became permeable to Ca^{2+} rather than to Cl^-, and that the peak value of the PD was thus related to the electrochemical equilibrium potential for Ca^{2+}.

FINDLAY (1961, 1962) found similar results for *Nitella*, except that for a change in $[Ca^{2+}]_o$ the change in the peak value of the action potential was about 35–40 mV, and he suggested that calcium was partly in the form of a monovalent complex, and that both this and the divalent calcium controlled the action potential. Both authors argued from the electrical effects of Ca^{2+} that the action potential in *Nitella* and *Chara* was brought about by an inward movement of calcium ions. No direct measurements of Ca^{2+} influx were made, as MULLINS (1962) pointed out. MULLINS suggested that the effects of $[Ca^{2+}]_o$ were most simply accounted for if during the action potential, the maximum change in PD, the membrane permeability to chloride, and consequently the chloride efflux increased with increasing $[Ca^{2+}]_o$. The peak value of the action potential would become more positive with increasing $[Ca^{2+}]_o$ as observed. The relation between peak value and $[Ca^{2+}]_o$ is not obvious on *a priori* grounds, and need not be the same for all species of *Nitella* and *Chara*.

The Action Potential at the Plasmalemma and Tonoplast. Apart from the obvious difference of time scale, the action potential in the vacuole of charophyte cells differs from that in nerve cells in having the recovery from the peak consisting of a fast phase followed by a slow phase, see Fig. *4*.6. GAFFEY and MULLINS (1958), on the basis of calculations of the diffusion time of potassium ions through the cell wall, claimed that the slow recovery phase was a depolarization which arose from the persistence of K^+ immediately outside the potassium-permeable plasmalemma. FINDLAY and HOPE (1964a) with rather thick-walled microelectrodes, but tip diameter less than 2 μm, were able to make shallow insertions into the flowing cytoplasm of *Chara corallina*, without stopping the streaming. They found that the action potential as measured in the vacuole actually consisted of two components, one across the plasmalemma and a slower one of smaller magnitude and longer time course, across the tonoplast (Fig. *4*.6). The peak of the action potential across the plasmalemma occurred earlier than the peak recorded in the vacuole. The action potentials across the separate membranes always occurred together; it was not possible to elicit an action potential separately at plasmalemma or tonoplast. The conductance of both membranes increased to maximum values at the respective action potential peaks. The plasmalemma conductance was 1.0 S m^{-2} in the resting state, and increased to about 40 S m^{-2}, whereas the tonoplast conductance went from 10 S m^{-2} to about 30 S m^{-2}.

Voltage-Clamp Analysis of the Action Potential. KISHIMOTO (1961) and FINDLAY (1961, 1962) independently and at about the same time applied the voltage-clamp technique to a study of the action potential in charophyte cells. In these experiments with *Nitella* species, the vacuolar PD was clamped. FINDLAY found that the membrane current flowing during a voltage-clamp above the threshold level consisted of an outward current beginning at the start of the clamp, followed within 250 ms by a transient inward current lasting up to 2 s and finally an outward current continuing for the remainder of the clamp. As the vacuolar PD was clamped at progressively more depolarized levels, the transient inward current decreased in magnitude and the continuing outward current increased. KISHIMOTO made similar observations. FINDLAY (1964) improved the voltage-clamp system (see *4*.1.4) and was able to clamp the PD across the plasmalemma alone. The peak of the transient inward current for the clamped plasmalemma PD occurred earlier than that for the clamped vacuolar PD. FINDLAY and HOPE (1964b) further analyzed the peak transient current as a function of the plasmalemma PD, in relation to the chloride and calcium hypotheses. They concluded that (a) depolarization of ψ_{co} above the threshold caused a transient increase of the permeability of the plasmalemma to Cl^-, with cytoplasmic chloride activity in the range 1–10 mM, and (b) the transient current is carried at least partly by Cl^-, and its magnitude depends on $[Cl^-]_o$, (c) the plasmalemma in its excited state becomes anion-permeable, as it does not distinguish between Cl^-, Br^- or NO_3^- in the internal medium, (d) the transient permeability to Cl^- and other anions is a function of $[Ca^{2+}]_o$, and that some Ca^{2+} must be present in the internal medium for excitation to occur, (e) the continuing outward clamp current could be carried by K^+ ions but that P_K of the plasmalemma decreases for several seconds from the start of the voltage clamp.

A simple and decisive test of the opposing calcium and chloride hypotheses was made by HOPE and FINDLAY (1964). They showed by the use of ^{45}Ca and ^{36}Cl as radioactive tracers that while there was no extra influx of Ca^{2+} during the action potential over and above the resting influx, there was an extra efflux of Cl^-. In voltage clamp experiments the integrated transient charge that flowed during clamps above threshold was accounted for by the Cl^- efflux during the corresponding times.

Factors Affecting the Action Potential. DOUGHTY (1973) found that the irradiation of cells of *Chara corallina* with 253.7 nm UV at an intensity of 4.7 W m^{-2} for 10 min prolonged the duration of the action potential at the plasmalemma but left the peak PD unaffected. At the tonoplast, the peak height of the action potential was diminished, and the duration was increased. These results can be explained by (a) a lengthening of the time course of P_{Cl} at the plasmalemma, but with the maximum value of P_{Cl} remaining unchanged or (b) a lengthening of the time courses of both P_{Cl} and P_K at the tonoplast, together with either a decrease in peak P_{Cl} or an increase in peak P_K to account for the decrease in peak potential.

ESCH et al. (1964) and ROTTINGER and HUG (1972) found that X-irradiation of *Nitella* cells caused an increase in the duration of the action potential.

F.J. BLATT (personal communication) has shown that in *Nitella flexilis* a change in the temperature from 28° C to 3° C changes the resting potential of the vacuole from −170 mV to −100 mV, while the peak of the action potential remains fairly constant, changing only from −45 mV to −35 mV. During the same change in temperature the rise and fall times of the action potential increased by about two orders of magnitude, while the general shape of the action potential remained unaltered. The rise and fall times plotted on a logarithmic scale against the reciprocal of the absolute temperature (Arrhenius plot) each yield two straight line segments intersecting at 13.5° C. For each of the rise and the fall, the slopes of the straight line segments were similar. This implies that the activation energies for the two processes are the same, with a phase transition at 13.5° C (cf. *4*.2.7.3). UMRATH (1934) obtained similar results for the dependence of rise-time with temperature, but with the change in slope at 15° C.

Because the rising phase of the action potential is brought about mainly by an efflux of chloride ions across the cell membranes, and the falling phase by an efflux of potassium ions, BLATT suggests that the near equality of the activation energies for rise and fall times implies that the processes determining Cl^- and K^+ fluxes are closely coupled, and perhaps mediated by the same ion or molecular complex.

Changes in Membrane Capacitance during the Action Potential. The capacitance of plasmalemma and tonoplast in series, considered as one "membrane" in cells of *Chara corallina,* decreases as the membrane is hyperpolarized from its resting PD and increases with depolarization (see *4*.2.3 above, and H.G.L. COSTER and D. HAREN, unpublished). During an action potential, the capacitance increases to about 3 times the resting value. However, during the action potential the capacitance is also a function of time-dependent variables apart from membrane PD. The fixed-charge membrane model (*4*.2.7) could account for the change in capacitance with membrane PD in the steady state. However, to account for the time-dependence of the capacitance during the action potential, the independent parameters in the fixed charge model, *viz*. the fixed-charged concentration and the dielectric constant would have to vary. At present there is no information about such changes.

2.4.2 *Acetabularia*

GRADMANN (1970) has recorded vacuolar action potentials of rather long duration, from 30 to 300 s, in the marine alga *Acetabularia crenulata*. The action potentials have an amplitude of about 120 mV and arise spontaneously or can be initiated by lowering the temperature, or by depolarizing the membrane. Replacement of the external sodium chloride with choline chloride does not affect the action potential, and GRADMANN concludes that the depolarization is caused by an efflux of chloride ions, together with potassium as the accompanying cation. There is no information about whether the action potential is propagated along the cell.

2.4.3 Higher Plant Cells; Mimosa, Dionaea, Drosera, Other Plants

Well-defined, propagated action potentials have been shown to occur in cells of the sensitive plants *Mimosa* (SIBAOKA, 1966), *Dionaea* (DIPALMA et al., 1961; JACOBSON, 1965; SIBAOKA, 1966; BENOLKEN and JACOBSON, 1970) and *Drosera* (WILLIAMS and PICKARD, 1972a, b; WILLIAMS and SPANSWICK, 1972). The action potentials, which normally arise as the result of mechanical stimulation of the plant, are closely associated with the subsequent movements.

In cells of the trap-lobes of *Dionaea*, the action potential has an amplitude of about 100 mV, the time of the rising phase is 0.1–0.2 s, and the total duration 0.6–1.0 s. The velocity of propagation of the action potential from one cell to another is in the range 60 to 170 mm s^{-1}. Parenchymal cells in the protoxylem and phloem of *Mimosa* have larger membrane potentials than neighboring cells, and generate and conduct action potentials with a rise time of about 0.5 s, and total duration of several seconds. The conduction velocity of the action potential is about 20–30 mm s^{-1}. A mechanical stimulus applied to the head of one of the tentacles of the *Drosera* leaf produces a depolarization of adjacent cells in the stalk of the tentacle. Action potentials of duration 3 to 15 s are initiated if the depolarization goes beyond a threshold level, and are propagated from cell to cell to the base of the tentacle with a velocity of about 5 mm s^{-1}. At present, the ionic basis of the action potentials of higher plant cells is poorly understood.

Other Plants. There have been many reports in the literature of various types of transient electrical behavior in cells and tissues. Most of these transients are not true action potentials. UMRATH (1959) has reviewed the evidence about propagated electrical fluctuations in plants such as *Cucumis, Cassia, Lathyrus, Phaseolus, Aeschynomene* and *Phylanthis*. More recently PICKARD (1973) has published an extensive review of the literature about electrical transients in a variety of plant cells. These transients often show considerable variability, and many of the results do not seem to be reproducible. Clearly further work on these problems is needed.

2.5 Testing for Active and Passive Transport

2.5.1 The Nernst Criterion

The PD is part of the driving force on ions. When considering whether ions are passively or actively transported, clearly the PD as well as the concentration gradient across the membrane must be taken into account. The basis of tests using the Nernst criterion or the flux-ratio criterion was discussed critically in Chap. *3*. The results of detailed examinations of algal cells, root cells and leaf cells in the light of these tests appear in Chapters devoted to them.

The almost universal acceptance of an externally-directed, active transport of sodium in plant cells is strongly dependent on the thermodynamically-based argument that, in an agreed steady-state and in the face of appreciable plasmalemma permeability to Na^+, the Nernst potential for Na^+ is much more positive than ψ_{co}. The cytoplasm concentration should rise to its equilibrium level, since there is a net driving force, $\psi_{co}-\psi_{Na}$, acting to cause a net influx. That this does not occur except in a very few instances (e.g. *Halicystis osterhoutii* (BLINKS and JACQUES, 1930) and possibly *Valoniopsis* sp.: G.P. FINDLAY, A.B. HOPE, M.G. PITMAN, F.A. SMITH and N.A. WALKER, unpublished data) is attributed to the active efflux pump. Only in isolated instances has evidence of such a pump been adduced with the aid, for example, of metabolic inhibitors (MACROBBIE, 1962; RAVEN, 1967; THOMAS, 1970). In a similar way, it has become accepted that anions, particularly the monovalent ones such as Cl^-, NO_3^- or $H_2PO_4^-$, must be actively transported inwards.

Very often passive and active transport occur independently of each other in the same membrane. Then, whether the ion species in question appears to be in electrochemical equilibrium or not depends on the relative contribution of such fluxes to the total unidirectional fluxes. Potassium is a good example of such an ion. Under some conditions it plainly behaves passively (some charophyte cells in Ca-free solutions, freshly-cut low-salt barley root segments) while under other conditions K^+ appears to be transported inwards or even outwards (pea and oat seedlings tissue in "10× standard nutrient solution"—HIGINBOTHAM et al., 1967) but the steady-state condition on which such a conclusion is based, together with comparison of ψ_K and ψ_M, was mostly not satisfied, as the authors recognized (but see also Part B, *3.3*.2).

2.5.2 Short-Circuited Preparations

Another method of revealing an active transport flux is to search for net ion-fluxes across a membrane or tissue under conditions where the electrochemical potential difference for the ion has been made zero. This is effected by maintaining the same concentration of that ion on either side together with a short-circuit of any PD remaining. The sum of the net flux of all ions actively transported is expected to agree with the size of the short-circuit current (SCC) once a steady-state has been reached. The large coenocytes *Halicystis* and *Valonia* have been treated in this way, the former having the vacuole perfused with sea water, while the latter tolerated only a change in the external medium to artificial vacuolar sap. Results from these experiments were variable and not altogether consistent, but active transport of sodium outwards, and potassium and chloride inwards was inferred in *Halicystis* (BLOUNT and LEVEDAHL, 1960). In *Valonia*, although considerations of the Nernst potential for Cl^- allows the conclusion that it is passively distributed between vacuole and sea water, a net influx seemed to persist under short-circuited, perfused conditions (GUTKNECHT, 1967).

In salt-excreting leaves of *Limonium*, HILL (1967) concluded from comparison of SCC with net fluxes of seven different ion species that Na^+ and Cl^- were both actively excreted, when 100 mM NaCl was the "salt load". Br^- and I^- substituted readily for Cl^-, and K^+, Cs^+ and Rb^+ substituted for Na^+, with increasing net flux in that order. In these experiments the fluxes in question were across leaf discs, almost the only path being *via* salt glands.

In later experiments (HILL, 1970) induction of the transport system was inferred from a lag in appearance of the SCC and open-circuit PD, on presenting the freshly cut leaf tissue with 100 mM NaCl. Part B, Chap. *5.2* is devoted to a detailed description of transport in salt glands, but see also Part B, *9.3.3.4*.

2.6 Electrogenic Effects—the PD as a Function of Metabolic Activity

The electrogenic influence is most obvious when the magnitude of the observed membrane PD is greater than the magnitude of any likely Nernst potential. Examples are given in Table *4.2*; see also *6.6.6.1*, *6.6.6.2*, *7.2.1.4* and Part B, *3.3.2*. An

Table *4.2*. Cells in which an electrogenic pump is inferred from a hyperpolarized state

Cells	ψ_K (mV)	ψ_M (mV)	Medium (mM)	Linkage	Ref.
Chara corallina	−160	−220	0.1 K^+, pH 7	PS[a]	HOPE (1965)
Neurospora crassa	− 80	−240	1 K^+	Resp[b]	SLAYMAN (1965)
Nitella translucens	−117	−162	0.5 K^+, pH 6	PS	SPANSWICK (1972b)
Beta vulgaris	−139	−209	0.6 K^+, pH 7.2	—	POOLE (1966)
Elodea densa	(−110)	−135	1 K^+, pH 5.8	PS	JESCHKE (1970)
Elodea canadensis	−172	−278	0.1 K^+, pH 7	PS	SPANSWICK (1972a)
Acetabularia mediterranea	− 93	−174	Sea water (10 K^+)	PS	SADDLER (1970)
Avena	− 90	−130	1 K^+	Resp	HIGINBOTHAM et al. (1970)
Pisum	− 55	−102	10 K^+	Resp	HIGINBOTHAM et al. (1970)
Riccia fluitans	−106	−200	1 K^+	PS	FELLE and BENTRUP (1974)

[a] Photosynthesis. [b] Respiration

electrogenic effect that tended to place the PD within the range of the Nernst potentials would be more difficult to detect but is possible in principle. Very clear connections between electrogenic effects and various metabolic pathways have been demonstrated in some instances, e.g. as a result of the sharp dependence of the membrane potential on the concentration of added inhibitor of oxidative metabolism. This was found in *Neurospora*, with N_3^- (SLAYMAN, 1965; 7.2.1.5) and pea epicotyl cells, with CN^- (HIGINBOTHAM et al., 1970). These workers did not rely on a reputed effect of the inhibitor; they confirmed that oxygen consumption in parallel experiments decreased in speeds and amounts corresponding to the depolarization of the plasmalemma. The electrogenic effect in both cases could be a proton efflux pump dependent on respiration. In *Neurospora*, a remarkable series of experiments (SLAYMAN et al., 1973) shows a close relationship between the extent of hyperpolarization (in which ψ_M is more negative than ψ_K, the most negative Nernst potential) and the concentration of ATP in the hyphae (Fig. *4*.7).

In other genera, the electrogenic pump is clearly photosynthesis-stimulated. This is so in *Nitella* (SPANSWICK, 1972b) and *Acetabularia* (SADDLER, 1970) while in *Ch. corallina* the hyperpolarized state often defies dark conditions and anaerobiosis but slowly declines in the presence of azide, an effect seemingly unrelated to respiration (RICHARDS and HOPE, 1974).

Whatever the metabolic link, in charophyte cells as well as in *Neurospora* the electrogenic effect is probably caused by extrusion of protons (SPANSWICK, 1972b, 1973; RICHARDS and HOPE, 1974).

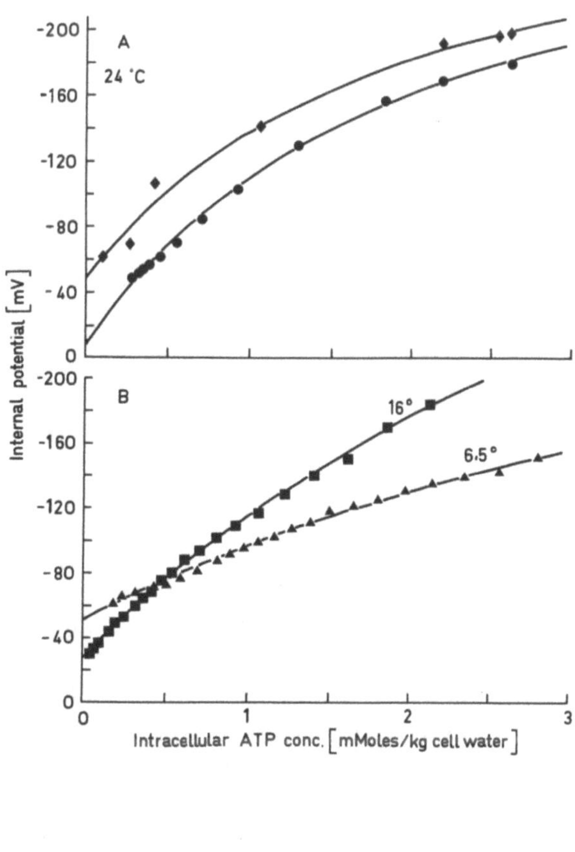

Fig. *4*.7. The membrane potential of hyphae of *Neurospora crassa* plotted against the corresponding concentration of ATP within the hyphae. The variation was obtained by additions of KCN to the medium, the determinations being made 30 s after the addition. (From SLAYMAN et al., 1973)

The electrical properties of electrogenic pumps have been discussed by FINKELSTEIN (1964) and RAPOPORT (1970), and in 3.5.4.2, and these notions applied to the situation in *N. translucens* by SPANSWICK (1972b, 1973). The pump is represented as having an EMF, and a conductance, as discussed in 4.2.2.1. In theory the rate of working of a proton efflux pump may be under control of external pH, and the membrane PD approaches the pump EMF the smaller the passive conductance (WALKER and SMITH, 1975). Therefore the PD is a function of pH even if the plasmalemma is not passively permeable to protons. The decrease in membrane conductance upon change to dark conditions is less easy to explain but may reflect changes in the thermodynamic coefficient linking active flux and driving force. Light/dark changes may also affect passive conductance.

2.7 Electrical Properties and Membrane Structure

2.7.1 Conductive Pores for K^+

The occurrence of true electroosmotic coupling in the plasmalemma of *Chara* and *Nitella* (TYREE and SPANNER, 1969; BARRY and HOPE, 1969) implies that conductance paths (probably not including the electrogenic pump in these instances) are such that water and $K^+(H^+?)$ interact during ion migration. Rather restricted, cation-selective paths would be expected to cause this effect. The non-independence of potassium influx and efflux, demonstrated on "passive" *Chara* and *Nitella* cells through measurement of flux-ratios (WALKER and HOPE, 1969) also suggests a restricted diameter of the conducting regions. Thirdly, the relatively high enthalpy of activation for K^+ migration, inferred from effects of temperature on PD and conductance in charophyte cells and in *Griffithsia* (HOGG et al., 1968b; HOPE and ASCHBERGER, 1970), again points to a measure of friction between K^+ ions and the membrane material and/or membrane water.

Taken together, the above results appear to eliminate the possibility of a few large (> 3 nm) pores as the passive conducting mechanism for ions.

2.7.2 The "Double Fixed-Charge Membrane Model"

Restricted, cation-selective conducting paths are not sufficient to account for "punch-through", the non-destructive increase in conductance occurring when the plasmalemma of charophytes is hyperpolarized. The only plausible explanation of this is, according to COSTER (1965, 1969, 1973a, b), that this membrane acts as a "p-n junction" in the style of semiconductors. P-type semiconductors have an excess of positive charge carriers (holes) while n-type semiconductors have an excess of electrons. By analogy, p-type membrane material could be a layer of negative fixed-charge, and n-type, an apposite layer of positive fixed-charge (Fig. 4.8). Calculations using this model are complicated but the predictions definite. The direction of punch-through, the PD at punch-through and its observed dependence on pH are all matched satisfactorily by judicious choice of several parameters in the model, such as the concentrations of fixed-charge and the dissociation constants.

The measurements of membrane capacitance referred to earlier have been used to test the validity of the model. The model predicts certain relationships between the capacitance and the frequency of an applied electric current. These are quite unexpected of a simple, parallel-plate capacitor, used at the beginning of this Chapter to explain membrane capacitance. Now, capacitance is vested in the depletion layer between the positive and negative fixed-charge regions and is expected to vary also with membrane PD since the width of the depletion layer is dependent on membrane PD. Experiments to test this have been done (see *4*.1.5.1; *4*.2.3).

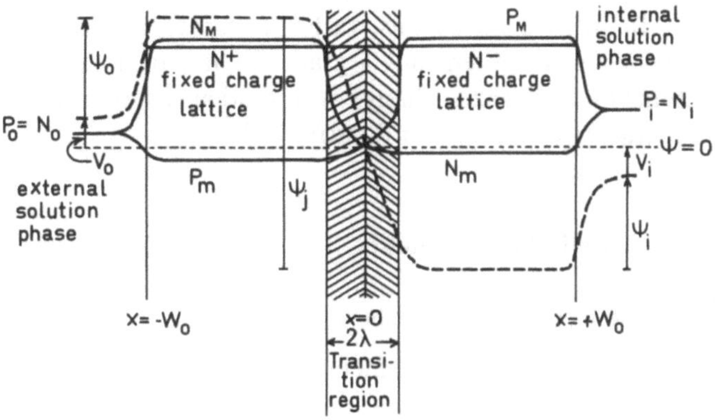

Fig. *4*.8. The profiles of phase boundary electric potentials (ψ), and concentrations of mobile ions in the double fixed-charge model of thickness $2\,W_o$. P_M is positive majority, N_M negative majority, N_m negative minority and P_m positive minority ion concentration. $N\pm$ are concentrations of fixed charge with respective lattices, which abut at $x=0$. The "membrane potential" is given by $V_i + V_o$. The transition region contains a space charge due to the withdrawal of mobile ions. (From COSTER, 1965)

How can this conclusion—that the plasma membrane has a structure that behaves like the model of Fig. *4*.8—be reconciled with the earlier conclusion about the nature of the conducting areas? One way is to propose that the restricted conductive areas are identical with some of the globular protein "floating in a sea of lipid" envisaged in the membrane model of SINGER and NICOLSON (1972). The positive and negative fixed charges would then be on appropriate side groups of the protein, and face the outside and the cytoplasm respectively. Other properties, as we postulated in *4*.2.2.1, can be given to yet other proteins in the membrane.

2.7.3 Temperature Effects and Membrane Phase Changes

If there are phase changes in membrane lipids (*1*.3.2) in the physiological temperature range, it might be expected that electrical properties might show corresponding discontinuities around the temperature of the phase changes. In retrospect, it was worth asking whether plant cell membranes exhibit such behavior, although at the time it was not known whether chiefly membrane lipid or protein was concerned with membrane permeability. Nor is it known with any certainty now, but the odds are in favor of the protein.

Nevertheless, changes in phase in the lipid component of membranes are accepted as affecting the rates of certain metabolic reactions plainly mediated by membrane-located enzymes (e.g. in mitochondria, see RAISON, 1972).

The evidence from measurements of electrical properties of *N. translucens* and *Griffithsia pulvinata* shows that permeability to K^+ or Na^+ is not discontinuous when an Arrhenius plot is made, in the temperature range 2–20° C (HOGG et al., 1968b; HOPE and ASCHBERGER, 1970). In *Chara corallina*, on the other hand, an indication of a change in slope in log $P_{K,Na}$ vs $1/T$ was obtained at 12° C (HOPE and ASCHBERGER, 1970). However, interpretation of changes in PD and conductance with temperature, in terms of changes in P_K and P_{Na} only, without consideration of the possible effects on the activity of an electrogenic pump, now appears unsatisfactory. In work such as this, rather accurate results are needed if the claim is to be made for a sudden, not a gradual change in slope in an Arrhenius plot. More convincing of a phase change is a sudden discontinuity in the rate at a particular temperature (see Fig. 1 in RAISON, 1972).

2.8 Unusual Measurements

2.8.1 The PD in Chloroplasts

The first reported measurements of an intra-organellar PD for plants seem to have been those of BULYCHEV et al. (1972). The unusually large chloroplasts in leaves of *Peperomia metallica* were penetrated by microelectrodes and the PD measured in light or dark conditions, relative to the cytoplasm, or *in vitro*, relative to a suspension medium. Two "states" or ranges of PD were observed, *in vivo*, with the chloroplast either positive or negative. Light-induced changes in PD were of two sorts, a slow, negative-going response, or a fast, positive-going response. The latter was particularly observed in isolated chloroplasts. The sign of the potential, and the direction of the light-induced change, was felt to be dependent on the location of the microelectrode tip. The fast, light-induced, positive changes may have been the response of intrathylakoid electrodes.

DAVIS (1974) used gametophytes from *Phaeocerus laevis*, also containing giant chloroplasts (Fig. 4.9) and made estimates of internal pH from the PD of antimony-covered microelectrodes. The intrachloroplast PD was measured using conventional microelectrodes. The latter observations fell once again into two groups; those obtained on the whole with finer-tipped electrodes were considered to be intrathylakoid measurements. Table 4.3 gives some of the values of PD found in the experiments

Table 4.3. PDs in chloroplasts, measured with respect to surrounding cytoplasm

Cells	$\Delta\psi$ (dark) (mV)	Effect of light	Ref.
Peperomia metallica	−60 or +15	10–20 mV −ve (slow, 60–180 s) *or* 10–30 mV +ve (fast, 10–50 ms)	BULYCHEV et al. (1972)
Phaeocerus laevis	−10 or	−35	DAVIS (1974)
	+10 (finer tips to microelectrodes)	+30	

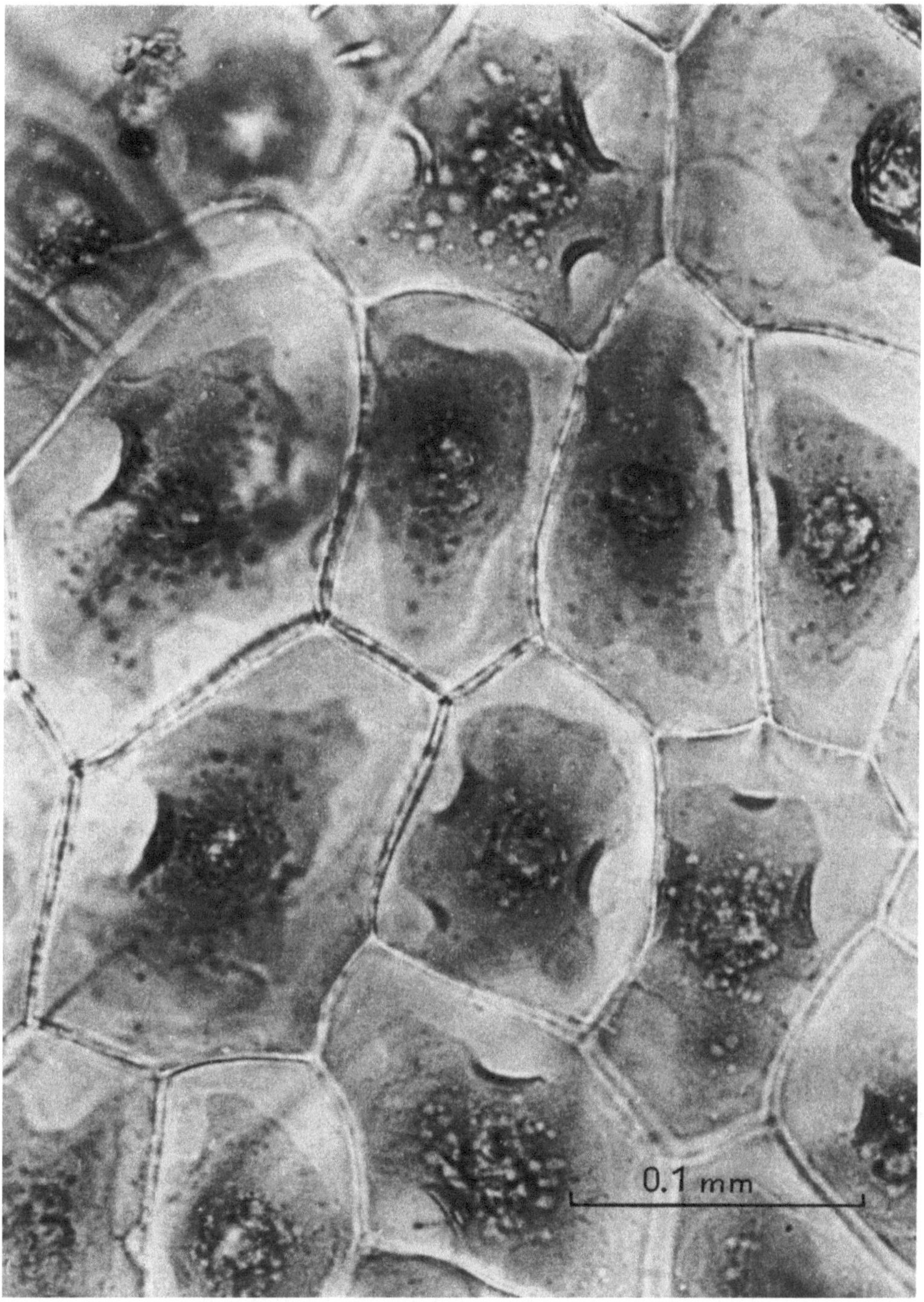

Fig. 4.9. Surface view of chloroplast in upper epidermal cells of gametophyte of *Phaeoceros laevis in vivo*. In some cells the plastid may be seen to curl away from the lateral walls. The large, dark granular mass in the center of each plastid is the pyrenoid body. The light areas are starch grains. By courtesy of R.F. DAVIS (Newark, New Jersey)

with chloroplasts. The accuracy of the estimates is not likely to be very high. Antimony pH electrodes are subject to errors from the presence of many surrounding substances. Also, when microelectrodes penetrate a membrane of small surface area, the inevitable current flow around the wound is relatively more likely to depolarize the membrane PD. This effect is discussed by LASSEN (1971) who also described means of extrapolating to the time of penetration in an attempt to find the membrane PD before the capacitance had been partly discharged by the wound current. Such techniques should prove useful in work with chloroplasts and even cells of higher plants.

2.8.2 *Chlorella*

This genus is the sole representative of the microalgae that has been used for measurement of intracellular PD. BARBER (1968) found that the cell interior was −40 mV with respect to the medium and that P_{Na}/P_K was 0.11, a value comparable to that for the plasmalemma of *Chara corallina*. Rather similar values for ψ_{co} were found by LANGMÜLLER and SPRINGER-LEDERER (1974), in a medium containing solely 1 mM KCl, for temperatures of 15–30° C.

2.8.3 Isolated Cytoplasmic Drops

Using a method originated by KAMIYA and KURODA (1965), INOUE et al. (1973) produced isolated drops of protoplasm from *Nitella* sp. and made measurements of transmembrane PD, conductance and excitability; these were followed by experiments to find the effect of proteases or lipases on the changes in PD and conductance following isolation of the drops. The drop surface acted as a rather depolarized and conductive membrane compared with the normal plasmalemma, exhibiting a PD of −50 to −100 mV (relative to an external solution containing 0.5 mM K^+, and other ions). The conductance was 3–25 S m^{-2} compared with 0.3–1 S m^{-2} in intact *Nitella* and *Chara* plasmalemmas. Brief (30–50 ms) action potentials were recorded.

2.9 Effects of Uncouplers and Antibiotics

2.9.1 Uncouplers

Substances such as DNP, CCCP, the gramicidins and nigericin render the "energy-coupling" membranes in mitochondria and chloroplasts more permeable to protons or in the case of nigericin allow an exchange of protons and metallic cations. Few tests have been made of such substances to discover their effects on the PD or conductance of plasma membranes. However, it has been found that CCCP will prevent light-stimulated hyperpolarization, acting rather like dark conditions in *Elodea* (JESCHKE, 1970). VREDENBERG (1972) observed that CCCP inhibited the light-induced depolarization occuring 1–2 min after light is switched on; concentrations of CCCP (*c.* 0.4 μM) could be found that had this effect while leaving the accompanying changes in membrane conductance unaffected.

2.9.2 Antibiotics

Nystatin, a polyene antibiotic that renders ultra-thin lipid membranes anion-permeable (CASS et al., 1970), depolarizes the plasmalemma of *Ch. corallina*. This may be attributed to an increase in P_{Cl} in the resting membrane (G.P. FINDLAY, A.B. HOPE and P.H. SYDENHAM, unpublished data) since large increases in passive chloride efflux accompany the depolarization. Curiously, in the nystatin-depolarized membrane the excitability decreases, probably because stimulation no longer elicits the usual increase in *transient* chloride permeability (see *4*.2.4 above). In view of the pronounced effect of nystatin, it is inferred that the plasmalemma of *Chara* contains sterols since this class of lipid is a necessary component of nystatin-sensitive membranes.

Valinomycin is a well-known macrocyclic polypeptide that complexes preferentially with K^+ to form a lipid-soluble Val-K^+ complex. When added to the bathing solution around *Ch. corallina*, up to 20 μM valinomycin has no detectable effects on electrical properties. Because of the relatively high resting permeability to K^+ under the conditions of these experiments (ψ_{co} was K^+-sensitive) it might be conjectured that the plasmalemma contains natural ionophores like valinomycin. The maximum discrimination in favor of K^+ over Na^+ reported for black lipid membranes is about 400 (MUELLER and RUDIN, 1967); it is interesting that the largest value for P_K/P_{Na} for plant cell plasmalemmas is close to this figure (*Griffithsia*, see *4*.2.1.1 above).

2.10 Visible, UV and Ionizing Radiations

Electrical measurements may give clues about the action of other agents besides the antibiotics and uncouplers.

2.10.1 Light

The PD in green cells has been shown to undergo complicated changes upon light/dark and dark/light transitions. These have not yet been fully interpreted. Apart from the stimulation of the electrogenic effect in *N. translucens* (but not in *Ch. corallina*), over a period of a few minutes, light may cause either depolarization (VREDENBERG, 1972) or hyperpolarization (SPANSWICK, 1972b), on initial application. Extremely rapid (ms) changed in PD were observed by VREDENBERG (1974) and interpreted as reflecting transient ion currents across the plasmalemma following light-driven ion currents out of the chloroplasts. The possibility of light receptors in the plasmalemma is a real one.

Transient changes in the intracellular PD in *Mnium* or in *Atriplex* leaf vesicles (bladders) were claimed to be connected with proton fluxes between cytoplasm and chloroplasts (LÜTTGE and PALLAGHY, 1969) hence causing changes in the gradient across the plasmalemma. However there is only a rather small translocation of protons across the chloroplast outer envelope, compared with that driven by light in isolated, class II chloroplasts (HELDT et al., 1973). The light-triggered transients in the PD may reflect the activity of electrogenic transport systems stimulated by chloroplast-exported substrate.

2.10.2 UV Radiation

A dose of 1,400 J m^{-2} of 254 nm UV had several effects on electrical properties of *Ch. corallina* (DOUGHTY and HOPE, 1973).

(i) at pH 5.5, depolarization and increase in the conductance of the plasmalemma were well correlated with an increase in passive chloride efflux

(ii) at pH 6.7, a secondary depolarization, i.e. following in time the depolarization postulated as due to induced chloride efflux, was consistent with an inhibition of the electrogenic pump.

Both effects (i) and (ii), and those on the action potential were almost fully reversed 30 min after the end of irradiation. The membrane targets were postulated to be sulphydryl and disulphide groups in membrane proteins.

2.10.3 Ionizing Radiation

A series of studies by ESCH et al. (1964), ESCH (1966) and HANSEN (1967) have shown that 12–50 krad doses of γ- or X-radiation depolarized the intracellular PD, probably ψ_{co}, in *Nitella flexilis.* An early interpretation, a decrease in potassium permeability, is probably wrong; the depolarization (and increase in membrane conductance) is more consistent with an increase in passive, resting chloride permeability.

References

ANDERSON, W.P., HENDRIX, D.L., HIGINBOTHAM, N.: Higher plant cell membrane resistance by a single intracellular electrode method. Plant Physiol. **53**, 122–124 (1974).

BARBER, J.: Measurement of the membrane potential and evidence for active transport of ions in *Chlorella pyrenoidosa.* Biochim. Biophys. Acta **150**, 618–625 (1968).

BARRY, P.H., HOPE, A.B.: Electro-osmosis in *Chara* and *Nitella* cells. Biochim. Biophys. Acta **193**, 124–128 (1969).

BENOLKEN, R.M., JACOBSON, S.L.: Response properties of a sensory hair excised from Venus' fly-trap. J. Gen. Physiol. **56**, 64–82 (1970).

BENTRUP, F.-W.: Die Morphogenese pflanzlicher Zellen im elektrischen Feld. Z. Pflanzenphysiol. **59**, 309–339 (1968).

BENTRUP, F.-W.: Elektrophysiologische Untersuchungen am Ei von *Fucus serratus:* Das Membranpotential. Planta **94**, 319–332 (1970).

BLINKS, L.R., JACQUES, A.G.: The cell sap of *Halicystis.* J. Gen. Physiol. **13**, 733–737 (1930).

BLOUNT, R.W., LEVEDAHL, B.H.: Active sodium and chloride transport in the single celled marine alga *Halicystis ovalis.* Acta Physiol. Scand. **49**, 1–9 (1960).

BRENNECKE, R., LINDEMANN, B.: A chopped-current clamp for current injection and recording of membrane polarisation with single electrodes of changing resistance. T.I.T.J. Life Sci. **1**, 53–58 (1971).

BRIGGS, G.E., HOPE, A.B., ROBERTSON, R.N.: Electrolytes and plant cells. Oxford: Blackwell 1961.

BULYCHEV, A.A., ADRIANOV, V.K., KURELLA, G.A., LITVIN, F.F.: Microelectrode measurements of the transmembrane potential of chloroplasts and its photoinduced changes. Nature **236**, 175–177 (1972).

CASS, A., FINKELSTEIN, A., KRESPI, V.: The ion permeability induced in thin lipid membranes by the polyene antibiotics nystatin and amphotericin B. J. Gen. Physiol. **56**, 100–124 (1970).

COLE, K.S.: Dynamic electrical characteristics of the squid axon membrane. Arch. Sci. Physiol. **3**, 253–258 (1949).

COLE, K.S., CURTIS, H.J.: Electrical impedance of *Nitella* during activity. J. Gen. Physiol. **22**, 37–64 (1939).

COLE, K.S., HODGKIN, A.L.: Membrane and protoplasm resistance in the squid giant axon. J. Gen. Physiol. **22**, 671–687 (1939).

COSTER, H.G.L.: A quantitative analysis of the voltage-current relationships of fixed charge membranes and the associated property of "punch-through". Biophys. J. **5**, 669–686 (1965).

COSTER, H.G.L.: The role of pH in the punch-through effect in the electrical characteristics of *Chara australis*. Australian J. Biol. Sci. **22**, 365–374 (1969).

COSTER, H.G.L.: The double fixed charge membrane. Low frequency dielectric dispersion. Biophys. J. **13**, 118–132 (1973a).

COSTER, H.G.L.: The double fixed charge membrane. Solution-membrane ion partition effects and membrane potentials. Biophys. J. **13**, 133–142 (1973b).

COSTER, H.G.L., SYRIATOWICZ, J.C., VOROBIEV, L.N.: Cytoplasmic ion exchange during rest and excitation in *Chara australis*. Australian J. Biol. Sci. **21**, 1069–1073 (1968).

CURTIS, H.J., COLE, K.S.: Transverse electric impedance of the squid giant axon. J. Gen. Physiol. **21**, 757–765 (1938).

DAVIS, R.F.: Photoinduced changes in electrical potentials and H^+ activities of the chloroplast, cytoplasm, and vacuole of *Phaeoceros laevis*. In: Membrane transport in plants (U. ZIMMERMANN, J. DAINTY eds.), p. 197–201. Berlin-Heidelberg-New York: Springer 1974.

DIPALMA, J.R., MOHL, R., BEST, W.J.R.: Action potential and contraction of *Dionaea muscipula* (Venus' flytrap). Science **133**, 878–879 (1961).

DOUGHTY, C.J.: The effects of ultraviolet radiation on the ionic relations of *Chara corallina*. Ph.D. Thesis, Flinders University of South Australia (1973).

DOUGHTY, C.J., HOPE, A.B.: Effect of ultraviolet radiation on the membranes of *Chara corallina*. J. Membrane Biol. **13**, 185–198 (1973).

ELLIS, G.W.: Piezoelectric micromanipulators. Science **138**, 84–91 (1962).

ESCH, H.E.: Dose-dependent gamma irradiation effects on the resting potential of *Nitella flexilis*. Radiation Res. **27**, 355–362 (1966).

ESCH, H.E., MILTENBURGER, H., HUG, O.: Die Beeinflussung elektrischer Potentiale von Algenzellen durch Röntgenstrahlen. Biophysik **1**, 380–388 (1964).

ETHERTON, B., HIGINBOTHAM, N.: Transmembrane potential measurements of cells of higher plants as related to salt uptake. Science **131**, 409–410 (1960).

EWART, A.J.: On the physics and physiology of protoplasmic streaming in plants. Oxford: Oxford University Press 1903.

FELLE, H., BENTRUP, F.-W.: Light-dependent changes of membrane potential and conductance in *Riccia fluitans*. In: Membrane transport in plants (U. ZIMMERMANN, J. DAINTY, eds.), p. 120–125. Berlin-Heidelberg-New York: Springer 1974.

FINDLAY, G.P.: Studies of action potentials in the vacuole and cytoplasm of *Nitella*. Australian J. Biol. Sci. **12**, 412–426 (1959).

FINDLAY, G.P.: Voltage-clamp experiments with *Nitella*. Nature **191**, 812–814 (1961).

FINDLAY, G.P.: Calcium ions and the action potential in *Nitella*. Australian J. Biol. Sci. **15**, 69–82 (1962).

FINDLAY, G.P.: Ionic relations of cells of *Chara australis*. VIII. Membrane currents during a voltage clamp. Australian J. Biol. Sci. **17**, 388–399 (1964).

FINDLAY, G.P.: Membrane electrical behaviour in *Nitellopsis obtusa*. Australian J. Biol. Sci. **23**, 1033–1045 (1970).

FINDLAY, G.P., HOPE, A.B.: Ionic relations of cells of *Chara australis*. VII. The separate electrical characteristics of plasmalemma and tonoplast. Australian J. Biol. Sci. **17**, 62–77 (1964a).

FINDLAY, G.P., HOPE, A.B.: Ionic relations of cells of *Chara australis*. IX. Analysis of transient membrane currents. Australian J. Biol. Sci. **17**, 400–411 (1964b).

FINDLAY, G.P., HOPE, A.B., PITMAN, M.G., SMITH, F.A., WALKER, N.A.: Ionic relations of marine algae. III. *Chaetomorpha*: membrane electrical properties and chloride fluxes. Australian J. Biol. Sci. **24**, 731–745 (1971).

FINDLAY, G.P., HOPE, A.B., WILLIAMS, E.J.: Ionic relations of marine algae. I. *Griffithsia*: membrane electrical properties. Australian J. Biol. Sci. **22**, 1163–1178 (1969).

FINDLAY, G.P., HOPE, A.B., WILLIAMS, E.J.: Ionic relations of marine algae. II. *Griffithsia*: ionic fluxes. Australian J. Biol. Sci. **23**, 323–338 (1970).

FINKELSTEIN, A.: Carrier model for active transport of ions across a mosaic membrane. Biophys. J. **4**, 421–440 (1964).

FRANK, K., BECKER, M.C.: Microelectrodes for recording and stimulation. In: Physical techniques in biological research (W.L. NASTUK, ed.), vol. V, part A, p. 89–143. New York-London: Academic Press 1964.

FUJITA, M.: Electrophysiological studies on *Nitella* cells, with special reference to electric resistance. Plant Cell Physiol. (Tokyo) **3**, 229–247 (1962).

FUJITA, M., MIZUGUCHI, K.: Excitation in *Nitella*, especially in relation to electric stimulation. Cytologia (Tokyo) **21**, 135–145 (1955).

GAFFEY, C.T.: Stimulation of action potentials with radiation in single cells of *Nitella gracilis*. Intern. J. Radiation Biol. **21**, 11–29 (1972).

GAFFEY, C.T., MULLINS, L.J.: Ion fluxes during the action potential in *Chara*. J. Physiol. (London) **144**, 505–524 (1958).

GOLDSMITH, M.H.M., FERNÁNDEZ, H.R., GOLDSMITH, T.H.: Electrical properties of parenchymal cell membranes in the oat coleoptile. Planta **102**, 302–323 (1972).

GRADMANN, D.: Einfluß von Licht, Temperatur und Außenmedium auf das elektrische Verhalten von *Acetabularia crenulata*. Planta **93**, 323–353 (1970).

GREENHAM, C.G.: The relative electrical resistances of the plasmalemma and tonoplast in higher plants. Planta **69**, 150–157 (1966).

GUTKNECHT, J.: Sodium, potassium and chloride transport and membrane potentials in *Valonia ventricosa*. Biol. Bull. **130**, 331–344 (1966).

GUTKNECHT, J.: Ion fluxes and short-circuit current in internally perfused cells of *Valonia ventricosa*. J. Gen. Physiol. **50**, 1821–1834 (1967).

GUTKNECHT, J.: Salt transport in *Valonia*: inhibition of potassium uptake by small hydrostatic pressures. Science **160**, 68–70 (1968).

HAAPANEN, L., SKOGLUND, C.R.: Recording of the ionic efflux during single action potentials in *Nitellopsis obtusa* by means of high-frequency reflectrometry. Acta Physiol. Scand. **69**, 51–68 (1967).

HANSEN, U.P.: Zusammenhänge zwischen den Strahlenbeeinflussungen der Membranspannung, der elektrischen Reizschwelle und des ohmschen Widerstandes der Zellmembran von *Nitella flexilis*. Atomkernenergie **12**, 447–462 (1967).

HARVEY, E.N.: Stimulation of cells by intense flashes of ultraviolet light. J. Gen. Physiol. **25**, 431–445 (1942a).

HARVEY, E.N.: Hydrostatic pressure and temperature in relation to stimulation and cyclosis in *Nitella flexilis*. J. Gen. Physiol. **25**, 855–863 (1942b).

HELDT, H.W., WERDAN, K., MILOVANCEV, M., GELLER, G.: Alkalisation of the chloroplast stroma caused by light-dependent proton flux into the thylakoid space. Biochim. Biophys. Acta **314**, 224–241 (1973).

HIGINBOTHAM, N., ETHERTON, B., FOSTER, R.J.: Mineral ion contents and cell transmembrane electropotentials of pea and oat seedling tissue. Plant Physiol. **42**, 37–46 (1967).

HIGINBOTHAM, N., GRAVES, J.S., DAVIS, R.F.: Evidence for an electrogenic ion transport pump in cells of higher plants. J. Membrane Biol. **3**, 210–222 (1970).

HIGINBOTHAM, N., HOPE, A.B., FINDLAY, G.P.: Electrical resistance of cell membranes of *Avena* coleoptiles. Science **143**, 1448–1449 (1964).

HILL, A.E.: Ion and water transport in *Limonium*. II. Short-circuit analysis. Biochim. Biophys. Acta **135**, 461–465 (1967).

HILL, A.E.: Ion and water transport in *Limonium*. IV. Delay effects in the transport process. Biochim. Biophys. Acta **196**, 73–79 (1970).

HILL, S.E.: Stimulation by cold in *Nitella*. J. Gen. Physiol. **18**, 357–367 (1935).

HODGKIN, A.L., HUXLEY, A.F.: A quantitative description of membrane current and its application to conduction and excitation in nerve. J. Physiol. (London) **117**, 500–544 (1952).

HODGKIN, A.L., HUXLEY, A.F., KATZ, B.: Measurement of current voltage relations in the membrane of the giant axon of *Loligo*. J. Physiol. (London) **116**, 424–448 (1952).

HODGKIN, A.L., RUSHTON, W.A.H.: The electrical constants of a crustacean nerve fibre. Proc. Roy. Soc. (London), Ser. B **133**, 444–479 (1946).

HÖRMANN, G.: Studien über die Protoplasmaströmung bei den Characeen. Jena: Fischer 1898.

HOGG, J., WILLIAMS, E.J., JOHNSTON, R.J.: A simplified method for measuring membrane resistances in *Nitella translucens*. Biochim. Biophys. Acta **150**, 518–520 (1968a).

HOGG, J., WILLIAMS, E.J., JOHNSTON, R.J.: The temperature dependence of the membrane potential and resistance in *Nitella translucens*. Biochim. Biophys. Acta **150**, 640–648 (1968b).

HOGG, J., WILLIAMS, E.J., JOHNSTON, R.J.: The membrane electrical parameters of *Nitella translucens*. J. Theoret. Biol. **24**, 317–334 (1969).

HOPE, A.B.: Ionic relations of cells of *Chara australis*. V. The action potential. Australian J. Biol. Sci. **14**, 312–322 (1961).

HOPE, A.B.: Ionic relations of cells of *Chara australis*. X. Effects of bicarbonate ions on electrical properties. Australian J. Biol. Sci. **18**, 789–801 (1965).

HOPE, A.B.: Ion transport and membranes. London: Butterworths 1971.

HOPE, A.B., ASCHBERGER, P.A.: Effects of temperature on membrane permeability. Australian J. Biol. Sci. **23**, 1047–1060 (1970).

HOPE, A.B., FINDLAY, G.P.: The action potential in *Chara*. Plant Cell Physiol. (Tokyo) **5**, 377–379 (1964).

HOPE, A.B., WALKER, N.A.: Ionic relations of *Chara australis* R.Br. IV. Membrane potential differences and resistances. Australian J. Biol. Sci. **14**, 26–44 (1961).

HOPE, A.B., WALKER, N.A.: The physiology of giant algal cells. Cambridge: C.U.P. 1975.

INOUE, I., UEDA, T., KOBATAKE, Y.: Structure of excitable membranes formed on the surface of protoplasmic drops isolated from *Nitella*. I. Conformation of surface membrane determined from the refractive index and from enzyme actions. Biochim. Biophys. Acta **298**, 653–663 (1973).

JACOBSON, S.L.: Receptor response in Venus' fly-trap. J. Gen. Physiol. **49**, 117–129 (1965).

JAFFE, L.F.: Localization in the developing *Fucus* egg and the general role of localizing currents. Advan. Morphogenesis **7**, 295–328 (1968).

JAFFE, L.F., ROBINSON, K.R., NUCCITELLI, R.: Transcellular currents and ion fluxes through polarising fucoid eggs. In: Membrane transport in plants and plant organelles (J. DAINTY, U. ZIMMERMANN, eds.), p. 226–233. Berlin-Heidelberg-New York: Springer 1974.

JESCHKE, W.D.: Lichtabhängige Veränderungen des Membranpotentials bei Blattzellen von *Elodea densa*. Z. Pflanzenphysiol. **62**, 158–172 (1970).

KAMIYA, N., KURODA, K.: Rotational protoplasmic streaming in *Nitella* and some physical properties of the endoplasm. In: Proc. of the 4th Int. Cong. on Rheology. Part 4, Symposium on Biorheology, p. 157–171. Brown University (U.S.A.): Wiley and Sons 1965.

KATZ, B.: Experimental evidence for a non-conducted response of nerve to subthreshold stimulation. Proc. Roy. Soc. (London), Ser. B **124**, 244–276 (1937).

KISHIMOTO, U.: Electrical characteristics of *Chara corallina*. Ann. Rept. Scient. Works Fac. Sci., Osaka Univ. **7**, 115–146 (1959).

KISHIMOTO, U.: Current voltage relations in *Nitella*. Biol. Bull. **121**, 370–371 (1961).

KISHIMOTO, U.: Current voltage relations in *Nitella*. Jap. J. Physiol. **14**, 515–527 (1964).

KISHIMOTO, U.: Voltage clamp and internal perfusion studies on *Nitella* internodes. J. Cell. Comp. Physiol. **66**, No. 3; Part II, 43–54 (1965).

KISHIMOTO, U.: Repetitive action potentials in *Nitella* internodes. Plant Cell Physiol. (Tokyo) **7**, 547–558 (1966a).

KISHIMOTO, U.: Action potential of *Nitella* internodes. Plant Cell Physiol. (Tokyo) **7**, 559–572 (1966b).

KISHIMOTO, U.: Response of *Chara* internodes to mechanical stimulation. Ann. Rept. Biol. Works Fac. Sci., Osaka Univ. **16**, 61–66 (1968a).

KISHIMOTO, U.: Electromotive force of *Nitella* membrane during excitation. Plant Cell Physiol. (Tokyo) **9**, 539–551 (1968b).

KISHIMOTO, U.: Characteristics of the excitable *Chara* membrane. Advan. Biophys. **3**, 199–226 (1972).

KOPAC, M.J.: Micromanipulators: principles of design, operation and application. In: Physical techniques in biological research (W.L. NASTUK, ed.), vol. V, part A, p. 193–233. New York-London: Academic Press 1964.

KOPPENHÖFER, E.: Ruhe- und Aktionspotential von *Nitella mucronata* (A. Braun) Miquel unter Normalbedingungen. Pflügers Arch. **336**, 289–298 (1972a).

KOPPENHÖFER, E.: Die Wirkung von Kupfer, TTX, Cocain und TEA auf das Ruhe- und Aktionspotential von *Nitella*. Arch. Ges. Physiol. **336**, 299–309 (1972b).

LANGMÜLLER, G., SPRINGER-LEDERER, H.: Membranpotential von *Chlorella fusca* in Abhängigkeit von pH-Wert, Temperatur und Belichtung. Planta **120**, 189–196 (1974).

LASSEN, U.V.: Measurement of membrane potential of isolated cells. In: Proc. of the 1st European Biophysics Cong. (E. BRODA, A. LOCKER, H. SPRINGER-LEDERER, eds.), vol. III, p. 13–22. Vienna: Verlag der Wiener Medizinischen Akademie 1971.

LASSEN, U.V., STEN-KNUDSEN, O.: Direct measurements of membrane potential and membrane resistance of human red cells. J. Physiol. (London) **195**, 681–696 (1968).
LÜTTGE, U., PALLAGHY, C.K.: Light-triggered transient changes of membrane potentials in green cells in relation to photosynthetic electron transport. Z. Pflanzenphysiol. **61**, 58–67 (1969).
LÜTTGE, U., ZIRKE, G.: Attempts to measure plasmalemma and tonoplast electropotentials in small cells of the moss *Mnium* using centrifugation techniques. J. Membrane Biol. **18**, 305–314 (1974).
MACROBBIE, E.A.C.: Ionic relations of *Nitella translucens*. J. Gen. Physiol. **45**, 861–878 (1962).
MARMONT, G.: Studies on the axon membrane. I. A new method. J. Cell. Comp. Physiol. **34**, 351–382 (1949).
MOORE, J.W., COLE, K.S.: Voltage clamp techniques. In: Physical techniques in biological research (W.L. NASTUK, ed.), vol. VI, p. 263–321. New York: Academic Press 1963.
MUELLER, P., RUDIN, D.O.: Development of $K^+ - Na^+$ discrimination in experimental bimolecular lipid membranes by macrocyclic antibiotics. Biochem. Biophys. Res. Commun. **26**, 398–404 (1967).
MULLINS, L.J.: Efflux of chloride ions during the action potential of *Nitella*. Nature **196**, 986–987 (1962).
NEWMAN, I.A.: Electric potentials and auxin translocation in *Avena*. Australian J. Biol. Sci. **16**, 629–646 (1963).
NEWMAN, I.A., BRIGGS, W.R.: Phytochrome-mediated electric potential changes in oat seedlings. Plant Physiol. **50**, 687–693 (1972).
ODA, K.: Resting and action potentials in *Chara braunii*. Sci. Rept. Tohoku Univ., Fourth Ser. **22**, 167–174 (1956).
ODA, K.: The electrical constants in *Chara braunii*. Sci. Rept. Tohoku Univ., Fourth Ser. **27**, 187–198 (1961).
ODA, K.: Polarised and depolarised states of the membrane in *Chara braunii*, with special reference to the transition between the two states. Sci. Rept. Tohoku Univ., Fourth Ser. **28**, 1–16 (1962).
OSTERHOUT, W.J.V.: Physiological studies of single plant cells. Biol. Rev. **6**, 369–411 (1931).
OSTERHOUT, W.J.V.: Nature of the action current in *Nitella*. I. General considerations. J. Gen. Physiol. **18**, 215–227 (1934).
OSTERHOUT, W.J.V.: Electrical phenomena in large plant cells. Physiol. Rev. **16**, 216–237 (1936).
OSTERHOUT, W.J.V.: Nature of the action current in *Nitella* V. Partial response and the all-or-none law. J. Gen. Physiol. **27**, 61–68 (1943).
PICKARD, B.G.: Action potentials in higher plants. Botan. Rev. **39**, 172–201 (1973).
PICKARD, W.F.: The spatial variation of plasmalemma potential in a spherical cell polarised by a small current source. Mathematical Biosciences **10**, 307–328 (1971).
PITMAN, M.G., MERTZ, S.M. JR., GRAVES, J.S., PIERCE, W.S., HIGINBOTHAM, N.: Electrical potential differences in cells of barley roots and their relation to ion uptake. Plant Physiol. **47**, 76–80 (1970).
POOLE, R.J.: The influence of the intracellular potential on potassium uptake by beetroot tissue. J. Gen. Physiol. **49**, 551–563 (1966).
RAISON, J.K.: The influence of temperature-induced phase changes on the kinetics of respiratory and other membrane-associated enzyme systems. Bioenergetics **4**, 559–583 (1972).
RAPOPORT, S.I.: The sodium-potassium exchange pump: Relation of metabolism to electrical properties of the cell. I. Theory. Biophys. J. **10**, 246–259 (1970).
RAVEN, J.A.: Light-stimulation of active ion transport in *Hydrodictyon africanum*. J. Exptl. Bot. **50**, 1627–1640 (1967).
RICHARDS, J.L., HOPE, A.B.: The role of protons in determining membrane electrical characteristics in *Chara corallina*. J. Membrane Biol. **16**, 121–144 (1974).
ROBINSON, G.R., SCOTT, B.I.H.: A new method of estimating micropipette tip diameter. Experientia **29**, 1033–1034 (1973).
ROTTINGER, E.M., HUG, O.: Effects of low energy X-rays on membrane potential, membrane resistance and action potential of *Nitella flexilis*. Radiation Res. **50**, 491–503 (1972).
SADDLER, H.W.D.: The membrane potential of *Acetabularia mediterranea*. J. Gen. Physiol. **55**, 802–821 (1970).
SCOTT, B.I.H.: Electric oscillations generated by plant roots and a possible feedback mechanism responsible for them. Australian J. Biol. Sci. **10**, 164–179 (1957).

SCOTT, B.I.H.: Feedback-induced oscillations of five-minute period in the electric field of the bean root. Ann. N.Y. Acad. Sci. **98**, 890–900 (1962).

SIBAOKA, T.: Action potentials in plant organs. Symp. Soc. Exp. Biol. **20**, 49–74 (1966).

SINGER, S.J., NICHOLSON, G.L.: The fluid mosaic model of the structure of cell membranes. Science **175**, 720–731 (1972).

SLAYMAN, C.L.: Electrical properties of *Neurospora crassa*. Respiration and the intra-cellular potential. J. Gen. Physiol. **49**, 93–116 (1965).

SLAYMAN, C.L., LONG, W.S., LU, C.Y-H.: The relationship between ATP and an electrogenic pump in the plasma membrane of *Neurospora crassa*. J. Membrane Biol. **14**, 305–338 (1973).

SPANSWICK, R.M.: Electrophysiological techniques and the magnitudes of the membrane potentials and resistances of *Nitella translucens*. J. Exptl. Bot. **21**, 617–627 (1970).

SPANSWICK, R.M.: Electrical coupling between cells of higher plants: A direct demonstration of intercellular communication. Planta **102**, 215–227 (1972a).

SPANSWICK, R.M.: Evidence for an electrogenic ion pump in *Nitella translucens*. I. The effects of pH, K^+, Na^+, light and temperature on the membrane potential and resistance. Biochim. Biophys. Acta **288**, 73–89 (1972b).

SPANSWICK, R.M., Electrogenesis in photosynthetic tissues. In: Ion transport in plants (W.P. ANDERSON, ed.), p. 113–128. London-New York: Academic Press 1973.

SPANSWICK, R.M., COSTERTON, J.W.F.: Plasmodesmata in *Nitella translucens;* structure and electrical resistance. J. Cell. Sci. **2**, 451–464 (1967).

SPYROPOULOS, C.S., TASAKI, I., HAYNARD, G.: Fractionation of tracer effluxes during action potential. Science **133**, 2064–2065 (1961).

STRUNK, T.H.: Correlation between metabolic parameters of transport and vacuolar perfusion results in *Nitella clavata*. J. Exptl. Bot. **22**, 863–874 (1971).

THOMAS, D.A.: The regulation of stomatal aperture in tobacco leaf epidermal strips. II. The effect of ouabain. Australian J. Biol. Sci. **23**, 981–989 (1970).

TYREE, M.T., SPANNER, D.C.: A reappraisal of thermodynamic transport coefficients in *Nitella* cell walls. Canad. J. Bot. **47**, 1497–1503 (1969).

UMRATH, K.: Untersuchungen über Plasma und Plasmaströmung an Characeen. IV. Potentialmessungen an *Nitella mucronata* mit besonderer Berücksichtigung der Erregungserscheinungen. Protoplasma **9**, 576–597 (1930).

UMRATH, K.: Die Bildung von Plasmalemma (Plasmahaut) bei *Nitella mucronata*. Protoplasma **16**, 173–188 (1932).

UMRATH, K.: Der Erregungsvorgang bei *Nitella mucronata*. Protoplasma **18**, 258–300 (1933).

UMRATH, K.: Der Einfluß der Temperatur auf das elektrische Potential, den Aktionsstrom und die Protoplasmaströmung bei *Nitella mucronata*. Protoplasma **21**, 329–334 (1934).

UMRATH, K.: Nature of the electric polarisability and excitability of *Nitella*. Protoplasma **34**, 469–483 (1940).

UMRATH, K.: Über Aktionsstrom und Stillstand der Protoplasmaströmung bei *Nitella opaca*. Protoplasma **42**, 77–82 (1953).

UMRATH, K.: Über die elektrische Polarisierbarkeit von *Nitella mucronata* und *Nitella opaca*. Protoplasma **43**, 237–252 (1954).

UMRATH, K.: Der Erregungsvorgang. In: Encyclopedia of plant physiology (W. RUHLAND, ed.), vol. XVII, pt. 1, p. 24–110. Berlin-Heidelberg-New York: Springer 1959.

VREDENBERG, W.J.: A method for measuring the kinetics of energy-dependent changes in the electrical membrane resistance of metabolising plant cells. Biochim. Biophys. Acta **274**, 505–514 (1972).

VREDENBERG, W.J.: Changes in transport determining electrical parameters of cell and chloroplast membranes associated with primary and associated photosynthetic reactions. In: Membrane transport in plants (U. ZIMMERMANN, J. DAINTY, eds.), p. 126–130. Berlin-Heidelberg-New York: Springer 1974.

WALKER, N.A.: Microelectrode experiments on *Nitella*. Australian J. Biol. Sci. **8**, 476–489 (1955).

WALKER, N.A.: The electric resistance of the cell membranes in a *Chara* and a *Nitella* species. Australian J. Biol. Sci. **13**, 468–478 (1960).

WALKER, N.A., HOPE, A.B.: Membrane fluxes and electric conductance in Characean cells. Australian J. Biol. Sci. **22**, 1179–1195 (1969).

WALKER, N.A., SMITH, F.A.: Intracellular pH in *Chara corallina* measured by DMO distribution. Plant Sci. Letters **4**, 125–132 (1975).

Williams, E.J., Bradley, J.: Voltage-clamp and current-clamp studies on the action potential in *Nitella translucens*. Biochim. Biophys. Acta **150**, 626–639 (1968).
Williams, E.J., Johnston, R.J., Dainty, J.: The electrical resistance and capacitance of the membranes of *Nitella translucens*. J. Exptl. Bot. **15**, 1–14 (1964).
Williams, S.E., Pickard, B.G.: Receptor potentials and action potentials in *Drosera* tentacles. Planta **103**, 193–221 (1972a).
Williams, S.E., Pickard, B.G.: Properties of action potentials in *Drosera* tentacles. Planta **103**, 222–240 (1972b).
Williams, S.E., Spanswick, R.M.: Intracellular recordings of the action potentials which mediate the thigmonastic movements of *Drosera*. Plant Physiol. **49**, suppl. 64 (1972).
Zimmermann, U., Steudle, E.: Bestimmung von Reflexionskoeffizienten an der Membran der Alge *Valonia utricularis*. Z. Naturforsch. **25b**, 500–504 (1970).

5. Measurement of Fluxes across Membranes

N. A. WALKER and M. G. PITMAN

1. Introduction

Other Chapters are devoted to the relationships between fluxes of ions and the driving forces that produce them—concentration differences, electric potential differences, flows of solvent or solute, metabolic flows. This Chapter is concerned with the problem of measuring such fluxes across cellular membranes, and with the related problem of determining the distribution of ions between the various cellular compartments.

The measurement of these fluxes and of the contents of inaccessible compartments relies entirely on tracer techniques; in practice the tracers are most often radioactive substances, usually isotopes of the ions in question. The existence of tracers makes it possible to measure the separate components of the net flux, which are here denoted by the words influx and efflux. These are defined in 5.2. Although as usually written the equations of irreversible thermodynamics refer only to net fluxes, the measurement of the separate components is in fact valuable in elucidating the mechanisms of ion movement.

Here we will not be directly concerned with mechanisms or with the related questions of energy supply and control; the techniques described are independent of these matters. The fluxes are estimated by allowing tissue to take up tracer, while following the uptake of tracer and/or the subsequent loss when the tissue is exposed to unlabeled solution. Generally the methods of measurement are relatively simple: the calculation of fluxes and of contents from the primary data is the major difficulty.

The procedure for calculating the fluxes of ions is based on the determination of rate constants for exchange of tracer between tissue and solution, and on the use of these rate constants to find values for the parameters of the chosen kinetic model. This approach is perhaps best exemplified by the measurements of fluxes in charophyte cells by the use of radio-active tracers (MACROBBIE and DAINTY, 1958a; DIAMOND and SOLOMON, 1959). Earlier examples of the techniques as applied to animal cells and to plant tissue are the works of HODGKIN and KEYNES (1955) on the *Sepia* axon and of MACROBBIE and DAINTY (1958b) on *Rhodymenia* fronds.

The theory of tracer measurements on compartmented systems is covered in SHEPPARD's book (1962), while the mathematical techniques appropriate to the extraction of parameter values from data are well introduced by MAGAR (1972). The plant literature is reviewed by MACROBBIE (1970a) and discussed in the books of BRIGGS et al. (1961), HOPE (1971), and HOPE and WALKER (1975).

2. Theoretical Basis of Flux Estimation

2.1 The Flux across a Membrane

The estimation of fluxes is based on the existence of radio-active isotopes, which, to the accuracy we require, are chemically indistinguishable from their stable isotopes, the substances under study.

Consider a system in which a membrane of area A separates two uniform (well-stirred) phases o and i, of large volumes. They contain concentrations $c_{o,j}$ and $c_{i,j}$ respectively of a chemical species j. If $\phi_{oi,j}$ and $\phi_{io,j}$ are the rates of arrival of particles of j in i from o, and in o from i, respectively, then the net flux of j from o to i is given by:

$$J_j = \phi_{oi,j} - \phi_{io,j} \tag{5.1}$$

where J and ϕ are both measured per unit area of membrane and expressed in mol $m^{-2} s^{-1}$. Suppose that at time zero, a quantity of a radio-active tracer for j is added to phase o, so that it contains d_o nuclei disintegrating per m^3 and s. Its specific activity $s_{o,j}$ is defined as equal to $d_o/c_{o,j}$. Then at time t the quantity of radioactivity in i, written as $Q^*_{i,j}$ will be given by:

$$Q^*_{i,j} = A\,t\,\phi_{oi,j}\,s_{o,j}. \tag{5.2}$$

The flux ϕ can be found from experimental values of t, s, Q^* and A.

In practice A may be impossible to determine with any degree of confidence in multicellular tissue, so that ϕ is then expressed in mol $g^{-1} s^{-1}$. It is convenient to express Q on the same basis as ϕ, per unit area of membrane or per unit weight of tissue; when this is done A no longer appears explicitly in Eq. (5.2) or related equations. We follow this practice from this point on.

More importantly, in all practical cases the volumes of the phases are not "large", so that Eq. (5.2) is replaced by a differential equation in which $s_{i,j}$ is no longer zero and both s_o and s_i are functions of t:

$$dQ^*_i = (\phi_{oi}\,s_o - \phi_{io}\,s_i)\,dt. \tag{5.3}$$

Such differential equations can be readily set up for systems of three or more compartments arranged in various ways. Solving sets of these equations is not generally possible except under restricted conditions but they are suitable for modeling on analog computers and for numerical solution with digital computers. The difficulties are for example in the need to assume the functional dependence of the fluxes on the concentrations and on the electric PDs, and in the objective fitting of model to data.

2.2 Two-Compartment Model

Perhaps the simplest model contains two compartments, representing the cell and the environment, with exchange between them limited by a membrane (Fig. 5.1). The differential equation for the tracer content of compartment i has already been given (Eq. (5.3)).

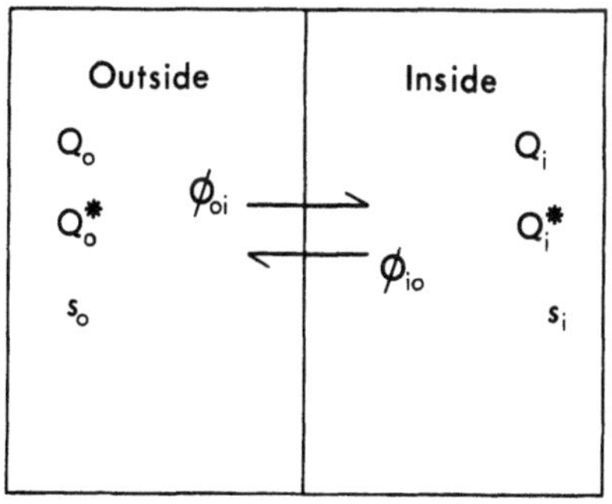

Fig. 5.1. Two-compartment system. Q_o, Q_i=chemical content of a particular ion; Q_o^*, Q_i^*=isotope content of that ion; s_o, s_i=specific activity; subscripts refer to outside and inside respectively. ϕ_{oi}, ϕ_{io}=fluxes into and out of the "inside". The volumes of o and i are not necessarily equal as shown here

If there is a steady state in which the chemical content Q_i is constant then $\phi_{oi}=\phi_{io}$ and $dQ_i^*=Q_i ds_i$, and one can write

$$ds_i=(\phi_{oi}/Q_i)(s_o-s_i)dt. \tag{5.4}$$

The solution to this for constant s_o is:

$$\ln[(s_o-s_i)/H]=-(\phi_{oi}/Q_i)t. \tag{5.5}$$

Where H is a constant of integration determined by the initial conditions. This Equation can be applied to both influx and efflux experiments with appropriate choice of H and of s_o.

For an unlabeled cell put into a radioactive solution of large volume and constant specific activity, s_o, so that at $t=0$, $s_i=0$, uptake is described by

$$\ln[(s_o-s_i)/s_o]=-(\Phi_{oi}/Q_i)t \tag{5.6}$$

or

$$s_i=s_o(1-e^{-k_1 t}) \tag{5.7}$$

where $k_1=\phi_{oi}/Q_i$. Fig. 5.2 shows the time course of s_i; Q_i^* is given by $s_i Q_i$.

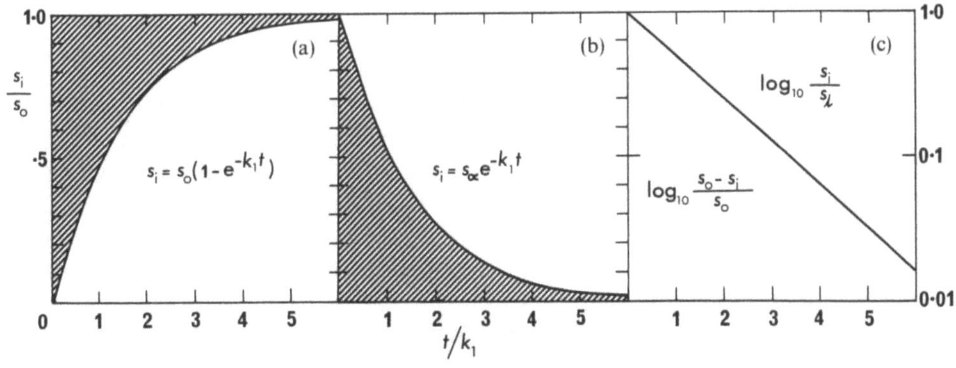

Fig. 5.2. (a) time course of increase in s_i (b) time course of decrease in s_i when put into solution where $s_o=0$ (c) semi-log plot of the shaded areas in 5.2 (a) and 5.2 (b). The left ordinate refers to (a) and (b), the right ordinate to (c)

Eq. (5.6) shows that a graph of $\ln(1-s_i/s_o)$ against time is linear with slope of $-\phi_{oi}/Q_o$; this is equivalent to plotting the logarithm of the shaded part of Fig. 5.2a against t.

The other common experimental situation is measurement of loss of tracer from a cell to an unlabeled solution. In this case s_o is kept very low and is assumed to be effectively zero, while s_i has some arbitrary value s_α at $t=0$. In this case Eq. (5.5) becomes:

$$\ln(s_i/s_\alpha) = -(\phi_{oi}/Q_i)\,t \tag{5.8}$$

or

$$s_i = s_\alpha e^{-k_1 t} \tag{5.9}$$

(see Fig. 5.2b). Again the shaded area when plotted semilogarithmically gives a straight line of slope $-k_1 = -\phi_{oi}/Q_i$.

2.2.1 Estimation of Fluxes

The exponential terms in Eq. (5.7) and in Eq. (5.9) have the same rate constant, equal to ϕ_{oi}/Q_i. Chemical analysis gives Q_i, so ϕ_{oi} $(=\phi_{io})$ can be calculated from k_1, determined either from uptake or efflux experiments. In making semi-log plots as in Fig. 5.2c note that Q_i^* can be plotted instead of s_i, and that if logarithms to base 10 instead of to base e are used, the slope is not $-k_1$ but $-k_1/2.303$.

In short-term experiments a single measurements of Q_i^* can be used to get ϕ_{oi} (Eq. (5.1)). The error is less than 4% if t is less than $0.07/k_1$, i.e. less than 0.1 of the half-time for the rise of s_i. This approach can of course also be used when ϕ_{oi} and ϕ_{io} are unequal, the error depending on the magnitude of ϕ_{io}.

In theory, a short-term measurement could be similarly used to get ϕ_{io} from the change in Q_i^* during elution. In practice, small changes in Q_i^* may not be very reliable, and it is often better to plot against time the successive values of $\ln(\Delta Q_i^*/t)$ and to extrapolate to zero time. Since $Q_o^* = s_\alpha Q_i e^{-k_1 t}$, $dQ_i^*/dt = -k_1 s_\alpha Q_i e^{-k_1 t}$; $\ln(dQ_i^*/dt)$ is proportional to $-k_1$, and extrapolates to $k_1 s_\alpha Q_i = \phi_{io} s_\alpha$.

2.3 Three Compartments in Series

Very few experimental systems behave as if they contained a single cellular compartment (but see Fig. 5.3). The exchange of tracer with many plant cells approximates better to a sum of two exponential terms than to one. Most plant cells have two major resistances to ion diffusion, the membranes at the outside of the cytoplasm, and between cytoplasm and vacuoles. Tracer exchange with a vacuolated cell is better represented by the model shown in Fig. 5.4. Below, 5.3 will deal with the cell wall and other extracellular compartments.

The three phases in series are often identified with solution, cytoplasm and vacuole and are referred to here by subscripts o, c and v. Mixing of ions within these phases is assumed instantaneous; only for substances of very high permeability (e.g. water) might diffusion inside the phases limit the rate of exchange.

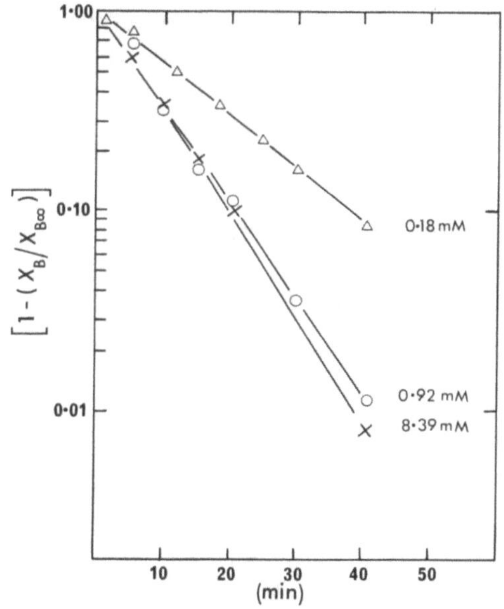

Fig. 5.3

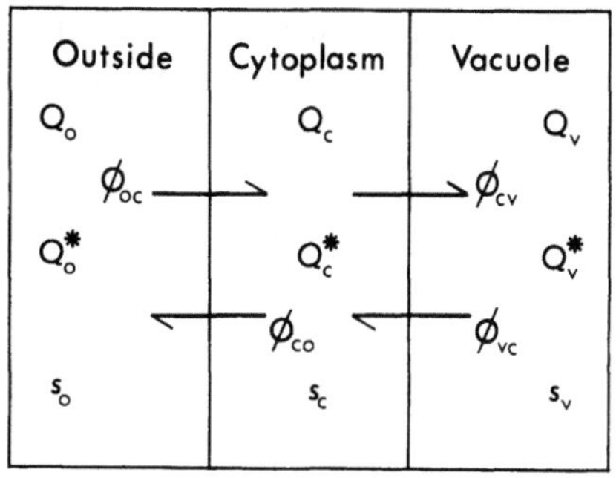

Fig. 5.4

Fig. 5.3. Tracer uptake used to measure K^+ flux in *Neurospora* hyphae. The graph is equivalent to Eq. (5.6), but X_B is used for the specific activity in the cell at time t and $X_{B\infty}$ is the maximum value of X_B reached. The slope gives ϕ_{oi}/Q_i from which ϕ_{oi} could be calculated. The numbers refer to different external K^+ concentrations. (From SLAYMAN and TATUM, 1965)

Fig. 5.4. Three-compartment system in which c refers to cytoplasm and v to vacuole. Q_o, Q_c, and Q_v are chemical contents in the solution, cytoplasm and vacuole; Q_o^*, Q_c^* and Q_v^* are amounts of isotope in these phases and ϕ_{oc}, ϕ_{cv} etc. are fluxes between phases as specified by the labels. s_o, s_c, s_v are specific activities in the three phases

At any moment the rates of change of tracer content of the two cellular phases are given by:

$$dQ_c^*/dt = \phi_{oc}\, s_o + \phi_{vc}\, s_v - (\phi_{co} + \phi_{cv})\, s_c \tag{5.10}$$

$$dQ_v^*/dt = \phi_{cv}\, s_c - \phi_{vc}\, s_v \tag{5.11}$$

and, if Q_T^* is written for $(Q_c^* + Q_v^*)$, then

$$dQ_T^*/dt = \phi_{oc}\, s_o - \phi_{co}\, s_c. \tag{5.12}$$

Explicit solutions for these differential equations can only be obtained for the steady state in which the values of ϕ, s and Q are constant, and hence $\phi_{oc} = \phi_{co}$, and $\phi_{cv} = \phi_{vc}$. Few plant systems approximate to this steady state, except perhaps non-growing cells in which Q_c and Q_v may change slowly. However it is convenient to consider this particular case, as the exact solution can be used to illustrate some points about change of radioactivity in the compartments of the cell that also hold for non-steady state conditions.

Analytical solutions for uptake and loss of tracer in the three phase model are discussed by PALLAGHY and SCOTT (1969). Briefly, uptake and efflux can be described as the sum of two exponential terms. For uptake

$$Q_T^* = A(1-e^{-k_1 t}) + B(1-e^{-k_2 t}). \tag{5.13}$$

After loading for a time T the specific activities will have increased in the tissue but will be less than external specific activity, s_o.

After the tissue is transferred to unlabeled solution at time T, the content is given by

$$Q_T^* = Ce^{-k_1 t} + De^{-k_2 t} \tag{5.14}$$

where $C = A(1-e^{-k_1 T})$ and $D = B(1-e^{-k_2 T})$.

The rate constants k_1 and k_2 are related to the fluxes and the contents by:

$$k_1 = \left(\frac{\phi_{oc}+\phi_{vc}}{2Q_c} + \frac{\phi_{vc}}{2Q_v}\right) + \left[\left(\frac{\phi_{oc}+\phi_{vc}}{2Q_c} + \frac{\phi_{vc}}{2Q_v}\right)^2 - \frac{\phi_{oc}\phi_{vc}}{Q_c Q_v}\right]^{\frac{1}{2}}, \tag{5.15}$$

$$k_2 = \left(\frac{\phi_{oc}+\phi_{vc}}{2Q_c} + \frac{\phi_{vc}}{2Q_v}\right) - \left[\left(\frac{\phi_{oc}+\phi_{vc}}{2Q_c} + \frac{\phi_{vc}}{2Q_v}\right)^2 - \frac{\phi_{oc}\phi_{vc}}{Q_c Q_v}\right]^{\frac{1}{2}} \tag{5.16}$$

and the multipliers A and B are given by:

$$A = s_o(\phi_{oc} - k_2 Q_T)/(k_1 - k_2), \tag{5.17}$$

$$B = s_o(k_1 Q_T - \phi_{oc})/(k_1 - k_2). \tag{5.18}$$

From the solutions, Q_T^* and the specific activities in each phase can be evaluated.

Note that the two terms in A and B are not equivalent to Q_c^* and Q_v^* respectively. The latter are given by:

$$Q_c^* = Q_c s_o \frac{(k_1 - \phi_{oc}/Q_c)(1-e^{-k_2 t}) - (k_2 - \phi_{oc}/Q_c)(1-e^{-k_1 t})}{k_1 - k_2},$$

$$Q_v^* = Q_v s_o \frac{k_1(1-e^{-k_2 t}) - k_2(1-e^{-k_1 t})}{k_1 - k_2}.$$

When there is sufficient difference between k_1 and k_2, the influx or efflux is clearly biphasic. Fig. 5.5 shows for this case how s_c and s_v change with time during an uptake period and a subsequent elution. The specific activity s_c rises or falls rapidly at rates determined mainly by k_1. After a quasi-steady state is reached, s_c increases more slowly, together with s_v. At this stage the rate of change of s_v and s_c is determined mainly by k_2. Note that there is a lag before s_v rises during uptake or falls during elution, because the movements of tracer out of compartment c depend on s_c.

Fig. 5.6 shows how the tracer content changes with time. The extent of the "shoulder" depends on the size of Q_c relative to Q_v and also on the ratio of ϕ_{oc} to ϕ_{cv}.

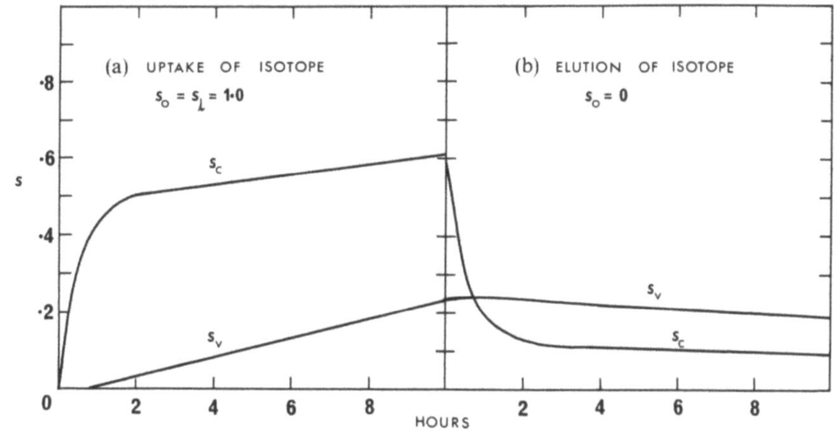

Fig. 5.5 a and b. Time course of change in s_c and s_v (relative to s_l) during uptake of tracer and subsequent elution. Calculated assuming $\phi_{oc}=\phi_{cv}=4$ μmol g_{FW}^{-1} h^{-1}; $Q_c=4$ μmol g_{FW}^{-1}; $Q_v=$ 70 μmol g_{FW}^{-1}

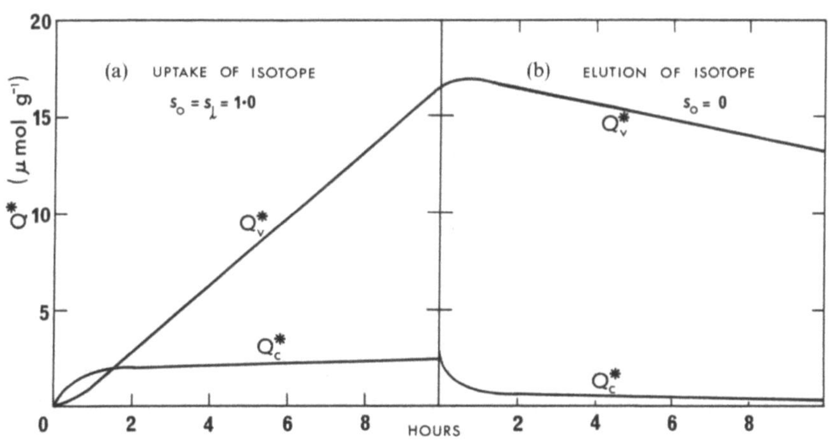

Fig. 5.6 a and b. Time course of change in Q_c^* and Q_v^* during uptake and elution, calculated as in Fig. 5.5

During elution, once s_c has reached a new quasi-steady level between s_v and s_o, Q_T^* can be approximated by a single exponential term ($D\,e^{-k_2 t}$). Hence $\ln(Q_T^*)$ tends to fall linearly with t, with a slope of $-k_2$; and the plot of $\ln(dQ_T^*/dt)$ against t also becomes linear, with the same slope as the plot of $\ln Q_T^*$ against t.

2.3.1 Estimation of Fluxes in the Steady State

Pallaghy and Scott (1969) showed that the observed values of C, D, k_1 and k_2 could be used to calculate the unknown fluxes and contents of the phases:

$$\phi_{oc} = \frac{1}{s_o}(A\,k_1 + B\,k_2) \tag{5.19}$$

$$Q_T = (A+B)/s_o \tag{5.20}$$

$$Q_c = \frac{(A\,k_1 + B\,k_2)^2}{s_o(A\,k_1^2 + B\,k_2^2)} \tag{5.21}$$

$$Q_v = Q_T - Q_c = \frac{AB(k_1 - k_2)^2}{s_o\,(Ak_1^2 + Bk_2^2)} \tag{5.22}$$

$$\phi_{vc} = \frac{k_1\,k_2\,Q_v\,Q_c}{\phi_{oc}} \tag{5.23}$$

where as before A and B are given by $C/(1-e^{k_1 T})$ and $D/(1-e^{-k_2 T})$ respectively. These calculations require the estimation of k_2 as well as k_1; in many experiments there is some uncertainty in k_2 since the loss of tracer does not decrease exponentially at long times. In many experimental situations certain approximations can be made to calculate the fluxes, as discussed below (5.4.2.1). We should note that the algebraic Eqs. (5.19)–(5.23) do not necessarily provide unbiased or efficient estimators of the parameters, in the statistical sense.

2.3.2 Estimation of Fluxes in the Non-Steady State

Many plant tissues such as storage tissues and low-salt roots show a net transport of ions. Though the Equations derived above cannot be used for this non-steady state, the general form of changes in specific activity is like that in Figs. 5.5 and 5.6. Initially there is rapid change in cytoplasmic specific activity and then a slow drift in s_c reflecting changes in s_v. In many of these tissues the volume and content of the cytoplasm is much smaller than that of the vacuole and the transient quasi-equilibration of the cytoplasmic phase occurs much more rapidly than changes in Q_v and in the fluxes into and out of the cell. A method of dealing with the quasi-steady state has been to assume that ϕ_{oc}, ϕ_{co}, ϕ_{cv} and ϕ_{vc} are not variable with time, and that the net flux J across each membrane is the same, so that Q_c is constant too.

Consider the initial quasi-equilibration of the cytoplasmic phase and assume that ds_v/dt is very small so that since

$$dQ_c^*/dt = \phi_{oc}\,s_o + \phi_{vc}\,s_v - (\phi_{cv} + \phi_{co})\,s_c \tag{5.24}$$

an approximate solution for the content during uptake is

$$Q_c^* = Q_c\frac{(\phi_{oc}\,s_o + \phi_{vc}\,s_v)}{\phi_{co} + \phi_{cv}}(1 - e^{-k_1 t}); \tag{5.25}$$

and during elution

$$Q_c^* = \frac{Q_c}{\phi_{co} + \phi_{cv}}(\phi_{vc}\,s_v + \phi_{oc}\,s_l\,e^{-k_1 t}), \tag{5.26}$$

where $k_1 = \dfrac{\phi_{co}+\phi_{cv}}{Q_c}$ and s_l is the specific activity of the *loading solution;* the specific activity of the solution during efflux is taken as zero. As t becomes large s_c tends to its quasi-steady value of

$$s_c = \frac{\phi_{oc}\, s_o + \phi_{vc}\, s_v}{\phi_{co}+\phi_{cv}}. \tag{5.27}$$

During the quasi-steady state the rate of change of content of tracer is

$$dQ_T^*/dt = \phi_{oc}\, s_o - \phi_{co}\, s_c = \frac{\phi_{oc}\,\phi_{cv}\, s_o - \phi_{vc}\,\phi_{co}\, s_v}{\phi_{co}+\phi_{cv}}. \tag{5.28}$$

When $s_o \gg s_v$ the net tracer influx tends to $\dfrac{\phi_{oc}\,\phi_{cv}\, s_o}{\phi_{co}+\phi_{cv}}$, and when $s_o = 0$, net tracer efflux tends to $\dfrac{\phi_{co}\,\phi_{vc}\, s_v}{\phi_{co}+\phi_{cv}}$.

Over long periods of elution, once the transient equilibration of Q_c^* is achieved, the rate constant k_2 can be calculated as

$$k_2 = -\frac{1}{Q_T^*}\frac{dQ_T^*}{dt} = \frac{1}{s_c\, Q_c + s_c\, Q_v}\,\frac{\phi_{co}\,\phi_{vc}\, s_v}{\phi_{co}+\phi_{cv}}$$

$$= \frac{\phi_{co}\,\phi_{vc}}{[\phi_{vc}\, Q_c + Q_v(\phi_{co}+\phi_{cv})]}. \tag{5.29}$$

Note that k_2 is not constant, since Q_v is increasing, and a graph of $\ln Q_T^*$ against t will not be linear. However, when Q_v is large, the change in k_2 during an experiment lasting only several hours should be small and the plot of $\ln Q_T^*$ against time becomes approximately linear.

Expressions can be derived from the equations to give estimates of the fluxes and content of the cytoplasmic phase; these solutions will be considered later (5.4.2).

An alternative approach to the non-steady state, as also to the steady-state problem is to calculate the tracer content as a function of time using analog or digital computing. The required parameter values are then found for best fit of the output to the results of experiments. Usually the fit has been carried out by eye. A digital computer may be used to solve the differential equations by numerical methods: in this case the program can provide for adjustment of the parameters for a least-squares fit to experimental data. A digital computer can also be used to produce dummy data representing the exchange of tracer in such a system. This data can then be used to test accuracy of approximations used to calculate fluxes from the exchange data. Unfortunately, these approaches have been used too rarely.

2.4 The Four-Compartment Model

In work on giant algal cells in particular it becomes necessary to consider a model with four compartments, arranged in a "Y" configuration as in Fig. 5.7. The four

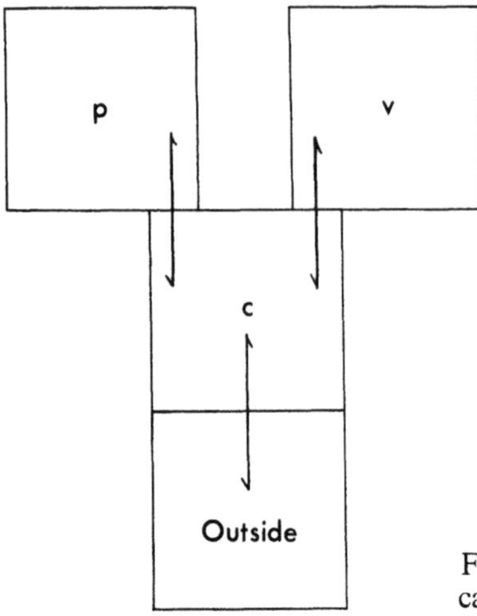

Fig. 5.7. Four-compartment model in which cytoplasm (*c*) can exchange with vacuole (*v*) or "plastids" (*p*)

compartments are labeled o, c, p and v but in any particular application their identification is of course part of the experimental problem.

The differential equations describing the behavior of the model of Fig. 5.7 are:

$$dQ_c^*/dt = \phi_{oc}\, s_o - (\phi_{co} + \phi_{cp} + \phi_{cv})\, s_c + \phi_{pc}\, s_p + \phi_{vc}\, s_v, \quad (5.30)$$

$$dQ_p^*/dt = \phi_{cp}\, s_c - \phi_{pc}\, s_p, \quad (5.31)$$

$$dQ_v^*/dt = \phi_{cv}\, s_c - \phi_{vc}\, s_v. \quad (5.32)$$

No explicit solution to these equations seems known to practitioners, even for the case of a steady state in which the fluxes ϕ and the contents Q are independent of time.

For this steady state, the only case used in practice, MACROBBIE has provided a description of the solution appropriate to an uptake experiment in which all the s but s_o are zero at zero time (MACROBBIE, 1969). Briefly, she states that if Q_c is sufficiently small compared with Q_p and Q_v, Q_c^* will rise quickly towards $Q_c\, s_o\, \phi_{oc}/(\phi_{cp} + \phi_{cv} + \phi_{co})$ with a rate constant given by:

$$k_1 \doteqdot (\phi_{co} + \phi_{cv} + \phi_{cp})/Q_c. \quad (5.33)$$

The vacuolar radio-activity, Q_v^*, will show an initial lag of about $1/k_1$ and then will rise slowly. If further the rate constant for exchange between c and p is much larger than that for exchange between c and v, then Q_p^* will rise with a rate constant given by:

$$k_2 \doteqdot (\phi_{pc}\, \phi_{cv})/(\phi_{cv} + \phi_{cp} + \phi_{oc})\, Q_p. \quad (5.34)$$

The steady state is easy to model on an analog computer, and several sets of experimental data are fitted in this way by HOPE and WALKER (1975). The curves shown in the Figures in 5.5.2 were produced by an iterative procedure on a desk calculator, which modeled the differential equations by a "book-keeping" method with one withdrawal per model second. Note that in the steady state $\phi_{co} = \phi_{oc}$.

The reader will notice a lack of sophistication in all of this, and will perhaps hope that future work may at least reach the level of least-squares fitting methods and quotable confidence limits.

3. The Effect of the Free Space on Flux Measurements

When tracers are used to measure the fluxes across cellular membranes, there is observed an uptake of tracer into extra-cellular regions of the tissue such as surface films of solution, intercellular spaces and cell walls. The term "free space" was introduced by BRIGGS (1957a) for those parts of a cell or tissue whose ions exchanged relatively rapidly with those in the surrounding solution. An extensive treatment of the properties of the free space is given in BRIGGS et al. (1961). Our concern here with the free space is with its effect on the measurement of membrane fluxes. Free space is equivalent to the "apoplast". (Part B, Chap. *1*)

It is possible to treat the free space as consisting of two components. One, called "water free space" (WFS), behaves as an extension of the external solution into the tissue: it contains ions at the same concentrations as in the bulk solution. Operationally it can be defined as the tissue water which is available to anions from the bulk solution, for a reason that will be made clear below.

The location of the WFS depends on the tissue and on the technique adopted. For single charophyte cells that have been blotted, the WFS is mostly in the cell wall; about 46 % of the water of the wall of *Chara* is accessible to the anion I^- (DAINTY and HOPE, 1959). These workers suggested that the WFS consisted of aqueous channels through the wall, with diameters greater than about 10 nm. In multi-cellular tissues that have been sliced and washed, the WFS includes the damaged cells at the cut surfaces, the intercellular spaces and possibly a part of the cell wall (by analogy with *Chara*). Usually the WFS in higher-plant tissue slices is about 20 % of the tissue water.

The second component of the free space is an ion-exchange system, the Donnan free space (DFS). It contains fixed carboxylic acid groups (on for example polymers of glucuronic acid, Part B, *1*.2.1.2) and so is a cation exchanger with a pK in the neighborhood of 2.8. In higher plant cells the exchange characteristics approximate to those of a Donnan system (for which see BRIGGS et al., 1961), but in charophyte cells DAINTY and HOPE (1961) found that a better model was the electric double-layer at a charged surface. In either case the DFS is a region electrically more negative than the bulk solution, with therefore higher equilibrium concentrations of cations (including H^+) and lower concentrations of anions. If we take K^+, Ca^{2+} and Cl^- as examples, their concentrations in the DFS (subscript w) are :

$$e^{(F\psi_w/RT)} = [K^+]_w/[K^+]_o = \{[Ca^{2+}]_w/[Ca^{2+}]_o\}^{\frac{1}{2}} = [Cl^-]_o/[Cl^-]_w \tag{5.35}$$

where ψ_w is, according to the model used, either the electric potential of the whole DFS or the potential at a particular distance from the charged surface. In the former, and simpler case, if there is only a uni-, uni-valent electrolyte present, ψ_w is given by:

$$e^{(F\psi_w/RT)} = \{a + (a^2 + 4[K^+]_o^2)^{\frac{1}{2}}\}/2[K^+]_o$$

where a is the concentration of fixed anions in the DFS. For *Chara* the ion-exchange measurements of DAINTY and HOPE (1959, 1961) give a as about 100 mM, so that a cell bathed in 1 mM solution of KCl has a wall potential of −116 mV, and the concentration of anions in the DFS is 10^{-2} mM. Thus anions are virtually absent from the DFS and their uptake into the free space can be used to estimate the WFS.

Table 5.1. Properties of free space exchange in various plants and tissues (blotted). Total content in the free space can be calculated from concentrations of ions in solution and the free space properties. (See BRIGGS et al., 1961)

Plant cell or tissue	WFS Volume ($cm^3\,g^{-1}$)	Fixed negative charges in DFS		Time for 50% exchange (exponential) (s)	
		Amount ($\mu mol\ g_{FW}^{-1}$)	Concentration (mM)		
Beetroot[a] slices (1 mm thick)	0.20	12	560	I^-, Cl^- = K^+ = Ca^{2+} =	120 720 2,640
Barley roots[b]	0.24	2.0	—	Cl^- = K^+, Na^+ =	66 132
Barley leaf[c] slices (0.9 mm)	0.21	3.5	300	Cl^- = Rb^+ =	96 130
Atriplex leaf slices[d]	—	16	—	—	—
Chara cell walls[e]	0.34	146	600	I^- = I^- = Na^+ = Ca^{2+} =	1 1 10 100
Chara cells (if wall is 0.058 of FW)	0.081	8.5	600	–	–

[a] BRIGGS et al. (1958b). [b] PITMAN (1965, 1972). [c] PITMAN et al. (1974). [d] OSMOND (1968). [e] DAINTY and HOPE (1959).

Table 5.1 gives some values for the number and concentration of exchange sites in the DFS of various tissues. Under normal conditions most of these exchange sites will be occupied by divalent ions, since the divalent distribution ratio is equal to the square of the univalent ion ratio (Eq. (5.35)). When there are simultaneously present ions of valence +, 2+ and −, the distribution ratios are given by a cubic equation: some numerical values are given in Table 5.2 to illustrate the dominant role played by divalent cations in determining ψ_w and the distribution ratios.

Table 5.2. Ion concentrations in bulk solution (o) and in DFS (w) with fixed anions of 100 mM

$[K^+]_o$ (mM)	$[Ca^{2+}]_o$ (mM)	$[Cl^-]_o$ (mM)	ψ_w (mV)	$[K^+]_w$ (mM)	$[Ca^{2+}]_w$ (mM)	$[Cl^-]_w$ (mM)
0.1	0.0	0.1	−173.4	100.0	0.0	0.0
1.0	0.0	1.0	−115.6	100.0	0.0	0.0
0.1	0.1	0.3	− 77.7	2.2	48.9	0.0
1.0	0.1	1.2	− 75.2	20.0	40.0	0.1
1.0	1.0	3.0	− 48.3	6.8	46.8	0.4
4.0	1.0	6.0	− 45.7	24.7	38.1	1.0
1.0	4.0	9.0	− 31.6	3.5	49.5	2.6

Table 5.3. Diffusion coefficients for various ions in the free space

Plant material	Ion	Diffusion coefficient ($m^2 s^{-1}$)	Ref.
Atriplex leaf slices	Na^+	1 to $3 \cdot 10^{-11}$	OSMOND (1968)
Barley roots	Na^+	3 to $4 \cdot 10^{-11}$	PITMAN (1965)
	Cl^-	$8 \cdot 10^{-11}$	PITMAN (1972)
Barley root (compressed)	Cl^-	$1 \cdot 10^{-10}$	CHARLEY and JENNY (1961)
	Fe^{3+}	2 to $4 \cdot 10^{-13}$	CHARLEY and JENNY (1961)
Beetroot slices	I^-	$6.9 \cdot 10^{-10}$	BRIGGS et al. (1958b)
	Na^+	$1.1 \cdot 10^{-10}$	BRIGGS et al. (1958b)
	Sr^{2+}	$3.7 \cdot 10^{-11}$	BRIGGS et al. (1958b)
Nitella cell walls	(KCl)	$8 \cdot 10^{-10}$	TYREE (1968)
Nitella cell walls	Cl	$5 \cdot 10^{-12}$	MAILMAN and MULLINS (1966)

Ions in the free space are thus present in, and diffuse through, regions of DFS and of WFS, whose geometrical relationship is not known. Accordingly "diffusion coefficients" of ions in the free space are rough operational values, not precisely known quantities, and are likely to depend on the particular experimental arrangement. The geometry of the charophyte cell wall is particularly simple, though here too the relationship of DFS to WFS is obscure; some values of diffusion coefficients in the charophyte wall are given in Table 5.3, and are compared with those in free solution. Of the values in Table 5.3, TYREE's (1968) were derived from measurements of rates of equilibration of cell wall preparations in KCl solutions, and independently from the activation energy derived from the temperature coefficient of electrical conductivity; while MAILMAN and MULLINS (1966) used the time course of appearance of chloride outside the cell wall after an action potential. The latter value is surprisingly low, and its explanation is not clear. It was attributed by HOPE and WALKER (1975) to a combination of occlusion and of exclusion from the DFS: they assumed that to pass from membrane to exterior, the ions must pass in series through DFS and WFS.

The exchange of anions between the cell wall and exterior is so rapid that it is rate-controlled by the unstirred layer of water (DAINTY and HOPE, 1961). For cation exchange the cell wall may delay the appearance of the new cation at the membrane with a halftime of 5 min (HOPE and WALKER, 1961). This halftime is derived from the time-course of the membrane PD after a rapid change of cation concentration in the external medium, corresponding to a diffusion coefficient of $3 \cdot 10^{-13}\ m^2\ s^{-1}$ in a uniform plane sheet. The low value again perhaps reflects rate-control by an unstirred layer whose effect is increased by the Donnan accumulation effect (see BRIGGS et al., 1961).

For the multicellular tissues of Table 5.3, diffusion coefficients were calculated from the time courses for tracer exchange between free space and solution. It was assumed that the tissue behaved as a simple, homogenous geometric shape for which solutions to the diffusion equations are readily available (disk, cylinder; BRIGGS et al.,

1961). Clearly the values are influenced by the layer of solution in damaged cells on the surface of the disks or by the tortuosity of the path in the tissue. Little can be deduced from them about the properties of cell walls for example, except that diffusion of polyvalent cations is slower than univalent cations. There is the further complication that the values were obtained for isotopic diffusion at constant total concentration of ions, and so may not predict net diffusion in a tissue. For example electrochemical activity gradients within the DFS could restrict the rate of diffusion of a salt to the net flux of the least abundant ion, even though the mobility of the ion (measured by isotopic exchange) was large.

It can be seen that the free space complicates the measurement of membrane fluxes in two ways. First, it takes up tracer cations from the bulk solution, setting the experimenter the task of separating this unwanted radioactivity from that inside the cell membranes, and second, it may introduce a significant delay in the appearance of tracer at the plasmalemma.

The difficulty of separating tracer taken up into the DFS from that inside the cells is seen at its most acute in the attempt to measure calcium influx into charophyte cells (SPANSWICK and WILLIAMS, 1965). It was not possible to wash out the free-space calcium, and although long uptake periods were used, the calcium in the DFS was indistinguishable from that in the cytoplasm. They were able to measure only the vacuole influx, Φ_{ov}, by a neat trick of vacuole separation. The measurement of univalent cation influxes in charophytes is much less difficult, and a reasonable approximation to be the influx ϕ_{oc} can be obtained with uptake periods of the order of 20 min followed by a similar wash-out period: at this point the radio-activity of the whole cell is largely intra-cellular. For anions the corresponding times are about 2 min. In each case the lower limit of uptake period is determined by the need to have a favorable ratio of intra- to extra-cellular radio-activity, and also by the timelags now to be discussed.

The exchange of tracer between cytoplasm and solution takes place through the free space, so that even if it were possible to measure the intracellular radio-activity precisely, other qualities in the equations already discussed are also in doubt. During the transition from, say, labeled to unlabeled solution there will be a delay in the appearance of the new specific activity at the cell membrane. This delay is, for charo-

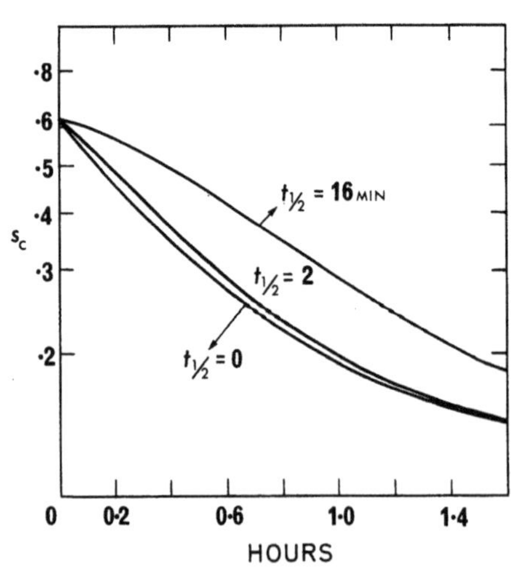

Fig. 5.8. Effect of free space exchange on s_c during elution based on computer simulation. The curves were calculated assuming $\phi_{oc}=\phi_{cv}=$ 2 µmol g_{FW}^{-1} h^{-1}, $Q_c=2$ µmol g_{FW}^{-1}, $Q_v=70$ µmol g_{FW}^{-1}. Free space exchange was set to have half exchange ($t_{\frac{1}{2}}$) in 16 min; in 2 min; or with no free space at all. Note the initial delay in decrease of s_c and change in slope of ln s_c, due to diffusion in the free space

phyte cells, a few seconds with anions, and 2 to 5 min with univalent cations, while for sections of higher-plant tissue it may be 5 to 10 min and more (see Table 5.1). Under steady-state conditions there will be a net uptake of tracer into the cells (or a net loss) so that the specific activity next to the membrane will be different from that in bulk solution. In most systems this difference can be assumed to be negligible, except perhaps in thick slices of tissue or in leaf discs (where diffusion takes place at the cut edge only).

Whether the free space seriously affects tracer elution measurements depends, in a particular tissue, on its geometry and on the relative sizes and rates of exchange of free space and cell interior (see Fig. 5.8). It is always a problem in such experiments that the calculation of the fluxes depends on an extrapolation "through" the free space contribution to the eluted radio-activity.

4. Practical Estimation of Fluxes from Tracer Elution Measurements

4.1 Experimental Procedures

The general procedure has been first to leave tissue in a labeled solution at high specific activity for a suitably long period so that specific activity increases adequately in the vacuole. One hopes that there is no change in fluxes. The tissue is then drained or blotted and put into an experimental set-up designed to allow rapid transfer of tissue from one solution to another so that tracer loss in relatively short periods can be determined. Time periods are usually 30–60s at the start of the elution, being increased to 60 min or more at later stages.

The loss of tracer from the tissue can be calculated from the radioactivity in the eluting solution at the end of each period.

It may be necessary to make allowance for tracer carried over from one sample to another in solution adhering to the tissue. In some cases allowance may also need to be made for the free space content where it is a large proportion of the total ion content of the solution.

By addition of the tracer loss for each period, values of Q_T^* are estimated as a function of time of elution. The tracer loss in each period divided by the length of the period also gives an estimate of average dQ_T^*/dt in that period, forming a series as a function of time. Chemical analysis of the tissue gives Q_T, though allowance must be made for the free space content. Q_T^* may be expressed as radioactivity per gram fresh weight or per unit area of cell surface. Instead, Q_T^*/s_l may be calculated which will have the units mol g_{FW}^{-1} or mol m^{-2}.

The basic procedure in analyzing the data is to plot ln Q_T^* and ln (dQ^*/dt) against t. The slope of the slow component of ln Q_T^* should be the same as the slope of $\ln(dQ_T^*/dt)$, so eventually the two lines should become parallel. If the lines are not parallel then the system is *not* behaving as a three-compartment system, as already discussed.

If this condition is satisfied, the slow component of Q_T^* can be extrapolated to $t=0$ and the difference between Q_T^* and this extrapolated value estimated (the shaded value in Fig. 5.9). Eq. (5.14) shows that this value should decrease exponentially with rate constant k_1 and so its logarithm should decrease linearly. Extrapolation to $t=0$

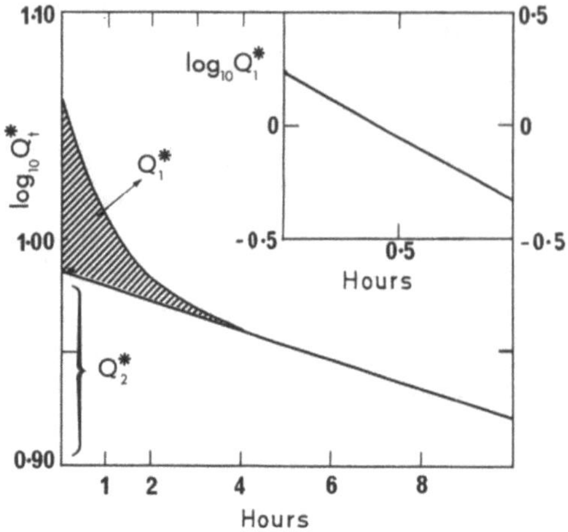

Fig. 5.9. Decrease in Q_T^* plotted semi-logarithmically. The shaded area (Q_1^*) is replotted in the insert. Calculated data assuming $\phi_{oc}=3$, $\phi_{cv}=1$ μmol g_{FW}^{-1} h^{-1}; $Q_c=4$, $Q_v=70$ μmol g_{FW}^{-1}

gives an estimate of the initial content of this phase $(Q_1^*)_o$. A similar extrapolation of the plot of dQ_1^*/dt gives the initial rate of loss from the phase, $(dQ_1^*/dt)_o$.

The data derived from this analysis are then:

Q_T = Content of the tissue (determined chemically)
$(Q_1^*)_o$ = Initial amount of isotope in "fast" component (see Fig. 5.9)
$(Q_2^*)_o$ = Initial amount of isotope in "slow" component (see Fig. 5.9)
$(dQ_1^*/dt)_o$ = Initial isotope efflux from fast component
$(dQ_2^*/dt)_o$ = Initial isotope efflux from slow component
Φ_{in}^* = Net influx of tracer measured in an uptake experiment
J = Chemically determined net uptake
k_1 = Rate constant of fast component
k_2 = Rate constant of slow component
s_l = specific activity of labeling solution prior to elution.

For convenience we can write $(dQ_1^*/dt)_o$ and $(dQ_2^*/dt)_o$ as "fluxes" with the symbols Φ_1^*, Φ_2^* respectively so that the initial isotope efflux from the cells, Φ_T^*, is equal to $\Phi_1^*+\Phi_2^*$. Note that Φ_1 and Φ_2 here are not equivalent to fluxes from the cytoplasm and vacuole, though $\Phi_T^*=\phi_{co}\,s_c$.

It is possible to apply a least-squares approach to the fitting of lines such as those in Fig. 5.9 (eg. VICKERY and BRUINSMA, 1973) but most workers have relied on their own intuition and a ruler. It must be remembered that while early points on the curve may be liable to error from free space, later points may have a disproportionately large error when they represent the (small) difference between vacuolar and total tracer content. It should be stressed that it is essential to compare values from slopes and intercepts of Q_T^* and (dQ_T^*/dt) as checks on consistency of the data with a three-compartment model.

4.2 Calculation of Fluxes

4.2.1 Steady State

Where there is no net uptake or loss of the ion from the tissue, the Equations of 5.2.3.1 can be used to calculate the fluxes and the contents of the compartments.

Many slowly growing algal cells approximate to this condition, for example charophyte cells, and those of *Chaetomorpha darwinii, Ulva lactuca, Valonia, Valoniopsis, Boergesenia,* and *Griffithsia.* The procedure for calculation is described by PALLAGHY and SCOTT (1969) or can be derived from equations given in 5.2.3.1. Note that Q_T can be established from the exchange data and also determined chemically, providing a check on the validity of the calculations.

Alternatively approximations may be used to estimate fluxes for tissues in both steady and non-steady state as discussed in 5.2.3.2. For this purpose it may be necessary to have measurements of the rates of tracer uptake as well as those of loss.

Provided the tissue has been loaded for a time $t > 5k^{-1}$ the cytoplasm will be in a quasi-steady state intermediate between s_l and s_v, and $s_c = (\phi_{oc} s_l + \phi_{vc} s_v)/(\phi_{co} + \phi_{cv})$ (PITMAN, 1963). It can be seen then that the following relationships exist between the measured quantities and the theoretical fluxes of Fig. 5.4. (Examination of Eqs. (5.15) and (5.16) shows that these relationships can only be derived when $Q_v \gg Q_c$.)

$$\Phi_T^* = \left(\frac{dQ_T^*}{dt}\right)_o = \phi_{co} s_c = \phi_{co} \frac{\phi_{oc} s_l + \phi_{vc} s_v}{\phi_{co} + \phi_{cv}}. \tag{5.36}$$

During the later stages of elution s_c is at a new quasi-steady value with $s_l = 0$, so

$$\Phi_2^* = \left(\frac{dQ_2^*}{dt}\right)_o = \frac{\phi_{co} \phi_{vc} s_v}{\phi_{co} + \phi_{cv}} \tag{5.37}$$

and therefore

$$\Phi_1^* = \Phi_T^* - \Phi_2^* = \frac{\phi_{co} \phi_{oc} s_l}{\phi_{co} + \phi_{cv}}. \tag{5.38}$$

Also

$$(Q_1^*)_o = \frac{1}{k_1} \left(\frac{dQ_1^*}{dt}\right)_o = \frac{Q_c s_l \phi_{oc} \phi_{co}}{(\phi_{co} + \phi_{cv})^2}, \tag{5.39}$$

$$k_1 = (\phi_{co} + \phi_{cv})/Q_c, \tag{5.40}$$

$$\Phi_{in}^* \doteqdot (\phi_{cv} \phi_{oc} s_o - \phi_{vc} \phi_{co} s_v)/(\phi_{co} + \phi_{cv}). \tag{5.41}$$

These Equations are quoted here in the form which distinguishes ϕ_{co} from ϕ_{oc}, etc., since they are applicable also to some cases of non-steady states.

Consider the following further cases for cells at flux equilibrium, i.e. $\phi_{oc} = \phi_{co}$ and $\phi_{cv} = \phi_{vc}$.

(a) $s_l = s_c = s_v$ at $t = 0$.

In this case $\phi_{oc} = \Phi_T^*/s_l$; $\phi_{cv} = \phi_{oc} \dfrac{\Phi_2^*}{\Phi_1^*}$; and $Q_c = k_1/(\phi_{co} + \phi_{cv})$. Note that Φ_{in}^* is not used, and so experimental measurement of Φ_{in}^* can be used as a check on validity of using this model.

(b) s_c and s_v both less than s_l.

If unlabeled tissue with the same history is available Φ^*_{in} can be estimated when s_v is very low and $\Phi^*_{in} \doteqdot \dfrac{\phi_{oc}\,\phi_{cv}\,s_o}{\phi_{co}+\phi_{cv}}$. Hence at the start of the elution, $s_v=\Phi_2\,s_o/\Phi^*_{in}$. Using this value of s_v, ϕ_{co}/ϕ_{cv} can be calculated from $\Phi^*_1\,s_v/\Phi^*_2\,s_l$. Writing this ratio as β, we can solve for ϕ_{cv} in Φ^*_1 since $\phi_{cv}=\dfrac{\Phi^*_1}{s_l}\cdot\dfrac{1+\beta}{\beta^2}$. Hence $\phi_{co}=\beta\,\phi_{cv}$; Q_c can be estimated from k_1; and $Q_v=Q_T-Q_c$. Note that Q_T must be estimated chemically.

An alternative approach is to use $(Q^*_2)_o/Q_T$ as an approximation to s_v and calculate Q_c as above, then use this value to calculate Q_v and Q^*_v, and hence a better estimate of s_v.

4.2.2 Non-Steady State

In this case we assume that $\phi_{oc}-\phi_{co}=\phi_{cv}-\phi_{vc}=J$. Approximations have been used to calculate the fluxes on the further assumption that J and Q_v change only slowly with time, and that the fluxes are constant over the early periods of the elution.

To calculate fluxes from the relationships given in Eqs. 5.36–41 it is important to know J or Φ^*_{in}, and to have an estimate of s_v (e.g. $(Q^*_2)_o/Q_T$).

Since $\phi_{oc}-\phi_{co}=\phi_{cv}-\phi_{vc}$, $\phi_{oc}+\phi_{vc}=\phi_{co}+\phi_{cv}$, hence $\phi_{co}=\Phi^*_1/s_l+\Phi^*_2/s_v$ and $\phi_{oc}=\phi_{co}+J$. Similarly $\phi_{vc}=\phi_{oc}(\Phi^*_2/\Phi^*_1)$, (s_l/s_v); $\phi_{cv}=\phi_{vc}+J$ and $Q_c \doteqdot (\phi_{co}+\phi_{cv})/k_1$.

Again, fresh values for s_v can be determined to refine the calculation. Note that as Φ^*_{in} has not been used it can be compared with values calculated from the fluxes (see Table 5.6).

An approach to estimating fluxes that is *not* valid for the three-compartment system was used for beetroot by BRIGGS et al. (1958b) and by VAN STEVENINCK (1962) and for leaf slices by JACOBY and DAGAN (1969). This calculation was derived by BRIGGS (1957b) assuming a two-compartment model, when it can be shown that the influx can be calculated from the rate of change of specific activity in the tissue and chemical determination of net influx (J).

4.3 Validity of Calculations

Since the approximations of 5.4.2 involve assumptions about the system, it is useful to test the validity of the analysis. One method is to use a digital computer to produce dummy data based on the model and then use the data as if they were experimental results to calculate fluxes. Table 5.4 shows some results for variation in net transport and the ratio of ϕ_{oc} to ϕ_{cv}. In general it shows good agreement between the calculated and the preset values for fluxes, except when $\phi_{cv} \gg \phi_{oc}$ so that Φ_1 was small. The estimates of Q_c are liable to be less satisfactory, since errors in ϕ_{oc} and ϕ_{cv} are doubled in calculation of Q_c. Clearly the simplified procedures outlined above can give reasonable results, though the assumption is made that exchange of the free space is very much faster than that of the cytoplasmic phase.

Table 5.5 contains data calculated in a similar way for "cells" separated from the solution by a free space. The rate constant for equilibration of the free space

Table 5.4. Values of fluxes and of Q_c calculated from dummy data derived from the three-compartment model by computer simulation. Q_v was set at 70 μmol g_{FW}^{-1} at the start of a 20 h uptake period. No free-space exchange. Fluxes in μmol g_{FW}^{-1} h^{-1}; Q_c in μmol g_{FW}^{-1}

J	ϕ_{oc}		ϕ_{cv}		Q_c	
	Input	Calc.	Input	Calc.	Input	Calc.
0	4.0	4.02	4.0	4.25	4.0	4.25
0	4.0	4.03	1.0	1.05	4.0	4.4
0	4.0	3.89	0.4	0.43	4.0	4.4
0	0.4	0.42	4.0	4.8	4.0	4.1
2	4.0	4.02	4.0	3.98	4.0	3.9
3.5	4.0	3.99	4.0	3.995	4.0	4.0

Table 5.5. Effect of varied time for half-exchange of the free space on estimation of fluxes; results of computer-simulated experiment. Q_c= 2.0 μmol g_{FW}^{-1}; $\phi_{oc}=\phi_{cv}=2$ μmol g_{FW}^{-1} h^{-1}.
Note the increasing error in the fluxes as the time for half-exchange of the free space increases

Time for half exchange of FS (s)	120	240	480	960
Estimated ϕ_{oc} (μmol $g_{FW}^{-1}h^{-1}$)	2.2	2.8	5	9
Estimated ϕ_{cv} (μmol $g_{FW}^{-1}h^{-1}$)	1.8	1.6	1.3	1.1

varied but fluxes into the "cells" were the same ($\phi_{oc}=\phi_{co}=\phi_{cv}=\phi_{vc}=2$ μmol g_{FW}^{-1} h^{-1}) and Q_c was 2 μmol g_{FW}^{-1}. The time for half-exchange of the cytoplasmic compartment was 21 min. The results show the increasing inaccuracy of the method as the rate constant of the free space was made closer to that of the cytoplasmic phase. It is clear that there may be systematic errors in estimation of fluxes due to the interaction between free space and non-free space exchange.

4.4 Estimation of Influxes from Uptake Measurements

The rate of tracer uptake (Φ^*_{in}) is equal to the difference between tracer influx and efflux, i.e. to ($\phi_{oc}\,s_o-\phi_{co}\,s_c$). One experimental approach is to measure Φ^*_{in} with short labeling periods in the hope that $\phi_{co}\,s_c \ll \phi_{oc}\,s_o$, so that Φ^*_{in}/s_o approximates to ϕ_{oc}. This approach seems to give a valid estimate of ϕ_{oc} when net uptake is large and ϕ_{co} consequently small, as in low salt barley roots (Part B, 3.1.2.5). Estimation of ϕ_{oc} in this way can be justified too when s_c is much less than s_o during a prolonged quasi-steady state i.e. when $\phi_{vc} \gg \phi_{oc}$, as in *Chara corallina*. Alternatively Φ^*_{in}/s_o will give a valid estimate of ϕ_{oc} if the period of labeling is much less than the time needed for s_c to rise to its quasi-steady level. A further practical problem is that allowance must be made for the free space content. This is usually reduced to a constant and predictable amount by rinsing the tissue for a standard short time in ice-cold solution.

An alternative approach is to measure tracer uptake over longer periods during which s_c has been mainly at its quasi-steady level. At this stage

$$\Phi^*_{in}=\phi_{cv}\,\phi_{oc}\,s_o/(\phi_{cv}+\phi_{co})$$

provided that $s_v \ll s_o$. The tracer content of the tissue (Q_T^*) is then made up of a constant Q_c^* and a rising Q_v^*; Φ_{in}^* can be measured as the slope of Q_T^*, but more usually tracer content of the tissue is measured following a long rinse. In this case free space and much of the cytoplasmic tracer is removed and that part of Q_c^* left in the tissue compensates for the lower transfer to the vacuole when s_c was less than its quasi-steady value (see 5.2.3 and Fig. 5.6). In this experimental situation the tracer content at the end of the long rinse is also approximately $\phi_{cv}\, \phi_{oc}\, s_o/(\phi_{cv}+\phi_{co})$. These approaches are discussed by CRAM (1969).

This problem of knowing which flux is estimated by measurements of tracer uptake is common to cell systems. A further complication occurs in roots where tracer taken up by the root may be re-exported from the cut end of the root *via* the xylem (5.5.3.2).

5. Applications of Flux Analysis

Results from flux analysis are given in later Chapters. Here we are concerned with the suitability of the models for calculation of fluxes in particular cases. The general problems are the number and identification of the compartments involved and, in tissues, the effect of heterogeneity of cell type.

5.1 Storage Tissue

Tracer exchange with beetroot slices (PITMAN, 1963) and carrot slices (CRAM, 1968) has been analyzed into a free space component, and fast and slow non-free space components. The free-space exchange was unaffected by temperature, but the remainder of the exchange was much slower at 2° C than at 25° C (Fig. 5.10). Free-space exchange may limit the usefulness of this approach. Thus for K^+ in beetroot the separation of the cytoplasmic from the free-space component was more satisfactory at 2° C: the free-space content of K^+ was 10 μmol g_{FW}^{-1} and the half-time for exchange

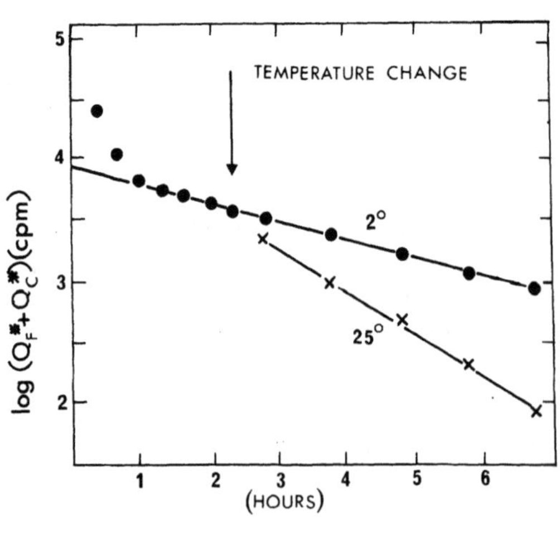

Fig. 5.10. Amount of isotope in the free space and eluted from the cytoplasmic plase (Q_1) plotted semi-logarithmically. Note effect of temperature on the cytoplasmic component [(●) at 2°C; (×) at 25°C]. (From PITMAN, 1963)

was 10 min; Q_c was about 2 to 3 μmol g_{FW}^{-1} and the half-time was 130 min at 2° C but only 50 min at 25° C.

Consideration was given to the location of these non-free-space components (see 8.2 and Fig. 8.4). For beetroot it was shown that the fast and slow components were both located in parenchyma cells and not in different cell types within the tissue, such as the vascular bundles (PITMAN, 1963). It was concluded that the fast component was some part of the cytoplasm and that the slow component was equivalent to the vacuole. In support of this view, it was found for both carrot and beetroot that the specific activity and content of the slow component were in agreement with the expected values based on the content of the tissue and volume of vacuoles. Similarly the content of the fast component was consistent with the volume of the cytoplasm.

Good evidence was provided by CRAM (1968) that the fast and slow components were arranged in series and hence that the fast component appeared to be equivalent to the whole of the cytoplasm bounded by the plasmalemma and tonoplast membranes. He measured the slow component of efflux of chloride after equilibration of the cytoplasmic phase with salt solution and then transferred the tissue to water, which reduced chloride efflux by eliminating exchange diffusion of Cl^-. When put back into KCl there was a transient burst of efflux which decreased with the rate constant of the cytoplasmic phase, showing that the specific activity of the cytoplasmic component had *increased* during the period in water. This behavior could only be explained by assuming a series arrangement of the phases. Support for this view has also been produced for beetroot (PITMAN, 1963).

A further test of the validity of using the 3-phase model was made by comparing observed values of tracer efflux during the later stages of elution with values calculated from the fluxes. Table 5.6 shows there was good agreement between the values.

Table 5.6. Observed and calculated values of apparent efflux (μmol g_{FW}^{-1} h^{-1}) for beetroot slices at 2° C. (Data from PITMAN, 1963)

$[KCl]_o$ (mM)	5	10	20	30	40
Observed efflux	0.18	0.14	0.13	0.13	0.13
Calculated efflux	0.16	0.16	0.16	0.17	0.15

5.2 Models for Giant Algal Cells

Work with giant algal cells has the advantage that it is—or should be—possible to measure directly the chemical concentrations and radio-activity of the main cellular compartments. If they can be identified with the kinetic compartments, the direct access gives the experimenter a powerful method which is not available for multicellular tissue.

For charophyte cells MACROBBIE and DAINTY (1958a) and DIAMOND and SOLOMON (1959), working with ^{42}K, ^{22}Na and ^{36}Cl, found the three-series-compartment to be an adequate representation of the cell. Both were able to identify the cell vacuole with the large, slowly-exchanging kinetic compartment, and to show that the faster compartment was the cytoplasm. DIAMOND and SOLOMON went further and showed that the two cellular compartments were in series.

These workers expected the chloroplasts to appear as a separately detectable compartment, and when this did not happen they investigated the exchange of ^{42}K between isolated chloroplasts and the cell sap in which they were suspended. Exchange was complete in a few seconds, and they concluded that the plastids were an indistinguishable part of the cytoplasm. As would be expected from the model, they found the time-course of s_v to rise from zero with an initial slope of zero (Fig. 5.11). This result is not mentioned as conclusive, but to show that even at this pioneering stage, the existence of the chloroplasts was not ignored. It is now clear that plastids isolated in cell sap are most unlikely to have intact envelopes. Indeed the direct measurement attempted by DIAMOND and SOLOMON (1959) is still not feasible with *Nitella* chloroplasts.

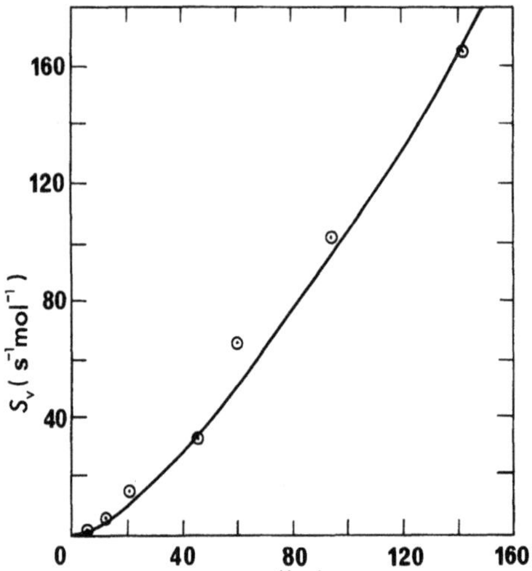

Fig. 5.11. Time course of specific activity of vacuolar K^+ in *Nitella axillaris*, re-drawn from DIAMOND and SOLOMON (1959). Each point is the mean of three determinations. The curve is calculated from the three-compartment model: the parameter values are given in Table 5.7

Most workers for the next decade used the model without testing it, although in work with Cl^- MACROBBIE (1964, 1966) used the check that the cytoplasm rate constant was independent of time.

In investigating the distribution of ^{36}Cl in cells of *Tolypella*, LARKUM (1968) found that he could detect the chloroplasts as a separate compartment. Using Cl^- with *Nitella* and *Tolypella*, MACROBBIE (1969) also found deviation from the predictions of the three-compartment model (Fig. 5.12). The discrepancy is that Q_v^* does not rise with the zero initial slope seen for example in Fig. 5.11, but with a slope apparently positive at zero time. This is perhaps more easily seen in a plot of F_v^* against time, where F_v^* is defined as Q_v^*/Q_T^*, Q_T^* being the radio-activity of the whole cell (Fig. 5.13). The same Figure shows that for K^+ in *Nitella* the effect is much less striking. In *Chara* both Cl^- and Na^+ exhibit marked deviations from three-compartment behavior (Fig. 5.14). In *Griffithsia*, K^+ shows similar deviations (Fig. 5.15), as was shown by FINDLAY et al. (1970).

Such results do not prove that F_v^* is indeed greater than zero at zero time, which would imply a direct pathway from exterior to vacuole, although this is possible. The alternative is that F_v^* rises from zero in a time shorter than the shortest uptake time in most cases, and much shorter than is predicted from the three-compartment

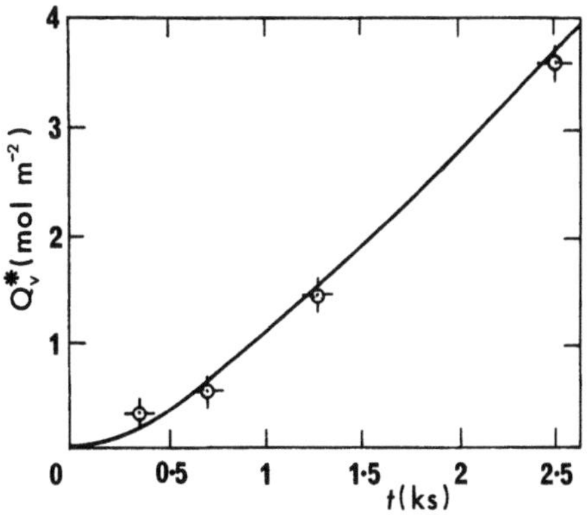

Fig. 5.12

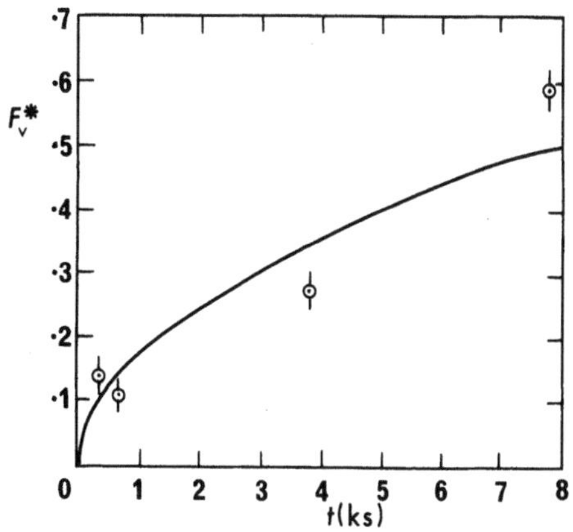

Fig. 5.13

Fig. 5.12. Time course of radio-activity in vacuole (^{36}Cl) in *Nitella translucens*, re-drawn from MACROBBIE (1969). Each point is the mean of 5–7 determinations. The curve is calculated from the "Y" four-compartment model: the parameter values are given in Table 5.7

Fig. 5.13. Time course of fraction of cellular radio-activity in vacuole (F_v^*) for ^{36}Cl in *Nitella translucens*, re-drawn from MACROBBIE (1969). Each point is the mean of 5 to 10 determination. The curve is calculated from the "Y" model: see Table 5.7 for the parameter values

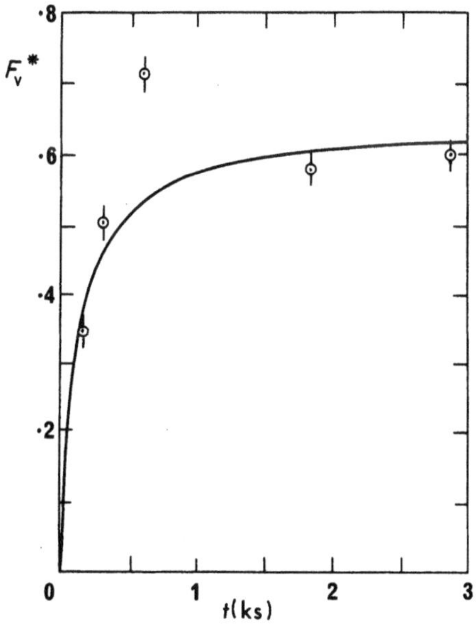

Fig. 5.14

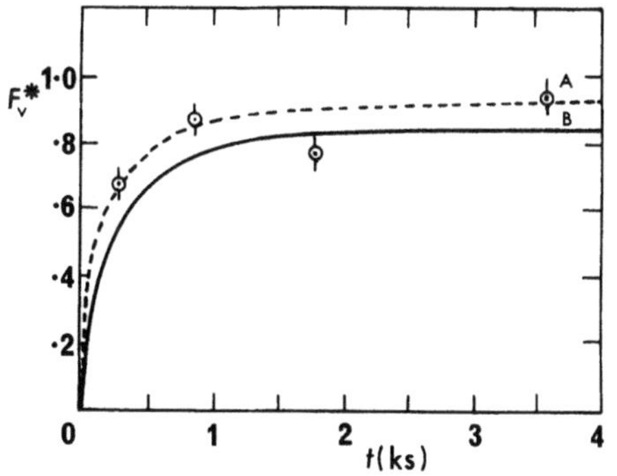

Fig. 5.15

Fig. 5.14. Time course of fraction of cellular radio-activity in vacuole (F_v^*) for ^{36}Cl in *Chara corallina*, re-drawn from HOPE and WALKER (1975). The points represent means of 10 determinations; the curve is calculated from the "Y" model, with values given in Table 5.7. (Data G.P. FINDLAY, A.B. HOPE and N.A. WALKER, unpublished)

Fig. 5.15. Time course of fraction of cellular radio-activity in vacuole (F_v^*) for ^{42}K in *Griffithsia monile*, re-drawn from FINDLAY et al. (1970). The points represent means of batches of cells; the curve is calculated from the "Y" model, with values given in Table 5.7

model. This model could in fact predict a rapid initial rise in F_v^* if one were free to make the cytoplasmic rate constant large by reducing Q_c or by increasing ϕ_{cv}: but Q_c can be got by direct methods and ϕ_{cv} has been measured in long-term uptake experiments. In any case the rapid rise to an intermediate value, followed by a steady further rise (Fig. 5.13) is not consistent with any three-compartment model.

These features can however be reproduced if we adopt the four-compartment model with the "Y" configuration already discussed. Curves derived from this model are displayed in Fig. 5.12–5.15, and the values of the parameters used are given in Table 5.7.

Table 5.7. Values of parameters used in calculating curves in figures

Figure number	Species used		Q_c (mmol m^{-2})	Q_p (mmol m^{-2})	Q_v (mmol m^{-2})	ϕ_{oc} (μmol $m^{-2}s^{-1}$)	ϕ_{cp} (μmol $m^{-2}s^{-1}$)	ϕ_{cv} (μmol $m^{-2}s^{-1}$)
5.11	*N. axillaris*		0.32	0	20	0.005	0.0	0.007
5.12	*N. translucens*		0.50	2.0	20	0.01	1.0	1.0
5.13	*N. translucens*		0.20	1.8	20	0.01	2.7	0.4
5.14	*Ch. corallina*		0.03	1.0	30	0.023	0.1	0.3
5.15	*G. monilis*	A	{1.00	17	150	2.5	0.9	10}
			3.00}	17	150	2.5	2.7	30}
		B	10	10	150	2.5	7.0	90

For *Griffithsia* the parameters were evaluated as follows: direct measurement gave $Q_v = 0.15$ mol m^{-2} and $\phi_{oc} = 2.5$ μmol $m^{-2} s^{-1}$; the flux analysis of FINDLAY et al. suggested $(Q_c + Q_p)$ 18 mmol m^{-2}; the later experimental points fix $\phi_{cv}/(\phi_{cv} + \phi_{cp})$ at about 0.9, and the early points fix the initial half-time t_1 at about 50 s. This determined the value of $Q_c(\phi_{cv} + \phi_{cp})$ at about 70 s. It was then determined by trial and error that if ϕ_{cv} were 10 μmol $m^{-2} s^{-1}$, the value calculated by FINDLAY et al. from the three-compartment model, the experimental points could be fitted at $\phi_{cp} = 0.9$ μmol $m^{-2} s^{-1}$, and $Q_c = 1$ mmol m^{-2}. This value for Q_c is merely one of a set of possible values among which the choice could only be made by determining ϕ_{cv} unambiguously. The value chosen is only 0.05 of the experimental value of $(Q_c + Q_p)$; this would seem implausible if it were not for electron micrographic evidence (cf. HOPE and WALKER, 1975, Plate 3) that almost all of the volume of the cytoplasm in *Griffithsia* is occupied by chloroplasts and mitochondria. Thus c might represent the continuous phase of the cytoplasm and p the plastids and mitochondria, though this is not in any way established.

For workers with *Nitella* and *Chara*, Cl^- has been the ion of interest. Since there are several methods by which the Cl^- content of the continuous phase of the cytoplasm may be estimated, the model may be tested by comparing the "direct" values $Q_{cyt, Cl}$ with the kinetic values of Q_c. This was first attempted by MACROBBIE (1969), who concluded that c could not represent the ground-plasm. Here we re-examine this question.

The flowing endoplasm of charophyte cells provides an approximation to the ground-plasm which can be collected by gentle centrifugation, for titration, or which can have a chloride electrode inserted into it *in vivo*. The results of such experiments are shown in Table 5.8; they are notable for their wide spread. Though generic differences must play some part in this, the different techniques seem to produce significantly different results too. One reason for this was demonstrated by KISHIMOTO and TAZAWA (1965), who showed that endoplasm centrifuged to one end of the

Table 5.8. Experimental values for the chloride content ($[Cl^-]_{cyt}$) in charophyte cytoplasm, and Q_{cyt}, assuming a cytoplasm 5 μm thick

Species	Reference	Method	$[Cl^-]_{cyt}$ (mM)	Q_{cyt} (mmol m^{-2})	Remarks
Ch. corallina	COSTER (1966)	Inserted chloride electrode *in vivo*	10	0.05	–
N. flexilis *Ch. corallina*	TAZAWA et al. (1974)	Cells slowly perfused with sorbitol + $CaCl_2$	20	<0.1	Value is for whole protoplasm including chloroplasts, Endoplasm must contain less
N. flexilis	LEFEBVRE and GILLET (1971)	Inserted chloride electrode *in vivo*	20	0.1	–
N. flexilis	KISHIMOTO and TAZAWA (1965)	Cells perfused with artificial sap containing no chloride, then centrifuged-endoplasm removed for titration	35	0.18	This method also gave low values for $[Na]_{cyt}$
N. flexilis	KISHIMOTO and TAZAWA (1965)	Cells centrifuged: endoplasm removed for titration	70	0.35	–
N. translucens	SPANSWICK and WILLIAMS (1964)	Cells centrifuged: endoplasm removed for titration	65	0.33	–
N. translucens	HOPE et al. (1966)	Cells centrifuged: endoplasm removed for titration	87	0.44	–

Nitella cell had Na^+ and Cl^- concentrations that depended strongly on those in the sap. If the ion being measured was removed from the sap by perfusion, its apparent concentration in the endoplasm was greatly reduced. The inference is that there is a considerable contamination of the centrifuged endoplasm with sap, which renders the results of SPANSWICK and WILLIAMS (1964) and of HOPE et al. (1966) suspect. Table 5.8 suggests that $[Cl]_{cyt}$ may be about 10–30 mM; Q_{cyt} is then about 60–180 μmol m^{-2}. The results of TAZAWA et al. (1974) suggest that the values may be lower still, and confirm the need for further work on this problem.

The estimates of Q_c from kinetic data range from 30 to 200 μmol m^{-2}, as shown in Table 5.7. The data on which these estimates are based do not however constrain the values very closely. Thus the *Chara* value of 30 μmol m^{-2} (Fig. 5.14) is determined only when the true slope of the curve between 0.5 and 3 ks is determined: the three points available do not determine the slope at all convincingly. Some of the results of MACROBBIE (e.g. Fig. 5.13) constrain ϕ_{cv} more closely, and so allow a better estimate of Q_c. It seems still not clear whether the lower estimates of Q_c are correct or reflect too low a choice of ϕ_{cv}. MACROBBIE has estimated the range of values of Q_c for one batch of *Nitella* cells as follows: the values of F_v^* were close to 0.13 at short uptake times, so that $k_1 = 0.02\ s^{-1}$, and $\phi_{cv}/(\phi_{cv}+\phi_{cp}) \doteqdot 0.13$; while cells with long times of uptake had values of F_v^* which gave k_2 values ranging from 2 to $12 \cdot 10^{-5}\ s^{-1}$. The Equations of 5.2.4 then yielded values of Q_c ranging from 30 to 170 $\mu mol\ m^{-2}$.

It seems possible to conclude that we have insufficient information in either Table 5.7 or 5.8; for the moment it is not an unreasonable assumption that c is the ground-plasm. The tentative conclusions we support then are that the four-compartment "Y" model is applicable to Cl^- in charophyte cells, and also to Na^+; it may or may not be necessary to fit the K^+ results. It is interesting that according to KISHIMOTO and TAZAWA (1965) the concentrations in the flowing cytoplasm of these ions are: [Na] = 2 mM; [Cl] = 35 mM; [K] = 120 mM. Thus, roughly, the "four-compartment" ions are those in low concentration in the cytoplasm.

This identification of the compartments in the "Y" model was rejected by MACROBBIE (1969, 1970b, 1971), who favored c being the endoplasmic reticulum and p being the cytoplasm and chloroplasts. Apart from the question of Q_c already discussed, she had two other grounds for this rejection.

The first was an apparent correlation between ϕ_{oc} and the rate constant k_2, in experiments in which F_v was constant (MACROBBIE, 1969: Fig. 8). The correlation shown in this figure is significant at the 0.05 level[1], and the explanation for it, suggested by MACROBBIE, was that in each cell ϕ_{pc} was determined by ϕ_{oc}. She suggested that this would be difficult to accommodate within the framework of the "Y" model. However, the "Y" model as discussed does not specify the values of the fluxes, nor the nature of the control systems that determine these values, and it is not invalidated by a finding that one flux appears to control another. Further, the nature of the correlation exhibited by Fig. 8 of MACROBBIE (1969) is not clear: the points represent the means of batches of cells under different conditions intended to alter ϕ_{oc}, as well as some batches under the same conditions. The origin of the significant correlation could thus be genetic, or physiological; but it is not yet shown that k_2 is determined directly by ϕ_{oc}. The observations, like those of a correlation between ϕ_{cv} and ϕ_{oc} (MACROBBIE, 1966) remain interesting and worth pursuing: their meaning depends on the compartment model adopted and on the ultimate source of the correlations, which needs to be critically identified.

The second ground for rejection of the "Y" model needs a Section of its own.

5.2.1 The Quantization of F_v^*

In several papers MACROBBIE (1970b, 1971, 1973) examined the distribution of values of F_v^* at zero time in individual cells of *Tolypella* and *Nitella*, using first ^{36}Cl and then both ^{36}Cl and ^{82}Br. She concluded that the distribution showed a degree of quantization, the values being clustered around integral multiples $n_i\alpha$ of a quantum α.

---

[1] By TUKEY's Quick Correlation Test (CONOVER, 1971).

In experimental batches of cells containing 17–42 individuals the values of α ranged from 0.035 to 0.395.

Since F_v^* can by definition range only between zero and 1, the values of n_i must satisfy

$$n_i \leqq 1/\alpha$$

and in practice n_i appeared to range between 1 and 6. If such a quantization exists its physical interpretation is thus obscure—physiological or genetic variables determine α, and after α is determined some mechanism would select an allowed integer. It is virtually impossible to imagine this happening within the framework of the "Y" model we have discussed, as MACROBBIE pointed out. She suggested that a vesicle transport model might produce the quantization, but no details of such a model have been worked out. In the 1970 paper consideration was given to the possibility that the observed values of F_v^* were affected by one or more action potentials (APs) at the tonoplast when the cells were cut open for sampling of the vacuole. This will be further considered after the data themselves have been discussed.

The data (original values of F_v^*) on which MACROBBIE's conclusions (1970b) are based are given as Fig. 4 by FINDLAY et al. (1971). They are not obviously distributed in a quantized way, although one experiment at least, identified as T, has its 17 values distributed in a noticeably clustered manner. FINDLAY et al. (1971) concluded that there was no valid statistical evidence for quantization. They also showed that in large experiments (of about 100 cells) *Chara* at least gave no sign of quanta, a conclusion also reached in an experiment with 100 *Tolypella* cells (G.P. FINDLAY, pers. comm.). WALKER (1973) has argued, on the basis of the run test and the empty cell test (BRADLEY, 1968), that only one, (T) of a total of 12 experiments of MACROBBIE shows a significant deviation from a null hypothesis of uniformly distributed data.

The existence of controversy suggests that at any rate the data do not force the observer to accept that rigid, integer quantization exists: nor is it *a priori* likely.

It remains possible that values of F_v^* are multi- or bimodally distributed, and this has led MACROBBIE to a reconsideration of the role of action potentials. When a cell of *Nitella* or *Chara* is cut open and vacuolar sap removed there occur one or more AP's (E.A.C. MACROBBIE and N.A. WALKER, unpublished). The charophyte AP has been shown to consist of membrane APs at both plasmalemma and tonoplast (HOPE and FINDLAY, 1964), and the magnitude and direction of that at the tonoplast are consistent with its being due to a transient increase in chloride permeability. If sufficiently large this will transfer radio-active chloride from cytoplasm to vacuole, although the net flux of Cl^- would be from vacuole to cytoplasm. If the cytoplasmic specific activity is very high, as the "Y" model predicts, this will measurably increase F_v^*, especially at short uptake times when Q_v^* is small.

We take the parameters to have the following values: $\phi_{oc}=23$ nmol m^{-2} s^{-1}, $\phi_{cp}=150$ nmol m^{-2} s^{-1}, $\phi_{cv}=270$ nmol m^{-2} s^{-1}; $Q_c=30$ µmol m^{-2}, $Q_p=1$ mmol m^{-2}, $Q_v=30$ mmol m^{-2}; transfer c−v = 10 µmol m^{-2} in two APs; then two APs will raise F_v^* as given below:

Uptake time	F_v^*	
	no AP	two APs
100 s	0.299	0.463
300 s	0.495	0.567

The assumed transfer is about 20 times the normal flux for the duration of the AP; it might be larger, but so might Q_c.

This sort of calculation suggests that if Q_c is low the inevitable APs will perhaps significantly increase F_v^*, while random variations in the number of APs might give it a distribution with several modes.

If F_v^* is affected by APs it will clearly be most in error at the shortest uptake times, so that the true value of k_1 will be lower than the data suggest and the true value of Q_c will be higher.

5.2.2 Vesicle Transport

The matters discussed in this Section can, at the risk of over-simplifying them, be condensed into two questions:

(a) are the observed kinetics consistent with the "Y" four-compartment model or do they demand some quite different model?

(b) given the "Y" model, is compartment c to be identified as the water phase of the cytoplasm, or as the interior of the endoplasmic reticulum?

At the moment the answers would appear to be:

(a) the established observations are consistent with the "Y" model, and

(b) so far compartment c is most probably the water phase of the cytoplasm.

These answers leave open the question of the mechanism of salt transport from cytoplasm to vacuole; either a molecular flux or an exocytotic vesicle flux would be quite consistent with the observed kinetics. It seems best to assume a molecular mechanism until it is disproved.

5.3 Fluxes in Cells of the Root

It has not been possible to investigate compartmentation of the cytoplasmic phase in higher plant cells, as has been done with giant algal cells. Separation of the cytoplasm from the vacuole in higher plant cells has not been practical and hence F_v^* cannot be measured. Concern with the validity of flux analysis in higher plant tissues has been more with the effect of the heterogeneity of the cells and of the organization of the tissue on the measurement of fluxes. The estimation of fluxes for cells of the root gives a good example of these types of difficulty, as they affect both uptake and efflux measurements.

5.3.1 Efflux Analysis

There is good evidence that cortical parenchyma cells show cytoplasmic and vacuolar components of isotope exchange, as does storage tissue (5.5.1). CRAM (1973) measured fluxes in isolated halves of the cortex of maize roots, so avoiding the complications caused by the stele. He found a clear separation of cytoplasmic from vacuolar components (Fig. 5.16).

Though the fast and slow components can be related to parenchyma cells there may be considerable variation in cell size within the tissue. The distribution of the

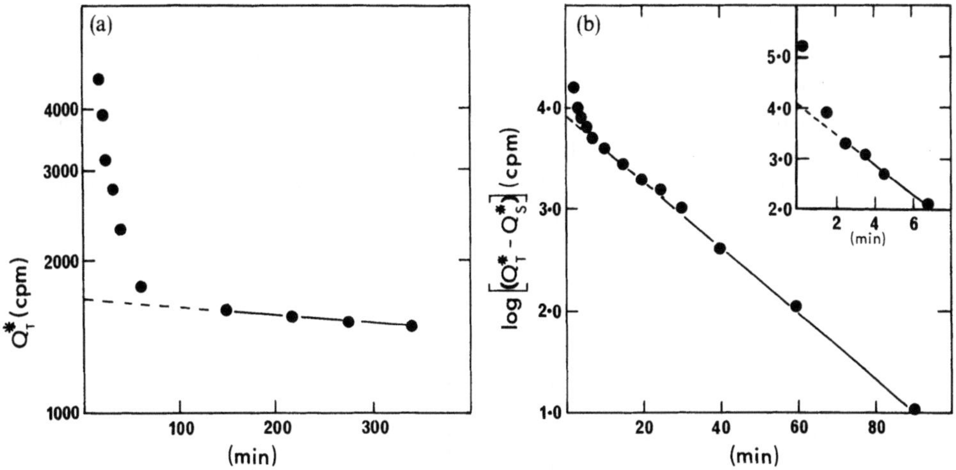

Fig. 5.16 a and b. Time course of loss of ^{36}Cl from maize root half-cortices plotted semilogarithmically and showing separation of various components. (a) Total content (Q_T^*), showing extrapolation of the slow component (Q_s^*). (b) Plot of ($Q_T^* - Q_s^*$) showing slope of exchange of the cytoplasmic component (Q_c^*). The inset shows a plot of log ($Q_T^* - Q_s^* - Q_c^*$) which is the free space exchange. (From CRAM, 1973)

Table 5.9. Distribution of cell diameters (transverse) in sections of barley root cortex

Cell diameter (μm)	Relative volume of tissue
0–20	40
20–25	10
25–30	11
30–30	21
40–50	9
50–65	9

transverse diameters of cells in the cortex of barley roots is given in Table 5.9. Even if the flux per unit of surface area of the cell is constant there will be large differences in uptake to cells on a weight basis and consequently in s_v. Also the flux per unit area may depend on the position of the cell along the length of the root (e.g. ESHEL and WAISEL, 1973).

A detailed analysis of isotope exchange in bean roots was made by PALLAGHY and SCOTT (1969). Over a period of 50 h they found decreases in the fluxes and in oxygen consumption, though the total K^+ content (Q_T) was steady. Comparison of observed and calculated values for Q_T (see 5.2.3.1 and 5.4.2) showed consistent differences that could be resolved if it was assumed there was a small net efflux from the roots. One result of this decrease in flux was that the plot of log Q_T against t was not straight but curved. However it was not thought that the estimation of fluxes was affected by this difficulty in determining a rate constant for the slow phase (k_2). They suggested that the drift in fluxes was due to decreases in permeability of the plasmalemma during ageing (see Part B, Chap. 8), and a change in proportion of "exchanging" and "non-exchanging" cells. (See also ETHERTON, 1967.)

Plant roots have another feature that makes it of doubtful validity to use the three-compartment model. A root takes in ions at the outer surface and secretes them through the xylem to the shoot (or to the solution when roots are excised). Hence, although the chemical content of the roots may be constant, there is in fact a large net flux of ions passing through the cells.

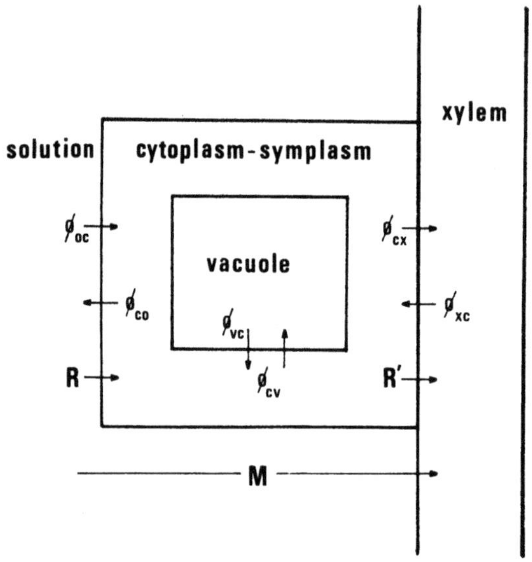

Fig. 5.17. Model for fluxes in relation to the organisation of the root. Symbols as in Fig. 5.4 except ϕ_{cx} is a flux to the xylem; R and R' are net fluxes $(=J)$ across the boundaries of cytoplasm and vacuole and "M" is net flux to the xylem outside of the symplast (this term is negligible for barley roots). (From PITMAN, 1971)

Fig. 5.17 shows a model for tracer exchange in the root that includes this xylem transport. The estimation of fluxes depends on measurements of tracer flux out of the surface of the root (ϕ_{co}) separately from that through the xylem (ϕ_{cx}). Experimentally, this separation can be achieved by setting up the roots in chambers so that the tracer from the cut end can be collected separately from that from the surface. The calculation of fluxes is simplified by using roots uniformly labeled with tracer, prepared by transferring low salt roots to 5 mM KCl solution labeled with ^{36}Cl for about 24 h. Since the initial Cl^- content was negligible, $s_v = s_c = s_l$ during uptake, and ϕ_{co} can be measured directly as Φ_T^*/s_l (PITMAN, 1971). Fig. 5.18 shows the time course of tracer from the cut end of the root compared with that from the surface (JESCHKE, 1973). It may be necessary to extrapolate the cytoplasmic component to some time later than $t=0$, since the stele originally contains ions at high specific activity which have to be swept out by the flow of xylem sap before the isotope efflux from the symplast can be measured. This lag in appearance of isotope can be estimated at about 5 min for barley roots (PITMAN, 1971).

Despite the regularity of exchange shown in Fig. 5.16 there have been observations of an anomalous increase in efflux with time, or of "shoulders" in the content of isotope during elution (PALLAGHY et al., 1970, maize; LÜTTGE and PALLAGHY, 1972, *Mnium*, bean root cortex). It has been suggested that this behavior is due to some other component of cell or tissue, but as yet no wholly acceptable hypothesis has been advanced.

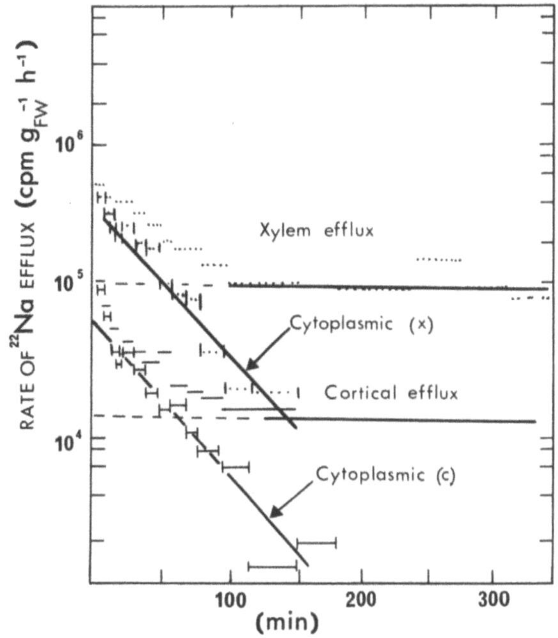

Fig. 5.18. Rate of ^{22}Na loss from excised barley roots, collected from the cut end [xylem efflux ($\phi_{cx} s_c$), ...] and from the surface of the root [cortical efflux ($\phi_{co} s_c$)———]. The slow components are subtracted to give the cytoplasmic components from the xylem (X|...|) and from the surface (cortex, C|—|). Note the two cytoplasmic components have the same rate constant. The roots were in 1 mM NaCl. (From JESCHKE, 1973)

5.3.2 Estimation of Fluxes from Uptake Measurements

The relationship between Φ^*_{in} and fluxes was discussed in general terms in 5.4.4. A particular problem with roots is the transport of tracer from the cut end, though its effect can be minimal in certain experimental situations. For periods up to 30 min after transfer from $CaSO_4$ to salt solutions, the transport from the xylem is negligible from low-salt roots, and tracer uptake seems to estimate ϕ_{oc} at concentrations less than 1 mM. At higher concentrations when ϕ_{oc} is larger the rate of rise of s_c to its quasi-steady level is very rapid and also ϕ_{co} may not be negligible so that $\phi_{co} s_c$ cannot be ignored. In this case $\Phi^*_{in}/s_o = \phi_{cv}\,\phi_{oc}/(\phi_{cv}+\phi_{co})$ which tends to ϕ_{cv} as ϕ_{oc} become very large. If $(\phi_{cv}+\phi_{oc}) = 20\ \mu mol\ g_{FW}^{-1}\ h^{-1}$ and $Q_c = 1\ \mu mol\ g_{FW}^{-1}$, the time for 50% rise in s_c is about 2 min. These values are reasonable for roots at 25° C-35° C in KCl solution of 10 mM and over, so that an extremely short period of measurement would be needed to estimate $\phi_{oc} s_o$ separately from $(\phi_{oc} s_o - \phi_{co} s_c)$. The use of short and long periods of measurement to compare ϕ_{oc} and ϕ_{cv} is discussed by CRAM and LATIES (1971).

When using roots grown in a nutrient solution or KCl solution the K^+ content of the roots is higher and there is commonly a large export from the xylem. When such roots are put into tracer, their tracer content rises rapidly at first but then more slowly as tracer is exported from the xylem. The rate of tracer uptake to the root may, however, be steady (Fig. 5.19). Generally,

$$\Phi^*_{in} = \phi_{oc}\, s_o - \phi_{co}\, s_c - \Phi^*_{ox}, \tag{5.36}$$

Φ^*_{in} will have a different significance, depending on the period over which it is measured. Again, over very short periods Φ^*_{in} should provide an estimate of $\phi_{oc}\, s_o$; over larger periods (up to 30 min in Fig. 5.19) Φ^*_{ox} is small and Φ^*_{in} is given by $(\phi_{oc}\, s_o -$

$\phi_{co}\, s_c$). At even later stages, when Φ^*_{ox} is appreciable, Φ^*_{in} becomes equal to $\phi_{cv}\,\phi_{oc}\cdot s_o/(\phi_{cv}+\phi_{co}+\phi_{cx})$. In most situations studied Φ^*_{in} at this stage approximates to $\phi_{cv}\, s_o$, but clearly the closeness of the approximation depends on the relative size of the fluxes (PITMAN, 1971; see also Part. B, Chap. *3.3*).

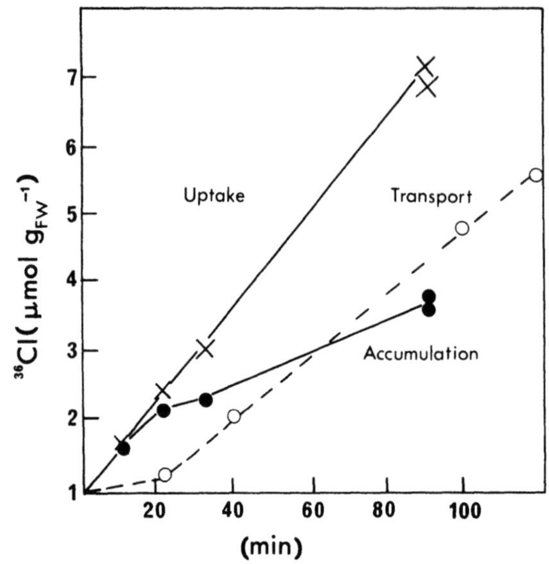

Fig. 5.19. Tracer uptake to barley roots (^{36}Cl) separated into export from the xylem (Transport) and content of the root (Accumulation). (From PITMAN, 1972)

6. Conclusions

Efflux analysis as applied to model systems is exact and can return values for fluxes in complicated systems, but it should be realized by now that its application to plant cells and tissues can only be an approximation. The main limitation of the approach is in knowing the compartmentation of the cell or organization of the tissue to which it is being applied. Despite these problems the method has yielded valuable information on the organization of *Chara* cells in relation to the transport and the only estimates of fluxes of ions across the tonoplast separate from the plasmalemma.

References

BRADLEY, J.V.: Distribution-free Statistical Tests. Englewood Cliffs: Academic Press 1968.

BRIGGS, G.E.: Some aspects of free space in plant tissue. New Phytologist **56**, 305–324 (1957a).

BRIGGS, G.E.: Estimation of the flux of ions into and out of the vacuole of a plant cell. J. Exptl. Bot. **8**, 319–322 (1957b).

BRIGGS, G.E., HOPE, A.B., PITMAN, M.G.: Exchangeable ions in beet disks at low temperature. J. Exp. Botany **9**, 128–141 (1958a).

BRIGGS, G.E., HOPE, A.B., PITMAN, M.G.: Measurement of ionic fluxes in red beet tissues using radioisotopes. Radioisotopes Sci. Res. Proc. Int. Conf. Paris **4**, 391–400 (1958b).

BRIGGS, G.E., HOPE, A.B., ROBERTSON, R.N.: Electrolytes and plant cells. Oxford: Blackwell 1961.

CHARLEY, J.L., JENNY, H.: Two-phase studies on availability of iron in calcareous soils. IV. Decomposition of iron oxide by roots, and Fe diffusion in roots. Agrochimica **5**, 99–107 (1961).

CONOVER, W.J.: Practical Nonparametric Statistics. New York: Wiley and Sons 1971.
COSTER, H.G.: Chloride in cells of *Chara australis*. Australian J. Biol. Sci. **19**, 545–554 (1966).
CRAM, W.J.: Compartmentation and exchange of chloride in carrot root tissue. Biochim. Biophys. Acta **163**, 339–353 (1968).
CRAM, W.J.: Short term influx as a measure of influx across the plasmalemma. Plant Physiol. **44**, 1013–1015 (1969).
CRAM, W.J.: Chloride fluxes in cells of the isolated root cortex of *Zea mays*. Australian J. Biol. Sci. **26**, 757–779 (1973).
CRAM, W.J., LATIES, G.G.: The use of short-term and quasi-steady influx in estimating plasmalemma and tonoplast influx in barley root cells at various external and internal chloride concentrations. Australian J. Biol. Sci. **24**, 633–646 (1971).
DAINTY, J., HOPE, A.B.: Ionic relations of cells of *Chara australis*. I. Ion exchange in the cell wall. Australian J. Biol. Sci. **12**, 395–411 (1959).
DAINTY, J., HOPE, A.B.: The Electric double layer and the donnan equilibrium in relation to plant cell walls. Australian J. Biol. Sci. **14**, 541–551 (1961).
DIAMOND, J.M., SOLOMON, A.K.: Intracellular compartments in *Nitella axillaris*. J. Gen. Physiol. **42**, 1105–1121 (1959).
ESHEL, A., WAISEL, Y.: Variations in uptake of sodium and rubidium along barley roots. Physiol. Plant. **28**, 557–560 (1973).
ETHERTON, B.: Steady state sodium and rubidium effluxes in *Pisum sativum* roots. Plant Physiol. **42**, 685–690 (1967).
FINDLAY, G.P., HOPE, A.B., WALKER, N.A.: Quantization of a flux ratio in charophytes. Biochim. Biophys. Acta **233**, 155–162 (1971).
FINDLAY, G.P., HOPE, A.B., WILLIAMS, E.J.: Ionic relations of marine algae. II. *Griffithsia:* ionic fluxes. Australian J. Biol. Sci. **23**, 323–338 (1970).
HODGKIN, A.L., KEYNES, R.D.: The potassium permeability of a giant nerve fibre. J. Physiol. **128**, 61–88 (1955).
HOPE, A.B.: Ion transport and membranes. London: Butterworth 1971.
HOPE, A.B., FINDLAY, G.P.: The action potential in *Chara*. Plant Cell Physiol. (Tokyo) **5**, 377–379 (1964).
HOPE, A.B., SIMPSON, A., WALKER, N.A.: The efflux of chloride from cells of *Nitella* and *Chara*. Australian J. Biol. Sci. **19**, 355–362 (1966).
HOPE, A.B., WALKER, N.A.: Ionic relations of *Chara australis*. IV. Membrane potential differences and resistances. Australian J. Biol. Sci. **14**, 26–36 (1961).
HOPE, A.B., WALKER, N.A.: The physiology of giant algal cells. Cambridge: Cambridge University Press 1975.
JACOBY, B., DAGAN, J.: Effects of age on sodium fluxes in primary bean leaves. Physiol. Plantarum **22**, 29–36 (1969).
JESCHKE, W.D.: K^+-Stimulated Na^+ efflux and selective transport in barley roots. In: Ion transport in plants (W.P. ANDERSON, ed.), p. 285–296. London-New York: Academic Press 1973.
KISHIMOTO, U., TAZAWA, M.: Ionic composition of the cytoplasm of Nitella flexilis. Plant Cell Physiol. (Tokyo) **6**, 507–518 (1965).
LARKUM, A.W.D.: Ionic relations of chloroplasts *in vivo*. Nature **218**, 447–449 (1968).
LEFEBVRE, J., GILLET, C.: Effects des cations externes sur l'activité des chlorures cytoplasmiques dosés par l'electrode Ag—AgCl introduite dans la cellule de *Nitella*. Biochim. Biophys. Acta **249**, 556–563 (1971).
LÜTTGE, U., PALLAGHY, C.K.: Unerwartete Kinetik des Efflux und der Aufnahme von Ionen bei verschiedenen Pflanzengeweben. Z. Pflanzenphysiol. **67**, 359–366 (1972).
MACROBBIE, E.A.C.: Factors affecting the fluxes of potassium and chloride ions in *Nitella translucens*. J. Gen. Physiol. **47**, 859–877 (1964).
MACROBBIE, E.A.C.: Metabolic effects on ion fluxes in *Nitella translucens*. II. Tonoplast fluxes. Australian J. Biol. Sci. **19**, 371–383 (1966).
MACROBBIE, E.A.C.: Ion fluxes to the vacuole of *Nitella translucens*. J. Exptl. Bot. **20**, 236–256 (1969).
MACROBBIE, E.A.C.: The active transport of ions in plant cells. Quart. Revs. Bioph. **3**, 251–294 (1970a).
MACROBBIE, E.A.C.: Quantized fluxes of chloride to the vacuole of *Nitella translucens*. J. Exptl. Bot. **21**, 335–344 (1970b).

MACROBBIE, E.A.C.: Vacuolar fluxes of chloride and bromide in *Nitella translucens*. J. Exptl. Bot. **22**, 487–502 (1971).
MACROBBIE, E.A.C.: Vacuolar ion transport in *Nitella*. In: Ion transport in plants. (W.P. ANDERSON, ed.), p. 431–446. London-New York: Academic Press 1973.
MACROBBIE, E.A.C., DAINTY, J.: Ion transport in *Nitellopsis obtusa*. J. Gen. Physiol. **42**, 335–353 (1958a).
MACROBBIE, E.A.C., DAINTY, J.: Sodium and potassium distribution and transport in the seaweed *Rhodymenia palmata*. Physiol. Plantarum **11**, 782–801 (1958b).
MAGAR, M.E.: Data analysis in biochemistry and biophysics. New York: Academic Press 1972.
MAILMAN, D.S., MULLINS, L.J.: The Electrical measurement of chloride fluxes in *Nitella*. Australian J. Biol. Sci. **19**, 385–398 (1966).
OSMOND, C.B.: Ion absorption in *Atriplex* leaf tissue. I. Absorption by mesophyll cells. Australian J. Biol. Sci. **21**, 1119–1130 (1968).
PALLAGHY, C.K., LÜTTGE, U., WILLERT, K. VON: Cytoplasmic compartmentation and parallel pathways of ion uptake in plant root cells. Z. Pflanzenphysiol. **62**, 51–57 (1970).
PALLAGHY, C.K., SCOTT, B.I.H.: The Electrochemical state of cell of broad bean roots. II. Potassium kinetics in excised root tissue. Australian J. Biol. Sci. **22**, 585–600 (1969).
PITMAN, M.G.: The determination of the salt relations of the cytoplasmic phase in cells of beetroot tissue. Australian J. Biol. Sci. **16**, 647–668 (1963).
PITMAN, M.G.: Ion exchange and diffusion in roots of *Hordeum vulgare*. Australian J. Biol. Sci. **18**, 541–545 (1965).
PITMAN, M.G.: Uptake and transport of ions in barley seedlings. I. Estimation of chloride fluxes in cells of excised roots. Australian J. Biol. Sci. **24**, 407–421 (1971).
PITMAN, M.G.: Uptake and transport of ions in barley seedlings. II. Evidence for two active stages in transport to the shoot. Australian J. Biol. Sci. **25**, 243–257 (1972).
PITMAN, M.G., LÜTTGE, U., KRAMER, D., BALL, E.: Free space characteristics of barley leaf slices. Australian J. Plant. Physiol. **1**, 65–75 (1974).
SHEPPARD, C.W.: Basic principles of the tracer method. New York: Wiley and Sons 1962.
SLAYMAN, C.W., TATUM, E.L.: Potassium transport in *Neurospora*. II. Measurement of steady-state potassium fluxes. Biochim. Biophys. Acta **102**, 149–160 (1965).
SPANSWICK, R.M., WILLIAMS, E.J.: Electric potentials and Na, K and Cl concentrations in the vacuole and cytoplasm of *Nitella translucens*. J. Exp. Botany **15**, 193–200 (1964).
SPANSWICK, R.M., WILLIAMS, E.J.: Calcium fluxes and membrane potentials in *Nitella translucens*. J. Exp. Botany **16**, 463–473 (1965).
STEVENINCK, R.F.M. VAN: Potassium fluxes in red beet tissue during its "lag phase". Physiol. Plantarum **15**, 211–215 (1962).
TAZAWA, M., KISHIMOTO, U., KIKUYAMA, M.: Potassium, sodium and chloride in the protoplasm of characeae. Plant Cell Physiol. (Tokyo) **15**, 103–110 (1974).
TYREE, M.T.: Determination of transport constants of isolated *Nitella* cell walls. Can. J. Botany **46**, 317–327 (1968).
VICKERY, R.S., BRUINSMA, J.: Compartments and permeability for potassium in developing fruits of tomato (*Lycopersicon esculentum* Mill.). J. Exp. Botany **24**, 1261–1270 (1973).
WALKER, N.A.: Discussion. In: Ion transport in plants (W.P. ANDERSON, ed.), p. 459–461. London-New York: Academic Press 1973.

II. Particular Cell Systems

6. Transport in Algal Cells

J.A. RAVEN

1. Introduction

The use of giant algal cells has been very important in recent advances in our understanding of the mechanism of solute and water transport in plants. This research has viewed giant algal cells as convenient material for the study of membrane properties *per se*; the involvement of solute transport in the life of the plant particularly with respect to nutrition, has been relatively neglected.

On the other hand, studies with unicellular (micro) algae have contributed to our knowledge of K^+/Na^+ regulation, the transport of metabolized solutes such as phosphate, nitrate and sugars, and the relation of transport to photosynthesis.

It is hoped in this Chapter to review not only the advances which the use of algal material has made to our understanding of membrane properties and functions, but also to consider how these advances contribute to our knowledge of the general metabolism, growth and development of algae.

Reviews which consider transport in algae include BRIGGS et al. (1961), BIEBL (1962), DAINTY (1962), EPPLEY (1962), GUILLARD (1962), STADELMANN (1962), SCOTT (1967), BUETOW (1968), GUTKNECHT and DAINTY (1968), KOTYK and JANACEK (1970), MACROBBIE (1970a, b), RAVEN (1970), WIESSNER (1970a, b), CORNER and DAVIES (1971), HOPE (1971), PITMAN (1971), RAVEN (1971c, d), GOLDMAN et al. (1972), RAVEN (1972a), HEALEY (1973), LÜTTGE (1973), SIMONIS and URBACH (1973), TYREE (1973), DROOP (1974), MACROBBIE (1974), O'KELLEY (1974) and RAVEN (1974a, b).

2. Algal Structure in Relation to Transport

Most of the work on solute transport and membrane properties of algae has involved the use of "plant-type" cells, i.e. having a large aqueous vacuole and a strong cell wall which is functional in osmoregulation, which are always hypertonic to the growth medium, and which are non-motile. DODGE (1973) has reviewed the ultrastructure of algae and considers (cf. FRITSCH, 1935, 1945) many organisms to be algae although they lack one or more of these "plant-type" features.

Quantitative electron microscopy shows that *Chlamydomonas* (Fig. 6.1; SCHÖTZ et al., 1972) and *Chlorella* (A.W. ATKINSON, 1972) have only about 10% of their cell volume occupied by (non-contractile) vacuoles. However, the tonoplast area in *Chlorella* (35% of the plasmalemma area) may not qualify these cells as non-vacuolate for the interpretation of flux data (ATKINSON et al., 1972). On the basis of light microscopy *Porphyra* has no central vacuole (EPPLEY, 1958), although electron microscopy shows that up to 20% of the cell volume can be occupied by small vacuoles (KAZAMA and FULLER, 1970). Eggs and zygotes of *Fucus* also have no central

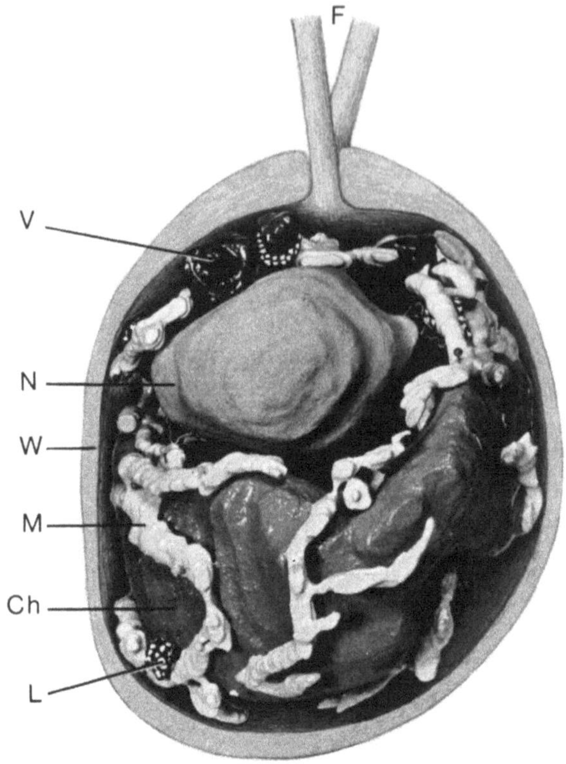

Fig. 6.1. Scale model of a (+) gamete cell of the unicellular alga *Chlamydomonas reinhardii* reconstructed from a series of electron micrographs. The model shows the spatial relations among the various organelles of the cell. The nucleus (*N*) is embedded in the cavity of the trough-shaped chloroplast (*Ch*). The mitochondria (*M*; 9 in this particular cell) are elongated, twisted and ramified "embracing" the other organelles. *V* are vacuoles, *L* lipid bodies, *F* the flagella. Magnification ca. × 15,000. Original photograph of the model by courtesy of Prof. Dr. F. SCHÖTZ (München; see SCHÖTZ et al., 1972). The cell wall (*W*) and the flagella (*F*) were drawn after electronmicrographs by Mrs. DORIS SCHÄFER to show how the cell wall plus the plasmalemma surround the cell

vacuole, although the cytoplasm contains many small membrane-bounded inclusions (QUATRANO, 1972).

Wall-less cells are common among flagellate algae, and also in flagellate or non-motile reproductive stages of multicellular algae whose vegetative cells are plant-like. Maintenance of non-spherical shape in these wall-less cells is by means of microtubules (BOUCK and BROWN, 1973; BROWN and BOUCK, 1973; PATENAUDE and BROWN, 1973). The possession of flagella means that a part of the plasmalemma is not covered by the cell wall (if present). The absence of a cell wall means that volume regulation cannot be of the conventional plant type. In sea water wall-less cells are isotonic with the medium (ALLEN et al., 1972) but require some volume control mechanism due to the presence of more impermeant solutes in the cell than in the medium. This may involve active Na^+ efflux as is the case in many animal cells (cf. ALLEN et al., 1972); active solute extrusion may also be involved in volume regulation of the cytoplasm *vis-a-vis* the vacuole in "plant-type" cells (GUTKNECHT and DAINTY, 1968). In fresh water, where for metabolic activity a cytoplasmic solute content equivalent to at least 20 mM KCl is needed, compared with 2–5 mM in the medium (GUILLARD, 1962; PROSSER and BROWN, 1962; LIVINGSTONE, 1963; POTTS and PARRY, 1964), the cell is hypertonic to the medium, and volume regulation occurs *via* contractile vacuoles (cf. ETTL, 1961). Mutant studies on the walled flagellate *Chlamydomanas* emphasize the role of contractile vacuoles rather than cell walls in volume regulation in fresh water flagellates (GUILLARD, 1960; DAVIES and PLASKITT, 1971).

Flagella are important in locomotion and prevention of sinking in planktonic forms (cf. SMAYDA, 1971) and also in disrupting unstirred layers around cells which may limit solute influx (KOCH, 1971; AARONSON, 1971).

The cell walls of some green microalgae of the order Chlorococcales contain a substance resembling sporopollenin which greatly reduces their permeability to solutes such as $NADH_2$ (ATKINSON et al., 1972b; SYRETT and THOMAS, 1973).

Most algae normally grow photoautotrophically, i.e. the major fluxes at the plasmalemma are of water and inorganic solutes. However, fluxes of organic solutes at the plasmalemma are important. Cell-wall synthesis, transfer of organic solutes to heterotrophic symbionts, and extracellular excretion are all important in the life of algae (FOGG, 1966; SIEBURTH, 1969; SMITH et al., 1969). A number of algae are obligate or facultative heterotrophs (VAN BAALEN and PURITCH, 1973; DROOP, 1974), and many photoautotrophs are vitamin auxotrophs (DROOP, 1974) and are thus dependent on transfer of soluble or particulate organic material into the cells.

As is discussed in *6*.4, most of the work on algal solute transport and membrane properties has involved either giant cells or unicellular micro-algae. Some algae form very large multicellular thalli with tissue differentiation rivalling that in higher plants (e.g. the *Phaeophyceae:* FRITSCH, 1945). DAVIES et al. (1973) point out that some of the organs of these algae, e.g. the haptera of *Laminaria*, could profitably be compared with organs of higher plants. In these plants some cells are non-photosynthetic, and even where meristems are photosynthetic the growth rate exceeds that which could be supported by local photosynthesis. Both symplastic and phloem-like transport occur in these algae; this may be involved in both the transfer of nutrients required in such differentiated plants, and the transfer of electrical or chemical information required to control differentiation (SPANSWICK, Part B, Chap. *2;* NICHOLSON and BRIGGS, 1972; FRASER and GUNNING, 1973; WALKER, 1974). However, it is clear that gradients occur within single cells, even when these are part of a symplast. Once the site of rhizoid production in a fucoid zygote has been determined, marked non-uniformities in ion fluxes over the cell surface appear (JAFFE et al., 1974); differences are also found in electrical behavior for different areas of membrane in regenerating *Acetabularia* (NOVAC and BENTRUP, 1972). Localized illumination of Characean internodes leads to localized effects on membrane phenomena, although the photoreceptor is chlorophyll in the chloroplast (BLINKS and CHAMBERS, 1958; VREDENBERG, 1969, E.A.C. MACROBBIE, H.D. JAYASURIYA, personal communication of experiments on halide influx; cf. WEST and PITMAN, 1967a; KNUTSEN, 1972).

Most of the work on transport in algae has concentrated on the plasmalemma and tonoplast. Relatively little is known of transport in algal organelles, with the exception of chloroplasts (e.g. STROTMANN and HELDT, 1969; LARKUM, 1968; KAHN, 1971; TRENCH et al., 1973). The relatively large fraction of the cytoplasm occupied by the chloroplast (often more than 40%: A.W. ATKINSON, 1972; SCHÖTZ et al., 1972; WAGNER, 1974), makes them important in considering the distribution of solutes and the significance of electrical measurements in the cytoplasm (*6*.5). In this respect a comparison of autotrophically and heterotrophically grown cells of algae (e.g. *Euglena*) with repressible chloroplast development (BUETOW, 1968) would be of interest.

It is hoped that this Chapter shows how algae can make good experimental material for a number of transport problems as well as for the biophysical approaches to which giant algal cells are so well suited.

3. Elemental Composition of Algae

Table *6*.1 gives the elemental composition of cells of *Euglena, Chlorella,* freshwater charophytes and *Hydrodictyon*. The first two are essentially non-vacuolate, while mature cells of charophytes and *Hydrodictyon* are more than 95% by volume vacuole. This is reflected in their composition; the N, S and P contents of the two microalgae are high compared with those of the coenocytes, while the K^+, Na^+, Ca^{2+}, Mg^{2+} and Cl^- contents are low, K^+, Na^+ and Cl^- are the major vacuolar ions in the coenocytes (Tables *6*.3 and *6*.4). The higher contents of N and P in the microalgae relative to C probably reflects the large quantity of polysaccharide wall materials in the coenocytes (ANDERSON and KING, 1961a, b).

Table 6.1. Ratio of elements in algal cells, taking C = 100 mol

	Euglena	*Chlorella*	*Nitella*	*Hydrodictyon*
C	100	100	100	100
N	9–16	10–13	4.5–6.0	8.4
S	0.3–1.5	0.3–0.6	0.34	1.3
P	0.5–1.5	0.3–1.9	0.23–0.6	0.23
K	0.2–0.3	0.5–1.5	1.5–3.0	3.26–5.3
Na	0.025	0.02–0.1	0.04–1.5	0.17–0.50
Ca	0.002–0.03	0.0–0.2	0.4	0.12–0.51
Mg	0.08–0.15	0.2–1.4	0.12	0.21
Cl	0.0–0.07	0.02	3.0	1.7

BLINKS and NIELSEN (1940), ANDERSON and KING (1961a, b). BARBER (1968b), COOK (1968), MACROBBIE (1970a), GOLDMAN *et al.* (1972), RAVEN, unpublished.

Thus in the formation of cell material equivalent to 100 moles of C, the other elements must enter the cell in the quantities shown in Table 6.1. At least in the microalgae (TAMIYA, 1966; COOK, 1968), the relative contents of the elements does not alter greatly during the growth cycle, i.e. the instantaneous *net* influxes of the elements must reflect their content in cell material. In the charophytes this is not so. Here the fraction of an internodal cell comprised of cytoplasm (and hence N, since most of the N is present as protein) decreases with expansion of the cell (MERCER and MERCER, 1971). Nevertheless, the quantities of elements shown in Table 6.1 must enter the cell during its development, either by direct influx or by translocation from more mature regions (WALKER, 1974): they must enter somewhere *via* the plasmalemma.

Of the elements listed in Table 6.1, K^+, Na^+, Ca^{2+}, Mg^{2+}, and Cl^- are mainly present as the osmotically active inorganic ion (exceptions include Mg^{2+} in chlorophyll and Cl^- in organic combination in *Ochromonas*, ELAVSON and VAGELAS, 1969). Organic solutes also contribute to the maintenance of osmotic pressure; in the cytoplasm many of these are intermediary metabolites (KANAZAWA, 1964; BASSHAM and KRAUSE, 1969 give values for *Chlorella*). As was discussed in 6.2, the osmolarity of cytoplasm must be more than the equivalent of 20 mM KCl. The solutes responsible for higher osmotic pressures may be related to the maintenance of isotonicity, turgor generation, or the storage of nutrients or energy, rather than being needed for metabolism *per se*. Osmoregulation and volume regulation, and the importance of turgor generation in the growth of walled cells, are briefly discussed in 6.2 (see also RAVEN, 1971c; see also *11*.3).

Nutrition of algae involves, in terms of transport, not only the uptake of essential elements required for growth (O'KELLEY, 1974), but also the active extrusion of solutes which are products of essential syntheses which would be toxic if allowed to accumulate in the cytoplasm; quantitatively significant examples are H^+ or OH^- (RAVEN and SMITH, 1974). Essential elements include not only C, H, O, N, S, P, Fe, Mg, etc, which are chemical constituents of essential organic molecules, but also K^+ which acts as an activator of a number of enzymes (EVANS and SORGER, 1966; SUELTER, 1970). This has not been well documented for algae, an exception being the thorough work of ANTIA et al. (1972) on threonine deaminase in marine planktonic algae. The absolute K^+ level in the cytoplasm of most algae (Table 6.4) is adequate for the maximum activation of K^+-requiring enzymes, i.e. 50 mM (EVANS and SORGER, 1966; SUELTER, 1970). However, in some freshwater algae (including *Enteromorpha intestinalis* growing in fresh water) the K^+ concentration is lower than this (RONKIN and BURETZ, 1960; GERLOFF and FISHBECK,

1969; Black, 1971); this must be the case in organisms with the minimum osmotic pressure consistent with growth (*6*.2).

The cell content of the elements (Table *6*.1) must be related to the net influx in relation to growth; in this connection the papers of Fuhs (1969) and of Gerloff and Fishbeck (1969) are of interest. These papers report the content of P, K^+, Na^+, Mg^{2+} and Ca^{2+} in various algae in relation to their growth rate. For the essential elements there is a minimum content per cell below which growth cannot occur. Above this level, increasing supply of the nutrient in the medium yields increasing cell content and growth rate, until the critical content is reached. This is the cell content of the element which gives maximum growth rate; cells can generally accumulate more of the element than this, but their growth rate is not thereby increased. This is illustrated in Fig. *6*.2.

These three domains of cell content (maintenance of viability; maintenance and growth; maintenance and luxury accumulation, with or without growth depending on the availability of other nutrients) are analagous to the concepts developed in relation to energy supply to cells (D.E. Atkinson, 1972). The extent of luxury accumulation varies with the element concerned. Thus P can be accumulated to more than three times the critical concentration, while K^+ can rarely be accumulated to more than twice this level (Gerloff and Fishbeck, 1969). This may reflect both the scarcity value of the element (P is much more likely to limit algal growth in nature than is K^+) and the ability to store the element in an innocuous form (e.g. P as polyphosphate, Ullrich, 1972; O'Kelley, 1974; any K^+ storage may lead to osmotic problems, since it cannot be converted to an osmotically inactive form). Luxury accumulation has probably been investigated inadvertently many times (e.g. Raven, 1971b). In this instance, it was found that solute accumulation was light-saturated at lower irradiances in the absence of CO_2 than in its presence. In the presence of CO_2 the influx of K^+ had a similar irradiance dependence as photosynthesis, while in its absence the surplus energy is used to accumulate elements whose "luxury accumulation" does not directly depend on C supply.

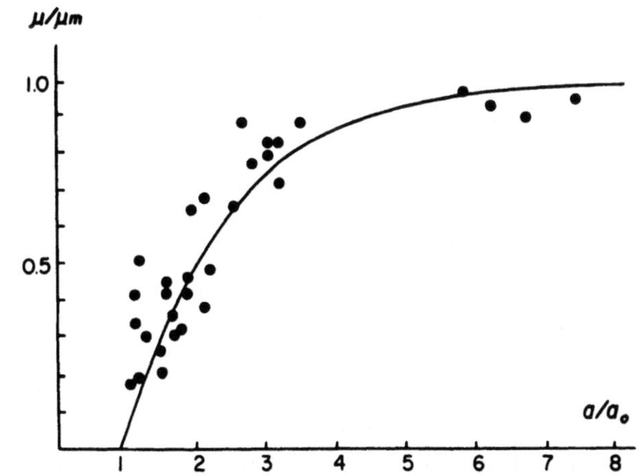

Fig. *6*.2. Growth rate of *Cyclotella nana* as a function of the total cell P. Chemostat experiment; growth expressed as the ratio of the observed growth rate (μ) to the maximum growth rate (μ_m) at saturating phosphorus. Phosphorus content expressed as the ratio of the observed content (a) to the minimum phosphorus content per cell (a_o). μ_m is 2 divisions per day, a_o is 0.9 mol P per 10^{15} cells. (From Fuhs, 1969)

4. Experimental Materials and Methods

4.1 Giant Algal Cells

4.1.1 General

The large size of these cells has enabled the experimenter to perform many manipulations which are not possible on smaller cells. The cells used here belong to a variety of taxonomic groups but most are green algae in the broad sense (ROUND, 1963, 1971): see Table *6*.2 and Fig. *6*.3. BLINKS (1936), UMRATH (1934) and GUTKNECHT and DAINTY (1968) mention genera of algal coencytes which have been used in electrophysiological research.

4.1.2 Methods

As described in Chap. *4* microelectrodes can be used to measure such electrical properties of cell membranes as potential difference, capacitance and resistance. Some of these measurements are difficult to interpret in that the location of the electrode tip is not known with certainty (*4*.1.2.1). Among the methods employed to cope with this difficulty in *Valonia* (UMRATH, 1938; VILLEGAS, 1966) are using "non-vacuolate" spores to measure the plasmalemma potential (GUTKNECHT, 1966), and recording the potential difference between the external solution and an electrode inside a cell as the latter electrode was pushed through the cytoplasm on the side at which it entered, on through the vacuole and into the cytoplasm on the other side of the cell (R.F. DAVIS, personal communication).

The use of ion-sensitive microelectrodes also allows measurement of electrochemical potential differences between compartments, and hence (if the electrical potential is known) their chemical activities. This technique has been applied to K^+, Na^+, Cl^- and H^+ (e.g. VOROBIEV et al., 1961; COSTER, 1965; VOROBIEV, 1967; ANDRIANOV et al., 1971).

Micro-pipettes inserted into cells have also been used for vacuolar perfusion (BLOUNT and LEVEDAHL, 1960; GUTKNECHT, 1967; KISHIMOTO and TAZAWA, 1965a; STRUNK, 1971), microinjection of tracers (WIDEMANN, 1968) and as micromanometers for the measurement and alteration of turgor (GREEN, 1968; STEUDLE and ZIMMERMANN, 1971, 1974; ZIMMERMANN and STEUDLE, 1974; *2*.4.2).

Physical separation of cell components is useful in determining the chemical concentration and radiochemical content of the various cell compartments. The most usual separation is into vacuole, cytoplasm (including all cell organelles) and cell wall (MACROBBIE, 1962; KISHIMOTO and TAZAWA, 1965a).

Compartmental analysis of tracer exchange (Chap. *5*) is not restricted in its use to the giant algal cells, but the use of these organisms means that it is possible to make an independent check on the method (MACROBBIE, 1970b).

The presence of high concentrations of solutes in the free space of algal cells, particularly cations in the Donnan free space of marine algae, presents difficulties in both chemical and radiochemical measurements. KESSELER (1964) discusses methods of replacing these solutes with others (different chemically or radiochemically) which will not interfere with the assay of solutes in the non-free space. In cells with short half-times of equilibrium of non-free space and bathing solution this method must be used with due regard to the respective half-times of exchange of the various compartments (SPANSWICK and WILLIAMS, 1965; SADDLER, 1970a; ROBINSON and JAFFE, 1973).

A number of experimental procedures are based on the use of the very long cylindrical cells of the Characean internodes (100 × 1 mm) or of *Acetabularia* with different areas of the cell exposed to different extracellular compartments. In this way various electrical measurements can be made (e.g. BLINKS and CHAMBERS, 1958; SMITH, 1970), as well as transcellular osmosis (KAMIYA and KURODA, 1956; DAINTY, *2*.4.2) and transcellular electro-osmosis measurements (FENSOM et al., 1967; MACROBBIE and FENSOM, 1969).

Ligation of cells after polarization of internal solutes by transcellular osmosis leads to production of cell fragments with higher or lower osmotic pressure than the original cells (KAMIYA and KURODA, 1956). Centrifugation followed by ligation can lead to the production of cell fragments which consist largely of cytoplasm (KISHIMOTO and TAZAWA, 1965a; KAMIYA and TAZAWA, 1966; SADDLER, 1970a). Cell turgor can be measured by the use of the "turgor balance" of TAZAWA (1957).

4.2 Microalgae

These are defined by their size, i.e. they are too small to be used for the microsurgery mentioned in *6*.4.1.2. The more widely used microalgae in transport studies are listed in Table *6*.2.

They have the advantage of a more rapid growth rate than the giant cells, so that large quantities of material are available for experiments. This, and the fact that they can be grown axenically, are particularly useful if metabolic or enzymic studies are to be made. Continuous culture can be used to assure uniformity of experimental material, while synchronous cultures can be used to obtain cells at particular stages of the life-cycle, so that transport studies can be made over the whole life cycle (e.g. MESZES et al., 1967; COOK, 1968; SOEDER, 1970; DOMANSKI-KADEN and SIMONIS, 1972; JEANJEAN, 1973b).

Chemostat cultures can be used to study the relationship of growth at low nutrient levels to transport kinetics (e.g. CORNER and DAVIES, 1971; FUHS et al., 1972; DROOP, 1973).

The small fraction of the cell volume of algae such as *Chlorella* which is occupied by the vacuole means that they have been used as non-vacuolate material (KANNAN, 1971), although this is probably only strictly true of the Cyanophyceae (DODGE, 1973).

A further advantage of the microalgae, which has been little applied to transport studies is the use of mutants. In *6*.2 some mutants of *Chlamydomonas* deficient in cell-wall synthesis and in contractile vacuole function were mentioned; little effort seems to have been devoted to producing mutants which have transport deficiencies, or to using the many photosynthetic mutants known in such algae as *Scenedesmus* and *Chlamydomonas* to study energy sources for active transport (BISHOP, 1973; HOPE et al., 1974).

A disadvantage in the use of the smaller micro-algae is in measuring electrical properties, although suspensions of *Chlorella* have been used to measure capacitance (HOPE, 1956) and intracellular microelectrodes have been used for measurements of electrical potential (BARBER, 1968b; LANGMÜLLER and SPRINGER-LEDERER, 1974). SKULACHEV (1971) reviews the use of the distribution of lipid-soluble cations (e.g. dimethyl dibenzyl ammonium) in estimating transmembrane potential difference in organelles. While these techniques are applicable to prokaryote and animal cells (HAROLD and PAPINEAU, 1972; CHRISTENSEN et al., 1973), they do not seem to be useful in eukaryotic algae (F.L. BENDALL, personal communication; E. KOMOR, personal communication; J.A. RAVEN, unpublished experiments).

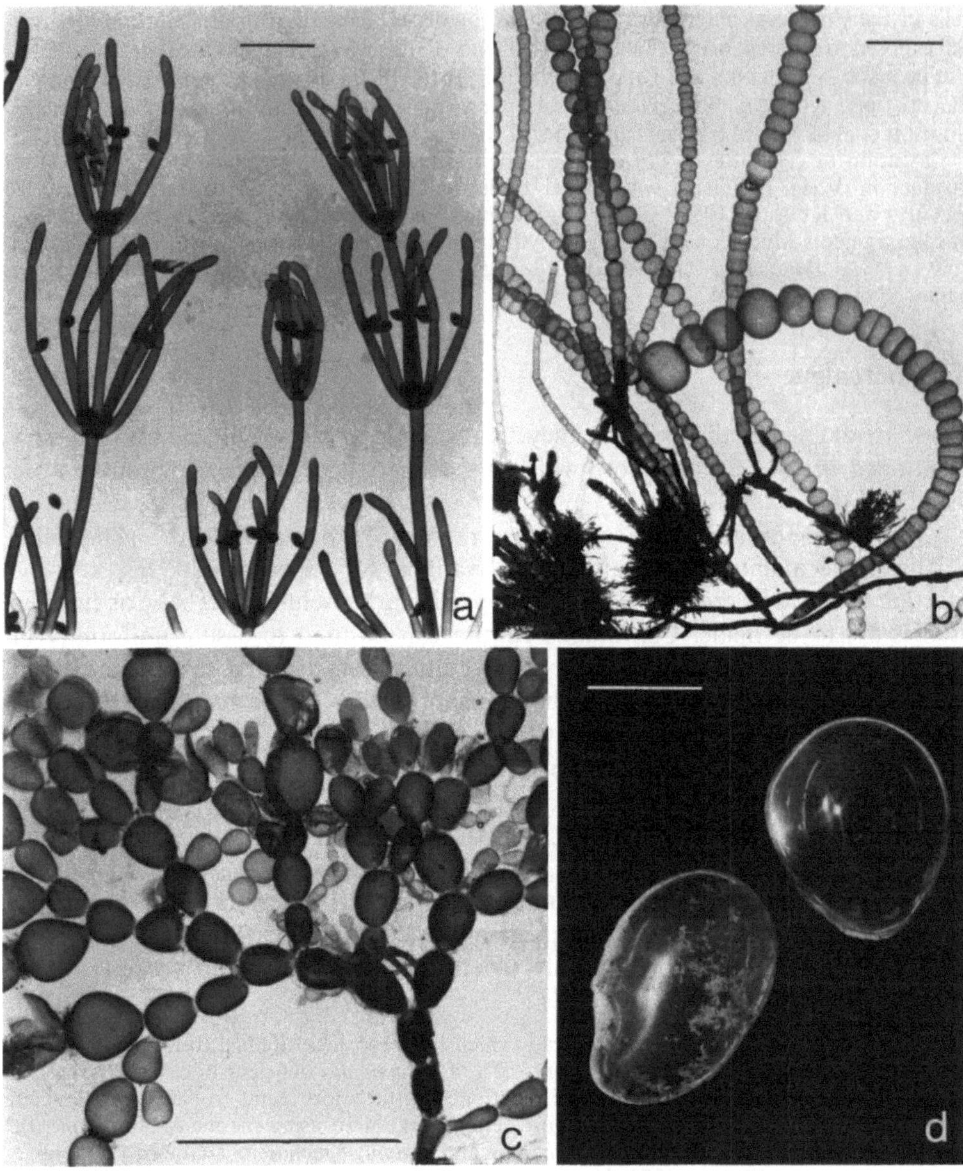

Fig. 6.3a–h. A selection of giant algal cells which are frequently used in transport studies (a) *Chara corallina*; (b) *Chaetomorpha darwinii*; (c) *Griffithsia monile*; (d) *Valonia* sp. (e) *Borgesenia* sp.; (f) *Acetabularia mediterranea*; (g) *Hydrodictyon africanum* net; (h) *Hydrodictyon africanum* larger single cells. Bars represent 1 cm. (a) to (e) by courtesy of B. LESTER, Sydney University; (f) is a photograph by Dr. SIGRID BERGER by courtesy of Max-Planck-Institut für Zellbiologie, Wilhelmshaven, Abteilung Dr. H.G. SCHWEIGER

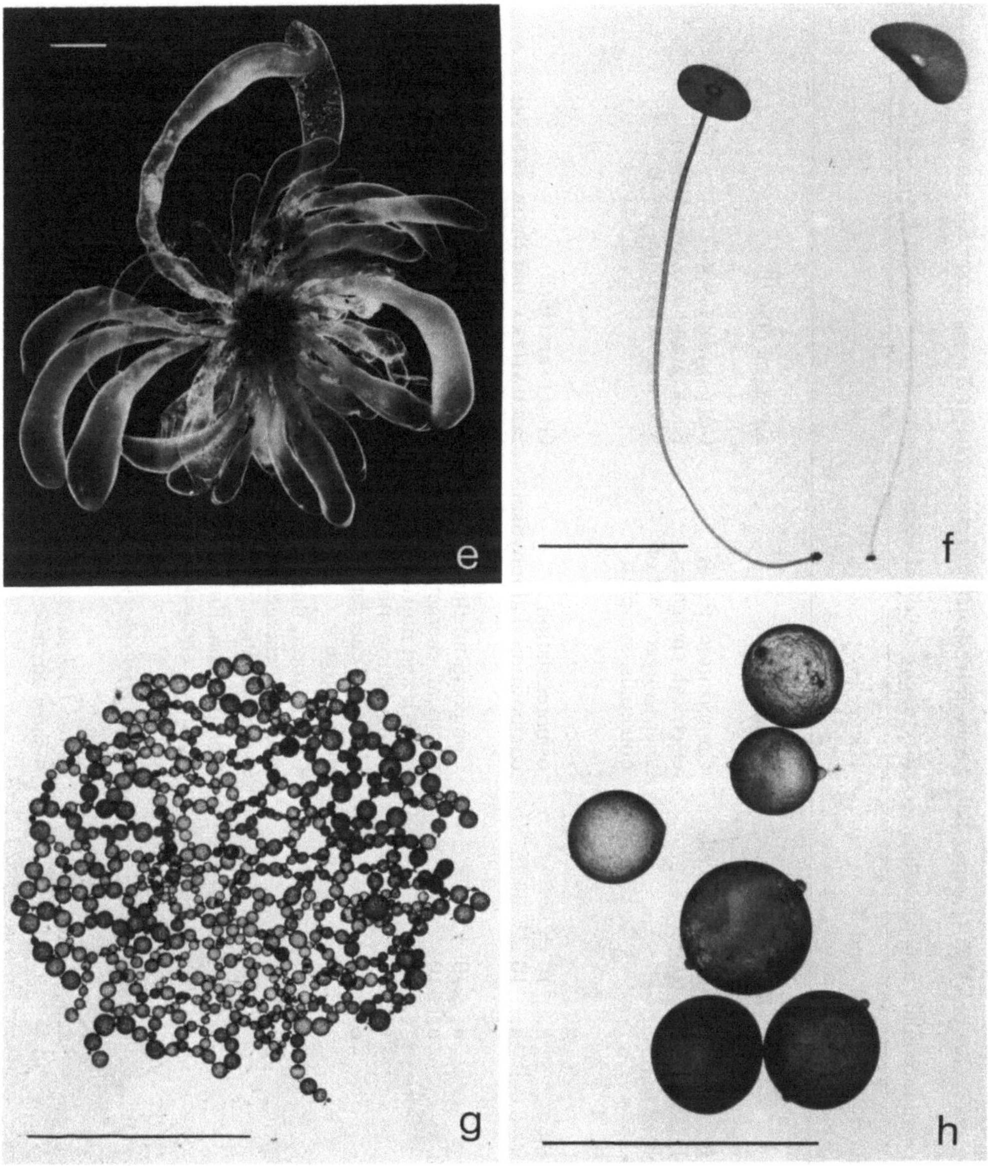

Fig. 6.3e–h. Legend see opposite page

4.3 Difficulties Associated with Metabolized Solutes and Weak Electrolytes

Free-space solutes pose a particular difficulty in measuring the intracellular concentration of metabolized solutes. In the case of inorganic phosphate there is evidence that its cytoplasmic concentration is regulated (the inorganic phosphate concentration is itself an important metabolic regulator: see RAVEN, 1974c). In the case of CO_2, glucose, nitrate and sulphate, it is likely that the time involved in removing the relevant solutes from the free space allows appreciable

Table 6.2. The taxonomy of the major algal genera used in transport studies, based on the scheme used in STEWART (1974)
Classes: Bac = Baccilariophyceae; Cha = Charophyceae; Chl = Chlorophyceae; Chr = Chrysophyceae; Cya = Cyanophyceae; Din = Dinophyceae; Eug = Euglenophyceae; Pha = Phaeophyceae; Rho = Rhodophyceae.
Habitat: F = freshwater (LIVINGSTONE, 1963); B = brackish (Intermediate between freshwater and seawater); S = seawater; S+ = more concentrated than sea water (up to 4000 mM).
Habit: u = unicellular; c = colonial; m = multicellular; uc = unicellular coenocyte; cc = colonial coenocyte; mc = multicellular coenocyte

Class	Order and Genus	Habitat (F, B, S, S+)	Habit and cell volume [m³]	Uses	Ref.
Bac	Baccilariales				
	Coscinodiscus	S	u; 10^{-10}	Composition	KESSELER (1967)
Cha	Charales				
	Chara,	F – B	mc; internodal cells	Composition, fluxes,	COLLANDER (1936)
	Nitella,	F – B	$\leqq 3 \cdot 10^{-7}$	electrical, energetics,	KAMIYA and TAZAWA (1966)
	Tolypella	F – B		micro-surgery, action	MACROBBIE (1970a, 1974)
	Nitellopsis	B		potentials, symplast	
	Lamprothamnium	B – S			
Chl	Chlamydomonadales				
	Chlamydomonas[a, c]	F – S+	u; $\leqq 10^{-15}$	osmoregulation,	GUILLARD (1960)
	Dunaliella[a, b, c]	S – S+	u; $\leqq 10^{-15}$	composition, halophytism	BOROWITZKA and BROWN (1974)
	Chlorococcales				
	Chlorella[a]	F (B, S)	u; $\leqq 10^{-15}$	Composition, fluxes,	BARBER (1968a, b)
	Ankistrodesmus[a]	F (B, S)	u; $\leqq 10^{-15}$	electrical, energetics	TAMIYA (1966)
	Scenedesmus[a]	F (B, S)	c; $\leqq 10^{-14}$	growth and nutrition	ULLRICH-EBERIUS (1973a, b)
	Hydrodictyon	F	$1\text{–}5 \cdot 10^{-8}$	Composition, fluxes,	RAVEN (1967a, 1974c)
			cc	electrical, energetics	RYBOVA et al. (1972)
				micro-surgery	
	Ulvales				
	Ulva	S	m; $\leqq 5 \cdot 10^{-15}$	Composition, fluxes,	BLACK (1971),
	Enteromorpha	(F, B) S		electrical, energetics,	BLACK and WEEKS (1972)
				osmoregulation	
	Zygnematales				
	Mougeotia	F	m; $\leqq 5 \cdot 10^{-13}$	Composition, fluxes,	STRAUSS (1967), UMRATH (1934)
	Spirogyra	F		electrical, energetics	WAGNER and BENTRUP (1973)
					WEISENSEEL and SMEIBIDL (1973)
	Cladophorales				
	Chaetomorpha	S	m; $\sim 10^{-8}$	as for *Hydrodictyon*	DODD et al. (1966)
					FINDLAY et al. (1971)
	Siphonocladales				
	Boergesenia[d]	S	uc; $\leqq 3 \cdot 10^{-7}$	as for *Hydrodictyon*;	BLINKS (1949, 1967, 1969)
	Valonia[d]	S		+ short-circuit	GUTKNECHT (1966, 1967)
	Valoniopsis[d]	S			HOPE (1971)
	Dasyclydales				
	Acetabularia[d]	S	uc; $0.5\text{–}1 \cdot 10^{-7}$	as for *Hydrodictyon*	SADDLER (1970a, b, c)

	Codiales				
	Codium[d]	S	uc; $\sim 10^{-8}$–10^{-7}	as for *Hydrodictyon*	GUTKNECHT and DAINTY (1968)
	Bryopsis[d]	S			
	Derbesiales				
	Halicystis[d, g]	S	uc; $\leqq 10^{-7}$	as for *Valonia*	GUTKNECHT and DAINTY (1968)
Chr	Ochromonadales				
	Ochromonas[a, b, c]	F	u; 10^{-16}–10^{-15}	Osmoregulation, composition	SCHOBERT et al. (1972)
Cya	Chroococcales				
	Anacystis[a]	F	u; $\sim 10^{-17}$	Composition, fluxes, energetics	DEWAR and BARBER (1973) SIMONIS et al. (1974)
Din	Peridiniales				
	Noctiluca[e]	S	u; $\sim 10^{-9}$	Composition, electrical (incl. AP)	ECKERT (1965) KESSELER (1966)
Eug	Euglenales				
	Euglena[a, b, c]	S	u; $3 \cdot 10^{-15}$	Composition, fluxes, nutrition	BUETOW (1968)
Pha	Desmarestiales				
	Desmarestia	S	m; $\sim 10^{-15}$–10^{-14}	Composition	KESSELER (1964)
	Fucales				
	Hormosira	S	m; $\sim 10^{-15}$–10^{-13}	Composition, regulation	BERGQUIST (1958)
	Pelvetia[a, b, f]	S	u; $5 \cdot 10^{-12}$	Composition, fluxes, electrical	ROBINSON and JAFFE (1973) QUATRANO (1972)
	Laminariales				
	Laminaria	S	m; $\sim 10^{-15}$–10^{-13}	Composition, symplast	BLACK (1950)
Rho	Bangiales				
	Porphyra[a]	S	m; $\leqq 10^{-13}$	Composition, fluxes, electrical	EPPLEY (1958, 1959, 1960) KAZAMA and FULLER (1970)
	Ceramiales				
	Griffithsia	S	mc; $\sim 10^{-8}$	as for *Hydrodictyon*	FINDLAY et al. (1969, 1970)
	Gigartinales				
	Gracilaria	S	m; $\sim 10^{-11}$	Composition, fluxes, electrical	GUTKNECHT (1965)
	Rhodymeniales				
	Rhodymenia	S	m; 10^{-15}–10^{-12}	Composition, fluxes, electrical	MACROBBIE and DAINTY (1958a)

[a] "non-vacuolate" (vacuole less than 20% of intracellular volume: see 6.2.). — [b] no cell wall. — [c] flagellate. — [d] Rhizoid on each cell (or coenocyte). — [e] obligate heterotroph. — [f] data apply to ovum and uncleaved zygote. — [g] see *6.5.4*: some species of *Bryopsis* and of *Halicystis* are gametophytes of *Derbesia* spp. i.e. their place in separate orders seems unjustified.

metabolism of the intracellular pools of these metabolites. The data reviewed by KESSLER (1964) suggests that a cytoplasmic concentration of 50 μM could be used in a few seconds of assimilation at the normal rate during separation of the cells from the nitrate-containing medium. These difficulties are exacerbated if the vacuole contains the metabolite under consideration. A method for overcoming this problem for inorganic phosphate is discussed by RAVEN (1974c); this method would not be applicable to a solute whose cytoplasmic content is not regulated as regards both minimum and maximum concentration.

In the absence of measurements of the internal concentration of metabolizeable solutes, it is possible to use the observed rate of assimilation, together with the K_m and V_m of the assimilation enzyme obtained from the alga concerned to compute the steady-state concentration of the solute in the cytoplasm. This value can then be substituted in the Nernst equation, and a decision can be reached as to whether active transport occurs.

Complications arise in the case of weak electrolytes. Here (e.g. phosphate, ammonia, carbon dioxide) electrochemical equilibrium is determined not only by the total concentration of the solute on each side of the membrane and the electrical PD, but also by the pH on each side of the membrane (which determines the concentration of the various ionic and molecular species of the solute) and the permeability of the membrane to the various species (ROOS, 1965; JACKSON et al., 1974). This is discussed in relation to indoleacetic acid transport in *Hydrodictyon africanum* by RAVEN (1975a).

5. Electrochemical Driving Forces in Algal Cells and Ionic Contents

5.1 Introduction

This Section considers the amounts of certain ions in algal cells in relation to the passive "driving forces" acting on the solutes due to concentration or electrical driving forces. In this way we can hope to find (thermodynamic) evidence that the ion has been actively transported into the cells, as described above in Chap. *3*. Other "metabolic" evidence from the use of treatments (e.g. inhibitors) which alter various aspects of energy metabolism will be considered below.

A point that should be made is that the overall transmembrane electrical potential is largely determined by the active and passive fluxes (and hence concentrations) of only a few ions (e.g. K^+, Cl^-, H^+). However, the potential generated in this way acts on the distribution of all other ions according to their charge and valency, concentration and permeability coefficients.

Measurement of the resting cell potential differences (ψ_{vo}) and concentration of K^+, Na^+ and Cl^- in tissue and solution are collected together in Table *6*.3. In some cases concentrations in the cytoplasm and vacuoles of giant algae have been determined and are given in Table *6*.4, together with measurements of ψ_{co} and ψ_{vc}.

5.2 Electrical Potential Differences between Compartments

Most of the measurements of electrical potential difference have been made on vacuolate cells, and most of the measurements are of the PD between the vacuole and the bathing solution (ψ_{vo}). This is generally inside-negative, the major exceptions being certain siphonaceous green algae (e.g. *Valonia, Chaetomorpha*) where the potential is generally slightly inside-positive (Tables *6*.3 and *6*.4).

Measurements of the potential difference between the cytoplasm (ψ_{co}) and the bathing solution are more difficult to make; the measured potential is invariably inside-negative in the plant's natural medium (Table *6*.4). This means that there is a large vacuole-positive PD across the tonoplast in *Valonia* and *Chaetomorpha*, with usually a smaller vacuole-positive PD in other algal cells.

5.3 Chloride

Cl^- is generally at a higher electrochemical potential in both the cytoplasm and the vacuole than in the bathing solution (Tables *6*.3 and *6*.4). However, in the cytoplasm of some marine algae the evidence is not very strong (GUTKNECHT and DAINTY, 1968; but see SADDLER, 1970a), and the internal concentration is low in the freshwater microalga *Chlorella* (SCHAEDLE and JACOBSEN, 1965; BARBER, 1968a). Thus there is, in general, active Cl^- transport at the plasmalemma of algae, and there may also be active transport at the tonoplast, both directed inwards. The tonoplast pump may be related to cytoplasmic volume regulation (GUTKNECHT and DAINTY, 1968; MACROBBIE, 1970a).

In *Valonia, Valoniopsis* and *Chaetomorpha* Cl^- is apparently close to electrochemical equilibrium in the vacuole which generally has a small positive potential with respect to the bathing solution (Table *6*.3). However, short-circuit studies on *Valonia ventricosa* (GUTKNECHT, 1967) suggest that Cl^- influx can be active. The electrochemical status of the Cl^- in the cytoplasm is also equivocal; GUTKNECHT (1966) and FINDLAY et al. (1971) suggest that Cl^- distribution between bathing medium, cytoplasm and vacuole is passive; this would require inhomogeneous Cl^- distribution in the cytoplasm. An even more inhomogeneous distribution is required if passive Cl^- distribution between cytoplasm and medium is to be compatible with the measurements of VILLEGAS (1966).

The high concentration of Cl^- salts in the chloroplasts of a number of giant algal cells (MACROBBIE, 1964, 1970a) suggests that the osmotic pressure of these organelles may be greater than that of the free cytoplasm and of the vacuole of these cells. This is borne out by the great osmotic tolerance of *Acetabularia* and *Codium* chloroplasts (SHEPPARD and BIDWELL, 1973; TRENCH et al., 1973).

A very interesting situation has been found in developing fucoid eggs (BENTRUP, 1970; ALLEN et al., 1972; WEISENSEEL and JAFFE, 1972; ROBINSON and JAFFE, 1973). Here the unfertilized egg lacks active Cl^- influx, and may even have active Cl^- extrusion. After fertilization the wall-less cell develops a cell wall, and the cell contents (formerly isotonic with sea water) become hypertonic to sea water. This is paralleled by the advent of an active Cl^- influx at the plasmalemma which brings about the increased internal concentration of solutes (probably in conjunction with active K^+ transport). The relationship of active Cl^- influx and the presence of cell walls has been discussed by DAINTY (1962).

5.4 Sodium and Potassium

K^+/Na^+ inside the cell is generally greater than that in the medium (Tables *6*.3, *6*.4, *6*.5). Marine siphonaceous green algae have a number of representatives with a low discrimination in favor of K^+ in their vacuoles (Tables *6*.3, *6*.4, *6*.5). At least in the case of *Bryopsis plumosa* the K^+/Na^+ in the cytoplasm is much higher than that in the vacuole (GUTKNECHT and DAINTY, 1968). In the case of both *Halicystis* and *Bryopsis* some species have a low K^+/Na^+ ratio in their vacuole, while other species have a high K^+/Na^+ ratio. It would be of interest to see if the K^+/Na^+ ratio in the alternate phase of the life-history of these algae (*Derbesia*

Table 6.3. Concentrations of K^+, Na^+, Cl^- in the solutions bathing algal cells and in the vacuole of these cells, and the potential difference between the bathing solution and the vacuole. An asterisk denotes that although the cell investigated has (as have all the cells in this Table) 50% or more of their cell volume occupied by vacuole, the cell is too small (see Table 6.2.) to allow separation of the vacuole by micro-dissection. The value of ψ_{vo} quoted in these cases may be in error in that the electrode could have been in the cytoplasm. The symbols in brackets represent the direction of active transport suggested by the Nernst equation; (e) means that the ion is close to electrochemical equilibrium between the phases specified; (i) means that there is an active influx into the vacuole (whole cell), (o) means that there is an active efflux from the vacuole (whole cell). A dagger (†) denotes that the vacuolar content has been estimated by flux analysis.

Further data on ψ_{vo} in various algae without corresponding ion content data may be found in UMRATH (1934) and BLINKS (1949). Values of ion content of algae are reviewed by COLLANDER (1936), KESSELER (1964) and GUTKNECHT and DAINTY (1968). See also Tables 6.4 and 6.5

Alga	Concentration in Solution (mM)			PD ψ_{vo} (mV)	Concentration in Vacuole (mM)			Ref.
	K^+	Na^+	Cl^-		K^+	Na^+	Cl^-	
Chara australis[a]	0.1	1.0	1.6	−160	70 (e)	50 (o)	110 (i)	HOPE and WALKER (1960)
Chara braunii	0.015	0.250	0.250	−173	128 (i)	3 (o)	124 (i)	ODA (1961)
Chara globularis	0.046	0.15	0.04	−181	65 (e)	66 (o)	112 (i)	GAFFEY and MULLINS (1958)
Lamprothamnium succintum	7	301	353	−100	164 (e)	368 (o)	532 (i)	KISHIMOTO and TAZAWA (1965b)
Nitella axilaris	0.075	0.2	1.2	−130 to −200	114 (e)	53 (o)	94 (i)	KISHIMOTO (1964) SAITO and SENDA (1973a)
Nitella clavata	0.1	3.0	4.1	−110	77 (i)	33 (o)	116 (i)	BARR (1965); BARR and BROYER (1964)
Nitella flexilis	0.1	0.2	1.3	−155	80 (i)	28 (o)	136 (i)	KISHIMOTO and TAZAWA (1965a)
Nitella translucens	0.1	1.0	1.3	−122	75 (i)	65 (o)	150–170 (i)	MACROBBIE (1962) SPANSWICK (1964), SPANSWICK and WILLIAMS (1965)
Nitellopsis obtusa	0.65	30	35	−122	113 (e)	54 (o)	206 (i)	MACROBBIE and DAINTY (1958a), FINDLAY (1970)
Tolypella prolifera	0.4	1.0	1.4	−155	80 (e)	28 (o)	136 (i)	SMITH (1968), LARKUM (1968)
Acetabularia mediterranea	10	470	550	−174	355 (o)	65 (o)	480 (i)	SADDLER (1970a)
*Bryopsis hypnoides**	11	467	483	− 50	280 (i)	345 (o)	705 (i)	BLINKS (1949), KESSELER (1964)
Bryopsis plumosa	10	483	563	− 50 to − 80	7 (o)	529 (o)	605 (i)	GUTKNECHT and DAINTY (1968), MUNDAY (1972)

Chaetomorpha darwinii	13	500	523	+ 5 to − 30	540 (i)	25 (o)	600 (e)	Dodd et al. (1966), Findlay et al. (1971)
Codium tomentosum	11	485	590	− 80	15 (o)	656 (o)	708 (i)	Gutknecht and Dainty (1968)
*Enteromorpha intestinalis**†	11	460	560	− 42	450 (i)	260 (o)	370 (i)	Black and Weeks (1972)
Halicystis osterhoutii	12	498	580	− 68	6.4 (o)	557 (o)	603 (i)	Blinks (1949), Gutknecht and Dainty (1968)
Halicystis ovalis	12	488	523	− 80	337 (e)	257 (o)	543 (i)	Blount and Levedahl (1960)
Hydrodictyon africanum	0.1	1.0	1.3	− 90	40 (i)	17 (o)	38 (i)	Blinks (1958), Raven (1967a)
Hydrodictyon petanaeforme[b]	0.15	1.3	0.9	− 55	76 (i)	4 (o)	55 (i)	Blinks (1949), Blinks and Nielsen (1940)
Hydrodictyon reticulatum[b]*	1.2	0.5	0.5	− 79	166 (i)	5 (o)	54 (i)	Janacek and Rybova (1966)
Mougeotia sp.*†	1	1	1	− 51	40–50 (e)	40–50 (e)	40–50 (i)	Wagner and Bentrup (1973) Weisenseel and Smeibidl (1973)
*Ulva lactuca**	10	470	–	− 60	300–410 (i)	40–52 (o)	–	West and Pitman (1967a) Aikman (1969)
Valonia macrophysa	13	500	523	+ 5 to − 10	509 (i)	113 (o)	622 (e)	Umrath (1938) Blinks (1949) Steward and Martin (1937)
Valonia ventricosa	12	508	596	+ 12	625 (i)	44 (o)	643 (e)	Gutknecht (1966) Aikman and Dainty (1966)
Valoniopsis sp.	10	500	580	− 5	50 (i)	620 (o?)	660 (e?)	Hope (1971)
Noctiluca miliaris	11	485	532	− 25	34 (e)	414 (o)	496 (i)	Eckert (1965) Kesseler (1966)
*Gracilaria foliifera**	11	471	532	− 81	660 (i)	66 (o)	462 (i)	Gutknecht (1965)
Griffithsia sp.	11	485	590	− 52 to − 54	500–600 (i)	30–90 (o)	600–650 (i)	Findlay et al. (1969)
Rhodymenia palmata	11	485	590	−165	(800) (i?)	(25) (o)	(600) (i)	MacRobbie and Dainty (1958b)

[a] Wood and Imahori (1965) have reduced *Ch. australis* to *Ch. corallina* var. *australis*. Kishimoto (1959) describes electrical characteristics of *Ch. corallina* sens. strict.
[b] Marchant and Pickett-Heaps (1972) suggest that *H. petanaeforme* is a form of *H. reticulatum*.

Table 6.4. Concentration of ions in the vacuole and cytoplasm of various algae. This table contains both algae which have a large central vacuole and which are large enough to allow separation of cytoplasm and vacuole and "non-vacuolate" algae (*Chlorella*, *Porphyra*: see 6.2 and Table 6.2). The cytoplasm contents for all but the charophytes include the chloroplasts; for the charophytes the values are for the streaming cytoplasm. MACROBBIE (1970a) tabulated further data on giant algal cells. Square brackets indicate estimates of activity using ion-sensitive microelectrodes. The symbols in brackets represent the direction of active transport suggested by the Nernst equation. (e) means electro-chemical equilibrium at the plasmalemma (cytoplasm column), or tonoplast (vacuole column); (i) means active influx (ϕ_{oc} or ϕ_{cv}); (o) means active efflux (ϕ_{co} or ϕ_{vc}). *Valonia macrophysa* (UMRATH, 1938) has a negative value of ψ_{co}, as with *V. ventricosa* and *Chaetomorpha*

Alga	Ion	Concentration (mM)			Potential difference (mV)		Ref.
		Solution	Cytoplasm	Vacuole	ψ_{co}	ψ_{vc}	
Chara australis	K^+	0.1	[115] (e)	70 [48] (e)			HOPE and WALKER (1960),
	Na^+	1.0	—	50 (o)			COSTER (1965),
	Cl^-	1.6	[10] (i)	110 (i)	−170	+18	VOROBIEV (1967),
	Ca^{2+}	0.5	—	2.6 (o)			COSTER and HOPE (1968),
	SO_4^{2-}	1.3	—	17 (i)			ROBINSON (1969a)
Nitella clavata	K^+	0.1	—	75–83 (i)	(ψ_{vo})		BARR (1965),
	Na^+	3.0	—	34–36 (o)	−120 to		BARR and BROYER (1964)
	Cl^-	4.1	—	120–124 (i)	−160		
	Ca^{2+}	1.3	—	12–19 (o)			
	Mg^{2+}	3.0	—	11–22 (o)			
	SO_4^{2-}	0.67	—	12–20 (i)			HOAGLAND and DAVIES (1923, 1929)
	$H_2PO_4^-$	0.0008	—	1.7–2.8 (i)			
	NO_3^-	1.8–1.9	—	3.3–5.3 (i)			
	H^+	10^{-6}–10^{-4}	—	10^{-3} (o)			
Nitella flexilis	K^+	0.1	125 (e)	80 (e)			KISHIMOTO and TAZAWA (1965)
	Na^+	0.2	5 (o)	28 (i)	−170	+16	
	Cl^-	1.3	36 (i)	136 (e)			
Nitella translucens	K^+	0.1	119 (i)	75 (e)			MACROBBIE (1962, 1964),
	Na^+	1.0	14 (o)	65 (i)	−140	+18	SPANSWICK and WILLIAMS (1965)
	Cl^-	1.3	65 (i)	150–170 (e)			
	Ca^{2+}	0.1	8 (o)	12 (e)			
Acetabularia mediterranea	K^+	10	400 (o)	355 (e)	−174	0	SADDLER (1970a)
	Na^+	470	57 (o)	65 (e)			
	Cl^-	550	480 (i)	480 (e)			
	SO_4^{2-}	55	—	8.8 (i)			
	$(COO)_2^{2-}$	0	—	110			

Bryopsis plumosa	K^+	10	210 (e)	7 (o)	− 70	0	GUTKNECHT and DAINTY (1968), MUNDAY (1972)
	Na^+	483	470 (o)	529 (i)			
	Cl^-	563	475 (i)	605 (i)			
Chaetomorpha darwinii	K^+	13	541 (i)	540 (i)	− 72 to − 75	+46 to +80	DODD et al. (1966), FINDLAY et al. (1971)
	Na^+	500	25 (o)	25 (i)			
	Cl^-	523	601 (e)	575 (e)			
Chlorella pyrenoidosa	K^+	6.5	108 (i)	–	− 40	–	BARBER (1968b) SHIEH and BARBER (1971)
	Na^+	1.0	1.1 (o)	–			
	Cl^-	1.0	1.3 (i)	–			
	H^+	$2 \cdot 10^{-4}$	$2 \cdot 10^{-4}$ (o)	–			
Halicystis ovalis	K^+	12	–	337 (e)	−80	0	HOLLENBERG (1932), BLOUNT and LEVEDAHL (1960)
	Na^+	488	–	257 (o)			
	Cl^-	523	–	543 (i)			
	SO_4^{2-}	29	–	1.4 (i)			
	H^+	10^{-5}	–	$3 \cdot 10^{-3}$ (o)			
Hydrodictyon africanum	K^+	0.1	(93) (i)	40 (e)	−116	+26	RAVEN (1967a, 1974c and unpublished)
	Na^+	1.0	(51) (o)	17 (i)			
	Cl^-	1.3	58 (i)	38 (e)			
	$H_2PO_4^-$	0.1	1–2 (i)	1–2 (e)			
	SO_4^{2-}	0.5	–	5–8 (i)			
	H^+	10^{-3}–10^{-5}	–	10^{-3} (o)			
Hydrodictyon petanae-forme	K^+	0.019	–	76 (i)	(ψ_{vo}) −55		BLINKS and NIELSEN (1940), BLINKS (1949)
	Na^+	1.3	–	4 (o)			
	Cl^-	0.90	–	55 (i)			
	Ca^{2+}	1.1	–	1.8 (o)			
	SO_4^{2-}	1.1	–	8.2 (i)			
	H^+	10^{-5}	–	$1–3 \cdot 10^{-3}$ (o)			
Valonia ventricosa	K^+	11	434 (i)	625 (i)	−71	+88	GUTKNECHT and DAINTY (1968)
	Na^+	485	40 (o)	44 (i)			
	Cl^-	590	138 (e)	643 (e)			
Pelvetia fastigiata (uncleaved walled zygotes)	K^+	11	400 (i)	–	−70 to −80		ALLEN et al. (1972)
	Na^+	485	20 (o)	–			
	Cl^-	590	322 (i)	–			
	Ca^{2+}	12	4 (o)	–			
	Mg^{2+}	57	25 (o)	–			
Porphyra perforata	K^+	11	480 (i)	–	−42		EPPLEY (1958), GUTKNECHT and DAINTY (1968)
	Na^+	485	51 (o)	–			
	Cl^-	590	81 (e)	–			
	NO_3^-	$1–10 \cdot 10^{-3}$	80–140 (i)	–			
	$H_2PO_4^-$	$1–10 \cdot 10^{-3}$	2.5 (i)	–			

Table 6.5. Extracellular and intracellular K^+, Na^+ and Cl^- values for some algae for which values of ψ_{vo} or ψ_{co} are not available; also tabulated is the K^+/Na^+ selectivity (K^+/Na^+ inside divided by K^+/Na^+ outside). (v) means vacuole, (c) means cytoplasm (including chloroplasts), (w) means whole cell.

Further data may be found in COLLANDER (1936), KESSELER (1964, 1965), STRAUSS (1967), HAYWARD (1970), LAQUERBE et al. (1970), GOLDMAN et al. (1972) and ONCUTT and PATTERSON (1973)

Alga	External concentration (mM)			Internal concentration (mM)			Selectivity	Ref.
	K^+	Na^+	Cl^-	K^+	Na^+	Cl^-		
Chara ceratophylla	1.4	68	80	69 (v)	148 (v)	232 (v)	23	COLLANDER (1936)
	0.6	31	36	61 (v)	126 (v)	208 (v)	25	
	0.13	0.81	0.78	85 (v)	99 (v)	194 (v)	5.4	
	0.04	0.21	0.13	77 (v)	84 (v)	176 (v)	4.8	
Chaetomorpha linum	8.8	410	435	743 (w)	44 (w)	762 (w)	785	KESSELER (1964)
Chlamydomonas moewusii	6.6	0.43	–	20 (c)	1.4 (c)	–	1	RONKIN and BURETZ (1960)
Chlamydomonas sp.	2.2	1,000	–	70 (c)	100 (c)	–	320	VAN AUKEN and MCNULTY (1969)
	–	–	2,750	–	–	90		
	2.2	5,000	–	110 (c)	800 (c)	–	310	
Chlamydomonas sp.	–	1,710	1,710	30–80 (c)	180–300 (c)	70 (c)		OKAMOTO and SUZUKI (1964)
Chlorella pyrenoidosa	3.0	4.0	–	142 (c)	4.3 (c)	–	44	SHIEH and BARBER (1971)
	0.04	7.0	–	28 (c)	54 (c)	–	91	
Codium fragile	8.8	410	435	16 (w)	475 (w)	495 (w)	1.6	KESSELER (1965)
Scenedesmus obtusiusculus	6.09	6.09	–	70 (c)	35 (c)	–	2	ERDEI and CSEH (1971)
	6.09	0.1	–	65 (c)	30 (c)	–	0.04	
	0.1	6.09	–	20 (c)	40 (c)	–	31	
Desmarestia viridis	8.8	410	435	279 (w)	111 (w)	89 (w)	110	KESSELER (1965)
Hormosira banksii	11	485	–	(650) (w)	(300) (w)	–	95	BERGQUIST (1958)
Coscinodiscus wailesii	9.5	463	525	461 (w)	125 (w)	345 (w)	180	KESSELER (1967)
Ceramium rubrum	8.8	410	435	515 (w)	31 (w)	575 (w)	775	KESSELER (1964)
Polysiphonia urceolata	8.8	410	435	644 (w)	96 (w)	583 (w)	314	KESSELER (1964)
Anacystis nidulans	21	1.3	–	149 (c)	5 (c)	–	18	DEWAR and BARBER (1973)

spp.: PAGE, 1970; RIETMA, 1971) resembles that in the corresponding *Halicystis* or *Bryopsis*. In the charophytes the intracellular K^+/Na^+ ratio decreases with cell age (MACROBBIE, 1962; BARR, 1965; STRAUSS, 1967). The occurrence of a higher K^+/Na^+ ratio in the cytoplasm than in the vacuole is also found in a number of algae with a high whole-cell K^+/Na^+ (Tables 6.3, 6.4, 6.5).

Electrochemical data (Tables 6.3 and 6.4) suggest active Na^+ extrusion at the plasmalemma of algal cells, while in a number of cases active Na^+ transport from the cytoplasm to the vacuole is also likely (e.g. freshwater coenocytes, marine coenocytes such as *Bryopsis plumosa*). Active Na^+ transport into the vacuole may be related to the regulation of cytoplasmic volume, or even to Na^+ extrusion *via* contractile vacuoles if both vacuoles and contractile vacuoles originated from the golgi apparatus (SCHNEPF and KOCH, 1967).

The situation with regard to K^+ transport is much less clear. In many cases (Tables 6.3 and 6.4) K^+ is close to electrochemical equilibrium at the plasmalemma, and it may well be passively distributed there. In a number of cases ψ_K is less than ψ_M for the plasmalemma, and active K^+ efflux at the plasmalemma is likely (e.g. SADDLER, 1970a, b; SPANSWICK, 1972). In rather more cases there is evidence for active K^+ influx at the plasmalemma (Tables 6.3 and 6.4). However, in number of cases, active K^+ influx is postulated on the basis of rather small excesses of ψ_K over ψ_M; here cytoplasmic inhomogeneity of K^+, or the use of the more appropriate K^+ activity rather than K^+ concentration (VOROBIEV, 1967) would weaken the thermodynamic argument for active K^+ influx (HIGINBOTHAM, 1973). K^+ appears to be passively distributed at the tonoplast of most algae, although in *Griffithsia, Chaetomorpha* and *Valonia* there is probably active K^+ transport from cytoplasm to vacuole; (Tables 6.3 and 6.4); this may be important in cytoplasmic volume regulation.

5.5 Metabolically Incorporated Ions

5.5.1 Phosphate and Sulfate

For phosphate and sulfate in giant algae there is evidence that net influx of the anion occurs even when the electrochemical activity in the vacuole is higher than that in the medium (Table 6.4; BLINKS, 1949; KESSELER, 1964; SMITH, 1966; GUTKNECHT and DAINTY, 1968; ROBINSON, 1969a; SADDLER, 1970a; RAVEN, 1974a, and unpublished experiments on sulfate in *Hydrodictyon africanum*). Thus active transport must occur somewhere between the bathing solution and the vacuole. In *Hydrodictyon africanum* it is highly likely that active $H_2PO_4^-$ transport occurs at the plasmalemma; the same is true of *Porphyra* (EPPLEY, 1958; GUTKNECHT and DAINTY, 1968).

If the inside-negative electrical potential found by BARBER (1968a, b) can be generalized to the cytoplasm of other green micro-algae and *Euglena*, then from measured internal and external concentrations the Nernst equation suggests active influx of sulfate (WEDDING and BLACK, 1960; KYLIN, 1964; VALLÉE and JEANJEAN, 1968a) and phosphate (MIYACHI and TAMIYA, 1961; BLUM, 1965; KANAI and SIMONIS, 1968; BASSHAM and KRAUSE, 1969; DOMANSKI-KADEN and SIMONIS, 1972). Gross inhomogeneities of cytoplasmic inorganic phosphate distribution have been ruled out as a complication in *Chlorella ellipsoidea*, where chloroplasts have been isolated and shown to have similar inorganic phosphate concentrations to the whole cell (MIYACHI and TAMIYA, 1961; HASE et al., 1963). In the examples quoted above, cytoplasmic phosphate concentrations are 1–8 mM, and cytoplasmic sulfate is about 1 mM. However, in two species of *Chlorella* no intracellular sulfate was detected (HASE et al., 1961; JOHNSTON and SCHMIDT, 1963). This may be due to the experimental artifact discussed in 6.4.3.

5.5.2 Nitrate

There is evidence from marine algae (and *Nitella clavata:* Table 6.4) that NO_3^- is present in the cytoplasm (*Porphyra*) or vacuole at a higher electrochemical potential than in sea water (JACQUES and OSTERHOUT, 1938; GUTKNECHT and DAINTY, 1968; FRAZISKET, 1973; cf. EPPLEY and ROGERS, 1970). In fresh-water algae there is often negligible NO_3^- measurable inside the cell when NO_3^- is known to be the N source for growth (e.g. BLINKS and NIELSEN, 1940; AHMED and MORRIS, 1968). Similar results were obtained using chlorate as an analog of NO_3^- (TROMBALLA and BRODA, 1971).

It is likely that the reduction of endogenous NO_3^- by the still active nitrate reductase system during the separation of cells from the medium can account for the apparent absence of NO_3^- from algal cells in the cases mentioned above (6.4.3). Nevertheless, it is likely that active NO_3^- transport can occur in these cells since cytoplasmic NO_3^- concentration should be at least 0.5 mM in order to account for *in vivo* nitrate reduction rates in terms of extractable enzyme activity (EPPLEY and ROGERS, 1970). In view of the very low nitrate levels found in many natural waters (around 1 μM: CORNER and DAVIES, 1971), this internal concentration could not be maintained by passive nitrate distribution, since net passive influx can only occur if the internal concentration is lower than the Nernst equilibrium concentration (cf. RAVEN, 1970). Thus it is likely that active transport of nitrate into cells of green microalgae occurs during growth. However, the reasoning is indirect, and would be weakened if it were discovered that the enzyme were not being extracted under optimal conditions. Further effort should be put into direct determinations of intracellular nitrate concentrations.

5.5.3 Bicarbonate

Bicarbonate is a weak electrolyte and also is metabolized. The complications related to nitrate, phosphate and sulfate transport studies are compounded for HCO_3^- by the requirement for knowing internal as well as external pH (6.4). As has been discussed by RAVEN (1970), photosynthesis at low external pH values involved entry of CO_2 as the undissociated molecule (see 6.7), while at high external pH values photosynthesis requires an active influx of HCO_3^- at the plasmalemma. Some algae appear to lack the ability to use HCO_3^- (RAVEN, 1970; 1974b; GANF, 1972). Measurements of the total internal and external inorganic carbon concentration, and of internal pH at a variety of external pH values are required in order to test these conclusions fully.

5.5.4 Ammonium

The complications associated with determining whether active or passive transport occurs for NH_4^+ resemble those encountered for HCO_3^-. As with HCO_3^-, at normal cytoplasmic pH values of about 7 (RAVEN and SMITH, 1973, 1974) the dissociated and associated form of NH_3 coexist. In a number of algae the internal total NH_3 concentration is greater than that outside (e.g. KESSELER, 1967; EPPLEY and ROGERS, 1970; PRIBIL and KOTYK, 1970). In most cases our knowledge of the pH and the electrical PD are insufficient to allow a decision to be made between active and passive distribution. In *Noctiluca,* where most data is available, there is the complication of NH_3 production from food in the holozoic organism. In *Valonia,* passive NH_3 distribution and internal and external pH values account for the observed distribution of NH_3 and NH_4^+ (OSTERHOUT and DORCAS, 1925). These experiments refer either to vacuolar NH_3 concentrations (the *Ditylum* value must be vacuolar, since such a concentration in the cytoplasm would repress nitrate reductase: EPPLEY and ROGERS, 1970), or involve very high external concentrations (PRIBIL and KOTYK, 1970).

Using the method applied to NO_3^- and HCO_3^-, it is clear that in *Ditylum brightwelli* the cytoplasmic ammonia plus ammonium concentration during steady-state assimilation must be much higher than that found in sea water. However, insufficient electrical, pH and permeability data are available to determine if this is due to passive or active transport.

5.5.5 Protons (and Hydroxyl Ions)

The measurement of internal, and particularly cytoplasmic, pH in algal cells is not easy. What evidence is available suggests that the cytoplasmic pH is about 6.5–7.0 (discussed by RAVEN and SMITH, 1973, 1974; see also *3*.3.4 and *12*.2.2), although DODD and BIDWELL (1971) suggest that the cytoplasmic pH is about 8 in *Acetabularia*. Marine algae which have values of ψ_{co} of −60 to −100 mV (see Tables *6*.3 and *6*.4) could have H^+ at electrochemical equilibrium between seawater and the cytoplasm at about pH 7. In either case H^+ distribution would involve active transport at the plasmalemma if, as RAVEN and SMITH (1973, 1974) have argued, the maintenance of cytoplasmic pH is important. Active H^+ extrusion at the plasmalemma is required if P_{H^+} is high enough for there to be appreciable H^+ fluxes in response to an inwardly-directed H^+ electrochemical gradient, and in growing cells with NH_4^+ as the nitrogen source entering the cell (RAVEN and SMITH, 1973, 1974). Thus active H^+ extrusion at the plasmalemma is common in algal cells. Active OH^- extrusion (H^+ influx) is required when NO_3^- or HCO_3^- are being assimilated at high pH, or when a finite P_{H^+} is combined with an inwardly directed electrochemical potential gradient for OH^- (RAVEN and SMITH, 1974).

Measurements of vacuolar pH in algal cells shows that the pH of the vacuole is usually less than 6, and in the extreme case of *Desmarestia* the pH is as low as 1. Combination with electrical potential data suggests that, in general, there must be active H^+ transport from the cytoplasm to the vacuole (e.g. OSTERHOUT, 1936; HOAGLAND and DAVIES 1929; GUTKNECHT and DAINTY, 1968; Tables *6*.3 and *6*.4).

5.6 Divalent Cations

The distribution of Ca^{2+} and Mg^{2+} suggests that their influx into algal cells is passive (Table *6*.4; OSTERHOUT, 1936; KESSELER, 1964; GOLDMAN et al., 1972). However, it is not clear that the low electrochemical activity of these ions in algal cells is due to low permeability alone, or requires an active efflux (WALKER, 1957; SPANSWICK and WILLIAMS, 1965). Large fluxes of Ca^{2+} must be involved in the intracellular production and subsequent extrusion of coccoliths (an organic matrix plus $CaCO_3$) in certain members of the Haptophyceae (PAASCHE, 1968).

6. Fluxes of Electrolytes

6.1 Introduction

Fluxes of electrolytes have been estimated for a few algae using efflux analysis as described elsewhere in Chap. *5* and in other cases from measurements of tracer uptake or net chemical change. A distinction between active and passive components is an important feature of interpreting these fluxes. Criteria distinguishing between active and passive fluxes are discussed in Chap. *3* (see also RAVEN, 1975b; *4.2.5*).

The two major techniques used to distinguish these various components are the manipulation of metabolism, which is thought to alter mainly active fluxes, and manipulation of electrochemical gradients which alters passive fluxes. However,

Table *6.6*a. Effects of treatments which alter rates of electron flow on active anion influx in *H. africanum*

Treatment			Electron flow occurring	Active anion influxes (tracer) in *H. africanum*	
Light wavelength (nm)	Gas phase	Inhibitor		Cl^-	$H_2PO_4^-$
400–730[a]	Air[c]	None	respiratory, cyclic and non-cyclic photosynthetic	100	100
None	Air[c]	None	respiratory	15	67
None	N_2[d]	None	None (fermentation)	6	17
None	Air[c]	Cyanide 1 mM[g]	None (fermentation)	7	15
None	Air[c]	Antimycin 100 μM[h]	None (fermentation)	9	18
700–730[b]	N_2[d]	None	cyclic photosynthetic	15	92
400–730[a]	N_2[d]	DCMU[i] 0.1 μM	cyclic photosynthetic	36	107
700–730[b]	N_2[d]	Antimycin 100 μM[h]	None (fermentation)	5	9
400–730[a]	N_2+5% CO_2[e]	Antimycin 100 μM[h]	non-cyclic photosynthetic (CO_2-linked)	86	78
400–730[a]	N_2+21% O_2[f]	Antimycin 100 μM[h] Cyanide 1 mM[g]	non-cyclic photosynthetic (O_2-linked, pseudocyclic)	75	65

[a] Absorbed by all photosynthetic pigments; permits the operation of both photosystems, i.e. both non-cyclic and cyclic electron transport and photophosphorylation can occur (phytochrome and other photoreceptors are not involved: RAVEN, 1969a, 1971e; WAGNER and BENTRUP, 1973).
[b] Absorbed by long wavelength form of chlorophyll a; permits only photosystem I to operate, i.e. cyclic but not non-cyclic electron transport and photophosphorylation.
[c] Allows all electron transport sequences to occur (CO_2 as acceptor for non-cyclic; O_2 as electron acceptor for respiration, O_2-linked non-cyclic).
[d] Permits only cyclic electron flow and phosphorylation (unless an electron acceptor like NO_3^- is present: ULLRICH, 1971, 1974; ULLRICH-EBERIUS, 1973b).
[e] Permits CO_2-linked non-cyclic; could also be cyclic, and O_2-requiring reactions (respiration, O_2-linked non-cyclic) from O_2 produced.
[f] Permits O_2-linked non-cyclic (pseudocyclic) and respiration; also cyclic, and CO_2-linked non-cyclic from CO_2 produced.
[g] Inhibits respiratory electron flow, and CO_2- and NO_3^--linked non-cyclic.
[h] Inhibits respiratory and cyclic.
[i] Inhibits photosystem II, and hence non-cyclic, but not cyclic.

(Data from RAVEN, 1967b, 1969a, b; 1971d, e, 1974d, and unpublished.)

Table 6.6b. Effects of treatments which alter rates of ATP synthesis on active anion influx in *H. africanum*. All treatments in saturating light (400–730 nm), air

Treatment	Effect on			Active anion influxes (tracer) in *H. africanum*	
	Rate of electron flow	Turnover of HEI of ATP synthesis	Rate of ATP synthesis	Cl^-	$H_2PO_4^-$
Control	Control	Control	Control	100	100
CCCP 5 μM [a]	Stimulated	Inhibited	Inhibited	73	13
DCCD 1 mM [b]	Inhibited	Inhibited (level is increased)	Inhibited	91	45
Ethionine 5 mM [c]	Inhibited	Inhibited (level is increased)	Inhibited	76	36

HEI = High energy intermediate.

[a] CCCP uncouples electron flow from ATP synthesis. Stimulated electron flow not seen with CO_2-linked non-cyclic *in vivo* because ATP is needed for CO_2 fixation.

[b] DCCD (energy transfer inhibitor) inhibits terminal steps of ATP synthesis, and hence inhibits coupled portion of electron flow.

[c] Ethionine (adenylate trapping agent) has somewhat similar effects to DCCD.

(Data from Raven, 1967b; 1969b, 1971e, 1974d.)

it is clear that manipulation of electrochemical gradients can also alter active fluxes, which are subject to control by chemical activity on both sides of the membrane (Chap. *11*) as well as by electrical potential in the case of electrogenic pumps. Conversely, alterations in metabolism can alter the "permeability coefficient" for passive ion movement, as well as the rate of exchange diffusion.

Tracer fluxes in the downhill direction, corrected for any exchange component, can be used in conjunction with measurements of the PD across the membrane and the concentration on each side, to compute "permeability coefficients" as described in Chap. *3*. The changes in metabolic state used in analyzing transport, and determining the chemical nature of the energy source, are brought about by the treatments shown in Table *6.6*.

The results shown there are consistent with all of the electron transport pathways known in green cells (respiratory, and photosynthetic of the cyclic type or the non-cyclic type linked to CO_2, NO_3^- or O_2 as electron acceptor) having the ability to support active transport. When more than one process can supply energy to active transport, relatively little is known of what the major energy source is under natural conditions in the light, when any of the pathways could be operative. Current evidence favors non-cyclic, probably with CO_2 as electron acceptor, being the major pathway. In some cases (e.g. Cl^- in many algae; see *6.6.2.5*) only non-cyclic can be used in the light; cyclic is apparently unavailable. The situation as regards the linkage between electron transport and active transport is unclear in some cases (e.g. Cl^- in some freshwater coenocytes: *6.6.2.5*), but in most cases ATP is the probable energy transmitter.

It must be emphasized that not all the properties of active transport processes deduced from experiments in which metabolism is altered need necessarily relate directly to the energy-coupling mechanism. Both metabolized and non-metabolized solutes have their active transport controlled by the requirement for that solute (*6*.3; *11*.3.2), and this kind of control may also be altered when energy metabolism is changed. Further, it is likely that the energetic state of the cell may influence the activity of the pump (the rate of action at constant substrate concentration) independently of changes in either the "solute requirement sensor" or the concentration of the direct energy source (e.g. ATP). The evidence for this is outlined in RAVEN (1974d). SMITH and RAVEN (1974) discuss further complications of interpretation. Despite these difficulties, the judicious use of the experimental procedures outlined in Table *6*.6 has led to some clarification of the energy sources for active transport in algae.

Exchange diffusion is a 1:1 chemical coupling of influx and efflux of a solute (antiport in the terminology of MITCHELL, 1966). For ions it has been invoked to explain parallel effects on both influx and efflux of a change in conditions. Most striking is the effect of removal of the ion concerned from the external medium; this frequently lowers the efflux in a manner not dependent on changes in the electrical component of the driving force. Parallel effects on influx and efflux of changes in temperature, illumination or rate of metabolism have also been used as evidence for exchange diffusion (e.g. HOPE et al., 1966; BARBER, 1968c; RYBOVA et al., 1972; RAVEN, 1974c). This phenomenon is also invoked to explain apparent excesses of passive ion fluxes over electrical conductance, and excess of tracer uphill fluxes over the energy available to support them. Exchange diffusion is commonly regarded as being mediated by a modified active transport system (HOPE et al., 1966; BARBER, 1968e; BARBER and SHIEH, 1973a). The involvement of energy here, as in passive transport, is in a non-stoichiometric or "triggering" role (cf. the irreversible thermodynamic definition of inter-active transport). In the case of active transport, energy metabolism must act in a stoichiometric way, but it also may be involved in an informational or triggering way (RAVEN, 1974d; SMITH and RAVEN, 1974).

6.2 Chloride Fluxes

6.2.1 Introduction

Values of chloride fluxes vary considerably between algae, as shown in Table *6*.7. Fluxes are generally low in freshwater species (up to 60 nmol m^{-2} s^{-1}), and in marine species can be up to 8 μmol m^{-2} s^{-1}.

6.2.2 Passive Fluxes

Tracer chloride efflux in algal cells is electrochemically downhill. Part of it is attributable to exchange diffusion (HOPE et al., 1966; RAVEN, 1974c). Voltage clamp studies have enabled investigations of the dependence of Cl^- efflux on electrical potential to be made; hyperpolarization increases the efflux while depolarization decreases it (e.g. HOPE et al., 1966; GUTKNECHT, 1967; COSTER and HOPE, 1968; KITASATO, 1968; GRADMANN et al., 1973; Fig. *6*.4). In both *Nitella translucens* and *Valonia ventricosa* (GUTKNECHT, 1967; WILLIAMS et al., 1972), tracer Cl^- influx is not voltage-sensitive in the range tested. In *Nitella* electrochemical evidence suggests a very small passive influx; in *Valonia* the situation is unclear (*6*.5.3).

Table 6.7. Tracer chloride, sulfate and phosphate fluxes at the plasmalemma and tonoplast in a number of algae living in freshwater or seawater

Genus	Ion (mM) (L=light, D=dark)		Flux (nmol m^{-2} s^{-1}) ϕ_{oc}	ϕ_{cv} (ϕ_{vc})	ϕ_{co}	Ref.
Chara australis	Cl^-, 1.3–2.3	L	1.6–35	~1,400	1–9	Coster and Hope (1968)
		D	1–10	~ 600	4–40	Findlay et al. (1969)
						Lannoye et al. (1970)
	SO_4^{2-}, 1.2	L	0.8	~ 100	$\ll \phi_{oc}$	Robinson (1969a)
		D	0.8–3.0	~ 36	$\ll \phi_{oc}$	
Nitella translucens	Cl^-, 1.3	L	5–30	1,000–2,000	2–7	Macrobbie (1962, 1964),
		D	0.5–1.8	300–400	17–40	Hope et al. (1966)
	SO_4^{2-}, 1.2	L	0.6	?	?	Robinson (1969a)
		D	0.5			
	$H_2PO_4^-$, 1.0	L	20–30	$\gg \phi_{oc}$	$\ll \phi_{oc}$	Smith (1966)
		D	10–20			
Acetabularia mediterranea	Cl^-, 550	L	2,000–8,000	?	2,900–8,500	Saddler (1970b, c)
		D	520			
	SO_4^{2-}, 55	L	1.7	?	$\ll \phi_{oc}$	
Chaetomorpha darwinii	Cl^-, 550	L	10,000	5,600	3,600	Findlay et al. (1971)
		D	6,500	900	800	
Chlorella pyrenoidosa	Cl^-, 1.0	L	1.2	?	~1.2	Barber (1968a, 1969)
		D	0.2		~0.2	
Hydrodictyon africanum	Cl^-, 1.3	L	10–30	500–1,000	3–20	Raven (1967a, 1969a, b, 1974c
		D	2–10	100–400	2–10	and unpublished)
	SO_4^{2-}, 1.3	L	0.5	20–40	<0.05	Raven (unpublished)
		D	0.3	5–20		
	$H_2PO_4^-$, 0.2	L	2–3	100–300	<0.2	Raven (1974a
		D	1–2	50–200		and unpublished)
Mougeotia sp.	Cl^-, 1.0	L	6	0.6 (1.3)	6.7	Wagner and Bentrup (1973)
		D	2	0.8 (1.8)	3.0	
Valonia ventricosa	Cl^-, 550	L	98	?	110	Gutknecht (1965)
			(Φ_{ov}185)		(Φ_{vo})	Aikman and Dainty (1966)
Pelvetia fastigiata (zygote >4 h after fertilisation)	Cl^-, 550	L or D	100	?	90	Robinson and Jaffe (1973)

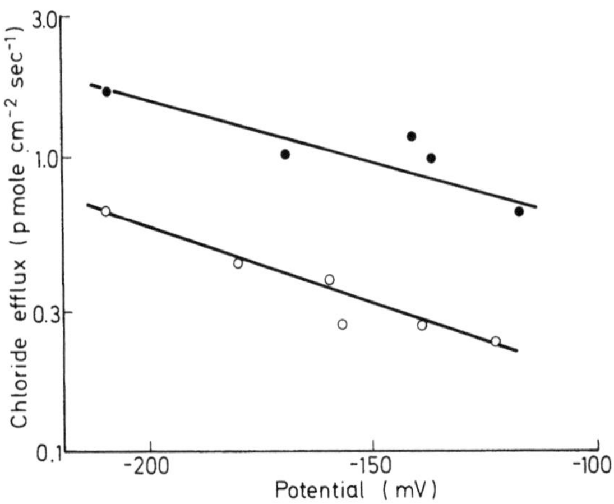

Fig. 6.4. Efflux of chloride from cell of *Chara australis*, as a function of membrane potential. Cell in chloride artificial pond water. (○) Efflux in light. (●) Efflux in dark after 8 h darkness. Membrane potential controlled by voltage clamping. (HOPE et al., 1966)

The effects of electrical potential on passive Cl^- fluxes in algae, like those on passive K^+ and Na^+ fluxes (6.6.3.2), do not quantitatively fit the simplest model for potential effects on independent ion fluxes (WALKER and HOPE, 1969).

6.2.3 Active Fluxes

It is likely that very little of the Cl^- influx in most algal cells (exceptions are noted in 6.5.3) is due to truly passive Cl^- influx; most is active influx or exchange diffusion. Comparison of effects of light on net fluxes, and tracer influx and efflux, are generally consistent with a "pump and leak" situation for Cl^-. Thus in *Chlorella, Acetabularia, Hydrodictyon africanum* and charophytes there is a net influx of Cl^- in the light, and a smaller net influx, or a net efflux, in the dark. While tracer influx is usually greatly light-stimulated, the efflux is less light-stimulated or even light-inhibited (TAZAWA, 1961; NAGAI and TAZAWA, 1962; SPEAR et al., 1969; Table 6.7). However, in a number of cases prolonged darkness does not alter the total electrolyte content (and, by implication, the Cl^- content) under conditions in which there is considerable net inward Na^+ and outward K^+ passive movement (SCOTT and HAYWARD, 1955; BERGQUIST, 1958). In *Valonia macrophysa* (SCOTT and HAYWARD, 1955) Cl^- distribution is probably passive; in the other cases (*Ulva, Hormosira*) any discrepancy between pump and leak must be smaller than for cations, either because the total solute pump (Cl^- plus cations) is preferentially powered when energy supply is low, or because the Cl^- efflux is low because the true P_{Cl} is low, despite the large driving force for passive Cl^- efflux. In *Acetabularia*, P_{Cl} falls under "low-energy" conditions and a new flux equilibrium with lower Cl^- content is established (SADDLER, 1970c; cf BARBER, 1969).

Net Cl^- influx is generally associated with net cation influx rather than the efflux of some metabolically produced organic anion. Tracer studies show a mutual dependence of a component of Cl^- influx on the presence of cations, and of a component of cation (K^+ and Na^+) influx on the presence of Cl^-. In *Chlorella* (BARBER, 1969) and *Nitella translucens* (SMITH, 1967a) the main cation is Na^+. Both K^+ and Na^+ are involved in *Hydrodictyon africanum* (RAVEN, 1968a) and *Enteromorpha intestinalis* (BLACK, 1971), while K^+ is the main cation involved in *Tolypella intricata* (SMITH, 1968), *Nitella flexilis* (NAGAI and TAZAWA, 1962; TAZAWA and NAGAI, 1966), *Chara corallina* (FINDLAY et al., 1969) and in zygotes of *Pelvetia* (ROBINSON and JAFFE, 1973).

This cation-Cl^- transport is completely light-dependent in some cases (SMITH, 1967a, 1968; RAVEN, 1968a; BARBER, 1969), but not in others (BLACK, 1971). The coupling of cations and Cl^- has been suggested to be an electrogenic or a chemical linkage (SMITH, 1967a; RAVEN, 1968a). Coupling *via* H^+ and OH^- fluxes (SMITH, 1970) has also been suggested.

6.2.4 Exchange Diffusion

Tracer Cl^- fluxes are greater than predicted from electrical conductance in a number of cases (e.g. at the tonoplast of *Griffithsia* and *Chaetomorpha:* FINDLAY et al., 1969; FINDLAY et al., 1971). Here some Cl^- transport by an electrically silent mechanism must be postulated, either exchange diffusion or some other chemically coupled antiport, or symport. Inhibition of Cl^- efflux upon removal of external Cl^- (HOPE et al., 1966; RAVEN, 1974c) is also consistent with an exchange diffusion component of Cl^- fluxes at the plasmalemma of freshwater coenocytes. In this situation, and also at the plasmalemma of *Griffithsia,* exchange diffusion makes the discrepancy between conductance calculated from tracer fluxes and conductance measured electrically even greater. Exchange diffusion has also been invoked to account for apparent inadequacy of metabolism to support tracer fluxes measured in the uphill direction (SADDLER, 1970b, c).

Exchange diffusion appears to account for 15 to 50% of the plasmalemma flux of Cl^- in *Griffithsia* (FINDLAY et al., 1970) and 85% and 35% respectively of the Cl^- fluxes at the plasmalemma and tonoplast in *Chaetomorpha* (FINDLAY et al., 1971).

6.2.5 Energy Sources for Active Transport

The energy source for the Cl^- pump in *Chlorella* (and possibly *Acetabularia*) appears to be ATP, which can be supplied by oxidative phosphorylation in the dark or by cyclic photophosphorylation in the light (BARBER, 1968d; SADDLER, 1970c; GRADMANN, 1970). In all other cases the light-stimulation appears to involve photosystem II of photosynthesis (MACROBBIE, 1964; RAVEN, 1967b, 1971a, e; SMITH, 1968; BLACK, 1971; LILLEY and HOPE, 1971; RYBOVA et al., 1972; SMITH and RAVEN,

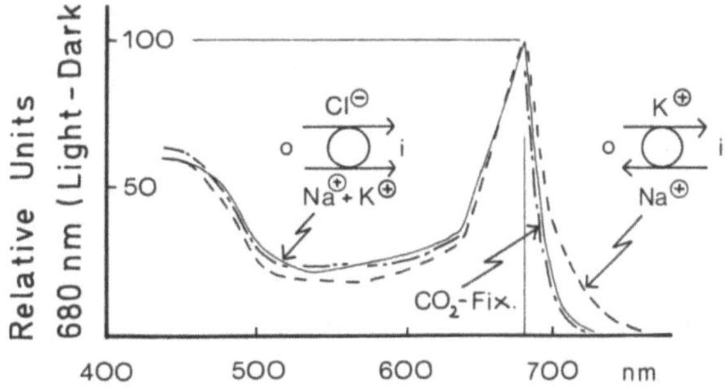

Fig. 6.5. Action spectra for active influxes of ^{36}Cl (——) and ^{42}K (----) in *Hydrodictyon africanum* in CO_2-free air. 0.5 Jm^{-2} s^{-1} of monochromatic light at each of the 15 wavelengths tested supplied by interference filters, half-band width 10 ± 2 nm. Results of RAVEN, 1969a, as modified by LÜTTGE (1973)

1974); although DCMU sensitivity is somewhat lower in N_2 than in air in *Hydrodictyon africanum* (RAVEN, 1971a), but not in *Chara corallina* (SMITH and RAVEN, 1974a). In other cases rigorous testing of the effect of inhibiting photosystem II under conditions (lack of O_2 and CO_2) which would not lead to inhibition, by "over-oxidation", of the cyclic electron transport pathway, has not been carried out. In some instances (e.g. *Nitella translucens*: MACROBBIE, 1965; *Hydrodictyon africanum:* RAVEN, 1967b, 1974d; *Hydrodictyon reticulatum:* RYBOVA et al., 1972; *Enteromorpha intestinalis:* BLACK, 1971) an internal control is available, in terms of the occurrence of some light-stimulated processes which *do not* require photosystem II under conditions in which Cl^- influx *does* need photosystem II.

Some of the algae which apparently need photosystem II for light stimulation of Cl^- influx appear to need ATP for active Cl^- transport (SMITH and WEST, 1969; LILLEY and HOPE, 1971), while others do not (MACROBBIE, 1965, 1966a; SMITH, 1967a, 1968; RAVEN, 1967b, 1969b, 1971e). The explanation for this difference is not clear; the complications of interpretation of the results are discussed by SMITH and RAVEN (1974). Similar confusion occurs over the nature of the dark Cl^- influx; in *Chara corallina* it is not inhibited by removal of O_2, and has been attributed to exchange diffusion (FINDLAY et al., 1969). In *Hydrodictyon africanum* (RAVEN, 1967b, 1969b: Table *6*.6) and in *Griffithsia* (LILLEY and HOPE, 1971) it is O_2-dependent, and may be active. This is in accord with the net Cl^- influx observed in darkness in *Nitella flexilis* (TAZAWA and NAGAI, 1966) when recovering from turgor lowering.

In view of the controversies over the energy source for Cl^- influx it is hardly surprising that there is no unanimity over its mechanism. Coupling to photosynthetic or respiratory ATP or reductant supply has been suggested (primary active transport: MACROBBIE, 1970a; SMITH and RAVEN, 1974). More distant relationships to metabolism, with the primary energy-transducing step at the cell membrane being active transport of H^+ have also been proposed (SMITH, 1970, 1972; RAVEN and SMITH, 1973; SMITH and RAVEN, 1974; RAVEN, 1974e). While an imposed inside-alkaline pH gradient across the plasmalemma has been shown to increase tracer Cl^- influx in *Chara corallina* (SMITH, 1972), and net Cl^- influx in *Hydrodictyon africanum* (measured as the difference between tracer influx and tracer efflux: RAVEN, 1974e, 1975b), it has *not* been proved that the energy for the Cl^- pump is provided solely by such a gradient.

In *Nitella translucens* the requirement for photosystem II for the Cl^- pump, and the relative insensitivity of tracer Cl^- influx to the uncoupler CCCP (MACROBBIE, 1965, 1966a; SMITH, 1967a) contrasts with the inhibition of the hyperpolarization attributed to the H^+ extrusion pump by CCCP, and the lack of a photosystem II requirement for its light stimulation (SPANSWICK 1974a, b). Thus if the hypothesis of SMITH is to apply to *Nitella translucens*, the requirement for photosystem II must be attributed to some (regulatory?) requirement other than the H^+ gradient, and the lack of CCCP inhibition of the Cl^- tracer influx in the light must be accounted for by increased exchange diffusion replacing an inhibited active transport. Certainly the tracer Cl^- efflux in *Nitella translucens* is stimulated by CCCP (MACROBBIE and FENSOM, 1969). Thus even if the H^+ pump is a *necessary* condition for the Cl^- influx in *Nitella translucens* it is not a *sufficient* condition. MACROBBIE (1974) discusses the criteria which must be fulfilled before this hypothesis can be considered proven. The situation with regard to the postulated role of vesicles in Cl^- transport is unsettled (MACROBBIE, 1973; WALKER and BOSTROM, 1973; WALKER, 1974). The chloride transport system can also transport bromide (HOAGLAND et al., 1928; MACROBBIE, 1962; WALKER and BOSTROM, 1973; RAVEN, unpublished experiments in *Hydrodictyon africanum*).

The influence of external Cl^- (Br^-) concentration on net influx is fitted by a simple rectangular hyperbola, although in some tracer experiments more complex relations are found (HOAGLAND et al., 1928; TAZAWA, 1961; MESZES et al., 1967; RAVEN, 1974c). Regulation of Cl^- transport in algae is in terms of homeostasis of Cl^- content, internal osmotic pressure or of turgor pressure (*11*.3.1.3.1).

6.3 Potassium and Sodium Fluxes

6.3.1 Introduction

Values of K^+ and Na^+ fluxes are given in Table *6.8*. Fluxes at the plasmalemma are generally rather lower than those of Cl^- in the same alga. Measurements of electrochemical potential (*6.5*) show that there is likely to be an active efflux of Na^+ at the plasmalemma of most algal cells; active Na^+ influx at the tonoplast is also quite common. There is less uniformity with K^+ fluxes; active K^+ influx at the plasmalemma is quite commonly found, although there are cases of electrochemical equilibrium and of apparent active K^+ influx at this membrane. Similarly, active K^+ influx as well as active K^+ efflux can occur at the tonoplast.

6.3.2 Passive and Exchange Diffusion Fluxes

Independent, passive fluxes of K^+ and Na^+ should respond to electrical and concentration gradients as predicted by the equations considered in *3*.5.3.1. The influence of electrical potential gradients have been investigated by the voltage clamp technique (e.g. BLOUNT, 1958; GUTKNECHT, 1967; KITASATO, 1968; WALKER and HOPE, 1969; MACROBBIE and FENSOM, 1969; WILLIAMS et al., 1972; RYAN and BARR, 1973; THAIN, 1973: Fig *6*.6). In the algae used, K^+ efflux and Na^+ influx are downhill, while there are active components of Na^+ efflux and (usually) K^+ influx; all of these fluxes respond in the (qualitatively) predicted manner. It is not clear whether the response of the partly active cation fluxes is by the passive component alone, or whether the active component is also involved. Quantitatively, these responses do not fit the simplest model (WALKER and HOPE, 1969; THAIN, 1973).

Since it is easier to control the external ion concentration, most of the work on concentration dependence of cation fluxes deals with changes in external concentration. Taking changes in electrical potential difference due to changes in ion concentration into account (preferably using empirical relationships derived from the voltage clamping experiments described above), in some cases there is a reasonable agreement between the concentration dependence of passive cation influx based on a simple model and the observed relationship (e.g. Na^+ influx in *Hydrodictyon africanum:* RAVEN, 1968a, and Fig. *6*.7). In other cases, the results can only be explained in terms of independent ion movement if the permeability coefficient for an ion is a function

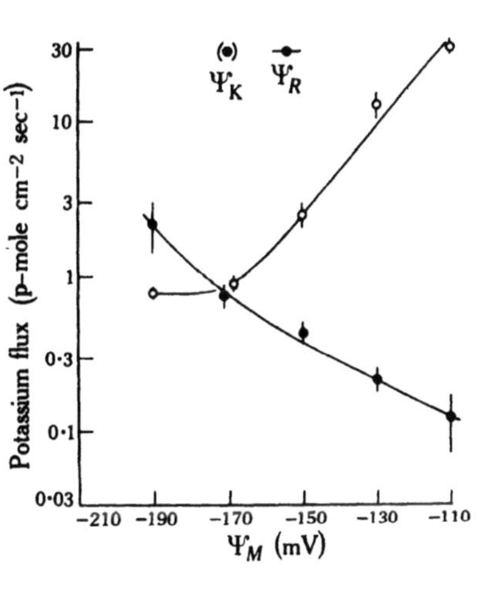

Fig. *6*.6. Mean influx (●, 10 cells) and efflux (○ 7 cells) of K^+ in *Ch. corallina* in APW, plotted on a logarithmic scale against vacuolar PD, ψ_M. Standard errors of the means (SEM) are shown by bars. The mean resting PD (ψ_R) with SEM is indicated. The mean Nernst PD (ψ_K) was -170 mV in the only three cells for which data is available. (From WALKER and HOPE, 1969)

Table 6.8. Tracer K^+, Na^+, Ca^{2+} and Mg^{2+} fluxes at the plasmalemma and tonoplast in a number of algae living in freshwater or seawater (see Table 6.2)

Genus	Ion (mM)	(L=light, D=dark)	Flux (nmol m^{-2} s^{-1}) ϕ_{oc}	ϕ_{cv} (ϕ_{vc})	ϕ_{co}	Ref.
Chara australis	K^+, 0.1–0.2	L	1.5–30	80–1950	6	Hope (1963),
		D	1.1–14			Findlay et al. (1969)
	Na^+, 1.0–2.0	L	1–9		1–8	Hope and Walker (1960)
		D	0.3–3		1–6	Findlay et al. (1969)
Nitella translucens	K^+, 0.1	L	5–17	20–1,250	4–40	MacRobbie (1962, 1964, 1966a, b)
		D	1–2			
	Na^+, 1.0	L	5–20	400–1,300	5	MacRobbie (1966b),
		D	1–5			Smith (1967a)
	Ca^{2+}, 0.1	L	0.5 (Φ_{ov})	?	$\ll \phi_{oc}$	Spanswick and Williams (1965)
Acetabularia mediterranea	K^+, 10	L	110–400	?	110–400	Saddler (1970b)
		D	110–400		110–400?	
	Na^+, 550	L	110–500	?	150–370	Saddler (1970b)
		D	110–500		35–100	
Chaetomorpha darwinii	K^+, 10	L or D	650–2,100	1,000–9,5000	650–2,100	Dodd et al. (1966)
	Na^+, 550	L or D	1,100	37	1,100	Dodd et al. (1966)
Chlorella pyrenoidosa	K^+, 6.5	L	10.3	?	9.2	Barber (1968a, d, e)
		D	1.8		3.3	
	Na^+, 1.0	L	1.1	?	1.4	Barber (1968c, 1969)
		D			0.12	
Hydrodictyon africanum	K^+, 0.1	L	5–15	200–400	4–12	Raven (1967a, 1968a,
		D	1–7	100–200	2–10	and unpublished)
	Na^+, 1.0	L	2–9	500–1,000	4–10	Raven (1967a, 1968a,
		D	1–7	200–800	1–5	and unpublished)
Mougeotia sp.	K^+, 1.0	L	1.5	1.2 (10)	10	Wagner and Bentrup (1973)
		D	5	1.5 (9)	8	
Valonia ventricosa	K^+, 10	L	890 (Φ_{ov})	?	860 (Φ_{vo})	Gutknecht (1965)
	Na^+, 550	D	360 (Φ_{ov})	?	330 (Φ_{vo})	Aikman and Dainty (1966)
Pelvetia fastigiata (zygote >4 h after fertilisation)	K^+, 10	L	330	?	290	Robinson and Jaffe (1973)
	Na^+, 550	L	68	?	68	
	Ca^{2+}	L	0.74	?	1.30	
	Mg^{2+}	L	4–11	?	(4–11)	
Griffithsia sp.	K^+, 10	L or D	500–3,800	3,600–9,900	500–3,800	Findlay et al. (1970)
	Na^+, 550	L	120–170	50–120	150	

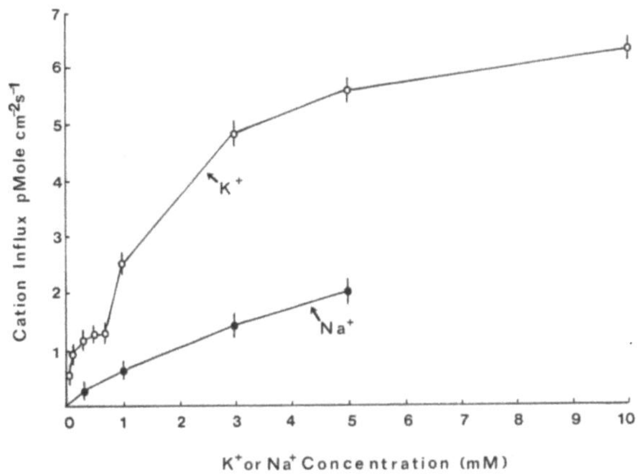

Fig. *6.7*. Concentration dependence of ^{42}K and ^{22}Na influx in *Hydrodictyon africanum*. Basic solution 0.1 mM KCl, 0.1 mM $CaCl_2$, 1 mM NaCl. For low concentrations of cations, some of the KCl or NaCl were omitted. For higher concentrations of cations, extra KCl or NaCl were added. 16° C, Fluorescent illumination 15 J m^{-2} s^{-1}. (J.A. RAVEN, unpublished; cf. RAVEN, 1967a, 1968)

of the concentration of that ion, e.g. K^+ influx in *Hydrodictyon africanum* (RAVEN, 1967a, 1974c and Fig. *6.7*) and *Nitellopsis obtusa* (MACROBBIE and DAINTY, 1958a), and Na^+ and K^+ influxes in *Nitella translucens* and *Chara australis* (HOPE and WALKER, 1960; HOPE, 1963; SMITH, 1967a; SPANSWICK, 1969; WILLIAMS and HOGG, 1970). Thus, as with electrical potential responses (cf. KITASATO, 1973), passive fluxes change with concentration in qualitative agreement with the expectation for independent ion movement.

There are a number of instances in which coupling of K^+ or Na^+ fluxes to some other ion flux is suspected. The possible chemical linkage of downhill K^+ and Na^+ influxes to Cl^- influx in a number of algae has been mentioned in *6.6.2.3*. Exchange diffusion of K^+ at the tonoplast of *Griffithsia* and *Chaetomorpha* has been suggested as necessary to account for the high electrical resistance of the membrane in comparison with measured (tracer) K^+ fluxes (FINDLAY et al., 1969, 1970; FINDLAY et al., 1971).

6.3.3 Active Fluxes

Much of the early work on monovalent cation transport at the plasmalemma of algal cells concentrated on K^+/Na^+ exchange mechanisms; recent work has tended to emphasise K^+/H^+ exchange, especially in growing cells (EPPLEY, 1962; GUTKNECHT and DAINTY, 1968; SMITH, 1970; MACROBBIE, 1970a, 1974).

Net cation fluxes have been studied in a number of marine algae recovering from treatments which have decreased their K^+/Na^+ ratio. These treatments generally involve decreased energy supply (prolonged darkness, or darkness in the absence of O_2: SCOTT and HAYWARD, 1955; EPPLEY, 1962; DODD et al., 1966). It appears from experiments on tracer fluxes in marine algae that such treatments decrease the "pump" more than the "leak" component of the fluxes (GUTKNECHT and DAINTY, 1968). The active processes which restore the K^+/Na^+ ratio are light-stimulated, and are slowed by inhibitors of energy metabolism; comparison with tracer experiments show that they act by inhibiting the pump component rather than by increasing the leak. Electrochemical data suggest that both the K^+ and Na^+ fluxes in the recovery phase are active (*6.5.4*). The data as to the stoichiometry of coupling and on the energy source are rather confused (EPPLEY, 1962).

Much of the recent work on K^+/Na^+ transport has involved tracer experiments, although some net flux measurements have been made on *Chlorella* (SCHAEDLE and JACOBSEN, 1965, 1966, 1967; SHIEH and BARBER, 1971), *Scenedesmus* (ERDEI and CSEH, 1971), *Chaetomorpha* (DODD et al., 1966) and charophytes (TAZAWA and NAGAI, 1966). These experiments show that the tracer efflux of Na^+ from normal cells of *Chlorella, Nitella translucens, Hydrodictyon africanum* and *Enteromorpha intestinalis,* and net efflux of Na^+ from "Na^+-rich" *Chlorella* and *Scenedesmus* require the presence of external K^+ (to a greater extent than expected from the potential changes at different K^+ levels) and in all cases except *Chlorella* and *Scenedesmus* both K^+ influx and Na^+ efflux are partially inhibited by ouabain (MACROBBIE, 1962, 1970a; RAVEN, 1967a, 1971a; BARBER, 1968c; BLACK, 1971; SHIEH and BARBER, 1971; ERDEI and CSEH, 1971). However in a number of cases the coupling of K^+ influx to Na^+ efflux seems not to occur. In *Chara corallina* neither addition of ouabain nor removal of external K^+ inhibits Na^+ efflux; (passive) K^+ influx is also not inhibited by ouabain (FINDLAY et al., 1969). In *Chlorella* (BARBER, 1968c, d), *Hydrodictyon reticulatum* (RYBOVA et al., 1972) and other instances listed by RAVEN (1971a) there is also no ouabain effect on K^+ influx or Na^+ efflux. In cells of *Nitella flexilis* whose vacuoles have been perfused with high-Na^+ solution the initial response of the cell is loss of NaCl (and a little KCl), followed by reaccumulation of KCl, with some K^+/Na^+ exchange as well (TAZAWA and NAGAI, 1966). This Na^+ loss is probably active (TAZAWA and KISHIMOTO, 1964). The cells die before complete recovery of the K^+/Na^+ ratio is achieved.

The high K^+ and Na^+ fluxes during restoration of the K^+/Na^+ ratio in Na^+-rich *Chlorella* and *Scenedesmus* are much greater than tracer fluxes in the steady state, or net K^+ fluxes required for growth (BARBER, 1968c, d, e; SHIEH and BARBER, 1971; ERDEI and CSEH, 1971; MESZES and CSEH, 1973). This may be due to the higher substrate (Na^+) concentration in the Na^+-rich cells. Experiments on charophytes and *Acetabularia* suggest that the Na^+ efflux is linearly related to internal Na^+ concentration (BARR, 1965; WIDEMANN, 1968; SADDLER, 1970b).

Net K^+ influx, whether related to Na^+ efflux (JACQUES and OSTERHOUT, 1935; EPPLEY, 1959; BARBER and SHIEH, 1972) or net salt accumulation (TAZAWA, 1961) exhibits a simple rectangular hyperbolic relationship to K^+ concentration. Tracer K^+ influx sometimes shows a simple hyperbolic relationship to K^+ concentration in the range tested (BARBER, 1968d), although in other cases a "dual isotherm" is found, even when there is little or no vacuole in the cells (RAVEN, 1967a, 1974c: Fig. 6.7; SPANSWICK, 1969; SADDLER, 1970b; KANNAN, 1971; SOLT et al., 1971). The interpretation of these effects in terms of passive, chemically coupled and active, and electrogenically coupled fluxes is difficult. In the case of net fluxes coupled to Na^+ efflux or Cl^- influx, limitation by the co- or counter-ion in determining the concentration dependence is a possibility.

The energy supply for K^+/Na^+ pumping appears to be ATP, based on responses to various inhibitors (MACROBBIE, 1970a, 1974; RAVEN, 1971e; SMITH and BARBER, 1971; BARBER and SHIEH, 1973b; RYBOVA et al., 1972). These papers also review data which suggest that, in the dark, the energy is supplied by oxidative phosphorylation, while in the light photophosphorylation is involved in energy supply to the light-stimulated flux (Fig. 6.5). In the absence of CO_2, O_2 and nitrate (ULLRICH, 1971, 1972) cyclic photophosphorylation is the main energy source. When CO_2 is present, non-cyclic photophosphorylation is also involved (RAVEN, 1971b); pseudocyclic photophosphorylation may also supply ATP to cation transport (RAVEN, 1969b). The situation with regard to Na^+ efflux in *Chara* and *Acetabularia* is unclear; here the flux is not always light-stimulated, and in *Acetabularia* is not inhibited by CCCP (HOPE and WALKER, 1960; FINDLAY et al., 1969; SADDLER, 1970b).

If ATP is indeed the energy source, it should be possible to isolate a plasmalemma fraction which contains a K^+, Na^+ activated ATPase which is distinct from other ATPases and from non-specific phosphatases, and which (at least in some cases) is inhibited by ouabain. A number of investigators have attempted to detect such ATPases (e.g. ATKINSON and POLYA, 1967; MIRSALIKOVA, 1968; MESZES and ERDEI, 1969; BATTERTON et al., 1972; SUNDBERG et al., 1973), although the correlation with the flux data is much less good than in the case of animal "transport ATPases" (see *10*.2 and *10*.5).

The role of the Na^+ extrusion pump during algal growth is not clear; Na^+ is not an obligate nutrient for many algae (O'KELLEY, 1974), but Na^+ influx may be linked to phosphate influx, and since microalgal cells contain less Na^+ than phosphate, this would involve an Na^+ efflux during growth (RAVEN, 1974c). The K^+/Na^+ ratio in charophytes decreases with cell age (MACROBBIE, 1962; BARR, 1965; STRAUSS, 1967); how this relates to K^+ and Na^+ active fluxes is not clear.

The specificity of cation transport has been investigated in a number of algae; however, it is not entirely clear whether the fluxes investigated are active or passive in a number of cases. In general Rb^+ and Tl^+ behave similarly to K^+, and may well be transported by a similar mechanism, although the fluxes are generally lower than those of K^+ (EPPLEY, 1959; SCHAEDLE and JACOBSEN, 1966, 1967; WEST and PITMAN, 1967b; FINDLAY et al., 1970; KANNAN, 1971; SOLT et al., 1971). The Na^+ extrusion pump may be related to H^+ extrusion, but does not deal with Li^+ (COLLANDER, 1939; EPPLEY, 1960). In *Chlorella* it appears that the coupled K^+/Na^+ (K^+/H^+) transport system can, under suitable conditions, bring about exchange diffusion of either K^+ (BARBER, 1968c) or Na^+ (BARBER and SHIEH, 1973a).

6.4 Metabolized Ions

6.4.1 Phosphate

Other anions have been studied in less detail than Cl^-; feedback effects from metabolic demand are important here, and complicate the picture as regards energy sources for uptake. As far as phosphate is concerned, fluxes are of the same order as, or lower than, those of Cl^- in algal coenocytes (Table *6*.7) and they are generally less light-stimulated (COOK, 1968; HEALEY, 1973; RAVEN, 1974d). In all cases investigated ATP seems to be the energy source for phosphate influx. In the dark oxidative phosphorylation supplies the ATP, while in the light it can be supplied by cyclic photophosphorylation in all cases except *Chara corallina* (references in SIMONIS et al., 1962; SIMONIS and URBACH, 1973; RAVEN, 1974d; SMITH and RAVEN, 1974b; SIMONIS et al., 1974; see Table *6*.6). In *Chara corallina* there may be no cyclic photophosphorylation, or at least no ATP from this process available outside the chloroplasts (SMITH and RAVEN, 1974). In *Hydrodictyon africanum* and *Anacystis nidulans* much of the ATP for phosphate influx comes from non-cyclic photophosphorylation when CO_2 is present in the light (RAVEN, 1974d; SIMONIS et al., 1974); in *Ankistrodesmus braunii* in the light NO_3^--linked non-cyclic photophosphorylation can power phosphate influx (ULLRICH, 1971, 1972, 1974; ULLRICH-EBERIUS, 1973b).

During growth of micro-algae on "ecological" phosphate concentrations it is likely that there is relatively little phosphate efflux (JEANJEAN and BLASCO, 1970); this is also true of *Hydrodictyon africanum* in 100 μM phosphate (RAVEN, 1974c). At higher phosphate concentrations in the external solution, phosphate efflux becomes significant (WEST, 1967; SOEDER, 1970), possibly in response to the higher internal inorganic phosphate levels in the cells at high external phosphate concentrations (DOMANSKI-KADEN and SIMONIS, 1972). Organic phosphate efflux may also occur (KUENZLER, 1970). The active phosphate influx is stimulated in response to phosphate deficiency; FUHS et al. (1972) showed that this was an effect on the K_m rather than the V_{max} of the system. The free inorganic phosphate content in the medium in equilibrium with low-phosphate algal cells is very low (about 0.001 μM; KUENZLER and KETCHUM, 1962). In the presence of high concentrations of phosphate in the medium algal cells can increase their phosphate content to two or three times the content (per cell) at which growth is saturated (FUHS, 1969; see *6*.3).

Phosphate influx in all of the investigated genera of the Chlorococcales (*Chlorella, Scenedesmus, Ankistrodesmus* and *Hydrodictyon*) is stimulated by Na^+ and Li^+ rather than K^+; (references in RAVEN, 1974c); this uniformity, as with the energy source, contrasts with the results obtained with Cl^- (*6*.6.2). The downhill entry of Na^+ cannot supply enough energy with the probable $Na^+/H_2PO_4^-$ of 1 to power the uphill phosphate influx (RAVEN, 1974c), although in view of the large free energy gradients against which phosphate can be transported (see above) association of phosphate influx with that of a cation present in the cell at a low electrochemical potential may be advantageous (RAVEN, 1975b).

The phosphate transport system can also transport arsenate, as judged by mutual competition between the two ions and the similar characteristics of their influxes; short-term experiments are required with arsenate which inhibits phosphorylation inside the cell (JEANJEAN and BLASCO, 1970; RAVEN, 1971e, 1974c, d and unpublished experiments on *Hydrodictyon africanum*). RAVEN (1974c) gives references on response of active phosphate transport in algae to external phosphate concentration. In some phytoplankton algae the half-saturation concentration for phosphate influx is as low as 1 μM (FUHS et al., 1972). The pH dependence of phosphate influx suggests that HPO_4^{2-} as well as $H_2PO_4^-$ can be actively transported in algae (ULLRICH-EBERIUS, 1973a; RAVEN, 1974c).

6.4.2 Sulfate

Sulfate influx has been less completely investigated; the energy source is less certain than that for phosphate influx. While phosphate influx is uniformly light-stimulated by a factor of 1.5–2 (see Tables *6*.6 and *6*.7, and RAVEN, 1974c, d), SO_4^{2-} influx can be slightly light-stimulated (KYLIN, 1964; VALLÉE and JEANJEAN, 1968a; PENTH and WEIGL, 1971; RAMUS and GRAVES, 1972) or the steady-state SO_4^{2-} influx can be the same in light and dark (COOK, 1968) with a transitory increase after the light is turned off (ROBINSON, 1969b). This stimulation in the "dark" occurs in photosystem I light, i.e. its suppression requires photosystem II (ROBINSON, 1969b). Even when there is no light stimulation the influx in the light can use energy from photosynthetic partial reactions (ROBINSON, 1969b). Where light stimulation occurs there is probably dependence on photosynthetic partial reactions rather than on respiratory reactions, but whether ATP is involved and which photosystem is required is unclear (KYLIN, 1964; PENTH and WEIGL, 1971). On balance it is likely that ATP is required.

As with phosphate, SO_4^{2-} influx is stimulated in SO_4^{2-}-deficient cells, the main effect being on K_m (VALLÉE, 1968; VALLÉE and JEANJEAN, 1968a). The feed-back signal seems to be some sulfate ester rather than internal inorganic SO_4^{2-} itself (VALLÉE and JEANJEAN, 1968b). Most evidence suggests that SO_4^{2-} influx greatly exceeds SO_4^{2-} efflux during net SO_4^{2-} transport, at least during recovery from SO_4^{2-} deficiency (WEDDING and BLACK, 1960; VALLÉE and JEANJEAN, 1968a, b; ROBINSON, 1969a). This may well be due, as with phosphate, to a low internal concentration rather than to a very low permeability (RAVEN, 1974c). The fact that phosphate and SO_4^{2-} are much further from flux equilibrium during growth than is Cl^- (Table *6*.7) is consistent with the lower tracer influxes of phosphate and SO_4^{2-} relative to Cl^- (Table *6*.7) in comparison with the cell content (Table *6*.1). In contrast to the situation with phosphate, SO_4^{2-} influx in *Ankistrodesmus* is stimulated by K^+ rather than Na^+ (ULLRICH-EBERIUS, 1973a).

The sulfate transport system is also able to transport chromate and selenate (KYLIN, 1967; VALLÉE and JEANJEAN, 1968a; VALLÉE, 1969; ROBINSON, 1969a). Here internal selenium is toxic by virtue of its conversion to seleno-analogs of sulfur-containing amino-acids (KYLIN, 1967; JEANJEAN, 1973a). The kinetics of algal SO_4^{2-} influx are discussed by ROBINSON (1969a). (See also *7*.6.)

6.4.3 Nitrate

Nitrate influx is strongly correlated with NO_3^- reductase activity in microalgae such as *Ankistrodesmus* (ULLRICH-EBERIUS, 1973b); this means that incorporation of its mechanism and energetics independent of those of nitrate reductase is difficult. Nitrate influx is increased in NO_3^--deficient cells (EPPLEY and ROGERS, 1970). The NO_3^- transport system can also transport chlorate (TROMBALLA and BRODA, 1971); this is less toxic in *Chlorella* than in higher plants, since the chlorate is reduced to Cl^- rather than the toxic chlorite and hypochlorite.

Relatively few experiments have been carried out on NO_3^- fluxes in algae in which there is a relatively high concentration of free NO_3^- in the cell. In *Nitella clavata* the accumulation of free NO_3^- is light-stimulated (HOAGLAND and DAVIES, 1923). In *Ditylum brightwellii*, NO_3^- net influx also exceeds the rate of NO_3^- reduction (EPPLEY and COATSWORTH, 1968). Here the light stimulation of NO_3^- reduction is much greater than that of NO_3^- influx, and the influx is much less sensitive to DCMU than is reduction. In *Ulva*, NO_3^- influx is not light-stimulated, but NO_3^- reduction is (FRAZISKET, 1973).

In these cases the close regulation of NO_3^- influx by the rate at which it is metabolized is not present, and it is possible to study influx subject only to the controls exerted on non-metabolized ions (energy supply, ionic and turgor regulation).

Within the concentration range studied, NO_3^- net influx in planktonic microalgae fits a rectangular hyperbola, with a very low half-saturation concentration of some 0.1–5 µM, depending on the algal species (CORNER and DAVIES, 1971).

No experiments appear to have been conducted with ^{15}N-nitrate (or ^{15}N-ammonium: *6.6.4.5*) in algae with the object of measuring bidirectional fluxes. This may be partly due to the experimental complexity of measuring non-radioactive isotopes, although metabolic and ecological studies have been made on algae using ^{15}N labeled NO_3^- and ammonium (e.g. CORNER and DAVIES, 1971).

6.4.4 Bicarbonate

What little evidence is available on the energy source for HCO_3^- influx suggests that it is similar, in *Hydrodictyon africanum*, to that for Cl^- influx, i.e. non-cyclic electron transport (RAVEN, 1970). FINDENEGG (1974) has shown that light-stimulation of Cl^- influx in *Scenedesmus* is correlated with the ability to use HCO_3^- in photosynthesis. However, the data on competition between HCO_3^- and Cl^- (RAVEN, 1968b, 1970) in *H. africanum* are not entirely consistent with the two ions being transported by the same mechanism.

6.4.5 Ammonium

Since the evidence that ammonium ion is actively transported in algae is equivocal (*6.5.5.4*), the interpretation of evidence that net ammonium influx in *Ditylum brightwellii* has a half-saturation constant of about 1 µM, is light-stimulated and is relatively insensitive to DCMU (compared with nitrate and nitrite reduction in the same organism) is difficult (EPPLEY and ROGERS, 1970).

6.4.6 Protons (and Hydroxyl Ions)

H^+ and OH^- net fluxes are major components of ion fluxes in growing algal cells (*6.4.6.*). Net H^+ efflux and, at high pH, net OH^- efflux, are active (*6.5.5.5*). There are many complications in interpreting changes of external pH in terms of H^+ or OH^- net flux across the membrane, since many other metabolic processes lead to changes in external pH (RAVEN and SMITH, 1974; BARR et al., 1974). Bidirectional (tracer) fluxes of H^+ or OH^- cannot be measured. Data on energy sources for the active H^+ extrusion pump in algae are relatively scant, and are largely based on the response of the hyperpolarization which is thought, in many cases, to reflect the properties of H^+ efflux (*6.6.6.2*). In *Nitella translucens* (SPANSWICK, 1972, 1974a, b; VREDENBERG and TONK, 1973) the pump is ouabain-insensitive, is stimulated by light, and (on the basis of inhibition by CCCP and DCCD) requires ATP. The use of inhibitors to investigate the metabolic linkage of the H^+ pump is even more risky than usual, since many uncouplers (e.g. CCCP) catalyze H^+ uniport in biological membranes, and many energy-transfer inhibitors (e.g. DCCD) inhibit the bacterial H^+ pump directly (RAVEN, 1971e; SPANSWICK, 1973, 1974a). The coupling to photosynthetic reactions is of interest, in that CO_2 inhibits when both photosystems are operative as well as when only photosystem I is operative (SPANSWICK, 1974a). This is consistent with a coupling to cyclic photophosphorylation alone, and (as SPANSWICK points out) an inhibition of cyclic photophosphorylation by CO_2 *via* "over-oxidation" when only photosystem I is operative (cf. RAVEN, 1971b). It also requires that cyclic photophosphorylation is inhibited when non-cyclic photophosphorylation is occurring (presence of CO_2 in white light) and that the ATP from this non-cyclic photophosphorylation is not available for the H^+ pump. SHIEH and BARBER (1971) showed that an active H^+ extrusion accompanied net K^+ and Na^+ fluxes in *Chlorella;* this probably has the same energy source (i.e. ATP from oxidative or cyclic photophosphorylation) as the other cation fluxes. Nothing appears to be known of the energy source or other characteristics of the active OH^- efflux which is found in some algae at high external pH values.

6.5 Permeability Coefficients of Cell Membranes for Ions

As discussed earlier in Chapter *3*, computation of ion permeability coefficients can be attempted from tracer flux measurements combined with measurements of the electrical PD, or from electrical measurements alone.

"Electrical" and "tracer" permeabilities are often different for the same ion in the same plant (see Table *6*.9) but relative permeabilities to different ions in the same plant are usually ranked in the same order by the two methods (e.g. FINDLAY et al., 1969, 1970). Despite uncertainties about the calculated values due to corrections for chemically coupled fluxes, and underestimation of tracer fluxes due to inadequate mixing of the cell-wall solution with the bulk solution (WALKER and HOPE, 1969), a number of features emerge from these results. P_K is generally greater than P_{Na}, and in turn P_{Cl} is even lower. The permeability of *Chlorella* plasmalemma is relatively low; this has been attributed to the large energy requirement for "pump and leak" maintenance of solute content in cells with a high surface/volume ratio (BARBER, 1968d; ROTHSTEIN, 1972). P_K for the plasmalemma of marine algae is generally higher than that in freshwater algae; the difference for Na^+ is less marked (possibly because the plasmalemma flux is more likely to be underestimated; PITMAN, 1971). This high P_K in marine algae might be related to the apparent increase in P_K in freshwater algae when the K^+ concentration is increased above 0.8–1 mM (KISHIMOTO, 1966; SPANSWICK, 1969).

In all cases the permeability of algal membranes is much higher than the permeability of lipid bilayers (Table *6*.9), suggesting that the algal membranes must contain some uniport-catalyzing agents (cf. HAYDON and HLADSKY, 1972). The occurrence of electro-osmosis in algal cell membranes is consistent with the occurrence of membrane channels which can carry both ions and water, although other explanations are not ruled out (BRIGGS, 1967; BARRY and HOPE, 1969; MACROBBIE and FENSOM, 1969).

The apparent P_{ion} of plant cell membranes varies with a variety of experimental conditions, as shown by both tracer and electrical experiments. However, as with the absolute values of P_{ion}, these two methods do not give quantitatively similar responses to environmental changes (compare the electrical data of HOGG et al., 1968, 1969, with the flux data of MACROBBIE, 1962, 1964 and SMITH, 1967a, for the effect of light and temperature on P_K and P_{Na} in *Nitella translucens*).

Table 6.9. Permeability to ions of algal cell membranes, in $m s^{-1}$. Calculated from tracer fluxes, concentrations and electrical potential differences, or "electrically" (see 4.2.2.2 or 3.5.3.1), where possible corrections have been made for non-independent ion fluxes

	Membrane	Method	Permeability coefficients $(m s^{-1}) \times 10^{11}$					
			K^+	Na^+	Cl^-	Ca^{2+} (Mg^{2+})	$H_2PO_4^-$	H^+
1	Plasmalemma freshwater Charophytes and *Hydrodictyon*	Tracer	50–700	10–100	0.2–4	40	2	–
2	Plasmalemma freshwater Charophytes	Electrical	3,000 to 10^4	1,000 to 2,000	–	–	–	$\leqq 10^4$
3	Tonoplast freshwater Charophytes and *Hydrodictyon*	Tracer	50–100	30–300	100 to 1,000	–	300	–
4	Plasmalemma *Chlorella*	Tracer, Electrical	20	2	–	–	–	–
5	Plasmalemma, *Chaetomorpha, Acetabularia, Griffithsia, Enteromorpha, Pelvetia*	Tracer, Electrical	500 to $4 \cdot 10^4$	6–100	60–300	1–2 (1–2)	–	–
6	Tonoplast, *Chaetomorpha*	Tracer, Electrical	20–30	20–30	800	–	–	–
7	Polar lipid bilayer	Tracer, Electrical	0.2–0.5	0.2–0.5	0.1	–	–	–

Rows 1 and 3: HOPE and WALKER, 1960; MACROBBIE, 1962; MULLINS, 1962; HOPE, 1963; FINDLAY and HOPE, 1964; BARR, 1965; SPANSWICK and WILLIAMS, 1965; HOPE et al., 1966; RAVEN, 1967a; SMITH, 1967a, 1968; LANNOYE et al., 1970; RAVEN, 1974c and unpublished.
Row 2: HOPE and WALKER, 1961; KITASATO, 1968; HOGG et al., 1969; WALKER and HOPE, 1969; LANNOYE et al., 1970; BROWN et al., 1973; GILLET and LEFÈVRE, 1973; SPANSWICK, 1973.
Row 4: BARBER, 1968b, 1968d.
Row 5: GUTKNECHT and DAINTY, 1968; FINDLAY et al., 1969, 1970; SADDLER, 1970b; BLACK, 1971; FINDLAY et al., 1971; ROBINSON and JAFFE, 1973.
Row 6: FINDLAY et al., 1971.
Row 7: PAGANO and THOMPSON, 1968; THOMPSON and HENN, 1970.

In general, P_{ion} is decreased by low temperatures (but see SADDLER, 1970b) and increased by illumination (see references in RAVEN, 1971c, and in 6.5.3; also HOPE and ASCHBERGER, 1970). While the temperature effect can be explained in terms of direct action on the membrane (cf. THORHAUG, 1971), the light effect cannot be so explained, since the action spectrum is invariably that of the photosynthetic pigments which do not occur in the plasmalemma of eukaryotes (references in RAVEN, 1969a, 1971c; WAGNER, 1974). Metabolic inhibitor effects can to some extent be explained in terms of direct action on the plasmalemma (MACROBBIE and FENSOM, 1969), the effects of other inhibitors, like those of light, are probably exerted *via* effects on metabolism in chloroplasts, mitochondria, etc. The nature of the metabolic signal which alters P_{ion}, and how P_{ion} is altered, remain unknown. The involvement of metabolism appears to be *via* oxidative reactions, and non-cyclic electron transport (and phosphorylation?) in photosynthesis (RAVEN, 1969a, 1971c). It is more likely that the metabolic effect is a triggering reaction (albeit dependent on continuous modifier availability: an on/off modification of, e.g., a polypeptide uniporter) than a stoichiometric relationship of metabolic energy to ion flux, which would take the passive flux into the realms of active transport as defined by irreversible thermodynamics (MACROBBIE, 1970a). These metabolic effects on passive fluxes probably reflect action on "carriers" or "pores" rather than on gross membrane structure (see also *6*.7.2.2.3).

6.6 Electrogenic Pumps

6.6.1 The Resting Potential

In some algae the resting potential of the plasmalemma is accounted for quite well by the Goldman equation (*3*.5.3.1): e.g. *Griffithsia* (FINDLAY et al., 1969), *Valonia* (GUTKNECHT and DAINTY, 1968), *Enteromorpha* (BLACK, 1971). Here the potential lies in the range of passive diffusion potentials for the major ions, and the effect of changes in the concentration of ions in the external solution can be accounted for in terms of the concentration-independent permeability values.

In other cases (e.g. many freshwater coenocytes) the resting potential often lies within the range permitted by diffusion potentials for the major ions, but the small changes in potential with variation of K^+ and Na^+ concentrations in the range 0.1–1.0 mM in solutions containing Ca^{2+} are difficult to explain in terms of the Goldman equation (e.g. FINDLAY and HOPE, 1964; KISHIMOTO et al., 1965; SPANSWICK et al., 1967). Flux experiments suggest that P_K and P_{Na} are concentration-dependent in this range (*6*.6.5). In the absence of Ca^{2+} the potential is more readily explained in terms of a Goldman-type equation (SPANSWICK et al., 1967); this may be related to the higher values of P_K and P_{Na} under these conditions (HOPE, 1963; SMITH, 1967a). While pH changes can alter electrical potentials in Ca^{2+}-containing solutions, control of the plasmalemma PD by H^+ diffusion is likely only at low pH values (KISHIMOTO and AKABORI, 1959; KISHIMOTO, 1959; ANDRIANOV et al., 1968; LANNOYE et al., 1970; STRUNK, 1971; SPANSWICK, 1973). Even at low pH the depolarizing effect of increasing H^+ concentration has been attributed to an increased P_{Na} rather than to H^+ diffusion *per se* (LANNOYE et al., 1970).

Thus attempts to explain the plasmalemma potential difference in freshwater coenocytes in "natural" (Ca^{2+}-containing) solutions solely in terms of diffusion gradients of inorganic ions have met with considerable difficulties.

6.6.2 Evidence for Electrogenic Pumps

Much more difficult to reconcile with explanation of resting potentials in terms of diffusion potentials are cases in which ψ_M is outwith the range of diffusion potentials of major ions. Specifically, there are a number of cases in which the resting potential is more negative than the K^+ equilibrium potential. In some cases this is by such a large amount that explanations in terms of an under-estimation of ψ_K because net K^+ influx depletes K^+ just outside the plasmalemma (WALKER and HOPE, 1969) are implausible.

Examples of such hyperpolarization are found for both marine and freshwater algae. The most spectacular example is *Acetabularia,* where ψ_K is about -90 mV and ψ_M is about -170 mV (SADDLER, 1970a, 1971; GRADMANN, 1970). This phenomenon is also seen to a smaller extent in various freshwater charophytes (e.g. HOPE, 1965; KISHIMOTO, 1966; VOLKOV and MISYUK, 1967; KITASATO, 1968; SPANSWICK, 1972; SAITO and SENDA, 1973a, b).

Other criteria for the existence of electrogenic pumps are less well founded, making depolarizing pumps and pumps causing small hyperpolarizations difficult to establish. The existence of a short-circuit current in the steady state in a complex membrane system (BLOUNT and LEVEDAHL, 1960; GUTKNECHT, 1967; STRUNK, 1971) is evidence for active ion transport, but does not necessarily indicate that the pump is electrogenic (GINZBURG and HOGG, 1967). The absence of such an effect does not mean that the pump is not electrogenic.

A final method for distinguishing electrogenic pump potentials from diffusion potentials is the rapidity with which inhibition of metabolism alters the electrical potential difference. From the Goldman equation, it is seen that inhibition of metabolism powering ion pumps should only affect the potential inasmuch as passive ion fluxes decrease ion gradients; in freshwater coenocytes this is a slow process ($t_{\frac{1}{2}}$ of days). If metabolism is inhibited by low temperature the diffusive PD component should be reduced slightly (directly proportional to the absolute temperature). However, as discussed in *6*.6.5, there are metabolic effects on permeabilities to ions, such that the ratios of ion permeabilities are altered by changes in light or temperature, and by other metabolic perturbations. This changes the PD predicted by the Goldman equation, as do changes in ion concentration in the cytoplasm such as might be caused by light-dependent redistribution of ions between the chloroplast and the rest of the cytoplasm (e.g. THROM, 1970).

The ion responsible for the hyperpolarization in *Acetabularia* is Cl^-. Reversible inhibition of the active Cl^- influx by Cl^- removal, low temperature treatment, darkening or addition of the inhibitors DNP or CCCP rapidly and reversibly depolarizes the plasmalemma potential (SADDLER, 1970c; GRADMANN, 1970). The residual potential can then be fitted to the Goldman equation with only passive terms. The properties of the hyperpolarization and flux suggests a linkage to cyclic photophosphorylation (SADDLER, 1970c; GRADMANN, 1970).

Cl^- removal from the medium surrounding *Halicystis ovalis* (BLINKS, 1940), *Nitella clavata* (BARR, 1965), *Chara corallina* (FINDLAY et al., 1969) and *Bryopsis plumosa* (MUNDAY, 1972) causes a rapid depolarization; this is consistent with the occurrence of an electrogenic pump. Results with other freshwater coenocytes (references in RAVEN, 1968a; LEFEVRE and GILLET, 1970; SPANSWICK, 1973, 1974a; Fig. *6*.8) do not, however, show such an effect. Thus electrogenic Cl^- influx is not well

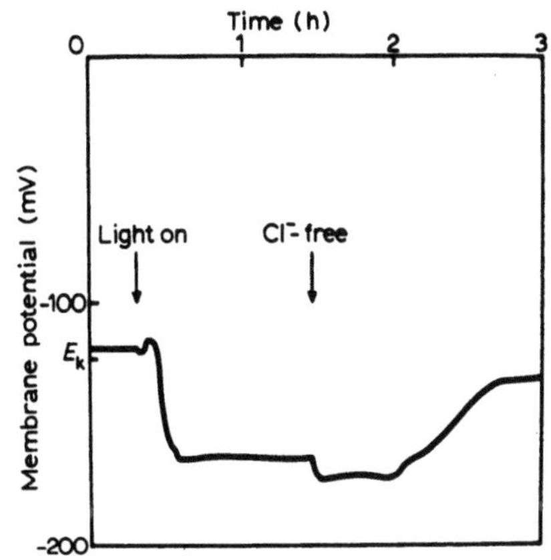

Fig. *6*.8. The membrane potential of *Nitella translucens* in artificial pond water plus 0.4 mM K_2SO_4, pH 6, in CO_2-free air. Initial reading in dark; arrows indicate illumination, and transfer to Cl^--free solution. (From SPANSWICK, 1974a)

established in fresh water coenocytes (cf. RAVEN, 1968a; SMITH, 1967a, 1970). PICKARD (1973) interprets the depolarization upon replacement of external sulphate by chloride in *Chara braunii* as being due to a coupled cation-Cl^- pump carrying excess cations (cf. 6.6.2.3). This seems in need of further investigation since direct measurement of cation-Cl^- ratio suggests the reverse (reviewed by MACROBBIE, 1970a).

HCO_3^- was proposed by HOPE (1965) as the electrogenically pumped ion in *Chara corallina*. While there is considerable evidence that HCO_3^- can be actively transported into algal cells (6.5.5.3; RAVEN, 1970), its involvement in electrogenesis is problematical (SPANSWICK, 1970). Much of the data on electrogenic pumps in freshwater coenocytes is currently interpreted in terms of active H^+ extrusion (see 6.6.4.6 and KITASATO, 1968; SPANSWICK, 1973, 1974a, b; VREDENBERG and TONK, 1973). However, the direct evidence for this view is not great. In part this stems from the difficulties in measuring net transmembrane H^+ fluxes, and the impossibility of measuring tracer H^+ fluxes (see RAVEN and SMITH, 1974; Chap. *12*). For example, THROM (1972) has investigated the correlation between light-induced net H^+ influx and the light-induced depolarization in *Griffithsia*. The different responses of the two effects to different inhibitory treatments suggests that the depolarization is not caused by an electrogenic OH^- extrusion; however, the external pH responses in light and dark are in the same direction as those induced by photosynthesis and respiration, and the contribution of these processes to the observed changes is unknown.

Difficulties in quantifying electrogenic pumps are numerous. In view of difficulties in interpretation of membrane conductance, it is clear that a direct measurement of flux from conductance measurements is not simple. The role of electrogenic pumps in altering passive fluxes has been mentioned earlier (6.6.2.2) and by RAVEN and SMITH (1974). SADDLER (1970b, c, 1971) has shown for *Acetabularia* that the correlation of passive ion fluxes with the electrogenic potential is not good; this suggests a voltage-dependent change in P_{ions}. It is clear that such a secondary coupling of passive fluxes to an electrogenic pump can have a metabolic dependence, although whether this is the same as that of the pump depends on how metabolism changes with ion permeability (6.6.5). Further, granted certain assumptions, it can have the concentration-dependence characteristic of an "ion pump" (BRIGGS, 1963). This underlines the importance of electrochemical measurements in determining active transport.

6.7 Strategies of Ion Transport

Ion transport in algal cells has both an osmoregulatory and a nutritional role (6.2 and 6.3). In general (6.5) anions are actively transported into plant cells, whether the ion is a nutrient or is involved in osmoregulation. An exception is OH^-, which is actively extruded at the plasmalemma during growth at high external pH. As regards cations, Na^+ and H^+ are usually extruded actively at the plasmalemma. Active K^+ transport shows much more variability.

In general (6.6) it appears that when nutrient anions are supplied at "ecological" concentrations, net influx and tracer influx are very similar, i.e. efflux is very small. This is certainly true of phosphate and sulfate; nitrate has not apparently been tested. This is not true of the major osmoregulatory ions (K^+, Na^+, Cl^-, Ca^{2+}, Mg^{2+}), where the bidirectional (tracer) fluxes generally exceed the influx (e.g. GUTKNECHT and DAINTY, 1968; SADDLER, 1970b, c; ROBINSON and JAFFE, 1973). Thus there is a "pump and leak" situation for these ions (data for Ca^{2+} and Mg^{2+} are rather sparse), and it appears that the "leak" is more readily checked (possibly by decreasing a metabolically controlled efflux) for total salt content than it is for K^+/Na^+ regulation (*6.6*). The partitioning of these tracer fluxes into active, exchange and passive components is often very complex (RAVEN, 1968a; FINDLAY et al., 1969; FINDLAY et al., 1970; FINDLAY et al., 1971; BLACK, 1971). At least as far as nutrient anions are concerned, control of content is *via* a control of influx rather than of the (usually very small) efflux (*6.6*). These considerations apply mainly to the plasmalemma; in vacuolate cells it is generally found that cations (H^+, Na^+, more rarely K^+) are actively transported from cytoplasm to vacuole (*6.5* and *6.6*). This, and possible active Cl^- transport inwards at the tonoplast, may be related to volume regulation of cytoplasm and vacuole.

Growth of algal cells, and maintenance of mature cells, requires that the fluxes of the various ions are co-ordinated. This is considered by CRAM (Chap. *11*). Suffice it to say here that a case has been made in some algae for a coupled K^+/Na^+ exchange at the plasmalemma and in many algae for a coupling of active Cl^- influx to that of K^+ and/or Na^+ at the plasmalemma. Influx of various cations (K^+, NH_4^+, Na^+) coupled to H^+ active efflux, and active influx of Cl^-, NO_3^- or HCO_3^- coupled to OH^- efflux (active or passive) has also been suggested (*6.6*). Patently, the coupled KCl or NaCl influxes could be the result of interactions of cation-H^+ and Cl^+—OH^- exchanges (SMITH, 1970, 1972). Alternatively, the observed interactions might relate to co-ordinated control of separate transport mechanisms, rather than coupled transport. Even where, in short-term experiments, such interactions do not appear to occur (SADDLER, 1970b, c), there must be some underlying regulation and co-ordination of the transport of nutrient and osmoregulatory ions, and of pH regulation.

The requirement for active transport rather than facilitated diffusion for the uptake of metabolized solutes has already been mentioned (*6.4* and *6.6*). It was suggested that active transport was required to keep the internal concentration high enough for an adequate rate of metabolism by the first enzyme of the solute assimilation pathway. Alternatives to active transport in this situation are increasing the concentration of enzyme, or modification of the enzyme to increase substrate affinity. However, the former requires extra capital investment in protein, while the latter can involve both additional enzymes and extra energy input to the metabolic sequence. Examples of the latter approach are the four-carbon acid pathway of photosynthesis, which appears not to occur in algae (*6.7*), and the glutamine synthetase – glutamine oxoglutarate amino transferase (oxido-reducing) pathway, alternative to the glutamate dehydrogenase pathway of ammonia assimilation which may be present in some blue-green algae (DHARMAWADENE et al., 1973; BROWN et al., 1973). The selective advantages of these three strategies for coping with the use of nutrients present at very low external concentrations are not readily quantified; however, active transport is a common solution.

In terms of net fluxes associated with growth, the most investigated ion fluxes are not those which are the quantitatively most important (*6.3*). Thus for growth of *Chlorella* in continuous illumination, using the data of Table *6.1*, of BARBER, 1968a, b, and of HOOGENHOUT and AMESZ (1965), it can be shown that the net influx of inorganic carbon during growth is some 300 nmol m^{-2} s^{-1}, that of ammonium or nitrate (and the associated net H^+ flux whose direction depends on the N source) is some 30 nmol m^{-2} s^{-1}, that of K^+, phosphate and sulfate is some 3 nmol m^{-2} s^{-1}, while Cl^- and Na^+ net influxes are only some 0.3 nmol m^{-2} s^{-1}. BARBER (1968a, b, c) has shown that tracer fluxes of K^+, Na^+ Cl^+ in the light are at least three times these net fluxes. However, in both *Scenedesmus* and *Chlorella* the net K^+ influx and Na^+ efflux associated with recovery from K^+ deficiency in "Na^+-rich" cells are larger than the net fluxes associated with growth, or the tracer fluxes in "normal" cells (BARBER, 1968c, d, e, 1969; ERDEI and CSEH, 1971; SHIEH and BARBER, 1971). The capacity of the K^+/Na^+ exchange in *Chlorella* is up to 200 nmol m^{-2} s^{-1} (SHIEH and BARBER, 1971).

For algae in which bicarbonate is the form in which inorganic carbon enters the cell, bicarbonate influx and the corresponding OH^- efflux are the largest net fluxes (see RAVEN, 1970, *12.3.1* and *12.5.2*). In many marine giant algal cells the tracer ion fluxes at the plasmalemma are greater than the net CO_2 and O_2 fluxes involved in photosynthesis (e.g. in *Griffithsia;* FINDLAY et al., 1970; LILLEY and HOPE, 1971).

In a number of algae, different strategies to those mentioned above are adopted to meet common transport problems. Perhaps it is not surprising that most of the anomalies should involve the most investigated ions K^+, Na^+ and Cl^-. In some algae, active Cl^- transport does not seem to occur (*6.5.3*). In *Porphyra* active nitrate influx appears to replace it in contributing anions to turgor generation. In *Valonia* and *Chaetomorpha*, Cl^- is the major turgor-generating anion, but is close to electrochemical equilibrium, with active K^+ transport at the tonoplast maintaining the high salt content of the vacuole. Active K^+ transport at the tonoplast is also found in *Griffithsia* which has a normal Cl^- pump (*6*.5).

In some green algal marine coenocytes the K^+/Na^+ ratio in the vacuole is very much lower than usual (*6*.4; Tables *6*.3, *6*.4), although in the only case investigated *Bryopsis plumosa* a more normal K^+/Na^+ ratio is found in the cytoplasm (RIGLER, quoted by GUTKNECHT and DAINTY, 1968). This anomaly might involve mainly a peculiarity of the tonoplast, where the frequently occurring Na^+ pump which moves Na^+ into the vacuole might be much faster than usual, and with normal K^+/Na^+ discrimination at the plasmalemma.

7. Transport of Non-Electrolytes

7.1 Neutral Inorganic Compounds

Important neutral compounds which enter or leave algal cells during growth and maintenance include CO_2, O_2, N_2, NH_3 and H_2O. CO_2 is the form in which inorganic carbon enters the plant cell at low pH (below pH 8–9: OSTERHOUT and DORCAS, 1925; RAVEN, 1970) and leaves it during respiration. O_2 is, conversely, lost during photosynthesis and taken up in respiration. N_2 is taken up during nitrogen fixation by certain Cyanophyceae, while NH_3, is the form in which ammonia enters at high pH (COOPER and OSTERHOUT, 1930) where there is an appreciable concentration of the undissociated form (cf. RAVEN and SMITH, 1974). H_2O is the solvent for intracellular solutes as well as participating in many chemical reactions. In all of these cases it is widely assumed that the fluxes are passive; what little evidence is available is consistent with the presence of a chemical potential gradient in the direction of the net flux (RAVEN, 1970), and that passive permeability is high enough to account for the observed flux with such gradients (see *6*.7.2.2).

While this assumption is, therefore, probably correct, it does lead to problems in terms of the retention of metabolites, e.g. CO_2 produced from HCO_3^- taken up by photosynthetic cells at high pH (RAVEN, 1970, 1972b), or of ammonia produced by reduction of nitrate or N_2. Problems of relative permeabilities of natural membranes to O_2 and N_2 arise if the Cyanophycean heterocyst is considered (STEWART, 1973) as a means of allowing N_2 to reach nitrogenase while keeping inhibitory O_2 out.

Active transport of CO_2 from the medium to the site of ribulose diphosphate carboxylase activity, *via* a dicarboxylate-monocarboxylate shuttle, is believed to occur in certain higher plants [four carbon acid (C_4) pathway of photosynthesis; transport aspects of this pathway are discussed by RAVEN (1972b)]. This pathway does not appear to occur in algae (RAVEN, 1970, 1974b). Its operation raises difficulties with respect to the retention of the pumped CO_2 due to its high permeability (RAVEN, 1972b). Aspects of CO_2 and NH_3 transport related to their nature as weak electrolytes is dealt with in *6*.6.4.4 and *6*.6.4.5.

7.2 Organic Compounds

7.2.1 General Considerations

Heterotrophic growth implies the ability to take up organic compounds by the cells concerned, and there is evidence that even obligate autotrophs can take up organic compounds although they cannot grow on them (VAN BAALEN and PURITCH, 1973; DROOP, 1974). However, little is known of the internal concentration of these compounds during steady-state growth conditions (cf. the difficulties with making these measurements: *6.4.3*). The involvement of active transport is not well understood in many cases (an exception is the incisive work of TANNER and KANDLER and their co-workers on active hexose transport in green microalgae: *6.7.2.3*).

Organic compounds make substantial contributions to the osmotic pressure of algal cytoplasm in the form of essential metabolic intermediates, and catalytic and informational macromolecules. Other organic solutes are energy and carbon reserves, and are involved in osmoregulation. Mannitol can reach concentrations approaching 10^3 mM in phaeophyceae such as *Laminaria*, where seasonal variation suggests that a major function is as a carbon and energy reserve (BLACK, 1950; WILLENBRINK and KREMER, 1973). In other algae a variety of organic solutes function in regulation of turgor (walled cells) or cell volume (wall-less cells) in response to varying external osmotic pressure as dealt with in *11.3*; see also RAVEN, 1971c.

7.2.2 Permeability of Membranes to Non-Electrolytes

7.2.2.1 General Aspects

DAINTY (*2.4.2*) discusses the contribution that work on algal cells has made to our knowledge of non-electrolyte permeability in plants. The agreement between the permeability values found for algal cells and those for model bimolecular leaflet membranes is good for water, monohydric aliphatic alcohols, glycerol, urea and indole acetic acid (THOMPSON and HENN, 1970; RAVEN, 1975a: experiments on indole acetic acid permeation in *Hydrodictyon africanum*). However, it is clear that the very low permeabilities of model membranes to metabolites like glucose and mannitol (less than 10^{-10}m s^{-1}: VREEMAN, 1966; WOOD et al., 1968) could not support the observed glucose fluxes in algae (e.g. SMITH, 1967b). Thus even where active transport has not been proved, the transport of sugars and other polar organic compounds such as amino-acids at nutritionally significant rates requires some modification to the basic lipid bilayer (DROOP, 1974).

7.2.2.2 Difficulties in Measuring Very High Permeability Coefficients

These arise because of the presence of unstirred layers of water on either side of the membrane. Within these layers transport is diffusive and when its thickness makes the *P* values for movement through the unstirred layers of the same order as that for membrane transport, determination of *P* for the membrane alone is very difficult. This problem is particularly severe for the common metabolic gases CO_2, O_2 and N_2, where various lines of evidence suggest that P_{gas} for the membrane *per se* is at least 10^{-3} m s^{-1} (RAVEN, 1970), i.e. several orders of magnitude higher than P_{gas} in the unstirred layers. P_{NH_3} is at least 10^{-6} m s^{-1}; this estimate may well be limited by unstirred layer effects (BARR et al., 1974).

7.2.2.3 Effects of External Factors and of Metabolism

Passive permeability to non-electrolytes is influenced by temperature and osmotic pressure. The only well-authenticated case of metabolic modification of passive permeability not attributable to uniports of the type involved in ion and hydrophilic neutral solute permeation (*6.6.5*) is the effect of phytochrome on water permeability in *Mougeotia* (WEISENSEEL and SMEIBIDL, 1973).

7.2.3 Active Transport of Organic Compounds

The best-investigated example of active transport of an organic solute in algae is that of glucose and glucose analogs in *Chlorella* (and, in less detail, *Ankistrodesmus, Scenedesmus* and *Bumilleriopsis*). (This is the work of TANNER and co-workers, and some aspects of it are discussed in Part B, *5.3*.1.1.2.)

The first evidence that active transport was involved in glucose influx came from the work of TANNER (1969) and KOMOR and TANNER (1971). Various analogs of glucose (2 deoxyglucose, 6 deoxyglucose, 3-0 methyl glucose) were actively transported into *Chlorella vulgaris* and other microalgae (TANNER et al., 1970; HAASS and TANNER, 1974). Mutual competition between glucose and its analogs for influx into *Chlorella,* suggested that all the sugars shared a common mechanism, although active transport of glucose itself has not been demonstrated (see *6*.4). The analogs are very little metabolized (KOMOR and TANNER, 1974a), and the free solute accumulates to up to 10^4 times the external concentration.

The transport system is inducible to a variable extent in different algae; in one strain of *Chlorella vulgaris* the ratio of transport activities induced/uninduced is over 400; in all the other algae tested this ratio was less than 5 (TANNER and KANDLER, 1967; TANNER et al., 1970; HAASS and TANNER, 1974). The induction of transport capacity requires protein synthesis; turnover of the transport system occurs at a variable rate depending on conditions, but activity of the transport system does not require concommitant protein synthesis (TANNER and KANDLER, 1967; HAASS and TANNER, 1974).

Energy for the active hexose transport system can be supplied by oxidative phosphorylation (in dark, aerobic conditions) or by cyclic photophosphorylation (in light, anaerobic conditions) (TANNER, 1969; KOMOR and TANNER, 1971; DECKER and TANNER, 1972; KOMOR et al., 1973). However, recent evidence suggests that some closely related compound rather than ATP itself is the immediate chemical energy source (KOMOR and TANNER, 1974a). On the basis of extra O_2 consumption when hexose was being transported it was computed that some 1.18 ATP (or ATP equivalent) were used per hexose transported (DECKER and TANNER, 1972; KOMOR et al., 1973).

The kinetics of transport are complex. Hexose influx can only occur if energy is supplied, either from metabolism or from a coupled hexose efflux (exchange diffusion). This latter mechanism halves the apparent ATP requirement for tracer hexose influx when there is hexose inside the cell which can be exchanged (KOMOR and TANNER, 1971; KOMOR et al., 1972; KOMOR et al., 1973; KOMOR et al., 1973b). In energy-depleted cells the transport mechanism does catalyze a substantial hexose efflux, unless the antibiotic nystatin is present (KOMOR et al., 1973a, b). The energy released by the hexose efflux system can be used only for hexose influx, and not for other energy-requiring processes in the cell (TANNER et al., 1974).

Recently evidence has been presented for an H^+-hexose symport supplying the energy to the hexose influx system (KOMOR, 1973; KOMOR and TANNER, 1974b). This is further discussed in Part B, *5.3*.1.1.2.3. This suggests that the observations concerning the energetic linkage of hexose transport could well relate to the H^+ extrusion pump which sets up the inwardly-directed H^+ free energy gradient. The stoichiometry of H^+ to hexose, together with the O_2/hexose data of DECKER and TANNER (1972) suggests that the H^+ pump is a type 1 ATPase (or ATP equivalent-ase) in the terminology of MITCHELL (1966).

Ions other than H^+ affect hexose transport in *Chlorella*, possibly *via* an effect on the electrical potential difference across the membrane (KOMOR and TANNER, 1974b). SMITH (1967b) could not detect any influence of the external ionic environment on glucose influx in *Nitella translucens;* the increased sucrose (but not glucose) influx in *Nitella* when a hyperpolarizing current was passed (FENSOM et al., 1967) could be related to a stimulation of H^+-sucrose antiport. In neither of these cases, nor in *Chlorella*, has active transport of glucose (or sucrose) *per se* been demonstrated; assimilation and influx appear to be closely related (see 6.4.3). In *Chlorella*, the energy sources for induction of the transport system, its functioning, and of glucose assimilation appear to be very similar, i.e. oxidative or cyclic photosynthetic phosphorylation (KANDLER and TANNER, 1966; TANNER and KANDLER, 1967; KOMOR et al., 1973). In *Nitella translucens*, glucose assimilation in short-term experiments appears to involve cyclic photophosphorylation; in longer-term experiments (more than 2 h) non-cyclic photophosphorylation is the major ATP source (SMITH, 1967b). In *Chara corallina*, which appears to lack cyclic photophosphorylation, light stimulation of glucose uptake and metabolism uses non-cyclic photophosphorylation (SMITH and RAVEN, 1974).

The only other well-substantiated case of active transport of organic compounds in algae is the work of KNUTSEN (1972) on uracil, and PETTERSON and KNUTSEN (1974) on guanine transport in *Chlorella fusca*. These compounds are not non-electrolytes since they have ionizeable groups (—OH, —NH_2); however, at the pH at which these experiments were performed the net charge on the solutes was near zero. At least in the case of guanine the concentration factor of more than 10^4 makes passive transport in response to pH or electrical gradients as a complete explanation most implausible. These active transport processes are chemically specific, stimulated by light and are inhibited by DNP. In contrast to the situation with glucose in *Chlorella*, these purines and pyrimidines are taken up by a mechanism which allows appreciable accumulation within the cells, even though the solutes are metabolized.

In other cases (WIESSNER, 1970a; DROOP, 1974) the influx of sugars and amino-acids, and possibly of fatty acids, in algal cells is too rapid to be explained in terms of purely passive diffusion through an unmodified membrane; however, whether the transport is by active transport or facilitated diffusion is unsettled.

8. Conclusions and Evolutionary Speculations

The algal cell, like all others, requires that the concentration of essential metabolites inside the cell be kept within a certain range. This implies the presence of a membrane which can allow (or actively transport) nutrients into the cell, which rids the cell of toxic byproducts (e.g. H^+), and which does not allow essential metabolites generated in metabolism to leak out; and some mechanism which prevents the cell from expanding to bursting point in either hypotonic or isotonic media. The data reviewed in this Chapter suggest that present-day algae have membranes (and accessory structures such as walls) which perform these functions. Here I set out to suggest possible evolutionary routes by which this state of affairs could have arisen.

The production of metabolites from the solutes on which the cell grows requires metabolic manipulations which produce either excess H^+ or OH^-; this must be lost from the cell in order to maintain intracellular pH at a suitable level for enzymes to work in. Elsewhere (RAVEN and SMITH, 1973, 1974; *12*.2.4) it has been argued that H^+ extrusion is probably the fundamental process in pH control; there is a less widespread requirement for an OH^- extrusion pump. Thus a cation-H^+ exchange pump was probably an early acquisition in cell evolution. Why is the cation so often K^+? Volume regulation in primitive, wall-less cells (to which the present-day Mycoplasmas may be similar) probably involved an Na^+ extrusion pump, which means a high K^+/Na^+ ratio inside the cell (see DAINTY, 1962). Once enzymes had become adapted to this environment, both H^+ and Na^+ extrusion were coupled to K^+ influx. This K^+/H^+ and K^+/Na^+ exchange is required to keep the cytoplasmic environment suitable (in terms of H^+ and K^+ content) for metabolism. When cell walls evolved, the osmotic reason for the Na^+ pump was removed, but it was still needed for nutritional reasons (DAINTY, 1962). Osmoregulation in wall-less cells in freshwater requires contractile vacuoles (*6*.2).

The evolution of the vacuole may have been related to a requirement for the photoautotrophic algal cell to spread out its cytoplasm as much as possible to intercept light (for photosynthesis) and solutes; padding the cell out with a vacuole is a way of doing this. The vacuole requires not only additional solute transport processes for its production, but also continued active solute transport in the steady state for its maintenance (*6*.2) (GUTKNECHT and DAINTY, 1968). The solute transported is often Na^+; this raises interesting possibilities as to the inter-relations of Na^+ transport at the plasmalemma and tonoplast and in contractile vacuoles (*6*.2, *6*.5, *6*.6). The origin of the (non-contractile) vacuole is still a matter of some dispute (DAINTY, 1968). The contractile vacuole may be related to the activity of dictyosomes (SCHNEPF and KOCH, 1967), and can be involved in the excretion of Na^+ as well as water (BRUCE and MARSHALL, 1965). This may inter-relate the two sorts of vacuole, especially since the presence of a non-contractile vacuole is often correlated with the development of a cell wall whose synthesis involves dictyosome activity, which supplants the osmoregulatory function of the contractile vacuole. Contractile vacuoles may be involved in "emptying" the non-contractile vacuole in gametogenesis (ETTL, 1961; HOFFMANN, 1973). However, origin of the non-contractile vacuole from endoplasmic reticulum or lysosomes is also a possibility.

Another early requirement was anion and neutral molecule pumps to supply nutrients and osmoregulatory compounds. The possible relationships of these pumps to H^+ and Na^+ pumps is discussed elsewhere (*12*.2 and *12*.5). In view of the widely different properties of the pumps for different anions in various algae (*6*.6.2, *6*.6.4), it may well be that H^+ gradients may not be directly involved in them all, although regulatory rather than energetic considerations may be involved in some of the observed differences.

References

AARONSON, S.: Particle aggregation and phagotrophy by *Ochromonas*. Arch. Mikrobiol. **92**, 39–44 (1973).

AHMED, J., MORRIS, I.: The effect of 2,4 dinitrophenol and other uncoupling agents on the assimilation of nitrate and nitrite by *Chlorella*. Biochim. Biophys. Acta **162**, 32–38 (1968).

AIKMAN, D.P.: Ionic relations of marine algae. Ph. D. Thesis, University of East Anglia (1969).

AIKMAN, D.P., DAINTY, J.: Ionic relations of *Valonia ventricosa*. In: Some contemporary studies in marine science (H. BARNES, ed.), p. 37–43. London: Allen and Unwin 1966.

ALLEN, R.D., JACOBSEN, L., JOAQUIN, J., JAFFE, L.J.: Ionic concentrations in developing *Pelvetia* eggs. Develop. Biol. **27**, 538–545 (1972).

ANDERSON, D.M.W., KING, N.J.: Polysaccharides of the Characeae. II. The carbohydrate content of *Nitella translucens*. Biochim. Biophys. Acta **52**, 441–449 (1961 a).

ANDERSON, D.M.W., KING, N.J.: Polysaccharides of the Characeae. III. The carbohydrate content of *Chara australis*. Biochim. Biophys. Acta **52**, 449–454 (1961 b).

ANDRIANOV, V.K., BULYCHEV, A.A., KURELLA, G.A., LITVIN, F.F.: Effect of light on the resting potential and cation (K^+, Na^+) activity in the vacuole sap of *Nitella* cells. Biofizika **16**, 1031–1036 (1971).

ANDRIANOV, V.K., VOROBIEV, I.A., KURELLA, G.A.: Study of the nature of resting potential in *Nitella* cells. 2. Medium pH effect on resting potential of *Nitella* cells. Biofizika **13**, 335–336 (1968).

ANITA, N.J., KRIPPS, R.S., DESAI, I.D.: L-threonine deaminase in marine planktonic algae. III. Stimulation of activity by monovalent inorganic cations and diverse effects from other ions. Arch. Mikrobiol. **85**, 341–354 (1972).

ATKINSON, A.W., JR.: Ultrastructural studies on *Chlorella*. Ph. D. Thesis, Queens College, Belfast. (1972).

ATKINSON, A.W., JR., GUNNING, B.E.S., JOHN, PC.L., MCCULLOUGH, W.: Dual isotherms of ion absorption. Science **176**, 694–695 (1972a).

ATKINSON, A.W., JR., GUNNING, B.E.S., JOHN, P.C.L.: Sporopollenin in the cell wall of *Chlorella* and other algae: Ultrastructure, chemistry, and incorporation of ^{14}C-acetate studied in synchronous cultures. Planta **107**, 1–32 (1972b).

ATKINSON, D.E.: The adenylate energy charge in metabolic regulation. In: Horizons in bioenergetics (A. SAN PIETRO, H. GEST, eds.), p. 83–96. New York: Academic Press 1972.

ATKINSON, M.R., POLYA, G.M.: Salt-stimulated adenosine triphosphatase from carrot, beet and *Chara australis*. Australian J. Biol. Sci. **20**, 1069–1086 (1967).

AUKEN, O.W., VAN, MCNULTY, I.B.: Intracellular concentrations of sodium, potassium and chloride in a halophilic *Chlamydomonas*. Plant Physiol. Lancaster **44**, 20 (abstr.) (1969).

BAALEN, C. VAN, PURITCH, W.H.: Heterotrophic growth in the microalgae. Crit. Rev. Microbiol. **2**, 229–255 (1973).

BARBER, J.: Light-induced uptake of potassium and chloride by *Chlorella pyrenoidosa*. Nature **217**, 876–878 (1968a).

BARBER, J.: Measurements of the membrane potential and evidence for active transport of ions in *Chlorella pyrenoidosa*. Biochim. Biophys. Acta **150**, 618–625 (1968b).

BARBER, J.: Sodium efflux from *Chlorella pyrenoidosa*. Biochim. Biophys. Acta **150**, 730–732 (1968c).

BARBER, J.: The influx of potassium into *Chlorella pyrenoidosa*. Biochim. Biophys. Acta **163**, 141–149 (1968d).

BARBER, J.: The efflux of potassium from *Chlorella pyrenoidosa*. Biochim. Biophys. Acta **163**, 531–538 (1968e).

BARBER, J.: Light-induced net uptake of sodium and chloride by *Chlorella pyrenoidosa*. Arch. Biochem. Biophys. **130**, 389–392 (1969).

BARBER, J., SHIEH, Y.J.: Net and steady-state cation fluxes in *Chlorella pyrenoidosa*. J. Exptl. Bot. **23**, 627–636 (1972).

BARBER, J., SHIEH, Y.J.: Sodium transport in sodium-rich *Chlorella* cells. Planta **111**, 13–22 (1973a).

BARBER, J., SHIEH, Y.J.: Effects of light on net Na^+ and K^+ transport in *Chlorella* and evidence for *in vivo* cyclic photophosphorylation. Plant Sci. Letters **1**, 405–411 (1973b).

BARR, C.E.: Na^+ and K^+ fluxes in *Nitella clavata*. J. Gen. Physiol. **49**, 181–197 (1965).

BARR, C.E., BROYER, T.C.: Effect of light on Na^+ influx, membrane potential and protoplasmic streaming in *Nitella*. Plant Physiol. Lancaster **39**, 48–52 (1964).

BARR, C.E., KOH, M.S., RYAN, T.E.: NH_3 efflux as a means for measuring H^+ extrusion in *Nitella*. In: Membrane transport in plants (U. ZIMMERMANN, J. DAINTY, eds.), p. 180–185. Berlin-Heidelberg-New York: Springer 1974.

BARRY, P.H., HOPE, A.B.: Electro-osmosis in *Chara* and *Nitella* cells. Biochim. Biophys. Acta **193**, 124–128 (1969).

BASSHAM, J.A., KRAUSE, G.H.: Free energy changes and metabolic regulation in steady-state photosynthetic carbon reduction. Biochim. Biophys. Acta **189**, 207–221 (1969).

BATTERTON, J.C., BOUSH, G.M., MATSUMARA, F.: D.D.T.: Inhibition of NaCl tolerance by the blue-green alga *Anacystis nidulans*. Science **176**, 1141–1143 (1972).

BENTRUP, H.W.: Elektrophysiologische Untersuchungen am Ei von *Fucus serratus*: Das Membranpotential. Planta **94**, 319–332 (1970).

BERGQUIST, P.L.: Evidence for separate mechanisms of Na^+ and K^+ regulation in *Hormosira banksii*. Physiol. Plantarum **11**, 760–770 (1958).

BIEBL, R.: Seaweeds. In: Physiology and biochemistry of algae (R.A. LEWIN, ed.), p. 799–815. New York: Academic Press 1962.

BISHOP, N.I.: Analysis of photosynthesis in green algae through mutation studies. Photophysiology **8**, 65–96 (1973).
BLACK, D.R.: Ionic relations of *Enteromorpha intestinalis* (L. LINK). Ph. D. Thesis, University of St. Andrews (1971).
BLACK, D.R., WEEKS, D.C.: Ionic relations of *Enteromorpha intestinalis*. New Phytologist **71**, 119–127 (1972).
BLACK, W.A.P.: Seasonal changes in weight and chemical composition of the common British Laminariaceae. J. Marine Biol. Assoc. U.K. **29**, 45–72 (1950).
BLINKS, L.R.: The effects of current flow in large plant cells. Cold Spring Harbor Symp. Quant. Biol. **4**, 34–42 (1936).
BLINKS, L.R.: The relation of bioelectric phenomena to ionic permeability and to metabolism in large plant cells. Cold Spring Harbor Symp. Quant. Biol. **7**, 204–215 (1940).
BLINKS, L.R.: The source of the bioelectric potentials in large plant cells. Proc. Natl. Acad. Sci. U.S. **35**, 566–575 (1949).
BLINKS, L.R.: Bioelectric phenomena in *Hydrodictyon africanum*. Science **127**, 1058 (1958).
BLINKS, L.R.: Bioelectric properties of *Boergesenia forbesi*. Science **156**, 535 (1967).
BLINKS, L.R.: The effect of protoplasmic acidity and of light on the bioelectric potential of *Valonia* and *Boergesenia*. Proc. Natl. Acad. Sci. U.S. **63**, 223 (1969).
BLINKS, L.R., CHAMBERS, D.H.: Effect of light on the bioelectric potential of *Nitella*. Science **128**, 1143 (1958).
BLINKS, L.R., NIELSEN, J.P.: The cell sap of *Hydrodictyon*. J. Gen. Physiol. **23**, 551–559 (1940).
BLOUNT, R.W.: A quantitative analysis of active ion transport in the single-celled alga *Halicystis ovalis*. Ph. D. Thesis, University of California, Los Angeles (1958).
BLOUNT, R.W., LEVEDAHL, B.H.: Active Na and Cl transport in the single celled marine alga *Halicystis ovalis*. Acta Physiol. Scand. **49**, 1–9 (1960).
BLUM, J.J.: Phosphate uptake by phosphate-starved *Euglena*. J. Gen. Physiol. **49**, 1125–1137 (1965).
BOROWITZKA, L.J., BROWN, A.D.: The salt relations of marine and halophilic species of the unicellular green alga *Dunaliella*. The role of glycerol as a compatible solute. Arch. Mikrobiol. **96**, 37–52 (1974).
BOUCK, C.B., BROWN, D.L.: Microtubule biogenesis and cell shape in *Ochromonas*. I. The distribution of cytoplasmic and mitotic microtubules. J. Cell. Biol. **56**, 340–359 (1973).
BRIGGS, G.E.: Rate of uptake of salts by plant cells in relation to an anion pump. J. Exptl. Bot. **14**, 191–197 (1963).
BRIGGS, G.E.: Electro-osmosis in *Nitella*. Proc. Roy. Soc. (London) Ser. B **168**, 22–26 (1967).
BRIGGS, G.E., HOPE, A.B., ROBERTSON, R.N.: Electrolytes and plant cells. Blackwell Scientific Publications 1961.
BROWN, C.M., BURR, V.J., JOHNSON, B.: Presence of glutamate synthase in fission yeasts and its possible role in ammonia assimilation. Nature New Biology **246**, 114–116 (1973).
BROWN, D.F., RYAN, T.E., BARR, C.E.: The effect of light and darkness in relation to external pH on calculated H^+ fluxes in *Nitella*. In: Ion transport in plants (W.P. ANDERSON, ed.), p. 141–152. London: Academic Press 1973.
BROWN, D.L., BOUCK, G.B.: Microtubule biogenesis and cell shape in *Ochromonas*. II. The role of nucleating sites in shape development. J. Cell Biol. **56**, 360–378 (1973).
BRUCE, D.L., MARSHALL, J.H.: Some ionic and bioelectric properties of the amoeba *Chaos chaos*. J. Gen. Physiol. **49**, 151–178 (1965).
BUETOW, D.E.: The biology of *Euglena*, vols. I and II. New York: Academic Press 1968.
CHRISTENSEN, H.N., CESPEDES, C.DE., HARDLOGRON, H.E., RONQUIST, G.: Energisation of amino acid transport, studied for the Ehrlich ascites tumor cell. Biochim. Biophys. Acta **300**, 487–522 (1973).
COLLANDER, R.: Der Zellsaft der Characeen. Protoplasma **25**, 201–210 (1936).
COLLANDER, R.: Permeabilitätstudien an Characeen. III. Die Aufnahme und Abgabe von Kationen. Protoplasma **33**, 215–257 (1939).
COOK, J.R.: The cultivation and growth of *Euglena*. In: The biology of *Euglena* (D.E. BUETOW, ed.), vol. I, p. 243–314. New York: Academic Press 1968.
COOPER, W.C., OSTERHOUT, W.J.V.: The accumulation of electrolytes. I. The entrance of ammonia into *Valonia macrophysa*. J. Gen. Physiol. **14**, 117–125 (1930).
CORNER, E.D.S., DAVIES, A.G.: Plankton as a factor in the nitrogen and phosphorus cycles in the sea. Advan. Marine Biology **9**, 101–204 (1971).

COSTER, H.G.L.: Chloride in cells of *Chara australis*. Australian J. Biol. Sci. **19**, 545–554 (1965).
COSTER, H.G.L., HOPE, A.B.: Ionic relations of cells of *Chara australis*. XI. Chloride fluxes. Australian J. Biol. Sci. **21**, 243–254 (1968).
DAINTY, J.: Ion transport and electrical potentials. Ann. Rev. Plant Physiol. **13**, 379–402 (1962).
DAINTY, J.: The structure and possible functions of the vacuole. In: Plant cell organelles (J.B. (J.B. PRIDHAM, ed.), p. 40–46. London: Academic Press 1968.
DAVIES, D.R., PLASKITT, A.: Genetical and structural analyses of cell-wall formation in *Chlamydomonas reinhardi*. Genet. Res. Camb. **17**, 33–43 (1971).
DAVIES, J.M., FERRIER, W.C., JOHNSTON, C.S.: The ultrastructure of the meristoderm cells of the hapteron of *Laminaria*. J. Marine Biol. Assoc. U.K. **53**, 237–247 (1973).
DECKER, M., TANNER, W.: Respiratory increase and active hexose uptake of *Chlorella vulgaris*. Biochim. Biophys. Acta **266**, 661–669 (1972).
DEWAR, M.A., BARBER, J.: Cation regulation in *Anacystis nidulans* Planta **113**, 143–155 (1973).
DHARMAWARDENE, M.W.N., HAYSTEAD, A., STEWART, W.D.P.: Glutamine synthetase of the nitrogen-fixing alga *Anabaena cylindrica*. Arch. Mikrobiol. **90**, 281–295 (1973).
DODD, W.A., BIDWELL, R.G.S.: The effect of pH on the products of photosynthesis with $^{14}CO_2$ by chloroplast preparations from *Acetabularia mediterranea*. Plant Physiol. Lancaster **47**, 779–783 (1971).
DODD, W.A., PITMAN, M.G., WEST, K.R.: Sodium and potassium transport in the marine alga *Chaetomorpha darwinii*. Australian J. Biol. Sci. **19**, 341–354 (1966).
DODGE, J.D.: The fine structure of algal cells, p. xii+261. London-New York: Academic Press 1973.
DOMANSKI-KADEN, J., SIMONIS, W.: Veränderungen der Phosphatfraktionen, besonders des Polyphosphats, bei synchronisierten *Ankistrodesmus braunii* Kulturen. Arch. Mikrobiol. **87**, 11–28 (1972).
DROOP, M.R.: Some thoughts on nutrient limitation in algae. J. Phycol. **9**, 264–272 (1973).
DROOP, M.R.: Heterotrophy. In: Algal physiology and biochemistry (W.D.P. STEWART, ed.), p. 530–559. Oxford: Blackwell Scientific Publications 1974.
ECKERT, R.: Bioelectric control of bioluminescence in the dinoflagellate *Noctiluca*. I. Specific nature of triggering events. Science **147**, 1140–1142 (1965).
ELAVSON, J., VAGELAS, P.R.: A new class of lipids: chlorosulfolipids. Proc. Natl. Acad. Sci. U.S. **62**, 957–963 (1969).
EPPLEY, R.W.: Sodium exclusion and potassium retention by the marine red alga *Porphyra perforata*. J. Gen. Physiol. **41**, 901–911 (1958).
EPPLEY, R.W.: Potassium-dependent sodium extrusion by cells of *Porphyra perforata*, a marine red alga. J. Gen. Physiol. **42**, 281–288 (1959).
EPPLEY, R.W.: Potassium accumulation and sodium efflux in lithium and magnesium sea waters. J. Gen. Physiol. **43**, 29–38 (1960).
EPPLEY, R.W.: Major cations. In: Physiology and biochemistry of algae (R.A. LEWIN, ed.), p. 255–266. New York: Academic Press 1962.
EPPLEY, R.W., COATSWORTH, J.L.: Uptake of nitrate and nitrite by *Ditylum brightwellii*, kinetics and mechanisms. J. Phycol. **4**, 151–156 (1968).
EPPLEY, R.W., ROGERS, J.N.: Inorganic nitrogen assimilation of *Ditylum brightwellii*, a marine plankton diatom. J. Phycol. **6**, 344–350 (1970).
ERDEI, L., CSEH, E.: Changes in the K^+ and Na^+ content of *Scenedesmus obtusiusculus*. Ann. Univ. Sci. Budap. Rolando Eotvos Nominatae Sect. Biol. **13**, 79–90 (1971).
ETTL, H.: Über pulsierende Vakuolen bei Chlorophyceen. Flora (Jena) **151**, 88–98 (1961).
EVANS, H.J., SORGER, G.J.: Role of mineral elements with emphasis on the univalent cations. Ann. Rev. Plant Physiol. **17**, 47–76 (1966).
FENSOM, D.S., URSINO, D.J., NELSON, C.D.: Determination of the relative pore size in living membranes of *Nitella* by the techniques of electro-osmosis and radioactive tracers. Canad. J. Botany **45**, 1267–1275 (1967).
FINDENEGG, G.R.: Beziehungen zwischen Carboanhydraseaktivität und Aufnahme von Bicarbonat und Chlorid bei der Photosynthese von *Scenedesmus obliquus*. Planta **116**, 123–131 (1974).
FINDLAY, G.P.: Membrane electrical behaviour of *Nitellopsis obtusa*. Australian J. Biol. Sci. **23**, 1033–1045 (1970).
FINDLAY, G.P., HOPE, A.B.: Ionic relations of *Chara australis*. IX. Analysis of transient membrane currents. Australian J. Biol. Sci. **17**, 400–411 (1964).

FINDLAY, G.P., HOPE, A.B., PITMAN, M.G., SMITH, F.A., WALKER, N.A.: Ionic fluxes in cells of *Chara corallina*. Biochim. Biophys. Acta **183**, 565–576 (1969).

FINDLAY, G.P., HOPE, A.B., PITMAN, M.G., SMITH, F.A., WALKER, N.A.: Ionic relations of marine algae. III. *Chaetomorpha:* membrane electrical properties and Cl^- fluxes. Australian J. Biol. Sci. **24**, 731–746 (1971).

FINDLAY, G.P., HOPE, A.B., WILLIAMS, E.J.: Ionic relations of marine algae. I. *Griffithsia:* membrane electrical properties. Australian J. Biol. Sci. **22**, 1163–1178 (1969).

FINDLAY, G.P., HOPE, A.B., WILLIAMS, E.J.: Ionic relations of marine algae. II. *Griffithsia:* ionic fluxes. Australian J. Biol. Sci. **23**, 323–328 (1970).

FOGG, G.E.: The extracellular products of algae. Oceanogr. Marine Biol. Ann. Rev. **4**, 195–212 (1966).

FRASER, T.W., GUNNING, B.E.S.: Ultrastructure of the hairs of the filamentous green alga *Bulbochaete hiloensis* (Nordst) Tiffany: an apoplastidic plant cell with a well developed golgi apparatus. Planta **113**, 1–19 (1973).

FRAZISKET, L.: Uptake and accumulation of nitrate and nitrite by reef corals. Naturwiss. **60**, 552 (1973).

FRITSCH, F.E.: The structure and reproduction of the algae, vol. I. Cambridge: Cambridge University Press 1935.

FRITSCH, F.E.: The structure and reproduction of the algae, vol. II. Cambridge: Cambridge University Press 1945.

FUHS, G.W.: Phosphorus content and the rate of growth in the diatoms *Cyclotella nana* and *Thalassiosira fluviatilis*. J. Phycol. **5**, 312–321 (1969).

FUHS, G.W., DEMMERELE, S.D., CARELLI, E., CHEN, M.: Characterisation of phosphorus-limited plankton algae (with reflections on the limiting nutrient concept). Limonol. Oceanogr. Special Symposia Volume **1**, 113–133 (1972).

GAFFEY, C.T., MULLINS, L.J.: Ion fluxes during the action potential in *Chara*. J. Physiol. (London) **144**, 505–524 (1958).

GANF, G.G.: The regulation of net primary production in Lake George, Uganda, East Africa. In: Productivity problems of freshwater (Z. ZAJAK and A. HILLBRICHT-ILKOWSKA, eds.), p. 693–708. Warsaw-Krakow: P.W.N. (Polish Scientific Publishers) 1972.

GERLOFF, G.C., FISHBECK, K.A.: Quantitative cation requirements of several green and blue-green algae. J. Phycol. **5**, 109–114 (1969).

GILLET, C., LEFEVRE, J.: Combined effect of potassium and bicarbonate ions on the membrane potential and electrical conductance of *Nitella flexilis*. In: Ion transport in plants (W.P. ANDERSON, ed.), p. 101–112. London: Academic Press 1973.

GINZBURG, B.Z., HOGG, J.: What does a short-circuit current measure in biological systems? J. Theoret. Biol. **14**, 316–322 (1967).

GOLDMAN, J.C., PORCELLA, D.B., MIDDLEBROOKS, E.J., TOERINZ, D.F.: The effect of carbon on algal growth—its relationship to eutrophication. Water Research **6**, 637–679 (1972).

GRADMANN, D.: Einfluß von Licht, Temperatur und Außenmedium auf das elektrische Verhalten von *Acetabularia mediterranea*. Planta **93**, 323–353 (1970).

GRADMANN, D., WAGNER, G., GLÄSEL, R.H.: Chloride efflux during light-triggered action potential in *Acetabularia mediterranea*. Biochim. Biophys. Acta **323**, 151–155 (1973).

GREEN, P.B.: Growth physics of *Nitella:* a method for continuous *in vivo* analysis of extensibility based on a micromanometer technique for turgor pressure. Plant Physiol. Lancaster **43**, 1169–1184 (1968).

GUILLARD, R.T.L.: A mutant of *Chlamydomonas reinhardii* lacking contractile vacuoles. J. Protozool. **7**, 262–268 (1960).

GUILLARD, R.R.L.: Salt and osmotic balance. In: Physiology and biochemistry of algae (R.A. LEWIN, ed.), p. 529–540. New York: Academic Press 1962.

GUTKNECHT, J.: Ion distribution and transport in the red marine alga *Gracilaria foliifera*. Biol. Bull. **129**, 495–510 (1965).

GUTKNECHT, J.: Sodium, potassium and chloride transport and membrane potentials in *Valonia ventricosa*. Biol. Bull. **130**, 331–344 (1966).

GUTKNECHT, J.: Ion fluxes and short-circuit current in internally perfused cells of *Valonia ventricosa*. J. Gen. Physiol. **50**, 1821–1834 (1967).

GUTKNECHT, J., DAINTY, J.: Ionic relations of marine algae. Oceanogr. Marine Biol. Ann. Rev. **6**, 163–200 (1968).

HAASS, D., TANNER, W.: Regulation of hexose transport in *Chlorella vulgaris*. Characteristics of induction and turnover. Plant Physiol. Lancaster **53**, 14–20 (1974).

HAROLD, F.M., PAPINEAU, D.: Cation transport and electrogenesis by *Streptococcus faecilis*. I. The membrane potential. J. Membrane Biol. **8**, 27–44 (1972).

HASE, E., MIHARA, S., TAMIYA, M.: Role of sulphur in the cell division of *Chlorella*, with special reference to the sulphur compounds appearing during the process of cell division. Plant. Cell Physiol. (Tokyo) **2**, 9–24 (1961).

HASE, E., MIYACHI, S., MIHARA, S.: A preliminary note on the phosphorus compounds in chloroplasts and volutin granules isolated from *Chlorella* cells. In: Microalgae and photosynthetic bacteria, p. 619–626. Tokyo: Japanese Society of Plant Physiologists 1963.

HAYDON, D.A., HLADSKY, S.B.: Ion transport across thin lipid membranes: a critical discussion of mechanisms in selected membranes. Quart. Rev. Biophysics **5**, 187–282 (1972).

HAYWARD, J.: Studies on the growth of *Phaeodactylum tricornatum* VI. The relationship to Na^+, K^+, Ca^{2+} and Mg.$^{2+}$ J. Marine Biol. Assoc. U.K. **50**, 293–299 (1970).

HEALEY, F.P.: Inorganic nutrient uptake and deficiency in algae. Crit. Rev. Microbiol. **3**, 69–113 (1973).

HIGINBOTHAM, N.: The mineral absorption process in plants. Bot. Rev. **39**, 15–70 (1973).

HOAGLAND, D.R., DAVIES, A.R.: Further experiments on the absorption of ions by plants, including observations on the effect of light. J. Gen. Physiol. **6**, 47–62 (1923).

HOAGLAND, D.R., DAVIES, A.R.: The intake and accumulation of ions by plant cells. Protoplasma **6**, 610–626 (1929).

HOAGLAND, D.R., DAVIES, A.R., HIBBARD, P.L.: The influence of one ion on the accumlation of another by plant cells, with special reference to experiments with *Nitella*. Plant Physiol. Lancaster **3**, 473–486 (1928).

HOFFMAN, L.R.: Fertilisation in *Oedegonium*. I. Plasmogamy. J. Phycol. **9**, 62–84 (1973).

HOGG, J., WILLIAMS, E.J., JOHNSTON, R.J.: The temperature dependence of the membrane potential and resistance in *Nitella translucens*. Biochim. Biophys. Acta **150**, 640–648 (1968).

HOGG, J., WILLIAMS, E.J., JOHNSTON, R.J.: Light intensity and the membrane parameters of *Nitella translucens*. Biochim. Biophys. Acta **173**, 64–66 (1969).

HOLLENBERG, G.J.: Some physical and chemical properties of the cell sap of *Halicystis ovalis* (Lyngb) Aresch. J. Gen. Physiol. **15**, 651–653 (1932).

HOOGENHOUT, H., AMESZ, J.: Growth rate of photosynthetic micro-organisms in laboratory cultures. Arch. Mikrobiol. **50**, 10–25 (1965).

HOPE, A.B.: The electrical properties of plant cell membranes. I. The electrical capacitance of suspensions of mitochondria, chloroplasts and *Chlorella* sp. Australian J. Biol. Sci. **9**, 53–66 (1956).

HOPE, A.B.: Ionic relations of cells of *Chara australis*. IV. Fluxes of potassium. Australian J. Biol. Sci. **16**, 429–441 (1963).

HOPE, A.B.: Ionic relations of cells of *Chara australis*. X. Effects of bicarbonate ions on electrical properties. Australian J. Biol. Sci. **18**, 789–801 (1965).

HOPE, A.B.: Ion transport and membranes. London: Butterworths 1971.

HOPE, A.B., ASCHBERGER, P.A.: Effects of temperature on membrane permeability to ions. Australian J. Biol. Sci. **23**, 1047–1060 (1970).

HOPE, A.B., LÜTTGE, U., BALL, E.: Chloride uptake in strains of *Scenedesmus obliquus*. Z. Pflanzenphysiol. **72**, 1–10 (1974).

HOPE, A.B., SIMPSON, A., WALKER, N.A.: The efflux of chloride from cells of *Nitella* and *Chara*. Australian J. Biol. Sci. **19**, 355–362 (1966).

HOPE, A.B., WALKER, N.A.: Ionic relations of cells of *Chara australis*. III Vacuolar fluxes of sodium. Australian J. Biol. Sci. **13**, 277–291 (1960).

HOPE, A.B., WALKER, N.A.: Ionic relations of cells of *Chara Australis* IV. Membrane potential differences and resistances. Australian J. Biol. Sci. **14**, 26–44 (1961).

JACKSON, H.J., SCHIOU, Y.-F., BANE, S., FOX, M.: Intestinal transport of weak electrolytes. Evidence in favour of a three-compartment system. J. Gen. Physiol. **63**, 187–213 (1974).

JACQUES, A.G., OSTERHOUT, W.J.V.: The kinetics of penetration. XI. Entrance of potassium into *Nitella*. J. Gen. Physiol. **18**, 967–985 (1935).

JACQUES, A.G., OSTERHOUT, W.J.V.: The accumulation of electrolytes. XI. Accumulation of nitrate by *Valonia* and *Halicystis*. J. Gen. Physiol. **21**, 767–773 (1938).

JAFFE, L.F., ROBINSON, K.R., NUCCITELLI, R.: Transcellular currents and ion fluxes through developing fucoid eggs. In: Membrane transport in plants (U. ZIMMERMANN, J. DAINTY, eds.), p. 226–233. Berlin-Heidelberg-New York: Springer 1974.

JANACEK, K., RYBOVA, B.: The effect of ouabain on the alga *Hydrodictyon reticulatum*. Cytologia (Tokyo) **31**, 199–202 (1966).

JEANJEAN, R.: The relationship between the rate of phosphate absorption and protein synthesis during phosphate starvation in *Chlorella pyrenoidosa*. F.E.B.S. Letters **32**, 149–151 (1973a).

JEANJEAN, R.: Mécanismes d'absorption des ions phosphates par les Chlorelles: étude de l'absorption par les cultures synchrones. Compt. Rend. **277**, 193–196 (1973b).

JEANJEAN, R., BLASCO, F.: Influence des ions arsenates sur l'absorption des ions phosphates par les Chlorelles. Compt. Rend. **270**, 1897–1900 (1970).

JOHNSTON, R.A., SCHMIDT, R.R.: Intracellular distribution of sulfur during synchronous growth of *Chlorella pyrenoidosa*. Biochim. Biophys. Acta **74**, 428–437 (1963).

KAHN, J.S.: Evidence for two-directional hydrogen ion transport in chloroplasts of *Euglena gracilis*. Biochim Biophys. Acta **245**, 144–150 (1971).

KAMIYA, N., KURODA, K.: Artificial modifications of the osmotic pressure of the plant cell. Protoplasma **46**, 423–436 (1956).

KAMIYA, N., TAZAWA, M.: Surgical operations on characean cells with special reference to cytoplasmic streaming. Ann. Rept. Biol. Works, Fac. Sci. Univ. Osaka. **14**, 95–106 (1966).

KANAI, R., SIMONIS, W.: Einbau von ^{32}P in verschiedene Phosphatfraktionen, besonders Polyphosphate, bei einzelligen Grünanlagen (*Ankistrodesmus braunii*) im Licht und im Dunkeln. Arch. Mikrobiol. **62**, 56–71 (1968).

KANAZAWA, T.: Changes in the amino acid composition of *Chlorella* cells during their synchronous life cycle. Plant Cell Physiol. (Tokyo) **5**, 333–354 (1964).

KANDLER, O., TANNER, W.: Die Photoassimilation von Glucose als Indikator für die Lichtphosphorylierung *in vivo*. Ber. Deut. Bot. Ges. **79**, 48–57 (1966).

KANNAN, S.: Plasmalemma: the seat of dual mechanisms of ion absorption in *Chlorella pyrenoidosa*. Science **173**, 927–929 (1971).

KAZAMA, F., FULLER, M.S.: Ultrastructure of *Porphyra perforata* infected with *Pythium marinum*, a marine fungus. Canad. J. Botany **48**, 2103–2107 (1970).

KESSELER, H.: Collection of cell sap, apparent free space and vacuole concentrations of the osmotically most important mineral components of some Helgoland marine algae. Helgoländer Wiss. Meeresuntersuch. **11**, 258–269 (1964).

KESSELER, H.: Turgor, osmotisches Potential und ionale Zusammensetzung des Zellsaftes einiger Meeresalgen verschiedener Verbreitungsgebiete. Bot. Goth. **3**, 103 (1965).

KESSELER, H.: Beitrag zur Kenntnis der chemischen und physikalischen Eigenschaft des Zellsaftes von *Noctiluca miliaris*. Veröff. Inst. Meeresforsch. Bremerh. Sonderbd. **2** (6. Meeresbiol. Symp.) 357–368 (1966).

KESSELER, H.: Untersuchungen über die chemische Zusammensetzung des Zellsaftes der Diatomee *Coscinodiscus wailesii*. (Baccilariophyceae, Centrales). Helgoländer Wiss. Meeresuntersuch. **16**, 262–270 (1967).

KESSLER, E.: Nitrate assimilation by plants. Ann. Rev. Plant Physiol. **15**, 57–72 (1964).

KISAIMOTO, U., NAGAI, R., TAZAWA, M.: Plasmalemma potential in *Nitella*. Plant Cell Physiol. (Tokyo) **6**, 519–528 (1965).

KISHIMOTO, U.: Electrical characteristics of *Chara corallina* I. The membrane potential. Ann. Rept. Sci. Works, Fac. Sci. Univ. Osaka **7**, 115–129 (1959).

KISHIMOTO, U.: Current voltage relation in *Nitella*. Jap. J. Physiol. **14**, 515–527 (1964).

KISHIMOTO, U.: Hyperpolarising response in *Nitella* internodes. Plant Cell Physiol. (Tokyo) **7**, 429–439 (1966).

KISHIMOTO, U., AKABORI, H.: Protoplasmic streaming of an internodal cell of *Nitella flexilis*. J. Gen. Physiol. **42**, 1167–1183 (1959).

KISHIMOTO, U., TAZAWA, M.: Ionic composition of the cytoplasm of *Nitella flexilis*. Plant Cell Physiol. (Tokyo) **6**, 507–518 (1965a).

KISHIMOTO, U., TAZAWA, M.: Ionic composition and the electric response of *Lamprothamnium succinctum*. Plant Cell Physiol. (Tokyo) **6**, 529–536 (1965b).

KITASATO, H.: The influence of H^+ on the membrane potential and ion fluxes in *Nitella*. J. Gen. Physiol. **52**, 60–87 (1968).

KITASATO, H.: Potassium permeability of *Nitella clavata* in depolarised state. J. Gen. Physiol. **62**, 535–549 (1973).

KNUTSEN, G.: Uptake of uracil by synchronous *Chlorella fusca*. Physiol. Plantarum **27**, 300–309 (1972).

KOCH, A.L.: The adaptive responses of *E. coli* to a feast and famine existence. Advan. Microbial Physiol. **6**, 147–217 (1971).

KOMOR, E.: Proton-coupled hexose transport in *Chlorella vulgaris*. F.E.B.S. Letters **38**, 16–18 (1973).

KOMOR, E., HAASS, D., KOMOR, B., TANNER, W.: The active hexose uptake system of *Chlorella vulgaris*. K_m-values for 6-deoxyglucose influx and efflux and their contribution to sugar accumulation. European J. Biochem. **39**, 193–200 (1973b).

KOMOR, E., HAASS, D., TANNER, W.: Unusual features of the active hexose uptake system of *Chlorella vulgaris*. Biochim. Biophys. Acta **266**, 649–660 (1972).

KOMOR, B., KOMOR, E., HAASS, D., TANNER, W.: The unusual hexose transport uptake system of *Chlorella vulgaris*. Hoppe-Seylers Z. Physiol. Chem. **354**, 225–226 (1973a).

KOMOR, E., LOOS, E., TANNER, W.: A confirmation of the proposed model for the hexose uptake system of *Chlorella vulgaris:* Anaerobic studies in the light and the dark. J. Membrane Biol. **12**, 89–99 (1973).

KOMOR, E., TANNER, W.: Characterisation of the active hexose transport system of *Chlorella vulgaris*. Biochim. Biophys. Acta **241**, 170–179 (1971).

KOMOR, E., TANNER, W.: The nature of the energy metabolite responsible for sugar accumulation in *Chlorella vulgaris*. Z. Pflanzenphysiol. **71**, 115–128 (1974a).

KOMOR, E., TANNER, W.: Proton movement associated with hexose transport in *Chlorella vulgaris*. In: Membrane transport in plants (U. ZIMMERMANN, J. DAINTY, eds.), p. 209–215. Berlin-Heidelberg-New York: Springer 1974b.

KOTYK, A., JANACEK, K.: Transport, principles and techniques, p. vii + 498. New York-London: Plenum Press 1970.

KUENZLER, E.J.: Dissolved organic phosphorus excretion by marine phytoplanton. J. Phycol. **6**, 7–13 (1970).

KUENZLER, E.J., KETCHUM, B.H.: Rate of phosphate uptake by *Phaeodactylum tricornutum*. Biol. Bull. **123**, 134–145 (1962).

KYLIN, A.: Sulphate uptake and metabolism in *Scenedesmus* as influenced by phosphate, carbon dioxide and light. Physiol. Plantarum **17**, 422–433 (1964).

KYLIN, A.: The uptake and metabolism of sulphate in *Scenedesmus* as influenced by citrate, carbon dioxide and metabolic inhibitors. Physiol. Plantarum **20**, 139–148 (1967).

LANGMÜLLER, G., SPRINGER-LEDERER, H.: Membranpotential von *Chlorella fusca* in Abhängigkeit von pH-Wert, Temperatur und Belichtung. Planta **120**, 189–196 (1974).

LANNOYE, R.J., TARR, S.E., DAINTY, J.: The effects of pH on the ionic and electrical properties of the internodal cells of *Chara australis*. J. Exptl. Bot. **21**, 543–551 (1970).

LAQUERBE, B., BUSSON, F., MAIGROT, M.: Sur la composition en elements mineraux de deux Cyanophycees. *Spirulina platensis* (Gorn) Geitler et *Spirulina geitleri* J. de Tori. Compt. Rend. **270**, 2130–2132 (1970).

LARKUM, A.W.D.: Ionic relations of chloroplasts *in vivo*. Nature **218**, 447–450 (1968).

LEFEVRE, J., GILLET, C.: Variations de la difference de potential electrochique des chlorures chez *Nitella* en presence de benzensulphonate. Experientia **26**, 482–483 (1970).

LILLEY, R. MCC., HOPE, A.B.: Chloride transport and photosynthesis in cells of *Griffithsia*. Biochim. Biophys. Acta **226**, 161–171 (1971).

LIVINGSTONE, D.A.: Chemical composition of rivers and lakes. In: Data of geochemistry. U.S. Geol. Surv. Profess. Papers **440-G**, 64 pp. (1963).

LÜTTGE, U.: Stofftransport der Pflanzen. Berlin-Heidelberg-New York: Springer 1973.

MACROBBIE, E.A.C.: Ionic relations of *Nitella translucens*. J. Gen. Physiol. **45**, 861–878 (1962).

MACROBBIE, E.A.C.: Factors affecting the fluxes of potassium and chloride ions in *Nitella translucens*. J. Gen. Physiol. **47**, 859–877 (1964).

MACROBBIE, E.A.C.: The nature of the coupling between light energy and active ion transport in *Nitella translucens*. Biochim. Biophys. Acta **94**, 64–73 (1965).

MACROBBIE, E.A.C.: Metabolic effects on ion fluxes in *Nitella translucens*. I. Active influxes. Australian J. Biol. Sci. **19**, 363–370 (1966a).

MACROBBIE, E.A.C.: Metabolic effects on ion fluxes in *Nitella translucens*. II. Tonoplast fluxes. Australian J. Biol. Sci. **19**, 371–383 (1966b).

MACROBBIE, E.A.C.: Active transport of ions in plant cells. Quart. Rev. Biophys. **3**, 251–294 (1970a).

MACROBBIE, E.A.C.: Fluxes and compartmentation in plant cells. Ann. Rev. Plant Physiol. **22**, 75–96 (1970b).
MACROBBIE, E.A.C.: Vacuolar ion transport in *Nitella*. In: Ion transport in plants (W.P. ANDERSON, ed.), p. 431–446. London: Academic Press 1973.
MACROBBIE, E.A.C.: Ion transport. In: Algal physiology and biochemistry (W.D.P. STEWART, ed.), p. 678–713. Oxford: Blackwell Scientific Publications 1974.
MACROBBIE, E.A.C., DAINTY, J.: Ion transport in *Nitellopsis obtusa*. J. Gen. Physiol. **42**, 335–353 (1958a).
MACROBBIE, E.A.C., DAINTY, J.: Sodium and potassium distribution and transport in the seaweed *Rhodymenia palmata* (L) Grev. Physiol. Plantarum **11**, 782–801 (1958b).
MACROBBIE, E.A.C., FENSOM, D.S.: Measurements of electroosmosis in *Nitella translucens*. J. Exptl. Bot. **20**, 466–484 (1969).
MARCHANT, H.J., PICKETT-HEAPS, J.D.: Ultrastructure and differentation of *Hydrodictyon reticulum*. VI. Formation of the germ. net. Australian J. Biol. Sci. **25**, 119–1213 (1972).
MERCER, M.J., MERCER, F.V.: Studies on the comparative physiology of *Chara corallina*. III. Nitrogen relations of internodal cell components during internodal cell expansion. Australian J. Bot. **19**, 1–12 (1971).
MESZES, G., CSEH, E.: Effect of light-dark transition treatments on the uptake of ^{42}K by *Scenedesmus obtusiusculus*. Bot. Kozlem-Bot. Publ. **60**, 121–132 (1973).
MESZES, G., ERDEI, L.: (Mg^{2+}, K^+, Na^+) activated ATPase activity and its properties in *Scenedesmus obliquus* alga cells. Acta Biochem. Biophys. Acad. Sci. Hung. **4**, 131–139 (1969).
MESZES, G., KRALOVANSKY, J., CSEH, E., BÖSZÖRMENYI, Z.: Ion and amino acid absorption by unicellular algae. Acta Biochem. Biophys. Acad. Sci. Hung. **2**, 239–251 (1967).
MIRSALIKOVA, N.M.: On the mechanism of active ion transport in cells of blue-green algae. Uzbeksk. Biol. Zh. **12**, 6–8 (1968).
MITCHELL, P.: Chemiosmotic coupling in oxidative and photosynthetic phosphorylation. Bodmin. Glynn Research Publication **66**/1 (1966).
MIYACHI, S., TAMIYA, M.: Distribution and turnover of phosphate compounds in growing *Chlorella* cells. Plant Cell Physiol. (Tokyo) **2**, 405–414 (1961).
MULLINS, L.J.: Efflux of chloride during the action potential of *Nitella*. Nature **196**, 986–987 (1962).
MUNDAY, J.C., JR.: Membrane potentials in *Bryopsis plumosa*. Botan. Marina **15**, 61–63 (1972).
NAGAI, R., TAZAWA, M.: Changes in resting potential and ion absorption caused by light in a single plant cell. Plant Cell Physiol. (Tokyo) **3**, 323–339 (1962).
NICHOLSON, N.H., BRIGGS, W.R.: Translocation of photosynthate in the brown alga *Nereocystis*. Amer. J. Bot. **59**, 97–106 (1972).
NOVAC, B., BENTRUP, F.W.: An electrophysiological study of regeneration in *Acetabularia mediterranea*. Planta **108**, 227–244 (1972).
ODA, K.: The electrical constants of *Chara braunii*. Sci. Rept. Tohoku Univ. 4th Ser. **27**, 187–198 (1961).
OKAMOTO, H., SUZUKI, Y.: Intracellular concentrations of ions in a halophilic strain of *Chlamydomonas*. I. Concentrations of Na^+, K^+ and Cl^- in the cell. Z. Allgem. Mikrobiol. **4**, 350–357 (1964).
O'KELLY, J.C.: Inorganic nutrients. in: Algal physiology and biochemistry (W.D.P. STEWART, ed.), p. 610–635. Oxford: Blackwell Scientific Publications 1974.
ONCUTT, D.H., PATTERSON, G.W.: Lipid and elemental composition of diatoms grown in chemically defined media. J. Phycol. **9**, 13s (1973).
OSTERHOUT, W.J.V.: The absorption of electrolytes in large plant cells. Botan. Rev. **2**, 283–315 (1936).
OSTERHOUT, W.J.V., DORCAS, M.J.: The penetration of carbon dioxide into living protoplasm. J. Gen. Physiol. **4**, 255–267 (1925).
PAASCHE, E.: Biology and physiology of Coccolithophorids. Ann. Rev. Mikrobiol. **22**, 71–88 (1968).
PAGANO, R., THOMPSON, T.E.: Spherical lipid bilayer membranes: electrical and isotopic studies of ion permeability. J. Mol. Biol. **38**, 41–58 (1968).
PAGE, J.Z.: Existence of a *Derbesia* phase in the life history of *Halicystis Osterhoutii* Blinks and Blinks. J. Phycol. **6**, 375–380 (1970).
PATENAUDE, R.D., BROWN, D.L.: Microtubule distribution in *Polytomella agilis*. J. Phycol. **9**, 7s (1973).

PENTH, B., WEIGL, J.: Anionen-Influx, ATP-Spiegel und CO_2-Fixierung in *Limnophila gratioloides* und *Chara foetida*. Planta **96**, 212–223 (1971).

PETTERSEN, R., KNUTSEN, G.: Uptake of guanine by synchronised *Chlorella fusca*. Characterisation of the transport system in autospores. Arch. Mikrobiol. **96**, 233–246 (1974).

PICKARD, W.F.: Does the resting potential of *Chara braunii* have an electrogenic component? Canad. J. Botany **51**, 715–724 (1973).

PITMAN, M.G.: Ion transport in plant cells. In: Intestinal absorption of metal ions, trace elements and radionuclides (S.C. SKORYNA, E.D. WALDRON, eds.), p. 115–133. Oxford: Pergamon Press 1971.

POTTS, W.T.W., PARRY, G.: Osmotic and ionic regulation in animals. London: Pergamon Press 1964.

PRIBIL, S., KOTYK, A.: Oscillatory transport of ammonium ions by *Scenedesmus quadricauda*. Biochim. Biophys. Acta **219**, 242–244 (1970).

PROSSER, C.L., BROWN, F.A.: Comparative animal physiology. Philadelphia: Saunders 1962.

QUATRANO, R.S.: An ultrastructural study of the determined site of rhizoid formation in *Fucus* zygotes. Exptl. Cell Res. **70**, 1–12 (1972).

RAMUS, J., GRAVES, S.T.: Incorporation of sulphate into the capsular polysaccharide of the red alga *Porphyridium*. J. Cell Biol. **54**, 399–407 (1972).

RAVEN, J.A.: Ion transport in *Hydrodictyon africanum*. J. Gen. Physiol. **50**, 1607–1625 (1967a).

RAVEN, J.A.: Light stimulation of active transport in *Hydrodictyon africanum*. J. Gen. Physiol. **50**, 1627–1640 (1967b).

RAVEN, J.A.: The linkage of light-stimulated Cl^- influx to K^+ and Na^+ influxes in *Hydrodictyon africanum*. J. Exptl. Bot. **19**, 233–253 (1968a).

RAVEN, J.A.: The mechanism of photosynthetic use of bicarbonate by*Hydrodictyon africanum*. J. Exptl. Bot. **19**, 196–206 (1968b).

RAVEN, J.A.: Action spectra for photosynthesis and light-stimulated ion transport processes in *Hydrodictyon africanum*. New Phytologist **68**, 45–62 (1969a).

RAVEN, J.A.: Effects of inhibitors on photosynthesis and the active influxes of K and Cl in *Hydrodictyon africanum*. New. Phytologist **68**, 1089–1113 (1969b).

RAVEN, J.A.: Exogenous inorganic carbon sources in plant photosynthesis. Biol. Rev. **45**, 167–221 (1970).

RAVEN, J.A.: Oubain-insensitive K influx in *Hydrodictyon africanum*. Planta **97**, 28–38 (1971a).

RAVEN, J.A.: Cyclic and non-cyclic photophosphorylation as energy sources for active K influx in *Hydrodictyon africanum*. J. Exptl. Bot. **22**, 420–433 (1971b).

RAVEN, J.A.: The effects of visible light on the influx and efflux of solutes in plant cells. Chemistry and industry, p. 859–856. London: Society for chemical Industry 1971c.

RAVEN, J.A.: Energy metabolism in green cells. Trans. Bot. Soc. Edinb. **41**, 219–225 (1971d).

RAVEN, J.A.: Inhibitor effects on photosynthesis, respiration and active ion transport in *Hydrodictyon africanum*. J. Membrane Biol. **6**, 89–107 (1971e).

RAVEN, J.A.: Endogenous inorganic carbon sources in plant photosynthesis. I. Occurrence of the dark respiratory pathways in illuminated green cells. New Phytologist **71**, 227–247 (1972a).

RAVEN, J.A.: Endogenous inorganic carbon sources in plant photosynthesis. II. Comparison of total CO_2 production in the light with measured CO_2 evolution in the light. New Phytologist **71**, 995–1014 (1972b).

RAVEN, J.A.: Photosynthetic electron flow and photophosphorylation. In: Algal physiology and biochemistry (W.D.P. STEWART, ed.), p. 391–423. Oxford: Blackwell Scientific Publications 1974a.

RAVEN, J.A.: Carbon dioxide fixation. In: Algal physiology and biochemistry (W.D.P. STEWART, ed.), p. 434–455. Oxford: Blackwell Scientific Publications 1974b.

RAVEN, J.A.: Phosphate transport in *Hydrodictyon africanum* New Phytologist **73**, 421–432 (1974c).

RAVEN, J.A.: Energetics of active phosphate influx in *Hydrodictyon africanum*. J. Exptl. Bot. **25**, 221–229 (1974d).

RAVEN, J.A.: Time course of chloride fluxes in *Hydrodictyon africanum* during alternating light and darkness. In: Membrane transport in plants (U. ZIMMERMANN, J. DAINTY, eds.), p. 167–172. Berlin-Heidelberg-New York: Springer 1974e.

RAVEN, J.A.: Transport of indoleacetic acid in plant cells in relation to pH and electrical potential gradients, and its significance for polar IAA transport. New Phytologist **74**, 163–172 (1975a).

RAVEN, J.A.: Transport at algal membranes. In: Botanical Proc. of 50th Anniv. Meeting. Soc. for Experimental Biol., Cambridge, July 1974 (N. SUNDERLAND, ed.). Oxford: Pergamon Press 1975b.

RAVEN, J.A., SMITH, F.A.: The regulation of intracellular pH as a fundamental biological process. In: Ion transport in plants (W.P. ANDERSON ed.), p. 271–278. London: Academic Press 1973.

RAVEN, J.A., SMITH, F.A.: Significance of hydrogen ion transport in plant. Canad. J. Botany **52**, 1035–1048 (1974).

RIETMA, M.: Life-history studies in the genus *Bryopsis* (Chlorophyceae) IV. Life-histories in *Bryopsis hypnoides* Lamx from different points along the European coasts. Acta Botan. Neerl. **20**, 291–298 (1971).

ROBINSON, J.B.: Sulphate influx in characean cells. I. General characteristics. J. Exptl. Bot. **20**, 201–211 (1969a).

ROBINSON, J.B.: Sulphate influx in characean cells. II. Links with light and metabolism in *Chara australis*. J. Exptl. Bot. **20**, 212–220 (1969b).

ROBINSON, K.R., JAFFE, L.F.: Ion movements in developing fucoid eggs. Develop. Biol. **35**, 349–361 (1973).

RONKIN, R.R., BURETZ, K.M.: Sodium and potassium in normal and paralysed *Chlamydomonas*. J. Protozool. **7**, 109–114 (1960).

ROOS, A.: Intracellular pH and intracellular buffering in the cat brain. Amer. J. Physiol. **209**, 1233–1246 (1965).

ROTHSTEIN, A.: Ion transport in micro-organisms. In: Metabolic pathways, 3rd ed. (L.E. HOKIN, ed.), p. 17–39. New York: Academic Press 1972.

ROUND, F.E.: The taxonomy of the chlorophyta. Brit. Phycol. Bull. **2**, 224–235 (1963).

ROUND, F.E.: The taxonomy of the chlorophyta. II. Brit. Phycol. J. **6**, 235–264 (1971).

RYAN, T.E., BARR, C.E.: Effect of an applied current on K^+ fluxes in *Nitella clavata*. Plant Physiol. Lancaster **51**, 16s (1973).

RYBOVA, R., JANACEK, K., STAVIKOVA, M.: Ionic relations of the alga *Hydrodictyon reticulatum*. The effects of light conditions and inhibitors. Z. Pflanzenphysiol. **66**, 420–432 (1972).

SADDLER, H.D.W.: The ionic relations of *Acetabularia mediterranea*. J. Exptl. Bot. **21**, 345–359 (1970a).

SADDLER, H.D.W.: Fluxes of sodium and potassium in *Acetabularia mediterranea*. J. Exptl. Bot. **21**, 605–616 (1970b).

SADDLER, H.D.W.: The membrane potential of *Acetabularia mediterranea*. J. Gen. Physiol. **55**, 803–821 (1970c).

SADDLER, H.D.W.: Spontaneous and induced changes in the membrane potential and resistance of *Acetabularia mediterranea*. J. Membrane Biol. **5**, 220–260 (1971).

SAITO, K., SENDA, M.: The light-dependent effect of external pH on the membrane potential of *Nitella*. Plant Cell Physiol. (Tokyo) **14**, 147–156 (1973a).

SAITO, K., SENDA, M.: The effect of external pH on the membrane potential of *Nitella* and its linkage to metabolism. Plant Cell Physiol. (Tokyo) **14**, 1045–1052 (1973b).

SCHAEDLE, M., JACOBSEN, L.: Ion absorption and retention by *Chlorella pyrenoidosa* I. Absorption of potassium. Plant Physiol. Lancaster **40**, 214–220 (1965).

SCHAEDLE, M., JACOBSEN, L.: Ion absorption and retention by *Chlorella pyrenoidosa* II. Permeability of the cell to sodium and rubidium. Plant Physiol. Lancaster **41**, 248–254 (1966).

SCHAEDLE, M., JACOBSEN, L.: Ion absorption and rentention by *Chlorella pyrenoidosa*. III. Selective accumulation of rubidium, potassium and sodium. Plant Physiol. Lancaster **42**, 953–958 (1967).

SCHNEPF, E., KOCH, W.: Golgi-Apparat und Wasserausscheidung bei *Glaucocystis*. Z. Pflanzenphysiol. **55**, 97–109 (1967).

SCHOBERT, B., UNTNER, E., KAUSS, H.: Isofloridosid und die Osmoregulation bei *Ochromonas malhamensis*. Z. Pflanzenphysiol. **67**, 385–398 (1972).

SCHÖTZ, F., BATHELT, H., ARNOLD, C.-G., SCHIMMER, O.: Die Architektur und Organisation der *Chlamydomonas*-Zelle: Ergebnisse der Elektronenmikroskopie von Serialschnitten und der daraus resultierenden dreidimensionalen Rekonstruktion. Protoplasma **75**, 229–254 (1972).

SCOTT, B.I.H.: Electric fields in plants. Ann. Rev. Plant Physiol. **18**, 409–418 (1967).

SCOTT, G.T., HAYWARD, M.R.: Sodium and potassium regulation in *Ulva lactuca* and *Valonia macrophysa*. In: Electrolytes in biological systems (A.M. SHANES, ed.), p. 35–64. Washington D.C.: American Physiological Society 1955.

SHEPPARD, D.C., BIDWELL, R.G.: Photosynthesis and carbon metabolism in chloroplast preparation from *Acetabularia*. Protoplasma **76**, 289–307 (1973).

SHIEH, Y.J., BARBER, J.: Intracellular sodium and potassium concentration and net cation movements in *Chlorella pyrenoidosa*. Biochim. Biophys. Acta **233**, 594–603 (1971).

SIEBURTH, J. McN.: Studies on algal substances in the sea. III. The production of extracellular organic matter by littoral marine algae. J. Exptl. Marine Biol. Ecol. **3**, 290–309 (1969).

SIMONIS, W., BORNEFELD, T., LEE, J., MAJUMDAR, K.: Phosphate uptake and photophosphorylation in the blue-green alga *Anacystis nidulans*. In: Membrane transport in plants (U. ZIMMERMANN, J. DAINTY, eds.), p. 220–225. Berlin-Heidelberg-New York: Springer 1974.

SIMONIS, W., KUNTZ, F.J., URBACH, W.: Probleme der Phosphataufnahme in Abhängigkeit von Licht und Dunkelheit bei *Ankistrodesmus*. Vortr. Gesamtgebiet. Botanik. Deut. Botan. Ges. **1**, 139–148 (1962).

SIMONIS, W., URBACH, W.: Photophosphorylation *in vivo*. Ann. Rev. Plant Physiol. **24**, 89–114 (1973).

SKULACHEV, V.P.: Energy transformation in the respiratory chain. Current Topics in Bioenergetics **4**, 127–190 (1971).

SMAYDA, T.J.: The suspension and sinking of phytoplankton in the sea. Oceanogr. Marine Biol. Ann. Rev. **8**, 353–414 (1971).

SMITH, D.C., MUSCATINE, L., LEWIS, D.H.: Carbohydrate movement from autotrophs to heterotrophs in parasitic and mutualistic symbioses. Biol. Rev. **44**, 17–90 (1969).

SMITH, F.A.: Active phosphate uptake by *Nitella translucens*. Biochim. Biophys. Acta **126**, 94–99 (1966).

SMITH, F.A.: The control of Na^+ uptake into *Nitella translucens*. J. Exptl. Bot. **18**, 716–731 (1967a).

SMITH, F.A.: Links between glucose uptake and metabolism in *Nitella translucens*. J. Exptl. Bot. **18**, 348–358 (1967b).

SMITH, F.A.: Metabolic effects on ion fluxes in *Tolypella intricata*. J. Exptl. Bot. **19**, 442–451 (1968).

SMITH, F.A.: The mechanism of chloride transport in characean cells. New Phytologist **69**, 903–917 (1970).

SMITH, F.A.: Stimulation of chloride transport in *Chara* by external pH changes. New Phytologist **71**, 595–601 (1972).

SMITH, F.A., RAVEN, J.A.: Energy-dependent processes in *Chara corallina*: absence of light stimulation when only photosystem one is operative. New Phytologist **73**, 1–12 (1974).

SMITH, F.A., WEST, K.R.: A comparison of the effects of metabolic inhibitors on chloride uptake and photosynthesis in *Chara corallina*. Australian J. Biol. Sci. **22**, 351–364 (1969).

SOEDER, C.J.: Zum Phosphathaushalt von *Chlorella fusca* Sh. et Kr. Arch. Hydrobiol., Suppl. **38**, 1–17 (1970).

SOLT, J., PASCHINGER, H., BRODA, E.: Die energieabhängige Aufnahme von Thallium durch *Chlorella*. Planta **101**, 242–250 (1971).

SPANSWICK, R.M.: The potassium absorption isotherm in *Nitella translucens*. International Botanical Congress XI. Abstract 206 (1969).

SPANSWICK, R.M.: The effects of bicarbonate ions and external pH. on the membrane potential and resistance of *Nitella translucens*. J. Membrane Biol. **2**, 59–70 (1970).

SPANSWICK, R.M.: Evidence for an electrogenic ion pump in *Nitella translucens*. I. The effects of pH, K^+, Na^+, light and temperature on the membrane potential and resistance. Biochim. Biophys. Acta **288**, 73–89 (1972).

SPANSWICK, R.M.: Electrogenesis in photosynthetic tissues. In: Ion transport in plants (W.P. ANDERSON ed.), p. 113–128. London: Academic Press 1973.

SPANSWICK, R.M.: Evidence for an electrogenic ion pump in *Nitella translucens*. II. Control of the light-stimulated component of the membrane potential. Biochim. Biophys. Acta **332**, 387–398 (1974a).

SPANSWICK, R.M.: Electrogenesis in photosynthetic tissues. In: Ion transport in plants (W.P. (1974b).

SPANSWICK, R.M., STOLAREK, J., WILLIAMS, E.J.: The membrane potential in *Nitella translucens*. J. Exptl. Bot. **18**, 1–16 (1967).

SPANSWICK, R.M., WILLIAMS, E.J.: Calcium fluxes and membrane potentials in *Nitella translucens*. J. Exptl. Bot. **16**, 463–473 (1965).

SPEAR, D.G., BARR, J.K., BARR, C.E.: Localisation of hydrogen ion and chloride ion fluxes in Nitella. J. Gen. Physiol. **54**, 397–414 (1969).

STADELMANN, E.J.: Permeability, In: Physiology and biochemistry of algae, p. 493–528. New York: Academic Press 1962.

STEUDLE, E., ZIMMERMANN, U.: Hydraulische Leitfähigkeit von *Valonia utricularis*. Z. Naturforsch. **26b**, 1302–1311 (1971).

STEUDLE, E., ZIMMERMANN, U.: Determination of the hydraulic conductivity and of reflection coefficients in *Nitella flexilis* by means of direct cell-turgor pressure measurements. Biochim. Biophys. Acta **332**, 399–412 (1974).

STEWARD, F.C., MARTIN, J.C.: The distribution and physiology of *Valonia* at the Dry Tortugas with special reference to the problem of salt accumulation in plants. Carnegie Inst. Wash. Publ. **475**, 87–170 (1937).

STEWART, W.D.P.: Nitrogen fixation. In: The biology of blue-green algae (N.G. CARR, D.A. WHITTON, eds.), p. 260–278. Oxford: Blackwell Scientific Publications 1973.

STEWART, W.D.P. (ed.): Algal physiology and biochemistry. Oxford: Blackwell Scientific Publications 1974.

STRAUSS, R.: Contribution à l'étude des alcalins et des alcalinoterreux chez les algues. Int. Rev. Ges. Hydrobiol. **52**, 465–478 (1967).

STROTMANN, H., HELDT, H.W.: Phosphate containing metabolites participating in photosynthetic reactions in *Chlorella pyrenoidosa*. In: Progress in photosynthetic research (H. METZNER, ed.), vol. III, p. 1131–1140, Tübingen: International Biological Union 1969.

STRUNK, T.H.: Correlation between metabolic parameters of transport and vacuolar perfusion results in *Nitella clavata*. J. Exptl. Bot. **22**, 863–874 (1971).

SUELTER, C.H.: Enzymes activated by monovalent cations. Science **168**, 789–795 (1970).

SUNDBERG, I., DOSKACIL, H.J., LEMBI, C.A.: Studies on the isolation of plasma membranes from filamentous green alga. J. Phycol. **9**, 21s (1973).

SYRETT, P.J., THOMAS, E.A.: The assay of nitrate reductase in whole cells of *Chlorella*: strain differences and the effect of cell walls. New Phytologist **72**, 1307–1310 (1973).

TAMIYA, H.: Synchronous cultures of algae. Ann. Rev. Plant Physiol. **17**, 1–26 (1966).

TANNER, W.: Light-driven active uptake of 3-0-methylglucose *via* an indicible hexose uptake system of *Chlorella*. Biochem. Biophys. Res. Commun. **36**, 278–283 (1969).

TANNER, W., GRÜNES, R., KANDLER, O.: Spezifität und Turnover der induzierten Hexoseaufnahme von *Chlorella*. Z. Pflanzenphysiol. **62**, 376–386 (1970).

TANNER, W., HAASS, D., DECKER, M., LOOS, E., KOMOR, B., KOMOR, E.: Active hexose transport in *Chlorella vulgaris*. In: Membrane transport in plants (U. ZIMMERMANN, J. DAINTY, eds.), p. 202–208. Berlin-Heidelberg-New York: York: Springer 1974.

TANNER, W., KANDLER, O.: Die Abhängigkeit der Adaptation der Glucoseaufnahme von der oxydativen und der photosynthetischen Phosphorylierung bei *Chlorella vulgaris*. Z. Pflanzenphysiol. **58**, 24–32 (1967).

TAZAWA, M.: Neue Methode zur Messung des osmotischen Wertes einer Zelle. Protoplasma **48**, 342–359 (1957).

TAZAWA, M.: Weitere Untersuchungen zur Osmoregulation der *Nitella*zelle. Protoplasma **53**, 227–258 (1961).

TAZAWA, M., KISHIMOTO, U.: Studies on *Nitella* having artificial cell sap. II. Rate of cyclosis and electrical potential. Plant Cell Physiol. (Tokyo) **5**, 45–59 (1964).

TAZAWA, M., NAGAI, R.: Studies on osmoregulation in *Nitella* internodes with modified cell saps. Z. Pflanzenphysiol. **54**, 333–344 (1966).

THAIN, J.F.: The flux-ratio equation for potassium in *Chara corallina*. In: Ion transport in plants (W.P. ANDERSON, ed.), p. 77–94. London: Academic Press 1973.

THOMPSON, T.E., HENN, F.A.: Experimental phospholipid model membranes. In: Membranes of mitochondria and chloroplasts (E. RACKER, ed.), p. 1–52. New York: van Nostrand-Reinhold 1970.

THORHAUG, A.: Temperature effects on *Valonia* bioelectric potentials. Biochim. Biophys. Acta **225**, 151–158 (1971).

THROM, G.: Die lichtabhängige Änderung des Membranpotentials bei *Griffithsia setacea*. Z. Pflanzenphysiol. **63**, 162–180 (1970).

THROM, G.: Einfluß von Entkopplern auf die lichtabhängige Änderung des Membranpotentials und auf den lichtabhängigen Netto-Protonen-Influx bei *Griffithsia setacea*. Arch. Protistenk. **114**, 308–329 (1972).

TRENCH, R.K., BOYLE, J.E., SMITH, D.C.: The association between chloroplasts of *Codium fragile* and the mollusc *Elysia viridis*. I. Characteristics of isolated *Codium* chloroplasts. Proc. Roy. Soc. (London) Ser. B **184**, 51–61 (1973).

TROMBALLA, H.W., BRODA, E.: Das Verhalten von *Chlorella fusca* gegenüber Perchlorat und Chlorat. Arch. Mikrobiol. **78**, 214–223 (1971).

TYREE, M.T.: Ion concentrations, resting potentials, membrane fluxes and permeabilities: plant cells, and electrical constants, plant cells. In: Biology data book 2nd ed., vol. II (P.L. ALTMAN, D.S. DITTMER, eds.), p. 1241–1243. Bethesda, Maryland: Federation of American Societies for Experimental Biology 1973.

ULLRICH, W.R.: Nitratabhängige nichtcyclische Photophosphorylierung bei *Ankistrodesmus braunii* in Abwesenheit von CO_2 und O_2. Planta **100**, 18–30 (1971).

ULLRICH, W.R.: Untersuchungen über die Raten der Polyphosphatsynthese durch die Photophosphorylierung bei *Ankistrodesmus braunii*. Arch. Mikrobiol. **87**, 323–339 (1972).

ULLRICH, W.R.: Die nitrat- und nitritabhängige photosynthetische O_2-Entwicklung in N_2 bei *Ankistrodesmus braunii*. Planta **116**, 143–152 (1974).

ULLRICH-EBERIUS, C.I.: Die pH-Abhängigkeit der Aufnahme von $H_2PO_4^-$, SO_4^{2-}, Na^+ und K^+ und ihre gegenseitige Beeinflussung bei *Ankistrodesmus braunii*. Planta **109**, 161–176 (1973a).

ULLRICH-EBERIUS, C.I.: Beziehungen der Aufnahme von Nitrat, Nitrit und Phosphat zur photosynthetischen Reduktion von Nitrat, Nitrit und zum ATP-Spiegel bei *Ankistrodesmus braunii*. Planta **115**, 25–36 (1973b).

UMRATH, K.: Über die Erregungsvorgänge bei *Spirogyra* und *Vaucheria* und über Potentialmessungen an Pflanzenzellen. Protoplasma **22**, 193–202 (1934).

UMRATH, K.: Über das elektrische Potential und über Aktionsströme von *Valonia macrophysa*. Protoplasma **31**, 184–193 (1938).

VALLÉE, M.: Le système de transport de SO_4^{2-} chez *Chlorella pyrenoidosa* et sa régulation. Variations des K_m des vitesses au cours d'une carence en sulfate. Compt. Rend. **266**, 1767–1768 (1968).

VALLÉE, M.: Le système de transport de sulfate chez *Chlorella pyrenoidosa* et sa régulation. IV. Études avec l'ion chromate. Biochim. Biophys. Acta **173**, 486–500 (1969).

VALLÉE, M., JEANJEAN, R.: Le système de transport de SO_4^{2-} chez *Chlorella pyrenoidosa* et sa régulation. I. Étude cinétique de la permeation. Biochim. Biophys. Acta **150**, 599–606 (1968a).

VALLÉE, M., JEANJEAN, R.: Le système de transport de SO_4^{2-} chez *Chlorella pyrenoidosa* et sa régulation. II. Recherches sur la régulation de l'entrée. Biochim. Biophys. Acta **150**, 607–617 (1968b).

VILLEGAS, L.: The electrical potential of the cytoplasm in *Valonia*. Abstr. Meeting. Int. Org. Pure Appl. Biochem. Vienna, abstr. 364 (1966).

VOLKOV, G.A., MISYUK, L.A.: Hyperpolarisation of the cytoplasmic membrane surface in the plant cell. Dokl.-Botan. Sci. Sect. (Engl. Transl.) **175**, 121–122 (1967).

VOROBIEV, L.N.: Potassium ion activity in the cytoplasm and the vacuole of cells of *Chara australis*. Nature **216**, 1325–1327 (1967).

VOROBIEV, L.N., KURELLA, G.A., POPOV, G.A.: Intracellular pH of *Nitella flexilis* at rest and during excitation. Biofizika **6**, 582–589 (1961).

VREDENBERG, W.J.: Light-induced changes in membrane potential of algal cells associated with photosynthetic electron transport. Biochem. Biophys. Res. Commun. **37**, 785–792 (1969).

VREDENBERG, W.J., TONK, W.J.M.: Photosynthetic energy control of an electrogenic ion pump at the plasmalemma of *Nitella translucens*. Biochim. Biophys. Acta **298**, 354–368 (1973).

VREEMAN, M.I.: Permeability of thin phospholipid films. III. Koninkl. Ned. Akad. Wetenschap. Proc. Ser. **B69**, 564–577 (1966).

WAGNER, G.: Light-dependent ion fluxes in *Mougeotia*: Control by photosynthesis, not by phytochrome. In: Membrane transport in plants (U. ZIMMERMANN, J. DAINTY, eds.), p. 186–191. Berlin-Heidelberg-New York: Springer 1974.

WAGNER, G., BENTRUP, F.W.: Lichtabhängige K^+ und Cl^- Flüsse bei *Mougeotia*. Ber. Deut. Bot. Ges. **86**, 365–369 (1973).

WALKER, N.A.: Ion permeability of the plasmalemma of plant cells. Nature **180**, 94–95 (1957).
WALKER, N.A.: Chloride transport to the charophyte vacuole. In: Membrane transport in plants (U. ZIMMERMANN, J. DAINTY, eds.), p. 173–179. Berlin-Heidelberg-New York: Springer 1974.
WALKER, N.A., BOSTROM, T.E.: Intercellular movement of Cl in *Chara*: a test for models of Cl influx. In: Ion transport in plants (W.P. ANDERSON ed.), p. 447–458. London: Academic Press 1973.
WALKER, N.A., HOPE, A.B.: Membrane fluxes and electric conductance in Characean cells Australian J. Biol. Sci. **22**, 1179–1195 (1969).
WEDDING, R.T., BLACK, M.K.: Uptake and metabolism of sulphate by *Chlorella*. I. Sulphate accumulation and active sulphate. Plant Physiol. Lancaster **35**, 72–80 (1960).
WEISENSEEL, M.H., JAFFE, L.J.: Membrane potential and impedance of developing fucoid eggs. Develop. Biol. **27**, 555–574 (1972).
WEISENSEEL, M.H., SMEIBIDL, E.: Phytochrome controls the water permeability in *Mougeotia*. Z. Pflanzenphysiol. **70**, 420–431 (1973).
WEST, I.C.: The uptake and metabolism of ions by illuminated algal cells. D. Phil. Thesis. Oxford University (1967).
WEST, K.R., PITMAN, M.G.: Ionic relations and ultrastructure in *Ulva lactuca*. Australian J. Biol. Sci. **20**, 901–914 (1967a).
WEST, K.R., PITMAN, M.G.: Rubidium as a tracer for potassium in the marine algae. *Ulva lactuca* L and *Chaetomorpha darwinii* (Hooker) Kuetzing. Nature **214**, 1262–1263 (1967b).
WIDEMAN, J.H.: Sodium efflux from microinjected *Nitella* cells. Biophys. J. **8**, abstr. 125 (1968).
WIESSNER, W.: Photometabolism of organic substrates. In: Photobiology of micro-organisms (P. HALDALL, ed.), p. 95–134. London: Wiley-Interscience 1970a.
WIESSNER, W.: Light effects on ion fluxes in microalgae. In: Photo-biology of micro-organisms (P. HALDALL, ed.), p. 135–164. London: Wiley-Interscience 1970b.
WILLIAMS, E.J., HOGG, J.: The (K^+-Na^+)-dependence of the membrane parameters of *Nitella translucens*. Biochim. Biophys. Acta **203**, 170–172 (1970).
WILLIAMS, E.J., MUNRO, C., FENSOM, D.S.: The influence of small applied electrical currents on Na, K and Cl fluxes in *Nitella translucens*. Canad. J. Bot. **50**, 2255–2263 (1972).
WILLENBRINK, J., KREMER, B.P.: Lokalisation der Mannitbiosynthese in der marinen Braunalge *Fucus serratus*. Planta **113**, 173–178 (1973).
WOOD, R.D., IMAHORI, K.: A revision of the characeae. First part, monograph of the characeae Weinheim: J. Cramer 1965.
WOOD, R.E., WIRTH, F.P., JR., MORGAN, H.E.: Glucose permeability of lipid bilayer membranes. Biochim. Biophys. Acta **163**, 171–178 (1968).
ZIMMERMANN, U., STEUDLE, E.: The pressure-dependence of the hydraulic conductivity, the membrane resistance and membrane potential during turgor pressure regulation in *Valonia utricularis*. J. Membrane Biol. **16**, 331–352 (1974).

7. Transport in Fungal Cells

D.H. Jennings

1. Introduction

This Chapter is concerned with those systems which are responsible for the transport of inorganic ions, amino acids and sugars into vegetative fungal cells. One of the virtues of fungi is the ease with which mutants can be obtained. Mutants have been used both to identify transport systems and to provide information about the way transport is regulated. With respect to this latter topic, a recurring theme in much of the information is feedback inhibition of transport either by the solute transported or by some compound closely related metabolically. The extensive use of mutants has of course meant that transport has been thought of in terms of transport proteins, with a consequent emphasis on transport kinetics.

The information about transport of inorganic ions is particularly relevant to those studying the same processes in green plants. The information about transport of organic molecules is important because it indicates the sorts of mechanisms which may be present in green plants or suggests how transport of these molecules into green plant cells might be studied.

It should be noted that transport processes have been studied in only a limited number of fungi. There is a great dearth of information about transport in Phycomycete and Basidiomycete species. This has been alluded to before (Jennings, 1973).

The following reviews on transport in fungi are available: cations (Jennings, 1973), amino acids (Oxender, 1972), sugars (Jennings, 1974), in yeast (Suomalainen and Oura, 1971); Armstrong, 1972) and in *Neurospora crassa* (Scarborough, 1973). Earlier work has been reviewed by Jennings (1963) and Burnett (1968).

2. Potassium-Hydrogen or -Sodium Exchange System: *Neurospora crassa* and *Saccharomyces cerevisiae*

This has been studied almost exclusively in two species; *Neurospora crassa* and baker's yeast, *Sachharomyces cerevisiae*. Both belong to the Ascomycetes and it is not unreasonable to suggest that the information about K^+ transport into these fungi be treated as a unity. There are obvious similarities between the systems from the two fungi. Be that as it may, the information is important since that for *N. crassa* is the first coherent account of an electrogenic pump in plant cells and demonstrates the role of ATP in cation transport, while the information for yeast indicates the source of the H^+ ions for exchange with K^+ and the possible nature of the carrier. The kinetic analysis of both systems provides the basis for an alternative interpretation of the dual isotherms observed with higher plant cells

(Part B, Chap. *3.2*; ARMSTRONG and ROTHSTEIN, 1967). There are some indications that ATP is involved in K^+-H^+ exchange in yeast but as yet a clear picture has not emerged (PEÑA et al., 1969, 1972).

2.1 *Neurospora crassa*

Potassium transport has been extensively studied in *N. crassa*, principally by C.L. and C.W. SLAYMAN and colleagues. There is now a convincing picture of the system in this fungus using the standard techniques for measurement of potassium uptake and exchange and cell potential. The data that we now have indicate that transport is due to an ATPase that differs from animal potassium-sodium-ATPases in that hydrogen ions play a major role. The system is summarized in Fig. *7.1*.

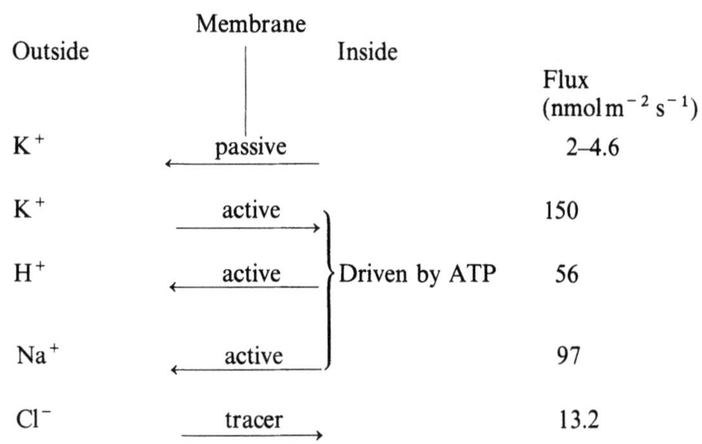

Composition (mM) (outside)	
H^+	$1.6 \cdot 10^{-3}$
K^+	30
Na^+	25
Cl^-	30
3,3 dimethyl-glutamate	20
glucose	0.55

Composition (mM) (inside)	
H^+	$3.9 \cdot 10^{-4}$
K^+	37
Na^+	152
P	300
$H_2PO_4^-$	12

Fig. *7.1*. A summary of those fluxes of ions which have been observed to occur when mycelium of *Neurospora crassa* of composition described above is put in medium whose composition is also given. (Data of SLAYMAN and SLAYMAN, 1968)

2.1.1 Flux Measurements

Uptake and flux measurements have been made using washed log-phase mycelium grown in shake culture but electrical measurements have been made using hyphae grown on cellophane over solid media. For the most part these latter measurement have been made using older hyphae of 20 μm diameter (C.L. SLAYMAN, 1965a, b) though there are some measurements with very young hyphae (SLAYMAN and SLAYMAN, 1962). The medium used by C.L. and C.W. SLAYMAN for growing the fungus was in all instances composed of sucrose/citrate plus appropriate inorganic salts. It should be noted that much of the mycelium used for uptake and flux studies may be without hyphal tips which burst during exposure to distilled water (ROBERTSON, 1959). Sealing of the septum after this treatment is almost certain to occur (TRINCI and COLLINGE, 1974). Further, although C.L. SLAYMAN (1965a) states that no obvious vacuoles are present in the larger hyphae used for electrical measurements, vacuolation is likely to have taken place (PARK and ROBINSON, 1967).

Table 7.1 gives the concentrations of potassium and sodium ions in mycelia grown under a variety of conditions. It needs to be emphasized that these results were obtained with batch culture; there is no information about ion content of mycelium grown under steady state conditions.

Table 7.1. Cation concentrations in mycelium of *Neurospora crassa* grown at 25° C in the presence of either 37 mM or 0.2 mM K^+ (SLAYMAN and SLAYMAN, 1968)

Type of mycelium	Average intracellular concentrations relative to cell water (mM)	
	K^+	Na^+
Normal cells, freshly harvested (16 h)	180	14
Low-K^+ cells, freshly harvested (16 h)	56	107
Low-K^+ cells, incubated after growth for 20 mins in K^+-free buffer (pH 5.8)	37	152
Low-K^+ cells, incubated in K^+-free buffer (pH 5.8) for 20 min then 30 mM KCl	181	28

Evidence that potassium transport is metabolically driven comes from use of the inhibitors sodium azide and DNP (LESTER and HECHTER, 1958; C.W. SLAYMAN and TATUM, 1965a). The former at 100 μM gives 80% inhibition of respiration and almost complete inhibition of net K^+ uptake. Loss of K^+ of any magnitude from the mycelium only occurs at higher concentrations of the inhibitor. The uptake system has a high degree of specificity for K^+, though it also transports Rb^+. The affinity for Rb^+ is 40% of that for K^+ at 5 mM concentration of each ion (C.W. SLAYMAN and TATUM, 1964). Little Na^+ is transported into the hyphae.

Net uptake of potassium by low-K^+/high-Na^+ mycelium at pH 5.8 is accompanied by net extrusion of Na^+ and H^+ ions (SLAYMAN and SLAYMAN, 1968). All three net fluxes are exponential with time and obey Michaelis kinetics as a function of external potassium (Table 7.2). The net flux of K^+ can certainly be regarded

Table 7.2. Maximal velocities and apparent Michaelis constants relative to external K^+, ($K_{m,K}$) for net cation fluxes in low potassium mycelium of *Neurospora crassa* (SLAYMAN and SLAYMAN, 1968)

Flux	V_{max} (nmol m^{-2} s^{-1})	$K_{m,K}$ (mM)
K^+ (influx)	150	11.8
Na^+ (efflux)	−97	11.6
H^+ (efflux)	−56	12.3

as being almost completely an active one. Measurements of loss of K^+ into distilled water and into buffer from azide-poisoned cells, and flux measurements at the minimum extra-cellular concentration allowing non-growing cells to remain in a steady state give very low values (Table 7.3). The fluxes of anions can also be discounted for the most part. SLAYMAN and SLAYMAN (1968) showed that tracer influx of chloride, sulphate and phosphate at pH 5.8 are at least an order of magnitude less than that of net K^+ flux (Table 7.4). (Note that the surface area of the hyphae is about $2.2\ m^2\ g_{cell\ water}^{-1}$.)

Table 7.3. Passive leak of potassium out of mycelium of *Neurospora crassa* determined by three different procedures (SLAYMAN and SLAYMAN, 1968)

Procedure	Flux ($nmol\ m^{-2}\ s^{-1}$)
Loss into buffer (pH 5.8) from mycelium poisoned with 1.0 mM sodium azide	2.0
Loss into distilled water from untreated mycelium	3.3
Unidirectional flux extrapolated to the minimum extracellular concentrations (0.05 mM) at which cells stay in the steady state	4.6

Table 7.4. Anion influxes into low potassium mycelium of *Neurospora crassa* (SLAYMAN and SLAYMAN, 1968, 1970)

Tracer anion	Concentration (potassium salt) (mM)	Initial anion influx ($nmol\ m^{-2}\ s^{-1}$)
pH 5.8		
^{36}Cl	30	13.2
$^{35}SO_4$	15	1.3
$^{32}PO_4$	27.4	3.9
pH 8.0		
^{36}Cl	30	14
$^{35}SO_4$	15	0.61
$^{32}PO_4$	15.5	3.7

Above pH 7, there are marked changes in the net fluxes (Fig. 7.2). Potassium uptake is reduced at pH 7 to 8 by roughly 50% of its maximum value at about pH 5, but Na^+ loss is reduced to a very much lower value ($10\ nmol\ m^{-2}\ s^{-1}$), which is sufficiently low to be ascribable to passive movement of the ion into the medium.

The difference between net K^+ influx and net Na^+ efflux above pH 7 has not been accounted for. The observed K^+-stimulated H^+ release at pH 8 was 15–30 $nmol\ m^{-2}\ s^{-1}$, which is only about a third of the difference between potassium and sodium net fluxes. Chloride, phosphate and sulphate tracer influxes at this pH are still low (Table 7.4). Uptake of bicarbonate has not been investigated. Loss of Mg^{2+} could also be important. SHERE and JACOBSON (1970) have shown that in *Fusarium oxysporium* ion balance in the mycelium when K^+ is absorbed is brought about by loss of Mg^{2+} as well as loss of H^+ and a gain in organic acids.

Above pH 8, K^+ uptake can be resolved into two distinct exponential components. There is a fast component (time constant = 1.2 min) which is matched quantitatively by a rapid loss of Na^+. This exchange can be attributed to ion exchange within the cell wall since it is comparatively insensitive to low temperatures and metabolic inhibitors.

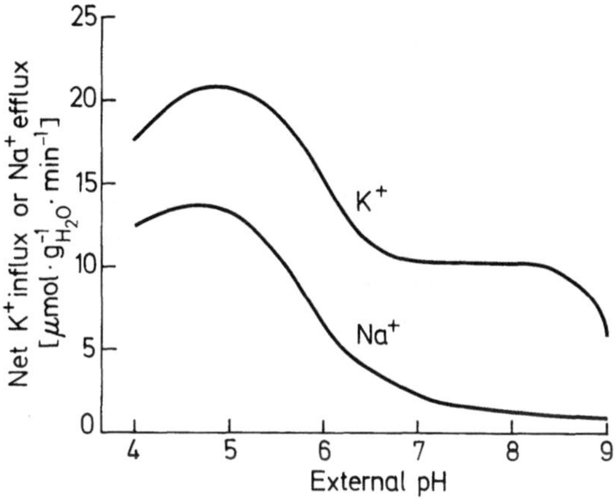

Fig. 7.2. Initital rates of net K^+ influx and Na^+ efflux in low potassium mycelium of *Neurospora crassa* in the presence of 30 mM KCl as a function of pH (SLAYMAN and SLAYMAN, 1970)

2.1.2 Potassium-Potassium Exchange

When mycelium harvested from log-phase growth is resuspended in 20 mM phosphate buffer at pH 5.8 and a concentration of K^+ (5 mM) which gives no change in mycelial K^+ content, K^+-K^+ exchange across the membrane can be demonstrated using ^{42}K (C.W. SLAYMAN and TATUM, 1965a). The mycelium acts as a single compartment with a half-time of about 5 min. This exchange is almost certainly brought about by the system responsible for the exchange of K^+ for H^+ or Na^+ ions. K^+-K^+ exchange has a V_{max} of 150 nmol m^{-2} s^{-1} which is the same as exchange of K^+ for Na^+ and H^+ ions. Further, a single gene can produce mycelium in which K^+-K^+ exchange and net K^+ uptake are equally affected (C.W. SLAYMAN and TATUM, 1965b; C.W. SLAYMAN, 1970). However the Michaelis constants for external K^+ for K^+-K^+ and K^+-Na^+,-H^+ exchange are quite different: 1.0 mM in the former instance and 11.8 mM in the latter. A similar situation is observed with red blood cells (GLYNN et al., 1970) but here all the evidence from biochemical studies (GLYNN et al., 1971) indicates that only a single transport system is involved.

2.1.3 Transport Kinetics

SLAYMAN and SLAYMAN (1970) have examined the kinetics of net K^+ uptake into the cytoplasm as a function of K^+ concentration and pH of the medium. At low pH (4.0–6.0), net flux is a simple exponential function of time which obeys Michaelis kinetics as a function of K^+ concentration. At high pH, K^+ uptake is more complex, obeying sigmoid kinetics. The data have been fitted satisfactorily by two different two-site models. In one, the transport system is thought to contain both a carrier site responsible for K^+ uptake and a modifier site: for an H^+ ion at low pH and for a K^+ ion at high pH. The other model postulates a transport consisting of multiple subunits, each with an active site for K^+, H^+ being an allosteric activator.

2.1.4 Electrical Measurements

Insertion of electrodes in mature hyphae, bathed in a sucrose/ionic medium of osmotic strength of 156 mOsm (C.L. SLAYMAN, 1965a) has shown that the internal potential can be large and negative, often exceeding −200 mV. This is more negative than the Nernst potential for any of the ions, indicating that it is not a diffusion potential. It is therefore sensible to conclude that the potential is in part being generated by an electrogenic pump.

The potential is sensitive to changing the concentration of K^+, Na^+ and Ca^{2+} in the external medium. Increasing the concentration of KCl or K_2SO_4 moves the internal potential towards zero, becoming less negative, at 45 mV per log unit over the range, 0.1–10 mM. Above 10 mM, the change of potential is greater: 77 mV per log unit. Sodium has a similar effect. Calcium at 1 mM diminishes the influence of an increasing concentration of K^+ on Na^+ such that the potential changes are only 17 and 9 mV per log unit respectively. Calcium exerts its effect when the K^+ concentration is below 20 mM.

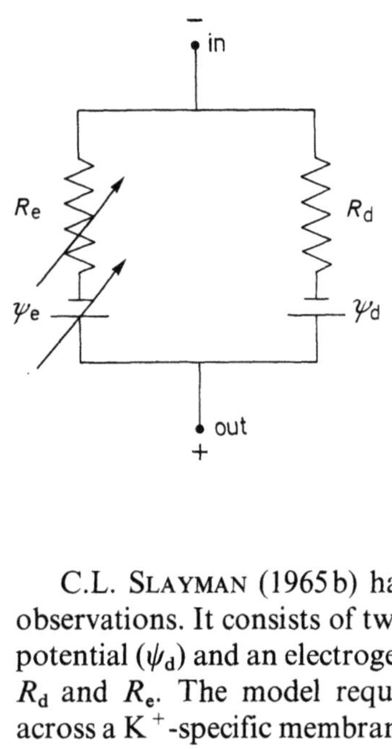

Fig. 7.3. The electrical circuit proposed by C.L. SLAYMAN (1965b) for the internal potential of *Neurospora crassa*. ψ_d, diffusion potential, ψ_e, electrogenic potential; R_d and R_e internal or intrinsic resistances of ψ_d and ψ_e respectively

C.L. SLAYMAN (1965b) has used the model shown in Fig. 7.3 to explain these observations. It consists of two distinct electromotive forces in parallel—a diffusion potential (ψ_d) and an electrogenic potential (ψ_e). They are linked through resistances R_d and R_e. The model requires only that R_d behave like an integral resistance across a K^+-specific membrane (relative to Na^+) in order that the internal potential should vary with log $[K]_o$ along a slope that is significantly less than the Nernst slope at low $[K]_o$ and significantly greater at high $[K]_o$. The decline in potential difference with increasing K^+ concentration would be made up of a decrease of ψ_d and a decrease in the fraction ψ_e which could be measured across the resistance network. The effect of calcium would be accounted for if Ca^{2+} simply increased R_d. Similar proposals have been put forward for *Nitella translucens* (SPANSWICK et al., 1967 and SPANSWICK, 1972, 1973 see also 4.2.6).

The ion causing electrogenicity is H^+. This ion remains the only likely candidate after all other possibilities have been removed in experiments where it was shown that the potential could not be changed by changing the anion composition of the medium nor by changing the Na^+ content of the mycelium (Table 7.5).

Table 7.5. Effects of various cations and anions on the membrane potential of *Neurospora crassa* (C.L. SLAYMAN, 1970)

Composition of cells	Ion tested	Control: Medium (mM)	Control: ψ_{cyt} (mV)	Testing conditions: Medium (mM)	Testing conditions: ψ_{cyt} (mV)
	NO_3^-	25 NaCl +37 KCl	-126 ± 4	25 $NaNO_3$ +37 KNO_3	-125 ± 3
	SO_4^{2-}	10 KCl	-145 ± 9	5 K_2SO_4	-149 ± 6
	$H_2PO_4^-$	25 NaCl +37 KCl	-126 ± 4	25 Na^+ +37 K^+ +phosphate at pH 5.9	-128 ± 7
	HCO_3^-	10 K+6.7 phosphate +1 $CaCl_2$ pH 6.9	-217 ± 6	10 $KHCO_3$+1 $CaCl_2$ +5% CO_2 pH 7.0	-212 ± 6
	Na^+	10 KCl +1 $CaCl_2$	-181 ± 4	10 NaCl+1 $CaCl_2$	-209 ± 4
Low-Na^+	–	10 KCl +1 $CaCl_2$	-174 ± 3	–	
Low-K^+	K^+	10 NaCl +1 $CaCl_2$	-165 ± 7	10 KCl+1 $CaCl_2$	-170 ± 7

All media contained the above constituents +2% sucrose or 1% glucose. Sodium in low-Na cells was estimated at 0.2 mM; potassium in Low-K^+ cells at 42 mM.

2.1.5 Effects of Inhibitors

SLAYMAN (1965b) has examined in some detail the effect of respiratory inhibitors on the electrical potential. Rapid changes are brought about by cyanide, azide, carbon monoxide, and DNP and anoxia. Thus 1 mM azide or DNP diminishes the potential from near -200 mV to near -30 mV within one minute, at a maximal rate of 20 mV s^{-1} (Fig. 7.4). The surface resistivity of the hyphae did not shift significantly from the control value of 0.47 Ωm^2 during this change, though over

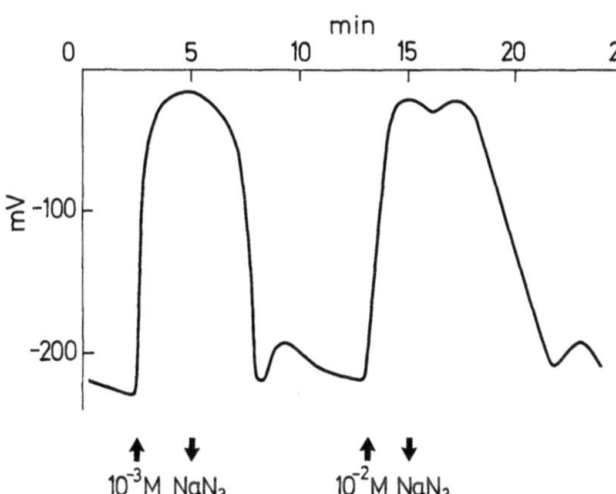

Fig. 7.4. The maximum response of the internal potential of *Neurospora crassa* to sodium azide (SLAYMAN, C.L., 1965b)

the following 5 min it rose two- to five-fold. The internal potential usually recovers within 10 min after the inhibitor has been removed. The effect of carbon monoxide can be reversed with light, specifically of the wavelengths 430 nm and 590 nm, known to decompose cytochrome-CO complexes in yeast (CASTOR and CHANCE, 1955). C.L. SLAYMAN et al. (1970) have shown that the decay in potential brought about by a metabolic inhibitor (in this case 1 mM cyanide) is paralleled by a drastic drop in the mycelial concentration of ATP (from 2.7 mM to 0.25 mM in 30 s) such that the voltage/time curve is superimposable upon the ATP/time curve with rate constants for both corresponding to a half-time of 3.7 s. The cell can be considered as acting as an ATP electrode (see Fig. *4*.7). With respect to the effect of inhibitors, it is of considerable interest to find that with DNP uncoupling of respiration (i.e. increase of O_2 uptake) takes place *before* an effect on the potential and on transport (Fig. *7*.5). Further information about adenine nucleotide levels in *N. crassa* has been given by C.L. SLAYMAN (1973).

The observations support the hypothesis that a major component of the membrane potential is electrogenic. It is reasonable to suppose that the potential that remains once an inhibitor has exerted its full effect is brought about by ion diffusion across the membrane.

One interesting result for comparison with other plant cells is the inhibition of potassium uptake by deoxycorticosterone (LESTER et al., 1958; LESTER and HECHTER, 1958; SLAYMAN and SLAYMAN, 1970). There is some disagreement about its effect on respiration in the presence of exogenous glucose but there is no doubt that the effect of deoxycorticosterone on respiration is very much less than its effect on K^+ uptake. It also has an effect on sugar and amino acid transport (LESTER et al., 1958) which is of interest in view of possible coupling between ion transport and transport of these compounds (see *7*.10).

Ouabain has no effect on the transport system (SLAYMAN and SLAYMAN, 1968).

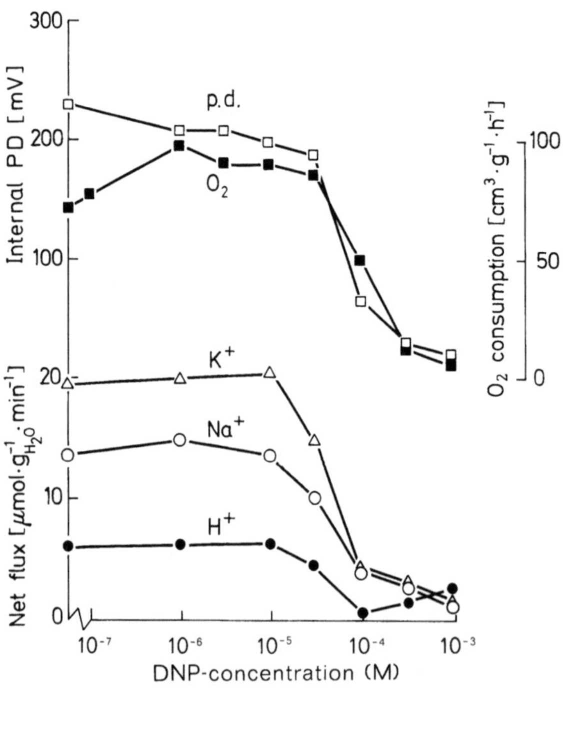

Fig. *7*.5. The effect of 2,4-dinitrophenol on the O_2 consumption and internal potential (data of C.L. SLAYMAN, 1965b) and on initial rates of net K^+ influx and net Na^+ and H^+ ion effluxes of low potassium mycelium (Data of SLAYMAN and SLAYMAN, 1970)

2.2 *Saccharomyces cerevisiae*

2.2.1 General Characteristics

Studies on alkaline ion uptake by *Saccharomyces cerevisiae* often have used commercially available baker's yeast, which previously had been washed and starved for a short time in distilled water. Some idea of the solute composition of such yeast can be gained from inspection of Table 7.6. In the great majority of studies only net uptake has been studied. The monovalent ion content of yeast can be readily changed by exposing the cells to aerated solutions of the appropriate salt. In this way, NH_4^+-rich (CONWAY and BREEN, 1945) and Na^+-rich yeast (CONWAY et al., 1954) can be produced (Table 7.6).

Table 7.6. Intracellular concentration of constituents in yeast

	CONWAY and ARMSTRONG (1961) (mM)	ROTHSTEIN (1960) (μmol g_{FW}^{-1})		CONWAY and ARMSTRONG (1961) (mM)	ROTHSTEIN (1960) (μmol g_{FW}^{-1})
K^+	261	150	Hexose P	9	120
Na^+	7	–	Meta P	5	–
Mg^{2+}	14	20	Amino acids	224	–
Ca^{2+}	9	3	Succinate	26	7
Mn^{2+}	–	0.2	Citrate	7	–
HCO_3^-	–	50	Keto acids	1	–
Phosphate	26	15	Ether soluble acids	10	15
Acid soluble P	–	40	Total molarity	602	420.2

Intracellular concentration (μmol g_{FW}^{-1}) of K^+ and ammonia–N in yeast after 4 days in either 0.2 M KCl or NH_4Cl (CONWAY and BREEN, 1945):

	Potassium yeast	Ammonia yeast
K^+	177	2.6
Ammonia–N	–	239

Intracellular composition (μmol g_{FW}^{-1}) of K^+ and Na^+ in yeast before and after exposure to 0.28 M glucose + 0.2 M sodium citrate (CONWAY and MOORE, 1954):

	Na^+	K^+
Before	2	128.1
After	141.6	4.1

Though commercial baker's yeast has been a convenient material, there are now genetically defined strains available. Since transport properties have been found to vary between different mutants (e.g. C.W. SLAYMAN, 1970) it seems preferable for the genetically defined strains to be used in transport studies. In some cases these strains may help clarify certain aspects of earlier research.

Net movement of K^+ into yeast is usually negligible in the absence of substrate, though some K^+ exchange can occur, particularly in aerobic conditions (CONWAY et al., 1954). In the presence of glucose, there is a rapid uptake of K^+ under both aerobic and anaerobic conditions (CONWAY and O'MALLEY, 1946; ROTHSTEIN and ENNS, 1946). Sodium loss from Na^+-rich yeast into 0.1 M KCl is markedly increased by glucose, as is also loss into distilled water and 0.1 M NaCl. However in this latter instance, the cause may be a decrease in cell volume (KOTYK and KLEINZELLER, 1959).

Under aerobic conditions, K^+ uptake can be brought about by a number of oxidisable substrates other than glucose (ØRSKOV, 1950), in particular ethanol, propanol and butanol, acetic, proprionic, pyruvic and lactic acids and acetaldehyde and proprionaldehyde.

Potassium is taken up in exchange for H^+ ions. Both ions move against large concentration gradients (ROTHSTEIN and ENNS, 1946). CONWAY and DUGGAN (1958) showed that the system is highly specific for K^+ and strongly discriminates in favor of K^+ when other alkali cations are present (Table 7.7). Only Rb^+ is transported to any appreciable extent by the same system. There is no evidence that when K^+ is taken into cells it is bound to an appreciable extent (CONWAY and ARMSTRONG, 1961). In yeast which is fermenting normally, it seems that K^+ is balanced inside the cell by succinate. If however the succinate level in the cell is reduced (by prolonged aeration before an experiment) K^+ uptake is associated with that of HCO_3^- (CONWAY and BRADY, 1950).

Two estimates (FOULKES, 1956; CONWAY and DUGGAN, 1958) arrive at around 0.1 μmol g_{FW}^{-1} for the amount of carrier present in yeast.

Table 7.7. The relative transport affinity of a cation for the yeast cell which was determined from the relation $(K^+)/(X) \times 100$, where (K^+) and (X) are the concentrations of potassium and the cation investigated, when the uptake of K^+ is depressed by 50%. Suspensions were made of 1 g of yeast in 5 to 200 cm^3 depending on the cations being examined, and the suspending fluid contained 5% glucose (CONWAY and DUGGAN, 1958)

Cation	Relative transport affinities	Cation	Relative transport affinities
K	100	Na	3.8
Rb	42	Li	0.5
Cs	7	Mg	0.5

2.2.2 Relationship of Potassium Uptake to Internal pH

CONWAY and DOWNEY (1950) showed that when yeast ferments glucose in the presence of 0.1 M KCl, the mean intracellular pH rises from an initial value of 5.8 to 6.35 after 30 min. ROTHSTEIN (1960) has suggested that the production of metabolic anions such as bicarbonate and succinate and the resulting change of cell pH to alkaline values may be the predominant factor in limiting K^+ uptake during glucose fermentation. The alternative view is that the limitation is brought about by the H^+ ion concentration in the cell acting directly on the influx.

Studies on net K^+ uptake during the oxidation of ethanol and related alcohols have indicated that the latter of these two alternatives is the more likely. RYAN (1967) found that K^+ uptake in the presence of ethanol in an unbuffered medium is three times that from a medium buffered at pH 7.4. The oxidation products of ethanol are acetic acid and acetaldehyde. Subsequent studies (RYAN et al., 1971) showed that, when yeast oxidizes ethanol at different pH values, the net uptake of K^+ corresponds to the amount of acetate which accumulates in the cell at each external pH value. As the pH of the medium is increased above pH 4.5, less acetate accumulates since more acetate is lost from the cell. When acetate or propionate is added to a yeast suspension oxidizing ethanol, the net uptake of K^+ increased due to uptake of these anions. Further, when a suspension has a 30:70 mixture of $CO_2:O_2$ bubbled through it, K^+ uptake is not affected between pH 4.5 and 7.0, since HCO_3^- is formed in solution and the uptake of this anion now replaces the acetate which has been lost.

It would seem from this that the oxidative process as such does not bring about net K^+ uptake; there must be an accompanying metabolic production of acid and its retention in the cell before uptake can occur. The hypothesis was tested by RYAN and RYAN (1972), using propan-2-ol, which is oxidized to acetone, not to acids. No net K^+ uptake occurs when this alcohol was oxidized by the yeast cell, but addition of proprionate to the medium or bubbling CO_2 through the medium brought about a considerable net uptake of K^+.

Although very little net K^+ uptake occurs with the oxidation of propan-2-ol, considerable K^+-K^+ exchange occurs. This exchange is affected by intracellular pH, the optimum rate being at 6.4. If yeast previously loaded with Na^+ (see 7.2.2.1) oxidizes propan-2-ol in the presence of KCl, a steady efflux of Na^+ and influx of K^+ occurs. The efflux of Na^+ is markedly inhibited by 10 mM proprionate due to the decrease of intracellular pH. The optimum intracellular pH for Na^+ efflux is the same for K^+-K^+ exchange, indicating the same system is responsible for both processes (cf. *N. crassa* 7.2.1.2).

RYAN and RYAN (1972) believe that K^+-H^+ exchange occurs *via* a different system, since the optimum intracellular pH for this process is 5.70, with cessation of activity at around 6.0, well below the optimum for K^+-Na^+ and K^+-K^+ exchange. However, this conclusion is not warranted, since it fails to take into account the possibility that H^+ competes with Na^+ and K^+ for the transport sites. The decrease in K^+-Na^+ and K^+-K^+ exchange below pH 6.4 could be due to competition between H^+ and Na^+ and K^+ ions and not due to an effect of pH on enzyme activity in the generally accepted sense.

It should be noted that, when considering the relationship between intracellular pH and K^+ transport, there is not total agreement that production in the medium of H^+ results in alkalinization of the interior. KOTYK (1963) using a dye-distribution method for measuring intracellular pH found that it decreases to pH around 4.0 in the presence of glucose, independently of whether or not K^+ is present in the external medium.

For a general discussion of the relations between ion transport and the regulation of cell pH see Chap. *12* and also *3*.3.4 and *6*.5.5.5.

2.2.3 The Redox Theory of Conway

It is clear that K^+-H^+ and K^+-Na^+ exchange depend upon metabolism. The redox theory of CONWAY (1951) has been an attempt to provide a mechanism for K^+-H^+ exchange. The essential features are as follows (Fig. *7*.6). A carrier in the membrane is reduced by metabolic reactions. The negatively charged catalyst combines with an inorganic cation which passes into the cell membrane. Here, with a final transport of electrons to oxygen (or appropriate electron acceptor) within the cell, there is loss of the cation into the cell interior. There is an equivalent amount of hydrogen ions liberated into the medium, so that the cell remains neutral.

CONWAY and KERNAN (1955) provided evidence in favor of this hypothesis. They examined the effect of redox dyes on the uptake of K^+. They used yeast fermenting under anaerobic conditions and dyes with redox potentials from +290 to −160 mV at 100 μM concentration. Without any dye, the redox potential of the system registered by a platinum electrode against a saturated calomel electrode was about +180 mV. When it was raised above this value, by the addition of dye there was increased potassium-hydrogen exchange. The reverse occurred when the redox potential was reduced and exchange was abolished at +100 mV, with almost no effect on CO_2 and alcohol production. There was a practically linear relation between the ratio of H^+ ion loss to the control value (without dye) and the redox potential of the system in the presence of a specific dye (Fig. 7.7). Thus transport of K^+ can be driven by an appropriate redox system, either electron

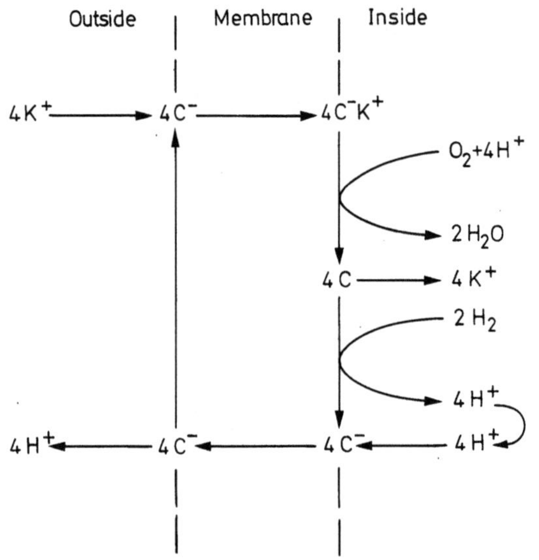

Fig. 7.6. A diagramatic representation of the redox theory of CONWAY as applied to yeast. See text for the description

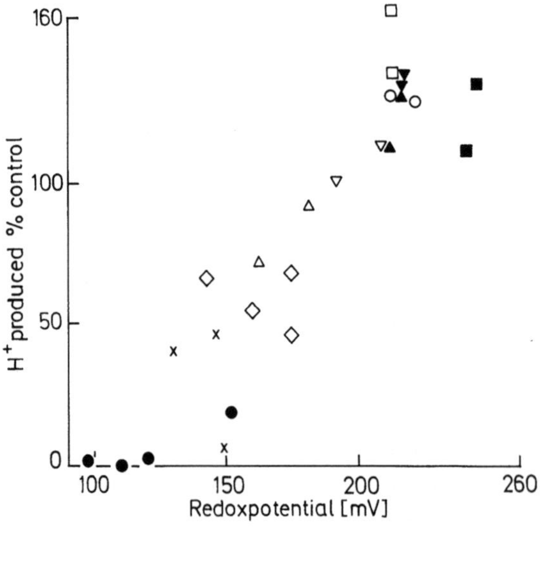

Fig. 7.7. Effect of redox dyes on the H^+ ion secretion by yeast fermenting in 40 mM K^+-H^+-succinate at pH 4.5 (CONWAY and KERNAN, 1955). Glucose added after expulsion of O_2. The following dyes were used (at 10^{-4} M): ● Nile blue, × Safranin, ◇ Neutral red, △ Benzyl-viologen, ▽ Phenol-1-indo-2,6-dichlorophenol, ▲ 1-naphthol-2-sodium sulphonate-indophenol, ○ 0-cresol-indo-2,6-dichlorophenol, ▼ Phenol-indo-2,6 dibromophenol, ■ 0-chloro-indo-2,6-dichlorophenol, □ 0-chlorophenol-indophenol

transfer to molecular oxygen or a suitable electron acceptor in the normal cell or a suitable redox dye outside the cell as just described.

Further support for and clarification of the ideas of Conway comes from the work of RYAN (1967) who studied the reduction of 125 μM methylene blue by yeast in the presence of ethanol. Rapid reduction occurs and since the dye does not penetrate the membrane, reduction must occur at the outer surface of the cell. The rate of reduction of methylene blue is strongly inhibited by potassium in the medium. Other ions also have an inhibitory effect and significantly the relative concentrations (Rb^+, Cs^+, Na^+, and Li^+) at 50% inhibition were found to be similar to the reciprocals of transport affinities of the same cations for the carrier (Table 7.8).

Table 7.8. Effect of various cations on the rate of reduction of methylene blue under anaerobic conditions (RYAN, 1967). Reciprocals of transport affinities calculated from data of CONWAY and DUGGAN (1958)

Cation	Concentration at 50% inhibition (mM)	Relative concentration at 50% inhibition	Reciprocals of transport affinities
K^+	0.4	1	1
Rb^+	1.1	2.8	2.4
Cs^+	3.9	9.8	14.0
Na^+	8.0	20.0	26.0
Li^+	13.6	34.0	200.0

The results are interpreted as follows. The alcohol: NAD oxidoreductase system reduces the K^+ carrier and in the absence of external K^+ passes both electrons and H^+ ions to the methylene blue. Potassium and other ions transported by the carrier compete with methylene blue for the carrier when it is in the reduced form. If this hypothesis is correct, rupture of the membrane should prevent potassium from exerting its effect. This was found to be the case when cells were treated with teepol, a cationic detergent, which ARMSTRONG (1957) showed has a cytolytic effect on yeast cells, and also when cells are frozen and thawed.

REILLY (1964, 1967), using a respiratory-deficient mutant of yeast, which unlike normal yeast cannot exchange potassium for sodium in the absence of glucose, showed that redox dyes at concentrations which did not affect the rate of fermentation had a different effect on K^+-Na^+ transport. Thus with normal yeast increase in the redox potential stimulated K^+ uptake but inhibited Na^+ loss. There was no effect on K^+ uptake and Na^+ loss by mutant cells. If dyes were having a non-selective effect such results would not be expected. From other studies, REILLY (1967) believes that redox dyes act primarily on a glutathione-like enzyme associated with the cytochrome system.

2.2.4 Kinetic Analysis of the System

ARMSTRONG and ROTHSTEIN (1964, 1967) from a detailed study of competition for uptake between various cations, proposed that the transport system possesses two cation binding sites for which cations compete. One site is concerned with transport and is highly specific for H^+, K^+ and Rb^+ ions (each competitively inhibiting

Table 7.9. Kinetic constants for cation uptake by *Saccharomyces cerevisiae* (ARMSTRONG and ROTHSTEIN, 1967)

	K_m (mM)	$K_{modifier}$ (mM)		K_m (mM)	$K_{modifier}$ (mM)
H^+	0.2	0.01–0.03	Rb^+	1.0	—
Li^+	27	19	Cs^+	7.0	1.3
Na^+	16	14.4	Mg^{2+}	500	4.0
K^+	0.5	1.6	Ca^{2+}	600	1.5

the transport of the other) and the other is a so-called "modifier" site. When certain cations are bound to this latter site, there is a reduction in the turnover of the carrier, the degree of reduction depending on which cation is bound and which cation is being transported into the cell.

Table 7.9 gives K_m values for the two sites. One can see from the data that H^+ has the lowest K_m for the transport site. Thus, at low pH (below 4.0) H^+ are transported into the cell in exchange one-for-one with K^+ causing enhanced efflux of this latter ion. The pump is thus reversed. Whether this is like reversal of the Na^+ pump in red blood cells (GARRAHAN and GLYNN, 1966) remains to be seen. If it is, there are implications for the biochemistry of the system.

Between pH 4.0–6.0 H^+ inhibits K^+ non-competitively by combining with the modifier site. The effect of H^+ can be reversed by high concentration of other cations and this is assumed to be due to displacement of H^+ from this site. The other cations listed in Table 7.9 as binding to the modifier site also non competitively inhibit K^+ transport.

Further work (BORST-PAUWELS et al., 1971; BORST-PAUWELS et al., 1973) has shown that there is also an activation site on the carrier. Table 7.10 summarizes the situation. When Rb^+ is being transported, the presence of either Rb^+ or K^+ or Na^+ on the activation site is stimulating. But when Na^+ is being transported only Rb^+ and K^+ stimulate; Na^+ now inhibits. The similarity of the half-maximal concentration of Na^+ which gives stimulation of Rb^+ transport and that which gives inhibition of Na^+ transport indicates that Na^+ is binding to the same site.

These effects were not observed by ARMSTRONG and ROTHSTEIN who were using higher concentrations for their studies.

Table 7.10. The activation or inhibition of Rb^+ or Na^+ transport by the presence of Rb^+, K^+ or Na^+ on the activation site of the carrier. Figure in brackets is the half-maximal concentration for the response. (BORST-PAUWELS et al., 1973)

Ion being transported	Effect of ion on activation site		
	Rb^+	K^+	Na^+
Rb^+	Activates (0.060)	Activates (0.018)	Activates (0.82)
Na^+	Activates (0.06)	Activates (0.02)	Inhibits (0.35)

3. Ammonium and Methyl Ammonium Transport

It is important to remember that uptake of ammonia may be via the undissociated molecule or the ammonium ion. As well as this, there is the complicating factor of subsequent metabolism once ammonia has entered the mycelium.

There is undoubted evidence that uptake of ammonia as the undissociated molecule occurs in *Scopulariopsis brevicaulis* (MORTON and MACMILLAN, 1954; MACMILLAN, 1956). The process appears to be passive since ammonia uptake is independent of the rate at which it is removed by assimilation. Ammonia is rapidly lost into the medium when mycelium, which has been accumulating ammonia, is returned to ammonia-free buffer solution. The loss is rapid, 50% of the ammonia being lost in 15 min. Further, respiratory inhibitiors have very little effect on the level of ammonia in the hyphae.

MACMILLAN (1956) measured the ammonia content and the internal pH of mycelia kept in a medium of constant ammonia composition but with varying pH. She found that the internal pH rises only slowly as the pH of the medium is changed from 5.0 to 9.0. Calculating the expected internal pH from the observed internal ammonia concentrations, she found very reasonable agreement between the observed pH and that calculated (Fig. 7.8) which provides favorable evidence for the diffusion hypothesis.

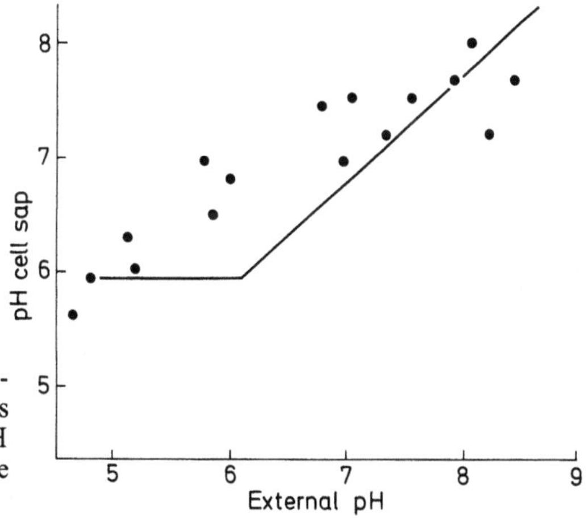

Fig. 7.8. Effect of shaking mycelium of *Scopulariopsis brevicaulis* in buffered solutions containing 0.4 mg cm^{-3} NH_3-N on the pH of the expressed sap. The line shows the calculated internal pH (MACMILLAN, 1956)

Evidence that the ammonium ion is absorbed is not quite so clear. CONWAY and DUGGAN (1958) have shown that the K^+-H^+ exchange system in *Saccharomyces cerevisiae* also transports ammonium ions, the affinity for the system being one-fifth of that for K^+. Ethylamine has an affinity one-tenth of K^+. It is highly likely that a similar sort of system must occur in other fungi since assimilation of ammonia is accompanied by loss of hydrogen ions into the medium (HOLLIGAN and JENNINGS, 1972).

There is however evidence for a specific permease. HACKETTE et al. (1970) showed that in *Penicillium chrysogenum* ammonia, methylamine and ethylamine are transported by a system, the K_m of which is respectively $2.5 \cdot 10^{-7}$ M, 10^{-5} M and

10^{-4} M. This system only occurs in nitrogen-starved mycelium. From experiments on the effects of refeeding substrate amounts of various nitrogen compounds on the uptake of methylamine, it is possible to show that repression of transport activity is brought about principally by asparagine and glutamine (Table 7.11). The rapidity of their action indicates that their effect is by feedback inhibition.

It is very interesting to find that the uptake of methylamine by this system shows a very similar sensitivity to pH as that shown by the K^+ transport system of *Neurospora crassa*. So the uptake of ammonia and the two amines may be as the positively charged ion. There is clearly scope for further investigation here.

In toto, the evidence indicates that in the fungi mentioned it is likely that there is a high affinity transport system for ammonia which at high concentrations of ammonia in the media may be obscured by diffusion.

Table 7.11. Effect of refeeding nitrogen-starved mycelium of *Penicillium chrysogenum* with various nitrogen compounds on the subsequent ability to absorb methylamine. Data selected from Table VI of HACKETTE et al. (1970)

Addition (0.01 M)	Subsequent transport rate (μmol g^{-1} min^{-1})	Addition (0.01 M)	Subsequent transport rate (μmol g^{-1} min^{-1})
None	8.5	L-asparagine	0.05
Ammonium chloride	1.0	L-glutamate	5.9
Potassium nitrate	8.5	L-glutamine	0.12
Methylamine HCl	1.8	Glycine	3.3
Urea	8.4	L-methionine	6.3
L-alanine	3.3	L-phenylalamine	7.4
L-aspartate	5.7	L-serine	1.0

4. Bivalent Cation Transport: *Saccharomyces cerevisiae*

Magnesium can enter the cells of *S. cerevisiae via* the K^+-H^+ exchange system (CONWAY and BEARY, 1958; CONWAY and GAFFNEY, 1966). Entry *via* this mechanism is slow and Mg^{2+} has a low affinity for the carrier.

There is another system which transports magnesium and other bivalent cations in the order of decreasing affinity Mg^{2+}, Co^{2+}, Zn^{2+}, Mn^{2+}, Ni^{2+}, Ca^{2+}, Sr^{2+} (FUHRMANN and ROTHSTEIN, 1968). Most of the information about this system has been obtained with ^{54}Mn which, once inside the cell, does not exchange across the membrane.

ROTHSTEIN et al. (1958) showed that Mn^{2+} is rapidly absorbed by yeast cells in the presence of phosphate, glucose and potassium. There appears to be no stoichiometric relationship between manganese uptake and K^+ and phosphate uptake. Phosphate is always absorbed more rapidly than Mn^{2+} and the rate of uptake of phosphate is unaffected by Mn^{2+}. However the absorption of Mn^{2+} can continue when all the phosphate is absorbed and, after a period of time, may exceed the amount of phosphate which is taken up. The absorption of Mn^{2+} occurs in the absence of K^+ (phosphate must be present) but is markedly stimulated in its presence.

But, whereas phosphate has a stimulatory effect over the whole range of concentrations, above 20 mM, K^+ ions are markedly inhibitory. While the stimulation of Mn^{2+} uptake at the lower K^+ concentrations is a direct effect of the stimulation of phosphate uptake, the inhibition at higher concentrations is ascribable to a large cation excess within the cells as the result of the high rate of K^+ uptake.

The absorption of Mn^{2+} can be isolated in time from the absorption of phosphate (JENNINGS et al., 1958). Cells pretreated with glucose, K^+ and phosphate, then resuspended in a K^+-free buffer are capable of absorbing Mn^{2+} without delay and without any reduction in the amount of Mn^{2+} absorbed compared with the quantity absorbed during a similar period of time when phosphate is present. However, as cells are allowed to stand, there is a loss of absorptive power which is dependent upon the metabolism of the cells and particularly dependent on the rate of glycolysis, for cells lose their absorptive power most rapidly in the presence of fermentable sugars.

By this approach, JENNINGS et al. were able to study the properties of the Mn^{2+} transport system in isolation from the process of phosphate uptake which is necessary for its initiation. Inhibitor studies with sodium acetate, redox dyes and arsenate caused parallel effects on Mn^{2+} and phosphate absorption. It was therefore postulated that both phosphate and Mn^{2+} are transferred into the yeast cell in an essentially irreversible manner, coupled in an unknown way to glycolysis at the glyceraldehyde-3-phosphate: NAD oxidoreductase step. Yeast cells normally absorb phosphate after a lag period of about 30 min, but, if pretreated with glucose and K^+, no delay occurs. It has been suggested therefore that a phosphate carrier is synthesized during glycolysis (Fig. 7.9). The manganese carrier is therefore postulated to be a phosphorylated compound formed in the cell membrane as a direct product of the phosphate-transferring system and therefore chemically related to it. This would account for the similar sensitivity of both phosphate and manganese uptake to the same inhibitors. As the ability of the yeast cell to absorb Mn^{2+} is transient, the synthesis of the Mn^{2+} carrier must be a reversible reaction. Its absence in cells undergoing glycolysis when phosphate is not in the medium eliminates glycolytic intermediates and since most of the phosphate which is absorbed can be accounted for as polyphosphated and glycolytic intermediates (JUNI et al., 1948; WIAME, 1949) the amount of carrier must be small.

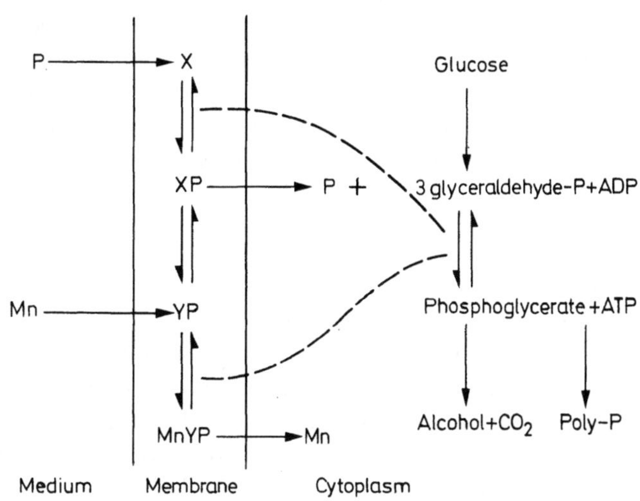

Fig. 7.9. A scheme for the absorption of Mn^{2+} and phosphate into the yeast cell. X and Y are postulated membrane components. (JENNINGS et al., 1958)

The properties of the carrier seem to be very similar to many phosphorylating enzymes. It is interesting to note that the analysis of isolated yeast plasma membranes (MATILE et al., 1967; NURMINEN and SUOMALAINEN, 1970) has revealed the presence of a Mg-activated ATPase; indeed this is the only ATPase which has been identified as being associated with these membranes.

5. Phosphate Transport

5.1 *Saccharomyces cerevisiae*

Although resting cells contain considerable amounts of phosphorus compounds, including inorganic phosphate, release of these compounds into the medium is negligible, nor does ^{32}P-orthophosphate exchange across the membrane (GOODMAN and ROTHSTEIN, 1957; SWENSON, 1960). A report (LEGGETT and OLSEN, 1964) that phosphate is accessible to 80% of the yeast cell can be explained in terms of the pretreatment given to the cells to exhaust intracellular metabolic substrate (SCHÖNHERR and BORST-PAUWELS, 1967). LEGGETT and OLSEN washed their cells with acid at pH 4.0 which leads to an increase in phosphate-accessible space. It is interesting that this increase can be prevented by DNP.

In reading the literature about phosphate uptake in yeast, it is necessary to be clear what system is under discussion. One should refer first to the paper by LEGGETT (1961) who has provided the most detailed kinetic analysis of phosphate uptake by yeast. Cells were pretreated for 60 min to establish steady-state rates of uptake which persisted for a further 60 min. During this steady-state period, uptake of ^{32}P was determined for periods of 30 to 300 s. The data was analyzed on the premise that active accumulation by the cells is considered to arise from initial combination with a specific carrier. Under steady-state conditions with negligible back-flow of phosphate out of the cell *via* the carrier, a single carrier system should exhibit first-order kinetics. A plot of rate of uptake against the ratio of uptake to phosphate concentration should give a straight line for such a carrier system. A curvilinear relationship results when there are two or more first-order reactions (two distinct transport systems) occurring simultaneously but independently (HOFSTEE, 1952). The curves can be resolved into linear components with a reasonable critical attitude toward precision and extent of the measurements and possible departure from first-order kinetics (see also Part B, *3.2.*2).

On this basis there appear to be three components of phosphate uptake. One component with a K_m of 10^{-6} M is strictly aerobic and is hypothesized from inhibitor studies to be associated with the oxidation and reduction of cytochrome b in the respiratory chain. Another component of K_m of 10^{-5} M is present under both aerobic and anaerobic conditions. It is inhibited by iodoacetate and sulphite suggesting that phosphate uptake is associated with the activity of glyceraldehyde-3-phosphate oxidoreductase. Finally, there is a component of phosphate uptake with a K_m of 10^{-3} M which is only effective at phosphate concentrations greater than 10^{-4} M in the presence of 0.3 M glucose. This component would have made an appreciable contribution to the phosphate uptake observed by GOODMAN and ROTHSTEIN (1957), who measured a K_m of about $5 \cdot 10^{-4}$ M and showed that uptake is associated with glycolysis (see also Part B, *3.3.*6.1).

BORST-PAUWELS (1962; BORST-PAUWELS et al., 1965) confirmed the presence of the component having a K_m of 10^{-5}M and showed that it occurred aerobically in the presence of ethanol and that arsenate could be transported by the same system. The arsenate which is absorbed is converted *via* arsenite to acid insoluble arsenic. However, not all arsenate is so changed, and the arsenate concentration rises such that feedback inhibition of transport takes place. HOLZER (1953) has produced results which indicate that phosphate can act in a similar way—phosphate uptake only proceeding when the concentration of inorganic orthophosphate is decreased below a certain level as a consequence of incorporation into organic form.

BORST-PAUWELS and JAGER (1969) have examined further the 10^{-5} M K_m phosphate transport system and have shown that, although absorption of phosphate is closely dependent upon glycolysis, it is possible to inhibit transport without inhibiting glycolysis, indicating that the coupling between the two processes is not tight. The kinetic data from these inhibitor studies is consistent with the following. Orthophosphate (and arsenate) binds to a primary acceptor located on the cell surface. The acceptor-phosphate complex then reacts with a hypothetical compound (Y) resulting in the accumulation of phosphate within the cell. Inhibitors of anaerobic phosphate uptake decrease the concentration of Y and so inhibit transport. These inhibitors include not only inhibitors of glycolysis but also uncouplers of oxidative phosphorylation, like DNP, isooctyl dinitrophenol and pentachlorophenol. If the hypothesis is correct, then Y is not ATP since inhibition of phosphate uptake by DNP is accompanied by only a slight decrease in the level of ATP in the cells. Nor does it appear from a comparison of the effect of DCCD and DNP on phosphate uptake that Y is some high-energy intermediate of oxidative phosphorylation (HUYGENS and BORST-PAUWELS, 1972).

On the other hand, the hypothesis may not be tenable in its entirety. Certainly the ATP concentration is reduced by 90% by 3 mM iodoacetate or 20 mM fluoride, suggesting that ATP could play the part of Y. But phosphate uptake is also inhibited by acetate (SAMSON et al., 1955; JENNINGS et al., 1958) and this effect, like the effect of DNP, is not immediately explicable in terms of the production of Y. Indeed further studies by BORST-PAUWELS and DOBBLEMAN (1972) have shown that a wide range of fatty acids inhibit phosphate uptake. This inhibition is determined by the degree of acidification of the cells and it can be therefore argued that the inhibition is not brought about by an interaction between the fatty acids and a particular protein molecule. In the light of this, BORST-PAUWELS and DOBBLEMAN (1972) suggest that a major component of the phosphate transport system is an exchange of phosphate for hydroxyl ions, a point which has been made earlier by GOODMAN and ROTHSTEIN (1957). Somewhat in line with these ideas BORST-PAUWELS and HUYGENS (1972) proposed that the effect of uncouplers and sodium azide may be *via* an effect on the cell membrane potential (but see 7.2.1.5).

There is no doubt however that phosphate uptake is closely connected with ATP metabolism since BORST-PAUWELS et al. (1962) showed that over a 10 s period ^{32}P labeled orthophosphate was incorporated into ATP more rapidly than any other compound. GTP, pyrophosphate and fructose 1–6 diphosphate were also rapidly labeled. After a further 5 s, rate of incorporation into ATP and GTP fell off and there was an increase in the rate of labeling of hexose phosphate. ROTHSTEIN (1963) and JUNG and ROTHSTEIN (1965) have shown that arsenate interacts with the phosphate transport system in a complicated manner.

5.2 *Thraustochytrium roseum*

Phosphate uptake in this organism, a non-filamentous obligate marine Phycomycete, deserves special mention, since sodium has been shown to stimulate the process (SIEGENTHALER et al., 1967a; SIEGENTHALER et al., 1967b). Experiments without added substrate showed that no other monovalent cation or any anion could bring about a stimulation of phosphate uptake. This observation suggests that unlike yeast, but similar to chlorococcal algae (ULLRICH-EBERIUS and SIMONIS, 1970; ULLRICH-EBERIUS and YINGCHOL, 1974; see also 6.6.4.1), ion balance during phosphate uptake in *T. roseum* is brought about by sodium ions.

6. Transport of Sulfate and Other Sulfur Compounds

The presence of a sulfur atom in the molecule of penicillin has meant that attention was directed early on to the transport of sulfur into moulds, particularly *Penicillium* sp. A recurring observation is the repression of sulfate transport by methionine. This has been shown to occur in *Penicillium chysogenum* (SEGEL and JOHNSON, 1961), *Saccharomyces cerevisiae* (MAW, 1963), *Penicillium notatum* and *Aspergillus nidulans* (TWEEDIE and SEGEL, 1970) and *Neurospora crassa* (MARZLUF, 1970a, 1972 see also 6.6.4.2). Other ions containing group VI elements can also often be transported by the sulfate transport system (Table 7.12).

Table 7.12. The relationship between various group VI anions and the sulfate transport system in various fungi

Fungus	Ions transported	Ions inhibiting transport of sulfate	References
Neurospora crassa			
Mechanism I (K_m=100 µM)	SO_4^{2-}	SeO_4^{2-}	MARZLUFF (1970a, b; 1972; 1973)
Mechanism II (K_m=10 µM)	CrO_4^{2-}	CrO_4^{2-}	VALLÉE and SEGEL (1972)
		MoO_4^{2-}	
		WO_4^{2-}	
Aspergillus nidulans, *Penicillium chrysogenum*, *P. notatum*	SO_4^{2-}, $S_2O_3^{2-}$, SeO_4^{2-}, MoO_4^{2-}		TWEEDIE and SEGEL (1970); SPENCER et al. (1968)
	[Distinct permeases for SO_3^{2-} and $S_4O_6^{2-}$ and choline-0-sulfate]		
Saccharomyces cerevisiae	SO_4^{2-}	$S_2O_3^{2-}$, SeO_4^{2-}	MAW (1963)

7. Amino Acid Transport

There is now a considerable body of evidence about amino acid transport in fungi. It is proposed to describe in detail only that information which appertains to the mycelium of *Neurospora crassa*. Much of the information that we have (Table 7.13) has been obtained by PALL (1969, 1970a, b, 1971). *N. crassa* was not the

Table 7.13. Major amino acid transport systems in *Neurospora crassa* (PALL, 1970b). The amino acids in this table are only a partial listing of the amino acids having affinity for the different transport systems. In most cases, amino acids with similar properties to those listed will also have affinity. Affinity constants (K_m or K_i) are all expressed in μM

System I: L-Neutral amino acids	System II: D- or L-basic, neutral and acidic amino acids	System III: L-Basic amino acids	System IV: D-or L-acidic amino acids
Amino acids transported and affinity constants (μM)			
L-Tryptophan (60)	L-Arginine (0.2)	L-Arginine (2.4)	L-Cysteic acid (7)
L-Leucine (110)	L-Phenylalanine (2)	L-Lysine (4.8)	L-Aspartic acid (13)
L-Phenylalanine (50)	D-Phenylalanine (25)		L-Glutamic acid (16)
	Glycine (7)		
	L-Aspartic acid (1200)		
Other amino acids showing affinity			
L-Valine	L-Lysine	L-Ornithine	D-Aspartic acid
L-Alanine	L-Leucine	L-Canavanine	D-Glutamic acid
Glycine	α-Aminoisobutyric acid	L-Histidine (low affinity)	
L-Histidine	β-Alanine		
L-Serine	L-Histidine		

first fungal species to be examined with respect to amino acid transport but it is the one about which we have the most comprehensive information. Much of what we have for other fungi is very similar to that for *N. crassa,* particularly with respect to the physical properties of the amino acids transported by any one system. Nevertheless, the reader should be clear that none of the systems has been described, in so far as we do not yet know the range of specificities for any one system. Thus reference to other amino acids showing affinity often does no more than indicate that these amino acids cause inhibition of transport into the cell of another amino acid. Often no investigation has been made to see whether or not the inhibitory amino acid is also taken into the cell. It is interesting to note that in the one higher plant tissue, carrot root discs, which has been examined in any detail with respect to competition between amino acids for entry into the cells, there do not seem to be distinct systems for particular types of amino acid (BIRT and HIRD, 1958).

7.1 *Neurospora crassa*

7.1.1 System I

This is specific for most aromatic and neutral L-amino acids, particularly L-tryptophan, L-leucine and L-phenylalanine. These amino acids can also be transported by system II. However, since this latter system can be inhibited by L-arginine, the activity of system I by itself can be easily assessed in the presence of 1mM arginine. L-citrulline is also transported by system I (SANCHEZ et al., 1972). System I is probably the same as the neutral amino acid system in germinated conidia (WILEY and MATCHETT, 1966, 1968).

System I is repressed by the presence of tryptophan (WILEY and MATCHETT, 1968) and phenylalanine (TISDALE and DE BUSK, 1970) in the growth medium.

WILEY (1970) has isolated a tryptophan-binding protein from germinated conidia by cold osmotic shock, the procedure being similar to that used for bacterial cells (HEPPEL, 1969). The germinated conidia so treated lose 90% of their ability to transport tryptophan without loss of viability. Alkaline phosphatases are lost at the same time. There are four lines of evidence that indicate that the tryptophan-binding protein is involved in transport of the amino acid: (i) it appears to be located near the cell surface in that no enzymes involved in tryptophan metabolism are also lost, (ii) a decreased capacity for binding tryptophan is observed in shock fluids from cells repressed for tryptophan uptake, (iii) the specificity of tryptophan binding is similar to that observed in the *in vivo* binding system, namely that binding phenylalanine, tryptophan and leucine, (iv) the dissociation constant for binding as measured by equilibrium dialysis, is approximately the same as K_m for tryptophan transport.

From all the data available, there is every reason for assuming that the tryptophan-binding protein is part of system I.

7.1.2 System II

This system has a broad range of specificity, transporting a variety of neutral or basic amino acids. The system is virtually absent in young, rapidly growing cells but appears in old mycelium in medium depleted of carbon. PALL (1969) has suggested that, in view of this and of the fact that the system has a low K_m (<50 µM) and broad specificity, system II has the role of a scavenger.

SANCHEZ et al. (1972), using appropriate mutants, have looked at this system further. Their data are consistent with regulation of system II both by the pool of ammonia and the pool of amino acids within the cell.

Histidine is principally taken up by system II. This can be demonstrated using mutants in which transport system for neutral and basic amino acids are absent (MAGILL et al., 1972). Using double mutants for these two latter systems, it has been possible to characterize system II further by studying the effect of selected compounds on histidine transport (MAGILL et al., 1973). The compounds used were selected L-α-amino acids, D-histidine, histidine analogues, β-alanine, carboxyl-modified amino acids, N-acyl amino acids and simple derivatives of imidazole. The positively charged α-amino group is necessary for binding; the negatively charged carboxyl group is however of less importance, since its replacement by a neutral carbonyl group does not completely abolish binding. The greatest latitude for binding was found in the side chain. Affinity for α-amino acids was uniformly high except for L-aspartic and L-glutamic acids, L-asparagine and L-proline.

7.1.3 System III

This is specific for basic amino acids. Its activity can be measured readily, since systems I, IV and V have no affinity for basic amino acids and system II is absent from rapidly growing mycelium. If necessary, activity can be measured by using L-lysine in the presence of 50 mM glycine which inhibits lysine uptake by system II but hardly by III.

7.1.4 System IV

This is specific for acidic amino acids and, like system III, its activity can be readily measured. L-aspartic acid can be used, with arginine at 1 mM, if necessary, to repress system II. There is little system IV activity in conidia or rapidly growing cells. The system is most active in mycelium which is starved of carbon, nitrogen or sulphur.

7.1.5 System V

This is specific for methionine and ethionine. Since the former amino acid is also transported by systems I and II, its uptake *via* system V can be estimated in the presence of 50 mM glycine which inhibits the activity of systems I and II. System IV is absent in growing mycelium, being only present in sulfur-starved mycelium. This suggests that it has a role scavenging sulfur (PALL, 1971). PALL (1971) has observed very little decline in system V even 3 h after CHM addition, indicating that the permease is relatively stable.

7.1.6 Transinhibition

We now know from studies on the uptake of amino acids in *N. crassa*, that regulation of transport is brought about by the concentration of amino acids within the hyphae (PALL, 1971). Such inhibition has been described (RING and HEINZ, 1966) from studies on amino acid uptake by *Streptomyces hydrogenans* and has been called transinhibition.

The transinhibition in the methionine transport system (PALL, 1971) and system I and II (PALL and KELLY, 1971) is system-specific. The ability of an amino acid to transinhibit a transport system is highly correlated with its affinity for the system. Amino acids with high affinity are effective transinhibitors, those with lower affinity are less effective transinhibitors.

PALL (1971) has indicated two possible ways in which transinhibition might occur. In the first, transinhibition is suggested to occur through the protein involved in transport having an allosteric binding site for the amino acids concerned. Binding of the amino acids could inhibit the activity of the protein and therefore transport. By this mechanism, the transport system would have two binding sites, one binding the amino acid prior to transport into the hyphae and another binding transinhibiting amino acids inside the hyphae.

On the other hand, and this more likely, transport and transinhibition may be determined by a single site. In this case, transinhibition will be caused by the binding of the appropriate amino acid to the active site of the carrier when that site is oriented towards the hyphal interior. PALL (1971) has proposed a model on this basis in which the binding of the amino acid to the carrier prevents the carrier returning to the outside of the membrane. The features of the various reactions involved in this model are shown in the diagram in Fig. 7.10.

The slowness of reaction 4 compared with 3 is equivalent to the carrier-amino acid complex being in a lower energy state when oriented to the inside of the hypha than when oriented towards the outside. When an amino acid which brings about transinhibition accumulates inside the hypha, the equilibrium between the

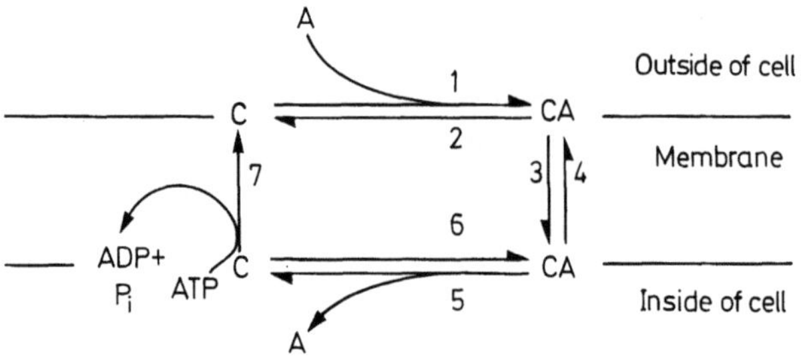

Fig. 7.10. A model for system specific transinhibition (PALL, 1971). *1* and *2* Allows *reversible* association of the amino acid with the carrier. *3* Brings about orientiation of the carrier from the outside to the inside of the membrane. *4* The reverse of *3* but very much slower. *5* and *6* Allows *reversible* dissociation of the amino acid from the carrier. *7* Brings about recycling of the carrier by a process involving ATP or some other energy-rich compound. *A* amino acid; *c* Protein carrier

Table 7.14. Properties of some transport systems present in *Penicillium* (HUNTER and SEGEL, 1971)

Transport system[a]	Conditions for development	V_{max} (µmol $g^{-1} min^{-1}$)	K_m (M)	Ref.
L-Methionine	Sulfur deficiency	1	10^{-5}	[1]
L-Cystine	Sulfur deficiency	1.2	$2 \cdot 10^{-5}$	[2]
L-Cysteine	Constitutive[b]	2.1	$1.4 \cdot 10^{-4}$	[2]
Choline-0-sulfate	Sulfur deficiency	3	$2 \cdot 10^{-4}$	[6]
Sulfate (thiosulfate, selenate, molybdate)	Sulfur deficiency	1.5	$1.7 \cdot 10^{-5}$	[4, 7]
Tetrathionate	Sulfur deficiency	3	$2.5 \cdot 10^{-5}$	[9]
Glutathione (reduced)	Sulfur deficiency	2.3[c]	$1.7 \cdot 10^{-5}$[c]	[8]
Ammonium[d]	Nitrogen deficiency	10	$2.5 \cdot 10^{-7}$[d]	[5]
L-α-Neutral and basic amino acids	Nitrogen or carbon deficiency	10	$2 \cdot 10^{-5}$	[2, 3]
L-α-Acidic amino acids	Nitrogen or carbon deficiency	10	$0.2–1 \cdot 10^{-4}$	
L-α-Basic amino acids	Constitutive[b]	1	$5 \cdot 10^{-6}$	
L-Arginine	Constitutive[b]	—[e]	—[e]	
L-Lysine	Constitutive[b]	—[e]	—[e]	
L-Proline	Nitrogen or carbon deficiency	—[e]	—[e]	

[a] The transport systems are designated in terms of their best substrate.
[b] "Constitutive" means that the transport system is active in mycelia without adding the substrate as an inducer and without incubating the mycelia in any nutrient-deficient media.
[c] Glutathione transport shows a second, apparently nonsaturable function at high substrate concentrations.
[d] The ammonium transport system was assayed with methylammonium-^{14}C as the substrate (K_m ca. 10^{-5}M). The K_m value for NH_4^+ listed is the experimentally determined K_i value for NH_4^+ as an inhibitor of methylammonium-^{14}C transport.
[e] Not determined, or undeterminable in wild-type mycelium.

[1] BENKO et al. (1967). [2] SKYE and SEGEL (1970). [3] BENKO et al. (1969). [4] YAMAMOTO and SEGEL (1966). [5] HACKETTE et al. (1970). [6] BELLENGER et al. (1968). [7] BRADFIELD et al. (1970). [8] SKYE, G.E. and SEGEL, I.H. Unpublished data. [9] KLEIN et al. (1970).

free carrier and carrier-amino acid complex is shifted towards the latter. Thus the concentration of free carrier is reduced. This leads to an inhibition of the energy-dependent recycling process. Recycling cannot occur *via* reaction 4 because of its slowness.

An important aspect of this model concerns reaction 4. The slowness of this reaction leads to transinhibition and also a low rate of exchange diffusion. The results of WILEY and MATCHETT (1968) show little or no exchange diffusion of tryptophan in *N. crassa*. On the other hand, if reaction 4 is fast, there will be no transinhibition but considerable exchange diffusion.

Of course, if amino acid transport is driven by ion gradients (see 7.10) an energy-dependent recycling process will not be involved. However, there is little problem about reconciling the above model with transport driven in this way. The only necessary assumption (and this is implicit in the above model) is that the unloaded carrier travels back to the outside of the membrane at a much higher rate than when it is loaded.

7.2 Other Fungi

Tables 7.14 and 7.15 give relevant information for *Penicillium chrysogenum*, and for *Saccharomyces* sp. respectively.

Table 7.15. Summary of studies on amino acid transport in *Saccharomyces cerevisiae* and *S. chevalieri*

Amino acid	Type	K_m (M)	Trans-inhibition	Ref.
S. cerevisiae				
S-adenosyl methionine	specific	$3.3 \cdot 10^{-6}$		MURPHY and SPENCE (1972)
amino-isobutyrate		$1.5 \cdot 10^{-4}$ $5.4 \cdot 10^{-3}$	Yes	KOTYK and ŘÍHOVÁ (1972)
arginine	basic	10^{-5}		GRENSON et al. (1966)
glutamate	acidic			JOIRIS and GRENSON (1969)
histidine	one high; one low affinity — both specific		Probably	CRABEEL and GRENSON (1970)
lysine	specific basic	$2.5 \cdot 10^{-5}$ $2 \cdot 10^{-4}$		GRENSON (1966)
methionine	specific narrow-specificity	$1.2 \cdot 10^{-6}$ $7.7 \cdot 10^{-4}$		GITS and GRENSON (1967)
—	general permease			GRENSON et al. (1970)
S. schevalieri				
proline	specific	$2.5 \cdot 10^{-5}$		SCHWENKE and MAGANA-SCHWENKE (1969); MAGANA-SCHWENKE and SCHWENKE (1969)

8. Monosaccharide Transport

8.1 *Saccharomyces cerevisiae*

8.1.1 Glucose Transport

VAN STEVENINCK and ROTHSTEIN (1965) showed that, while glucose is normally transported into yeast cells by a system of high affinity and relatively high specificity, when the metabolic poison iodoacetate is present, glucose enters by facilitated diffusion, the system then having broader specificity transporting also sorbose and galactose (Table 7.16). When glucose transport is occurring by the high specificity system, there is a drop in the binding of Ni^{2+} and Co^{2+} by the cell surface, and transport is inhibited by uranyl ions (UO_2^{2+}) at relatively low concentrations. On the other hand, when facilitated diffusion is occurring, there is no change in the binding of Ni^{2+} and Co^{2+} and only relatively high concentrations of UO_2^{2+} in the medium inhibit transport. Earlier studies (ROTHSTEIN, 1954; VAN STEVENINCK and BOOIJ, 1964) showed that Ni^{2+} and Co^{2+} and UO_2^{2+} at low concentrations bind to polyphosphate groups at the cell surface. VAN STEVENINCK and ROTHSTEIN therefore conceived the sugar transport system in yeast to be functioning as shown in Fig. 7.11.

Table 7.16 Summary of the interaction between Ni^{2+} and UO_2^{2+} and sugar transport processes in *Saccharomyces cerevisiae* (VAN STEVENINCK and ROTHSTEIN, 1965)

Transport Process	Ni^{2+} binding	Ni^{2+} effect on sugar transport	UO_2^{2+} effect on sugar transport	Kinetic parameter		Specificity
				V_{max}	K_m	
Facilitated diffusion[a]	No change	None	Inhibited by relatively higher concentration (0.3 mM)	Low	High	Broad
Active transport[b]	Reduced by 70%	Inhibited to maximum of 70%	Inhibited by relatively lower concentration (0.1 mM)	High	Low	Narrow

[a] Sorbose in idoacetate-poisoned normal or galactose-induced cells. Glucose in iodoacetate-poisoned cells. Galactose in uninduced cells ± iodoacetate. Galactose in induced, iodoacetate-poisoned cells.
[b] Glucose, mannose or fructose in normal cells. Galactose in induced cells.

VAN STEVENINCK (1969) showed in iodoacetate-poisoned cells (fermentation blocked) that there is a rapid active glucose uptake of about 5 μmol g^{-1}. No free glucose could be recovered from the yeast cells and it was found that the transported sugar had been phosphorylated. The amount of sugar phosphorylated was independent of cellular ATP concentration, even when it was less than 0.1 μmol g^{-1}, an amount only sufficient to phosphorylate 20% of the glucose transported. Parallel studies (VAN STEVENINCK, 1968) showed that, when 2-deoxyglucose is taken up by yeast, it is recovered inside the cells partly as the free sugar and partly as 2-

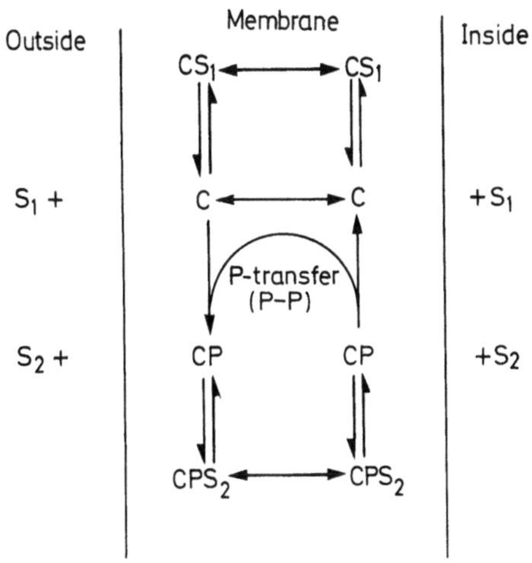

Fig. 7.11. A model proposed by VAN STEVENINCK and ROTHSTEIN (1965) of the glucose transport systems in yeast. C refers to the carrier, S_1 to a sugar transported by facilitated diffusion, S_2 to a sugar actively transported and P-P to polyphosphate groups

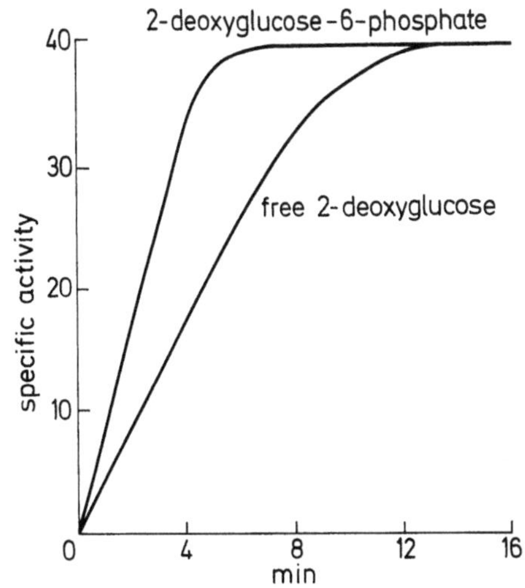

Fig. 7.12. Specific activity of intracellular free 2-deoxyglucose and 2-deoxyglucose-6-phosphate after (^{14}C)-deoxyglucose pulsing. The preincubation conditions were: 3% yeast, 1% ethanol and 3 mM deoxyglucose (unlabeled) at 25° C for 20 min. 0.06 M (^{14}C)-deoxyglucose was then added to the suspension (VAN STEVENINCK, 1968)

deoxyglucose-6-phosphate. Pulsing experiments with ^{14}C-deoxyglucose showed that deoxyglucose phosphate behaves as the precursor of intracellular free deoxyglucose (Fig. 7.12). This means that the transfer step itself involves the phosphorylation of the sugar. The source of phosphate is believed to come from polyphosphate formed in turn from the bulk polyphosphate of the cell *via* ATP (HAROLD, 1966). The process, one of group translocation, is postulated (VAN STEVENINCK and DAWSON, 1968) to function as follows (S=sugar; C=carrier; $(PP)_n$=polyphosphate):

$$\begin{array}{l} S+C \searrow \\ \qquad\qquad (PP)_{n-1}+S\text{-phosphate-C}\rightarrow \text{Transmembrane transport} \rightarrow S\text{-phosphate} \\ (PP)_n \nearrow \\ \uparrow \\ \text{metabolic link} \end{array}$$

The same type of process brings about the transport of α-methylglucoside (VAN STEVENINCK, 1970) and also galactose (see 7.8.1.2) in those cells which have been induced to transport this latter sugar into the cell under the influence of metabolism.

8.1.2 The Inducible Galactose Transport System

Normally yeast cells cannot metabolize galactose but can be induced to do so. When this happens, both a transport system and the necessary metabolic enzymes are synthesized (DOUGLAS and HAWTHORNE, 1964, 1966; MORTIMER and HAWTHORNE, 1966). VAN STEVENINCK and ROTHSTEIN (1965), VAN STEVENINCK and DAW-

SON (1968) and VAN STEVENINCK (1972) have shown that the induction process leads to a "permease" which catalyzes the binding of galactose to the carrier, changing it from one which in uninduced cells brings about facilitated diffusion, to one bringing about group translocation in a way similar to that for glucose.

The actual mechanism of transport of galactose into induced cells is not clear. Evidence has been presented (DEIERKAUF and BOOIJ, 1968; VAN STEVENINCK and DAWSON, 1968; VAN STEVENINCK, 1972) that the sugar is phosphorylated during transport, phosphatidyl glycerol-phosphate being involved. However CIRILLO and his co-workers (CIRILLO, 1968b; KUO et al., 1970; KUO and CIRILLO, 1970) believe that galactose transport always occurs *via* facilitated diffusion both before and after induction. Further work is required here. Attention should be paid to the effect of pH on transport, since before induction it is pH insensitive whilst after it is highly pH dependent, having two optima at pH 4.5 and 6.0 (VAN STEVENINCK and DAWSON, 1968).

8.2 *Rhodotorula gracilis*

Monosaccharide transport in *Rhodotorula gracilis* has been studied in some detail (HOFER, 1970, 1971a, b; HOFER and KOTYK, 1968; KOTYK and HOFER, 1965). The transport system brings about accumulation of monosaccharides and the mechanism is tightly coupled to metabolism. This means that when metabolism is inhibited (anaerobic conditions) or uncoupled (DNP) the sugar concentration inside the cell does not even reach diffusion equilibrium with the external medium. However, under these conditions when net movement is inhibited, there is still exchange of material (as demonstrated by the movement of ^{14}C-sugars from preloaded cells) across the membrane.

The phenomenon is explained (HOFER, 1971b) in terms of a carrier which exists in two forms with differing affinity for monosaccharides (Fig. 7.13). The normal form has a high affinity which is lowered metabolically at the inner surface. Only the low affinity form can diffuse across the membrane in the free state. The high affinity form must combine with a sugar molecule before it can move. When a carrier

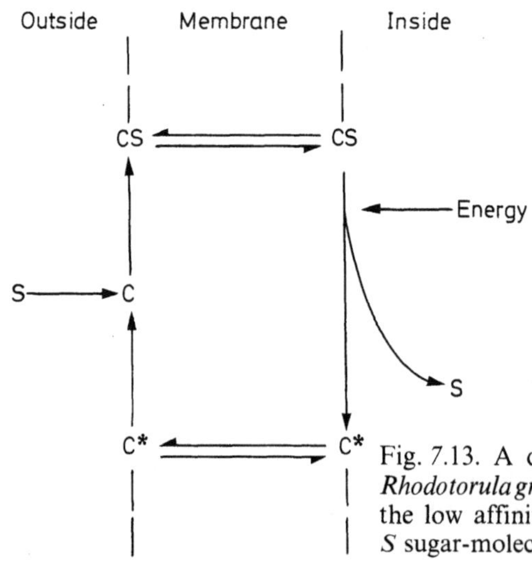

Fig. 7.13. A diagramatic representation of sugar transport in *Rhodotorula gracilis*. Adapted from HOFER (1971b). *C** represents the low affinity form, *C* the high affinity form of the carrier; *S* sugar-molecule; for description see text

in the low affinity from reaches the outer surface, it converts spontaneously into the high affinity form, remaining there until it combines with a sugar molecule. Therefore in the absence of metabolism, we can envisage the following. Cells will initially and for a very brief period, be able to move sugar molecules across their membranes. However, once carriers reach the other side of the membrane they cannot return unless they still have sugar molecules attached to them. Therefore no free carriers will return to the outer surface and so net movement of sugar will cease. On the other hand, if labeled sugar molecules are already present inside the cell, they can exchange with unlabeled molecules on carriers coming from outside, causing a movement of radioactivity into the medium, the so-called *exchange* flux.

There are some similarities between this system in *R. gracilis* and the hexose transport system in *Chlorella vulgaris* but in the latter system DNP inhibits exchange (KOMOR et al., 1972; see also *6.7.2.3* and Part B, *5.3.1.1.2.2* and 3).

8.3 *Neurospora crassa*

There are possibly two distinct transport systems for glucose in the hyphae of *N. crassa* (SCARBOROUGH, 1970a, b; SCHNEIDER and WILEY, 1971a, b, c). The two systems have K_m's of around 10–70 µM (high affinity system) and 8–25 mM (low affinity system). The high affinity system is repressed by growth of the fungus in glucose. When this occurs, only the low affinity system is operative and this brings about the facilitated diffusion of 3-0-methylglucose and presumably glucose into the hyphae. Active transport of sorbose is also repressed.

It appears that it is a metabolic product of glucose, not glucose itself, which represses the high affinity system, since 3-0-methyl glucose has no effect. Further, although CHM can interrupt derepression, this protein synthesis inhibitor has no effect on the rate of transport by the mycelium to which the inhibitor is presented. This is unlike the tryptophan transport system in the same fungus (WILEY, 1970) which turns over with a half-time of 15 min, synthesis being inhibited by CHM (WILEY and MATCHETT, 1966) (see *7.7.1.1*).

8.4 Specificity of Transport Systems

Table *7.17* gives information about the substrate specificity of the D-glucose transport system in *Saccharomyces cerevisiae* (CIRILLO, 1968a) and *Aspergillus nidulans* (MARK and ROMANO, 1971). From such data, it is possible to say what part of the D-glucopyranose structure is required for binding to the transport protein. It is interesting to see that *S. cerevisiae* and *A. nidulans* possess transport systems with a pattern of broad similarity with respect to specificity. In general, it can be said that alterations of a substituent group at any one of the six carbon positions can be tolerated. Except for methylation of the anomeric hydroxyl group (α-methylglucoside), no single change completely abolishes binding activity. The hydroxyl group at C-2 does not contribute to binding, since 2-deoxyglucose has the same affinity as glucose. Multiple changes (L-xylose and L-sorbose) bring about complete loss of activity. The notable exception is fructose which has a relatively high affinity for the D-glucose transport system in yeast. This does not readily fit into the pattern which has emerged from studies with aldose sugars but it may be related to the fact that D-fructose in aqueous solution exists in the furanose form to the extent of 30%. However, in *A. nidulans*, fructose has negligible affinity for the glucose transport system and appears to be transported by a highly specific carrier (MARK and ROMANO, 1971).

An equally interesting situation exists for mannitol which has been shown by MAXWELL and SPOERL (1971) to have a high affinity for the D-glucose transport system in yeast. The structural requirements of the D-glucose carrier in yeast makes it difficult to understand why this should happen.

Table 7.17. The comparative affinities of various sugars for the glucose transport systems of *Saccharomyces cerevisiae* and *Aspergillus nidulans*. (Data from CIRILLO, 1968a, and MARK and ROMANO, 1971, respectively)

Sugar	*S. cerevisiae*	*A. nidulans*
D-Glucose	1	1
2-Deoxy-D-glucose	1	1
D-Fructose	0.2	<0.0012
D-Galactose	0.13	0.055
D-Mannose	0.1	0.046
1,5-Anhydro-D-glucitol	0.1	0.04
3-0-Methyl-D-glucose	0.02	0.48
D-Xylose	0.013	0.016
D-Fucose	0.02	<0.0012
D-Arabinose	0.02	<0.0012
L-Sorbose	0.05	<0.0012
L-Glucose	0.0025	<0.0012
α-Methyl-D-glucoside	0.0025	<0.0012

8.5 Control of Monosaccharide Transport

A recent study by JENNINGS and AUSTIN (1973) indicates that the rate of sugar transport in fungi may be controlled by the concentration of metabolites within the mycelium. Using *Dendryphiella salina*, JENNINGS and AUSTIN showed that the non-metabolized sugar 3-0-methyl glucose is actively transported into hyphae and when this takes place, the concentration of the other soluble carbohydrates (predominantly mannitol and arabitol) adjusts so that the *total* soluble carbohydrate, which includes the 3-0-methyl glucose absorbed, remains unchanged. The drop in mannitol and arabitol concentration is accompanied by an increase in polysaccharide hydrolyzable with 0.5 M sulfuric acid. JENNINGS and AUSTIN suggest that the simplest hypothesis to explain their observations is that the transport of each molecule of sugar into the mycelium is accompanied by the concomitant conversion of either a mannitol or an arabitol molecule into polysaccharide or some other insoluble compound. Of the two sugar alcohols, mannitol is the more likely candidate. JENNINGS and AUSTIN also point out that the ability of the mycelium to maintain a constant concentration of soluble carbohydrates, whilst absorbing carbohydrates from the external medium, will mean that there is little fluctuation of the hyphal osmotic pressure. Inadequate regulation of this can lead to bursting of the hyphae or changes in their morphology (ROBERTSON, 1959; BARTNICKI-GARCIA and LIPPMAN, 1972). Similarly the control of amino acid uptake by amino acids already present in the medium could be viewed as being in part a consequence of the need to control hyphal osmotic pressure (see Chap. *11*).

Acetate has been shown to inhibit α-thioethylglucoside transport in *Saccharomyces cerevisiae* (OKADA and HALVORSON, 1964b), utilization of a whole range of both mono- and disaccharides (but not sucrose) and polyhydric alcohols, by *Aspergillus nidulans* (ROMANO and KORNBERG, 1968, 1969) and 3-0-methyl glucose transport by *Dendryphiella salina* (JENNINGS and AUSTIN, 1973). ROMANO and KORNBERG produced evidence from studies with mutants that the effect of acetate requires the presence in the cells of an enzyme that catalyzes the formation of acetyl-coenzyme A from acetate, though their conclusion that glucose transport involves phosphorylation is now thought to be incorrect (BROWN and ROMANO, 1969; JENNINGS and AUSTIN, 1973). The inhibition of glucose transport in *Candida* 107 by n-decane and other alkanes has been explained in similar terms (GILL and RATLEDGE, 1973). Here it is thought that there is an accumulation of fatty acids and fatty acyl-CoA esters which might inhibit transport either directly or through an accumulation of acetyl-CoA.

There is evidence that control of sugar transport in *Saccharomyces cerevisiae* may be brought about by glucose-6-phosphate. KOTYK and KLEINZELLER (1967) found that the apparent Michaelis constant (K_m) for glucose is higher aerobically (17.4 mM) than anaerobically (6.7 mM) or aerobically in the presence of 0.5 mM DNP (5.6 mM). They postulated that under aerobic conditions, there is an increase of citrate and ATP which inhibit phosphofructokinase activity (SALAS et al., 1965) leading to an increase in concentration of glucose-6-phosphate. AZAM and KOTYK (1969) have been able to show that efflux of D-xylose from yeast cells is inhibited by high intracellular concentrations of glucose-6-phosphate. The pentose is transported by the same carrier as D-glucose and fructose. It is difficult to relate this latter finding to the change in K_m, but it does indicate that glucose-6-phosphate does interact with the transport system. BECKER and BETZ (1972), as part of studies on the oscillations in the redox state of nicotinamide adenine nucleotides and glycolytic intermediates in *Saccharomyces carlbergensis*, have implicated feedback inhibition of glucose transport by glucose-6-phosphate as being a major part of the phenomenon.

9. Uptake of Di- and Trisaccharides

Maltose transport in *Saccharomyces cerevisiae* is by a specific system which after induction brings about accumulation of the sugar (ROBERTSON and HALVORSON, 1957; HARRIS and THOMPSON, 1961; GORTS, 1969). OKADA and HALVORSON (1964a and b) using the non-metabolized analog, α-thioethyl-D-glycopyranoside (αTEG), have shown that induction changes the transport system from one bringing about facilitated diffusion to one bringing about active transport. The K_m for influx changes from 50 mM to 1.8 mM, efflux is unchanged. A similar position seems to hold for maltose transport in *S. carlsbergensis* (DE KROON and KONINGSBERGER, 1970).

Other disaccharides, e.g. sucrose, and trisaccharides, e.g. melibiose, require surface hydrolysis to their constituent monosaccharides before they can be utilized by *S. cerevisiae* (FRIIS and OTTOLENGHI, 1959a, b). The same is true for sucrose utilization by *Chaetonium globosum* (WALSH and HARLEY, 1962), trehalose by *Myrothecium verrucaria* (MANDELS and VITOLS, 1967) and maltose by *Mucor rouxii* (REYES and RUIS-HERRERA, 1972).

10. Ion and Proton Movements Accompanying Organic Solute Absorption

10.1 *Saccharomyces carlsbergensis* and *cerevisiae*

EDDY et al. (1970) showed that when glycine is present in the external medium buffered to pH 7.0 with *tris* citrate, uptake is relatively small. The rate of uptake however increases five times when the pH is reduced to pH 4.5. On the other hand when potassium was present at pH 7.4, the rate of uptake was reduced by 80%. Kinetic analysis indicated that potassium acts as a non-competitive inhibitor binding to a single site. Absorption of glycine causes 2–3 equivalents of potassium to leave the cells, though later work (EDDY and NOWACKI, 1971) indicated 1–2 equivalents as more likely. All the effects are less evident when glucose is also present.

EDDY et al. (1970) examined the phenomenon further, using cells in which the levels of phosphorylated intermediates were reduced to a low level by starvation in the presence of antimycin and 2-deoxyglucose. In these cells, glycine can be concentrated up to 200 times at pH 4.5 and it is unlikely that the amino acid is adsorbed on cellular constituents to any significant extent. Replacement of potassium in the cell prevents amino acid uptake but not when cell metabolism is not inhibited.

EDDY and NOWACKI (1971) have examined the stoichiometry of the processes concerned. Addition of glycine to cells at pH 4.5 whose metabolism is inhibited, immediately stimulates proton uptake. About 2 extra equivalents of hydrogen are taken up per molecule of amino acid over the first 2–4 min, and after 20 min about 1–2 equivalents. In the absence of metabolic inhibitors, around 2 extra equivalents of hydrogen per molecule of amino acid are absorbed with the loss from the cell of an equal number of equivalents of potassium. Under similar conditions when Na^+ is the major intracellular cation, only one extra equivalent of hydrogen is absorbed. When yeast has been fed with glucose, almost all the protons which are absorbed with the amino acid are extruded again.

The conclusions from all this are that the protons and, in certain circumstances, K^+, are co-substrates on the amino acid carrier. EDDY and NOWACKI in the analysis of their data used the analogy of the roles of Na^+ and K^+ as co-substrates in certain mammalian systems (HEINZ, 1972). In essentials the mechanism proposed envisages that the amino acid is transported into the cell as a consequence of a proton gradient across the membrane, this gradient being maintained by the K^+-H^+ exchange pump (see *7.2.2*). There is some doubt however about how K^+ interacts with the amino acid carrier. (See also *3.3.3*; *6.7.2.3*.)

Other amino acids are also absorbed by the same mechanism (EDDY et al., 1970; EDDY and NOWACKI, 1971). Analysis of mutant strains of *Saccharomyces cerevisiae* first described by GRENSON and HENNAUT (1971) indicate that the fast absorption of glycine, L-citrulline and methionine *via* a general amino acid permease is associated with uptake of two extra equivalents of protons per molecule amino acid absorbed. The slower absorption of L-methionine, L-proline and possibly L-arginine through their specific permeases is associated with uptake of one proton equivalent. L-canavanine and L-lysine were also absorbed together with 1–2 equivalents of protons (SEASTON et al., 1973).

Saccharomyces carlsbergensis behaves similarly and SEASTON et al. showed that maltose is absorbed by this yeast along with two to three proton equivalents. Uptake of protons is faster in the presence of α-methyl-glucoside and sucrose, but not in the presence of glucose, galactose and 2-deoxyglucose. A strain of *Saccharomyces fragelis* absorbed extra protons in the presence of lactose.

10.2 Other Fungi

HUNTER and SEGEL (1973) have presented evidence that a proton gradient can drive amino acid transport in *Penicillium chrysogenum*.

ALLAWAY and JENNINGS (1970a, b, 1971) have presented evidence that glucose transport in *Dendryphiella salina* is affected by potassium and sodium ions and they indicated that sodium ions might stimulate transport. It should be noted that this fungus comes from marine habitats, so involvement of sodium is not entirely unexpected.

References

ALLAWAY, A.E., JENNINGS, D.H.: The influence of cations on glucose uptake by the fungus *Dendryphiella salina*. New Phytologist **69**, 567–579 (1970a).

ALLAWAY, A.E., JENNINGS, D.H.: The influence of cations on glucose transport and metabolism by, and the loss of sugar alcohols from, the fungus *Dendryphiella salina*. New Phytologist **69**, 581–593 (1970b).

ALLAWAY, A.E., JENNINGS, D.H.: The effect of cations on glucose utilisation by, and on the growth of, the fungus *Dendryphiella salina*. New Phytologist **70**, 511–518 (1971).

ARMSTRONG, W.McD.: Surface-active agents and cellular metabolism. I. Effect of cationic detergents on the production of acid and carbon dioxide by baker's yeast. Arch. Biochem. Biophys. **71**, 137–147 (1957).

ARMSTRONG, W.McD.: Ion transport and related phenomena in yeast and other microorganisms. In: Transport and accumulation in biological systems (E.J. HARRIS, ed.), p. 407–445. London and Baltimore: Butterworths 1972.

ARMSTRONG, W.McD., ROTHSTEIN, A.: Discrimination between alkali metal cations by yeast. I. Effect of pH on uptake. J. Gen. Physiol. **48**, 61–71 (1964).

ARMSTRONG, W.McD., ROTHSTEIN, A.: Discrimination between alkali metal cations by yeast. II. Cation interactions in transport. J. Gen. Physiol. **50**, 967–988 (1967).

AZAM, F., KOTYK, A.: Glucose-6-phosphate as regulator of monosaccharide transport in baker's yeast. F.E.B.S. Letters **2**, 333–335 (1969).

BARTNICKI-GARCIA, S., LIPPMAN, E.: The bursting tendency of hyphal tips of fungi: presumptive evidence for a delicate balance between wall synthesis and wall lysis in apical growth. J. Gen. Microbiol. **73**, 487–500 (1972).

BECKER, J.-U., BETZ, A.: Membrane transport as controlling pacemaker of glycolysis in *Saccharomyces carlsbergensis*. Biochim. Biophys. Acta **274**, 584–597 (1972).

BELLENGER, N., NISSEN, P.W., WOOD, T.C., SEGEL, I.H.: Specificity and control of choline-0-sulfate transport in filamentous fungi. J. Bacteriol. **96**, 1574–1585 (1968).

BENKO, P.V., WOOD, T.C., SEGEL, I.H.: Specificity and regulation of methionine transport in filamentous fungi. Arch. Biochem. Biophys. **122**, 783–804 (1967).

BENKO, P.V., WOOD, T.C., SEGEL, I.H.: Multiplicity and regulation of amino acid transport in *Penicillium chrysogenum*. Arch. Biochem. Biophys. **129**, 498–508 (1969).

BIRT, L.M., HIRD, F.J.R.: Kinetic aspects of the uptake of amino acids by carrot itssue. Biochem. J. **70**, 286–292 (1958).

BORST-PAUWELS, G.W.F.H.: The uptake of radioactive phosphate by yeast. I. The uptake of phosphate by yeast compared with that by higher plants. Biochim. Biophys. Acta **65**, 403–406 (1962).

BORST-PAUWELS, G.W.F.H., DOBBLEMAN, J.: The mechanism of inhibition of anaerobic phosphate uptake by fatty acids in yeast. Biochim. Biophys. Acta **290**, 348–354 (1972).

BORST-PAUWELS, G.W.F.H., HUYGENS, P.L.M.: Comparison of the effect of acidic inhibitors upon anaerobic phosphate uptake and dinitrophenol extrusion by metabolising yeast cells. Biochim. Biophys. Acta **288**, 166–171 (1972).

BORST-PAUWELS, G.W.F.H., JAGER, S.: Inhibition of phosphate and arsenate uptake in yeast by monoiodoacetate, fluoride, 2,4-dinitrophenol and acetate. Biochim. Biophys. Acta **172**, 399–406 (1969).

BORST-PAUWELS, G.W.F.H., LOEF, H.W., HAVINGA, E.: The uptake of radioactive phosphate by yeast. II. The primary phosphorylation products. Biochim. Biophys. Acta **65**, 407–411 (1962).

BORST-PAUWELS, G.W.F.H., PETER, J.K., JAGER, S., WIJFFELS, C.C.B.M.: A study of the arsenate uptake by yeast cells compared with phosphate uptake. Biochim. Biophys. Acta **94**, 314–316 (1965).

BORST-PAUWELS, G.W.F.H., SCHNETKAMP, P., VAN WELL, P.: Activation of Rb^+ and Na^+ uptake into yeast by monovalent cations. Biochim. Biophys. Acta **291**, 274–279 (1973).

BORST-PAUWELS, G.W.F.H., WOLTERS, G.H.J., HENRICKS, J.J.G.: The interaction of 2,4-dinitrophenol with anaerobic Rb^+ transport across the yeast cell membrane. Biochim. Biophys. Acta **225**, 269–276 (1971).

BRADFIELD, G., SOMERFIELD, P., MEYN, T., HOLBY, M., BABCOCK, D., BRADLEY, D., SEGEL, I.H.: Regulation of sulfate transport in filamentous fungi. Plant Physiol. **46**, 720–727 (1970).

BROWN, C.E., ROMANO, A.H.: Evidence against necessary phosphorylation during hexose transport in *Aspergillus nidulans*. J. Bacteriol. **100**, 1198–1203 (1969).

BURNETT, J.H.: Fundamentals of mycology. London: Arnold 1968.

CASTOR, L.N., CHANCE, B.: Photochemical action spectra of carbon monoxide-inhibited respiration. J. Biol. Chem. **217**, 453–465 (1955).

CIRILLO, V.P.: Relationship between sugar structure and competition for the sugar transport systems in baker's yeast. J. Bacteriol. **95**, 603–611 (1968a).

CIRILLO, V.P.: Galactose transport in *Saccharomyces cerevisiae*. I. Non-metabolised sugars as substrates and inducers of the galactose transport system. J. Bacteriol. **95**, 1727–1731 (1968b).

CONWAY, E.J.: The biological performance of osmotic work. A redox pump. Science **113**, 270–273 (1951).

CONWAY, E.J., ARMSTRONG, W.McD.: The total intra-cellular concentration of solutes in yeast and other plant cells and the distensibility of the plant-cell wall. Biochem. J. **81**, 631–639 (1961).

CONWAY, E.J., BEARY, M.E.: Active transport of magnesium across yeast cell membrane. Biochem. J. **69**, 275–280 (1958).

CONWAY, E.J., BRADY, T.G.: Biological production of acid and alkali. I. Quantitative relations of succinic and carbonic acids to the potassium hydrogen exchange in fermenting yeast. Biochem. J. **47**, 360–369 (1950).

CONWAY, E.J., BREEN, E.J.: An "ammonia"-yeast and some of its properties. Biochem. J. **39**, 368–371 (1945).

CONWAY, E.J., DOWNEY, M.: An outer metabolic region of the yeast cell. Biochem. J. **47**, 347–355 (1950).

CONWAY, E.J., DUGGAN, P.F.: A cation carrier in the yeast cell wall. Biochem. J. **69**, 265–274 (1958).

CONWAY, E.J., GAFFNEY, H.M.: The further preparation of inorganic cationic yeasts and some of their chief properties. Biochem. J. **101**, 385–391 (1966).

CONWAY, E.J., KERNAN, R.P.: The effect of redox dyes on the active transport of hydrogen, potassium and sodium ions across the yeast cell membrane. Biochem. J. **61**, 32–36 (1955).

CONWAY, E.J., MOORE, P.T.: A sodium yeast and some of its properties. Biochem. J. **57**, 523–528 (1954).

CONWAY, E.J., O'MALLEY, E.: The nature of the cation exchanges during yeast fermentation with the formation of 0.02 N H ion. Biochem. J. **40**, 59–67 (1946).

CONWAY, E.J., RYAN, H., CARTON, E.: Active transport of sodium ions from the yeast cell. Biochem. J. **58**, 158–167 (1954).

CRABEEL, M., GRENSON, M.: Regulation of histidine uptake by specific feedback inhibition of two histidine permeases in *Saccharomyces cerevisiae*. European J. Biochem. **14**, 197–204 (1970).

DEIERKAUF, F.A., BOOIJ, H.L.: Changes in the phosphatide pattern of yeast cells in relation to active carbohydrate transport. Biochim. Biophys. Acta **150**, 214–225 (1968).

DOUGLAS, H.C., HAWTHORNE, D.C.: Enzymatic expression and genetic linkage of genes controlling galactose utilization in *Saccharomyces*. Genetics **49**, 837–844 (1964).

DOUGLAS, H.C., HAWTHORNE, D.C.: Regulation of genes controlling synthesis of the galactose pathway enzymes in yeast. Genetics **54**, 911–916 (1966).

EDDY, A.A., BACKEN, K., WATSON, G.: The concentration of amino acids by yeast cells depleted of adenosine triphosphate. Biochem. J. **120**, 853–858 (1970).

EDDY, A.A., INDGE, K.J., BACKEN, K., NOWACKI, J.A.: Interactions between potassium ions and glycine transport in the yeast *Saccharomyces carlsbergensis*. Biochem. J. **120**, 854–852 (1970).

EDDY, A.A., NOWACKI, J.A.: Stoichiometrical proton and potassium ion movements accompanying the absorption of amino acids by the yeast *Saccharomyces carlsbergensis*. Biochem. J. **122**, 701–711 (1971).

FOULKES, E.C.: Cation transport in yeast. J. Gen. Physiol. **39**, 687–704 (1956).

FRIIS, J., OTTOLENGHI, P.: Localisation of invertase in a strain of yeast. Compt. Rend. Trav. Lab. Carlsberg **31**, 259–271 (1959a).

FRIIS, J., OTTOLENGHI, P.: Localisation of melibiase in a strain of yeast. Compt. Rend. Trav. Lab. Carlsberg **31**, 272–281 (1959b).

FUHRMANN, G., ROTHSTEIN, A.: The transport of Zn^{2+}, Co^{2+} and Ni^{2+} into yeast cells. Biochim. Biophys. Acta **163**, 325–330 (1968).

GARRAHAN, P.J., GLYNN, I.M.: Driving the sodium pump backwards to form adenosine triphosphate. Nature **211**, 1414–1415 (1966).

GILL, C.O., RATLEDGE, C.: Inhibition of glucose assimilation and transport by n-decane and other n-alkanes in *Candida* 107. J. Gen. Microbiol. **75**, 11–22 (1973).

GITS, J., GRENSON, M.: Multiplicity of the amino acid permeases in *Saccharomyces cerevisiae*. III. Evidence for a specific methionine—transporting system. Biochim. Biophys. Acta **135**, 507–516 (1967).

GLYNN, I.M., HOFFMAN, J.F., LEW, V.L.: Some "partial" reactions of the sodium pump. Phil. Trans. Roy. Soc. London, Ser. B **262**, 91–102 (1971).

GLYNN, I.M., LEW, V.L., LUTHI, U.: Reversal of the potassium entry mechanisms in red cells with and without reversal of the entire pump cycle. J. Physiol. (London) **207**, 371–391 (1970).

GOODMAN, J., ROTHSTEIN, A.: The active transport of phosphate into the yeast cell. J. Gen. Physiol. **40**, 915–923 (1957).

GORTS, G.P.M.: Effects of glucose on the activity and the kinetics of the maltose uptake system and of α-glucosidase in *Saccharomyces cerevisiae*. Antonie van Leeuwenhoek, J. Microbiol. Serol. **35**, 233–234 (1969).

GRENSON, M.: Multiplicity of the amino-acid permeases in *Saccharomyces cerevisiae*. II. Evidence for a specific lysine transporting system. Biochim. Biophys. Acta **127**, 339–346 (1966).

GRENSON, M., HENNAUT, C.: Mutation affecting activity of several distinct amino acid transport systems in *Saccharomyces cerevisiae*. J. Bacteriol. **105**, 477–482 (1971).

GRENSON, M., HOU, C., CRABEEL, M.: Multiplicity of the amino acid permeases in *Saccharomyces cerevisiae*. IV. Evidence for a general amino acid permease. J. Bacteriol. **103**, 770–777 (1970).

GRENSON, M., MOUSSET, M., WIAME, J.M., BECHET, J.: Multiplicity of the amino acid permeases in *Saccharomyces cerevisiae*. I. Evidence for a specific arginine-transporting system. Biochim. Biophys. Acta **127**, 325–338 (1966).

HACKETTE, S.L., SKYE, G.E., BURTON, C., SEGEL, I.H.: Characterisation of an ammonium transport system in filamentous fungi with methyl ammonium-^{14}C as the substrate. J. Biol. Chem. **245**, 4241–4250 (1970).
HAROLD, F.M.: Inorganic polyphosphates in biology: Structure, metabolism and function. Bacteriol. Rev. **30**, 772–794 (1966).
HARRIS, G., THOMPSON, C.C.: The uptake of nutrients by yeasts. III. The maltose permease system of a brewing yeast. Biochim. Biophys. Acta **52**, 176–183 (1961).
HEINZ, E.: Na-linked transport of organic solutes, p. 201. Berlin-Heidelberg-New York: Springer 1972.
HEPPEL, L.E.: The effect of osmotic shock on release of bacterial proteins and on active transport. J. Gen. Physiol. **54**, 95s–109s (1969).
HOFER, M.: Mobile membrane carrier for monosaccharide transport in *Rhodotorula gracilis*. J. Membrane Biol. **3**, 73–82 (1970).
HOFER, M.: Transport of monosaccharides in *Rhodotorula gracilis* in the absence of metabolic energy. Archiv. Mikrobiol. **80**, 50–61 (1971a).
HOFER, M.: A model of the monosaccharide uphill transporting cell membrane system in yeast. J. Theoret. Biol. **33**, 599–603 (1971b).
HOFER, M., KOTYK, A.: Tight coupling of monosaccharide transport and metabolism in *Rhodotorula gracilis*. Folia Microbiol. (Prague) **13**, 197–204 (1968).
HOFSTEE, B.H.J.: On the evaluation of constants V_m and K_m in enzyme reactions. Science **116**, 329–333 (1952).
HOLLIGAN, P.M., JENNINGS, D.H.: Carbohydrate metabolism in the fungus *Dendryphiella salina*. II. The influence of different carbon and nitrogen sources on the accumulation of mannitol and arabitol. New Phytologist **71**, 583–594 (1972).
HOLLIGAN, P.M., JENNINGS, D.H.: Carbohydrate metabolism in the fungus *Dendryphiella salina*. IV. Acetate assimilation. New Phytologist **72**, 315–319 (1973).
HOLZER, H.: Zur Penetration von Orthophosphat in lebenden Hefezellen. Biochem. Z. **324**, 144–155 (1953).
HUNTER, D.R., SEGEL, L.H.: Acidic and basic amino acid transport systems of *Penicillium chrysogenum*. Arch. Biochem. Biophys. **144**, 168–183 (1971).
HUNTER, D.R., SEGEL, I.H.: Effect of weak acids on amino acid transport by *Penicillium chrysogenum:* Evidence for a proton or charge gradient as the driving force. J. Bacteriol. **113**, 1184–1192 (1973).
HUYGENS, P.L.M., BORST-PAUWELS, G.W.F.H.: The effect of N,N′-dicyclohexyl carbodiimide on anaerobic and aerobic phosphate uptake by baker's yeast. Biochim. Biophys. Acta **283**, 234–238 (1972).
JENNINGS, D.H.: The absorption of solutes by plant cells. Edinburgh-London: Oliver and Boyd 1963.
JENNINGS, D.H.: Cations and filamentous fungi: Invasion of the sea and hyphal functioning. In: Ion transport in plants (W.P. ANDERSON, ed.), p. 323–335. London: Academic Press 1973.
JENNINGS, D.H.: Sugar transport into fungi: an essay. Trans. Brit. Mycol. Soc. **62**, 1–24 (1974).
JENNINGS, D.H., AUSTIN, S.: The stimulatory effect of the nonmetabolised sugar 3-0-methyl glucose on the conversion of mannitol and arabitol to polysaccharide and other insoluble compounds in the fungus *Dendryphiella salina*. J. Gen. Microbiol. **75**, 287–294 (1973).
JENNINGS, D.H., HOOPER, D.C., ROTHSTEIN, A.: The participation of phosphate in the formation of a 'carrier' for the transport of Mg^{++} and Mn^{++} ions into yeast cells. J. Gen. Physiol. **41**, 1019–1026 (1958).
JOIRIS, C.R., GRENSON, M.: Specificité et regulation d'une perméase des acides amines dicarboxyliques chez *Saccharomyces cerevisiae*. Arch. Intern. Physiol. Biochim. **77**, 154–156 (1969).
JUNG, C., ROTHSTEIN, A.: Arsenate uptake and release in relation to the inhibition of transport and glycolysis in yeast. Biochem. Pharmacol. **14**, 1093–1112 (1965).
JUNI, E., KAMEN, M.D., REINER, J.M., SPIEGELMAN, S.: Turnover and distribution of phosphate compounds in yeast metabolism. Arch. Biochem. Biophys. **18**, 387–408 (1948).
KLEIN, W.L., DAHMS, A.S., BOYER, P.D.: The nature of the coupling of oxidative energy to amino acid transport. Federation Proc. **29**, 341 (1970).

KOMOR, E., HAASS, D., TANNER, W.: Unusual features of the active hexose uptake system of *Chlorella vulgaris*. Biochim. Biophys. Acta **266**, 649–660 (1972).

KOTYK, A.: Intracellular pH of baker's yeast. Folia Microbiol. (Prague) **8**, 27–31 (1963).

KOTYK, A., HOFER, M.: Uphill transport of sugars in the yeast *Rhodotorula gracilis*. Biochim. Biophys. Acta **102**, 410–422 (1965).

KOTYK, A., KLEINZELLER, A.: Movement of sodium and cell volume changes in a sodium-rich yeast. J. Gen. Microbiol. **20**, 197–212 (1959).

KOTYK, A., KLEINZELLER, A.: Affinity of the yeast membrane carrier for glucose and its role in the Pasteur effect. Biochim. Biophys. Acta **135**, 106–111 (1967).

KOTYK, A., ŘÍHOVÁ, L.: Transport of α-amino butyric acid in *Saccharomyces cerevisiae:* Feedback control. Biochim. Biophys. Acta **288**, 280–389 (1972).

KROON, R.A. DE, KONINGSBERGER, V.V.: An inducible transport system for α-glucosides in protoplasts of *Saccharomyces carlsbergensis*. Biochim. Biophys. Acta **204**, 590–609 (1970).

KUO, S.-C., CHRISTENSEN, M.S., CIRILLO, V.P.: Galactose transport in *Saccharomyces cerevisiae*. II. Characteristics of galactose uptake and exchange in galactokinaseless cells. J. Bacteriol. **103**, 671–678 (1970).

KUO, S.-C., CIRRILO, V.P.: Galactose transport in *Saccharomyces cerevisiae*. III. Characteristics of galactose uptake in transferaseless cells: Evidence against transport-associated phosphorylation. J. Bacteriol. **103**, 679–685 (1970).

LEGGETT, J.E.: Entry of phosphate into yeast cells. Plant Physiol. **36**, 277–284 (1961).

LEGGETT, J.E., OLSEN, R.A.: Anion absorption by baker's yeast. Plant Physiol. **39**, 387–390 (1964).

LESTER, G., HECHTER, O.: Dissociation of rubidium uptake by *Neurospora crassa* into entry and binding phases. Proc. Natl. Acad. Sci. U.S. **44**, 1141–1149 (1958).

LESTER, G., STONE, D., HECHTER, O.: The effects of deoxycorticosterone and other steroids on *Neurospora crassa*. Arch. Biochem. Biophys. **75**, 196–214 (1958).

MACMILLAN, A.: The entry of ammonia into fungal cells. J. Exptl. Bot. **7**, 113–126 (1956).

MAGANA-SCHWENCKE, N., SCHWENCKE, J.: A proline transport system in *Saccharomyces chevalieri*. Biochim. Biophys. Acta **173**, 313–323 (1969).

MAGILL, C.W., NELSON, S.O., D'AMBROSIO, S.M., GLOVER, G.I.: Histidine uptake in mutant strains of *Neurospora crassa via* the general transport system for amino acids. J. Bacteriol. **113**, 1320–1325 (1973).

MAGILL, C.W., SWEENEY, H., WOODWARD, V.H.: Histidine uptake in strains of *Neurospora crassa* with normal and mutant transport systems. J. Bacteriol. **110**, 313–320 (1972).

MANDELS, G.R., VITOLS, R.: Constitutive and induced trehalose mechanisms in the spores of the fungus *Myrothecium verrucaria*. J. Bacteriol. **93**, 159–167 (1967).

MARK, C.G., ROMANO, A.H.: Properties of the hexose transport systems of *Aspergillus nidulans*. Biochim. Biophys. Acta **249**, 216–226 (1971).

MARZLUF, G.A.: Genetical and biochemical studies of distinct sulphate permease species in different developmental stages of *Neurospora crassa*. Arch. Biochem. Biophys. **138**, 254–263 (1970a).

MARZLUF, G.A.: Genetic and metabolic controls for sulphate metabolism in *Neurospora crassa:* Isolation and study of chromate-resistent and sulfate transport-negative mutants. J. Bacteriol. **102**, 716–721 (1970b).

MARZLUF, G.A.: Control of the synthesis, activity and turnover of enzymes of sulfur metabolism in *Neurospora crassa*. Arch. Biochem. Biophys. **150**, 714–724 (1972).

MARZLUF, G.A.: Regulation of sulfate transport in *Neurospora* by transinhibition and by inositol depletion. Arch. Biochem. Biophys. **156**, 244–254 (1973).

MATILE, PH., MOOR, H., MÜHLETHALER, K.: Isolation and properties of the plasmalemma in yeast. Arch. Mikrobiol. **58**, 201–211 (1967).

MAW, G.A.: The uptake of inorganic sulphate by brewer's yeast. Folia Microbiol. (Prague) **8**, 325–332 (1963).

MAXWELL, W.A., SPOERL, E.: Mannitol uptake by *Saccharomyces cerevisiae*. J. Bacteriol. **105**, 753–758 (1971).

MORTIMER, R.K., HAWTHORNE, D.C.: Genetic mapping in *Saccharomyces*. Genetics **53**, 165–173 (1966).

MORTON, A.G., MACMILLAN, A.: Assimilation of nitrogen from ammonium salts and nitrate by fungi. J. Exptl. Bot. **5**, 232–252 (1954).

MURPHY, J.T., SPENCE, K.D.: Transport of S-adenosylmethionine in *Saccharomyces cerevisiae*. J. Bacteriol. **109**, 499–504 (1972).
NURMINEN, T., SUOMALAINEN, H.: The lipolytic activities of the isolated cell envelope fractions of baker's yeast. Biochem. J. **118**, 759–763 (1970).
OKADA, H., HALVORSON, H.O.: Uptake of α-thioethyl-D-glucopyranoside by *Saccharomyces cerevisiae*. I. The genetic control of facilitated diffusion and active transport. Biochim. Biophys. Acta **82**, 538–546 (1964a).
OKADA, K., HALVORSON, H.O.: Uptake of α-thioethyl-D-glucopyranoside by *Saccharomyces cerevisiae*. II. General characteristics of an active transport system. Biochim. Biophys. Acta **82**, 547–555 (1964b).
ØRSKOV, S.L.: Experiments with substances which make baker's yeast absorb potassium. Acta Physiol. Scand. **20**, 62–78 (1950).
OXENDER, D.L.: Amino acid transport in micro-organisms. In: Metabolic pathways (L.E. HOKIN, ed.), vol. 4, p. 133–185. New York-London: Academic Press 1972.
PALL, M.L.: Amino acid transport in *Neurospora crassa*. I. Properties of two amino acid transport systems. Biochim. Biophys. Acta **123**, 113–127 (1969).
PALL, M.L.: Amino acid transport in *Neurospora crassa*. II. Properties of a basic amino acid transport system. Biochim. Biophys. Acta **203**, 139–147 (1970a).
PALL, M.L.: Amino acid in *Neurospora crassa*. III. Acidic amino acid transport. Biochim. Biophys. Acta **211**, 513–520 (1970b).
PALL, M.L.: Amino acid transport in *Neurospora crassa*. IV. Properties and regulation of a methionine transport system. Biochim. Biophys. Acta **233**, 201–214 (1971).
PALL, M.L., KELLY, K.A.: Specificity of transinhibition of amino acid transport in *Neurospora*. Biochem. Biophys. Res. Commun. **42**, 940–947 (1971).
PARK, D., ROBINSON, P.M.: A fungal hormone controlling internal water distribution normally associated with cell ageing in fungi. Symp. Soc. Exptl. Biol. **21**, 323–336 (1967).
PEÑA, A., CINCO, G., GÓMEZ PUYOU, A., TUENA, M.: Studies on the mechanism of the stimulation of glycolysis and respiration by K^+ in *Saccharomyces cerevisiae*. Biochim. Biophys. Acta **180**, 1–8 (1969).
PEÑA, A., CINCO, G., GÓMEZ PUYOU, A., TUENA, M.: Effect of the pH of the incubation medium on glycolysis and respiration in *Saccharomyces cerevisiae*. Arch. Biochem. Biophys. **153**, 413–425 (1972).
REILLY, C.: The active transport of potassium and sodium ions in respiratory-deficient mutants of *Saccharomyces cerevisiae*. Biochem. J. **91**, 447–452 (1964).
REILLY, C.: The effect of certain redox dyes on glutathione and on potassium accumulation in normal and mutant yeast. Folia Microbiol. (Prague) **12**, 495–499 (1967).
REYES, E., RUIZ-HERRERA, J.: Mechanism of maltose utilization by *Mucor rouxii*. Biochim. Biophys. Acta **273**, 328–335 (1972).
RING, K., HEINZ, E.: Active amino acid transport in *Streptomyces hydrogenans*. I. Kinetics of uptake of α-aminoisobutyric acid. Biochem. Z. **344**, 446–461 (1966).
ROBERTSON, J.J., HALVORSON, H.O.: The components of maltozymase in yeast and their behaviour during deadaptation. J. Bacteriol. **73**, 186–198 (1957).
ROBERTSON, N.F.: Experimental control of hyphal branching and branch forms in hyphomycetous fungi. J. Linn. Soc. London **56**, 207–211 (1959).
ROMANO, A.H., KORNBERG, H.L.: Regulation of sugar utilization by *Aspergillus nidulans*. Biochim. Biophys. Acta **158**, 491–493 (1968).
ROMANO, A.H., KORNBERG, H.L.: Regulation of sugar uptake by *Aspergillus nidulans*. Proc. Roy. Soc. (London), Ser. B **173**, 475–490 (1969).
ROTHSTEIN, A.: Enzyme systems of the cell surface involved in the uptake of sugars by yeast. Symp. Soc. Exptl. Biol. **8**, 165–201 (1954).
ROTHSTEIN, A.: Regulation of the inorganic ion content of cells. Ciba Found. Study Group **5**, 53–64 (1960).
ROTHSTEIN, A.: Interactions of arsenate with the phosphate transporting system of yeast. J. Gen. Physiol. **40**, 1075–1085 (1963).
ROTHSTEIN, A., ENNS, L.H.: The relationship of potassium to carbohydrate metabolism in baker's yeast. J. Cell. Comp. Physiol. **28**, 231–252 (1946).
ROTHSTEIN, A., HAYES, A.D., JENNINGS, D.H., HOOPER, D.C.: The active transport of Mg^{++} and Mn^{++} into the yeast cell. J. Gen. Physiol. **41**, 585–594 (1958).

RYAN, H.: Alcohol dehydrogenase activity and electron transport in living yeast. Biochem. J. **105**, 137–143 (1967).

RYAN, H., RYAN, J.P., O'CONNOR, W.H.: The effect of diffusible acids on potassium ion uptake by yeast. Biochem. J. **125**, 1081–1085 (1971).

RYAN, J.P., RYAN, H.: The role of intracellular pH in the regulation of cation exchanges in yeast. Biochem. J. **128**, 139–146 (1972).

SALAS, M.L., VINUELA, E., SALAS, M., SOLS, A.: Citrate inhibition of phosphofructokinase and the Pasteur effect. Biochem. Biophys. Res. Commun. **19**, 371–376 (1965).

SAMSON, F.E., KATZ, A.M., HARRIS, D.L.: Effects of acetate and other short-chain fatty acids on yeast metabolism. Arch. Biochem. Biophys. **54**, 406–423 (1955).

SANCHEZ, S., MARTINEZ, L., MORA, J.: Interactions between amino acid transport systems in *Neurospora crassa*. J. Bacteriol. **112**, 276–284 (1972).

SCARBOROUGH, G.A.: Sugar transport in *Neurospora crassa*. J. Biol. Chem. **245**, 1694–1698 (1970a).

SCARBOROUGH, G.A.: Sugar transport in *Neurospora crassa*. II. A second glucose transport system. J. Biol. Chem. **245**, 3985–3987 (1970b).

SCARBOROUGH, G.A.: Transport in *Neurospora*. Intern. Rev. Cytol. **34**, 103–122 (1973).

SCHNEIDER, R.P., WILEY, R.P.: Kinetic characteristics of the two glucose transport system in *Neurospora crassa*. J. Bacteriol. **106**, 479–486 (1971a).

SCHNEIDER, R.P., WILEY, R.P.: Regulation of sugar transport in *Neurospora crassa*. J. Bacteriol. **106**, 487–492 (1971b).

SCHNEIDER, R.P., WILEY, W.R.: Transcription and degradation of messenger ribonucleic acid for a glucose transport system in *Neurospora*. J. Biol. Chem. **246**, 4784–4789 (1971c).

SCHÖNHERR, O.TH., BORST-PAUWELS, G.W.F.H.: Investigation into the permeability of yeast cells to phosphate. Biochim. Biophys. Acta **135**, 787–790 (1967).

SCHWENCKE, J., MAGANA-SCHWENCKE, N.: Derepression of a proline transport system in *Saccharomyces chevalieri* by nitrogen starvation. Biochim. Biophys. Acta **173**, 302–312 (1969).

SEASTON, A., INKSON, C., EDDY, A.A.: The absorption of protons with specific amino acids and carbohydrates by yeast. Biochem. J. **134**, 1031–1043 (1973).

SEGEL, I.H., JOHNSON, M.J.: Accumulation of intracellular inorganic sulfate by *Penicillium chrysogenum*. J. Bacteriol. **81**, 91–106 (1961).

SHERE, S.M., JACOBSON, L.: Mineral uptake in *Fusarium oxysporium* f. sp. vasinfecta. Physiol. Plantarum **23**, 51–62 (1970).

SIEGENTHALER, P.A., BELSKY, M.M., GOLDSTEIN, S.: Phosphate uptake in an obligately marine fungus: a specific requirement for sodium. Science, **155**, 93–94 (1967a).

SIEGENTHALER, P.A., BELSKY, M.M., GOLDSTEIN, S., MENNA, M.: Phosphate uptake in an obligately marine fungus. II. Role of culture conditions, energy sources and inhibitors. J. Bacteriol. **93**, 1281–1288 (1967b).

SKYE, G.E., SEGEL, I.H.: Independent regulation of cysteine and cystine transport in *Penicillium chrysogenum*. Arch. Biochem. Biophys. **138**, 306–318 (1970).

SLAYMAN, C.L.: Electrical properties of *Neurospora crassa*. Effects of external cations on the intracellular potential. J. Gen. Physiol. **49**, 69–92 (1965a).

SLAYMAN, C.L.: Electrical properties of *Neurospora crassa*. Respiration and the intracellular potential. J. Gen. Physiol. **49**, 93–116 (1965b).

SLAYMAN, C.L.: Movement of ions and electrogenesis in microorganisms Amer. Zoo. **10**, 377–392 (1970).

SLAYMAN, C.L.: Adenine nucleotide levels in *Neurospora*, as influenced by conditions of growth and metabolic inhibitors. J. Bacteriol. **114**, 752–766 (1973).

SLAYMAN, C.L., LU, C.Y-H., SHANE, L.: Correlated changes in membrane potential and ATP concentrations in *Neurospora*. Nature **226**, 274–276 (1970).

SLAYMAN, C.L., SLAYMAN, C.W.: Measurement of membrane potentials in *Neurospora*. Science **136**, 876–877 (1962).

SLAYMAN, C.L., SLAYMAN, C.W.: Net uptake of potassium in *Neurospora*. Exchange for sodium and hydrogen ions. J. Gen. Physiol. **52**, 424–443 (1968).

SLAYMAN, C.W.: Net potassium transport in *Neurospora*. Properties of a transport mutant. Biochim. Biophys. Acta **211**, 502–512 (1970).

SLAYMAN, C.W., SLAYMAN, C.L.: Potassium transport in *Neurospora*. Evidence for a multisite carrier at high pH. J. Gen. Physiol. **55**, 758–786 (1970).

SLAYMAN, C.W., TATUM, E.L.: Potassium transport in *Neurospora*. I. Intracellular sodium and potassium concentrations and cation requirements for growth. Biochim. Biophys. Acta **88**, 578–592 (1964).

SLAYMAN, C.W., TATUM, E.L.: Potassium transport in *Neurospora*. II. Measurement of steady-state potassium fluxes. Biochim. Biophys. Acta **102**, 149–160 (1965a).

SLAYMAN, C.W., TATUM, E.L.: Potassium transport in *Neurospora*. III. Isolation of a transport mutant. Biochim. Biophys. Acta **109**, 184–193 (1965b).

SPANSWICK, R.M.: Evidence for an electrogenic ion pump in *Nitella translucens* 1. The effect of pH, K^+, Na^+, light and temperature on the membrane potential and resistance. Biochim. Biophys. Acta **288**, 73–89 (1972).

SPANSWICK, R.M.: Electrogenesis in photosynthetic tissues. In: Ion transport in plants (W.P. ANDERSON, ed.), p. 113–128. London: Academic Press 1973.

SPANSWICK, R.M., STOLAREK, J., WILLIAMS, E.J.: The membrane potential of *Nitella translucens*. J. Exptl. Bot. **18**, 1–16 (1967).

SPENCER, B., HUSSEY, E.C., ORSI, B.A., SCOTT, J.M.: Mechanism of choline-O-sulphate utilization in fungi. Biochem. J. **106**, 461–499 (1968).

STEVENINCK, J. VAN: Transport and transport-associated phosphorylation of 2-deoxy-D-glucose by yeast. Biochim. Biophys. Acta **163**, 386–394 (1968).

STEVENINCK, J. VAN: The mechanism of transmembrane glucose transport in yeast. Evidence for phosphorylation, associated with transport. Arch. Biochem. Biophys. **130**, 244–252 (1969).

STEVENINCK, J. VAN: The transport mechanism of α-methylglucoside in yeast. Evidence for transport-associated phosphorylation. Biochim. Biophys. Acta **203**, 376–384 (1970).

STEVENINCK, J. VAN: Transport and transport-associated phosphorylation of galactose in *Saccharomyces cerevisiae*. Biochim. Biophys. Acta **274**, 575–583 (1972).

STEVENINCK, J. VAN, BOOI , H.L.: The role of polyphosphates in the transport mechanism of glucose in yeast cells. J. Gen. Physiol. **48**, 43–60 (1964).

STEVENINCK, J. VAN, DAWSON, E.C.: Active and passive galactose transport in yeast. Biochim. Biophys. Acta **150**, 47–55 (1968).

STEVENINCK, J. VAN, ROTHSTEIN, A.: Sugar transport and metal binding in yeast. J. Gen. Physiol. **49**, 235–246 (1965).

SUOMALAINEN, H., OURA, E.: Yeast nutrition and solute uptake. In: The yeasts (A.H. ROSE, J.S. HARRISON, eds.), vol. 2, p. 3–74. London-New York: Academic Press 1971.

SWENSON, P.A.: Leakage of phosphate compounds from ultravioletirradiated yeast cells. J. Cell. Comp. Physiol. **56**, 77–91 (1960).

TISDALE, J.H., DE BUSK, A.G.: Developmental regulation of amino acid transport in *Neurospora crassa*. J. Bacteriol. **104**, 689–697 (1970).

TRINCI, A.J.P., COLLINGE, A.J.: Occlusion of the septal pores of damaged hyphae of *Neurospora crassa* by hexagonal crystals. Protoplasma **80**, 57–68 (1974).

TWEEDIE, J.W., SEGEL, I.H.: Specificity of transport processes for sulfur, selenium, and molybdenum anions by filamentous fungi. Biochim. Biophys. Acta **196**, 95–106 (1970).

ULLRICH-EBERIUS, C.I., SIMONIS, W.: Der Einfluß von Natrium- und Kaliumionen auf die Phosphataufnahme bei *Ankistrodesmus braunii*. Planta **93**, 214–226 (1970).

ULLRICH-EBERIUS, C.I., YINGCHOL, Y.: Phosphate uptake and its pH-dependence in halophytic and glycophytic algae and higher plants. Oecologia **17**, 17–26 (1974).

VALLÉE, M., SEGEL, I.H.: Sulphate transport by protoplasts of *Neurospora crassa*. Microbios **4**, 21–31 (1971).

WALSH, J.H., HARLEY, J.L.: Sugar absorption by *Chaetomium globosum*. New Phytologist **61**, 299–313 (1962).

WIAME, J.M.: The occurrence and physiological behaviour of two metaphosphate fractions in yeast. J. Biol. Chem. **178**, 919–929 (1949).

WILEY, W.R.: Tryptophan transport in *Neurospora crassa*: a tryptophan binding protein released by cold osmotic shock. J. Bacteriol. **103**, 656–662 (1970).

WILEY, W.R., MATCHETT, W.H.: Tryptophan transport in *Neurospora crassa*. I. Specificity and kinetics. J. Bacteriol. **92**, 1698–1705 (1966).

WILEY, W.R., MATCHETT, W.H.: Tryptophan transport in *Neurospora crassa*. II. Metabolic control. J. Bacteriol. **95**, 959–966 (1968).

YAMAMOTO, L.A., SEGEL, I.H.: The inorganic sulfate transport system of *Penicillium chrysogenum*. Arch. Biochem. Biophys. **114**, 523–538 (1966).

8. Transport in Cells of Storage Tissues

R.J. POOLE

1. Introduction

Storage tissues have long been a favored experimental material for studies of transport in the higher plant cell. Slices of these tissues provide a more or less homogeneous bulk source of material which may be treated as individual cells, unperturbed by long-distance transport to other parts of the plant, or by photosynthesis or growth. Although these are cells removed from their natural environment, their transport properties, including kinetics (OSMOND and LATIES, 1968) and regulation mechanisms (CRAM, 1973), are comparable to those of other plant material. Until the recent advent of cell suspension cultures, no other plant material appeared to offer such advantages for the study of the cell *per se*.

Cell suspensions have the advantage of providing cells of known nutritional history and without the large diffusion paths of tissue slices. Wound effects and abnormal developmental changes are avoided too. The suspensions vary between clumps of several cells to single cells. Pioneering work on use of tissue culture for transport is due to STEWARD (eg. STEWARD and MILLAR, 1954). Cell suspensions have been used for flux measurements, for example, by HELLER et al., 1973; BENTRUP et al., 1973 and THOIRON et al., 1974. An extensive study of ion uptake to cells of tissue culture has been made by MOTT and STEWARD (eg. 1972). The system clearly has many potential uses, though it has the limitation that it may contain cells in various stages of development.

Cells of tobacco pith have been grown on a liquid medium when they form filaments with cell division occurring throughout the filament, as in certain algae (FILNER, 1969). These cells have been used in studies of nitrate uptake and control of production of nitrate reductase (HEIMER and FILNER, 1971), and of SO_4^{2-} uptake (HART and FILNER, 1969). (See also *11*.3.2.2)

A special characteristic of storage tissue cells is their capacity for a progressive development of metabolism and transport, which is initiated by slicing the dormant tissue, and continues over a period of several days. This ageing phenomenon is not only of interest in itself as a developmental system (see Part B, *8*.3) but also permits the experimental manipulation of the transport properties of the cells (e.g. POOLE, 1971 b).

The cells of storage tissues are highly vacuolated: according to PITMAN (1963) the cytoplasm of beet occupies about 5% of the tissue volume, whereas the vacuoles occupy about 70% of the volume. The remaining space in the tissue slice is extracellular. Although their ion content (Table *8*.1) depends on the history of the plant, the cells are always far from salt saturation. Even when the tissue is at its most active stage, it requires at least 4 days of continuous uptake to reach flux equilibrium with a constant external ion concentration (SUTCLIFFE, 1952; MACDONALD et al., 1960; CRAM, 1973). (Such long-term experiments are, of course, facilitated by the food reserves of storage tissues.) If, on the other hand, the tissue is allowed to exhaust the ions in the external solution, extremely high accumulation ratios, up to $5 \cdot 10^5$ (VAN STEVENINCK, 1961), may be attained.

Table 8.1. Ion content of storage tissues: some typical valu s

Tissue	K^+	Na^+	Cl^-	Ca^{2+}	Mg^{2+}
			(μmol g_{FW}^{-1})		
Red beet (initial)[a,b,c]	50	11–28	5–28	2.5–10	2.5
Red beet (maximal)[d,e]	200–300	($K^+ + Na^+$)	90–110		
Carrot (initial)[b,f]	60–85	11–28	19	4.5	3
Carrot (maximal)[b,f]	150–190	($K^+ + Na^+$)	102–113		
Potato (initial)[b,g]	50–80	1	7–70	0.5–5.5	2.5–10
Potato (maximal)[h]			99		

[a] PITMAN, 1964; [b] MACDONALD et al., 1960; [c] The K:Na ratio may vary greatly (VAN STEVENINCK, 1972; cf. KYLIN, 1973); [d] PITMAN, 1963; [e] HURD, 1958; [f] CRAM, 1973; [g] STEWARD, 1933, meq l^{-1} in expressed sap; [h] STEWARD, 1933, $Cl^- + Br^-$ in expressed sap.

Because of their convenient geometry and relatively homogeneous structure, storage tissues have been used for studies of extracellular diffusion and free space properties. This topic is discussed in Chap. 5 (see 5.3 and 5.5.1, Tables 5.1 and 5.3).

In addition to these common features of storage tissues, certain properties of more limited occurrence have proved interesting. These include the striking effect of pH on transport in some storage tissues, and the preferential uptake of sodium in *Beta*, a halophilic genus. Experimental results with specific tissues will be discussed below.

2. Material

As illustrated in Fig. 8.1, plant storage tissues have diverse origins, and, even within a single species (*Beta vulgaris*), the storage tissue may develop mainly from the root (sugar beet) or mainly from the hypocotyl (red beet). Nevertheless, storage tissues have common features related to their function. They all consist predominantly of large, thin-walled, highly-vacuolate parenchyma cells. Vascular tissue (xylem and phloem) makes up about 2% of the tissue (e.g. TURNER, 1938).

Anatomically, sugar beet and red beet have the most heterogeneous appearance; these have concentric rings of vascular bundles, typical of the Chenopodiaceae, associated with small-celled parenchyma. There are, thus, layers of tissue composed entirely of large-celled parenchyma (50–80 μm in diameter) alternating with layers composed of small-celled parenchyma (10–30 μm in diameter) and vascular tissue. It is possible to prepare slices containing only large-celled parenchyma, and such slices are found to have transport properties similar to those of slices cut from the vascular region. Thus PITMAN (1963) has shown that the rapidly-exchanging "cytoplasmic" fraction of potassium is present in both types of slice, and is therefore not an artifact due to the presence of vascular tissue or of cells of various sizes. Changes in ion selectivity during ageing of the slices are also found in both regions of beet tissue (POOLE, 1971 b).

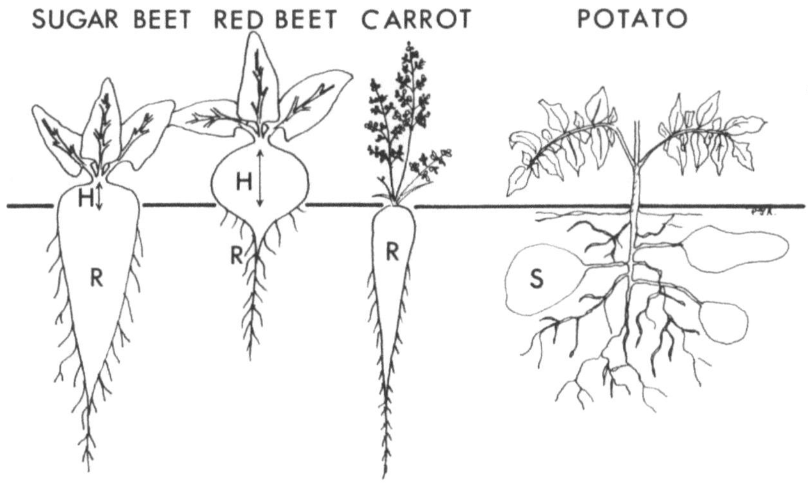

Fig. 8.1. Origin of storage tissues. *H* hypocotyl, *R* root, *S* stem

In experiments with carrot tissue, disks may be cut from the central region (xylem parenchyma), the outer region (phloem parenchyma), or the intermediate cambial region. In all cases, parenchyma cells about 50 μm in diameter make up the bulk of the slice, and, as in red beet, different regions of the tissue show similar patterns of transport (HURD, 1959).

Although potato is composed of a very homogeneous large-celled parenchyma, with a small amount of scattered vascular tissue, this material does not develop in a homogeneous way after slicing (see below). Potato thus seems a less favorable material for transport studies than beet or carrot. Although other species with storage tissues have received attention from time to time as experimental material, we will consider primarily those tissues (beet, carrot and potato) for which we have a relatively detailed picture of transport properties.

2.1 Preparation of Tissue

Disks of tissue, usually 1 mm in thickness, are cut with a cork borer and tissue slicer. In order to ensure adequate diffusion of oxygen and ions to all cells, it is desirable to slice the tissue as thinly as possible. On the other hand, the thinner the slices, the greater the proportion of cells damaged by cutting. Since a layer of dead cells about 0.1 mm thick is formed on each surface, it is not practicable to cut slices less than about 0.5 mm thick.

2.2 Diffusion in Slices of Storage Tissue

It is important to know to what extent metabolism and transport may be limited by the rate of diffusion of oxygen and ions into the tissue slice. Whenever there are net movements of substances into or out of the tissue (or an exchange of isotope for unlabeled ions) the concentration (or specific activity) at the surface of the cells will be different from that in the bulk of the external solution. BRIGGS and ROBERTSON (1948) show how the concentration of a substance may be calculated for any point in the tissue. First, it is necessary to find the diffusion coefficient of the substance in the tissue. This may be determined by measuring the transport between solutions in two external compartments separated by a disk of tissue. It is also necessary to know the relationship between the net uptake or loss of the substance by the cells, and its concentration at the cell surface. If the rate of uptake of oxygen, or an ion, is assumed to be independent of concentration and uniform throughout the tissue, its concentration at the centre of a tissue slice, $c_{\frac{1}{2}}$, is given by

$$c_{\frac{1}{2}} = c_o - J \cdot l^2 / 8D \tag{8.1}$$

where l is the thickness of the slice, and J is the net uptake.

For oxygen, D, the diffusion coefficient in the tissue, may be very much increased by the intercellular air spaces which occupy 1 or 2% of the volume of the tissue. The air spaces are not always continuous, however, and do not, of course, extend across the unstirred layer of solution around the disk. The width of this unstirred layer will depend on the experimental conditions, but it is unlikely to be less than the width of the surface layer of dead cells (approx. 0.1 mm). The unstirred layer of solution may thus add significantly to the length of the diffusion path for oxygen. Rates of oxygen uptake in storage tissue slices at 25° C reach a maximum of about 200 $mm^3\ g_{FW}^{-1}\ h^{-1}$, or about $11 \cdot 10^{-5}$ mg $cm^{-3}\ s^{-1}$ (MACDONALD and DEKOCK, 1958). Application of Eq. 8.1 indicates that the concentration of oxygen at the centre of a 1 mm slice of actively-respiring storage tissue would be about 50% of its concentration in the air-equilibrated external solution.

It would seem, then, that respiration in storage tissue slices may not be too seriously limited by the rate of diffusion of oxygen. Nevertheless, the development of these tissues, i.e., the increase in metabolism which takes place after slicing, is dependent on the thickness of the slices (STILES and DENT, 1947; LATIES, 1962; MACDONALD, 1968). This is particularly striking in potato slices, where cell division and other cytological changes are displayed only by the superficial layer of cells (STEWARD et al., 1932) and rates of oxygen uptake and ion accumulation are related to the surface area rather than the volume of the tissue. In slices of other storage tissues, no such localization of cytological or metabolic activity has been observed. With red beet, SKELDING and REES (1952) found that an initial lag in the development of transport in thicker slices disappeared with time, and the rate of ion accumulation per cell became independent of thickness.

The pathway for diffusion of ions in tissue is, of course, much more restricted than that for oxygen. Thus BRIGGS and ROBERTSON (1948) concluded that, in potato slices, the limitation of ion accumulation to the superficial layers of cells resulted not only from the more active development of these cells, but also from a very low diffusion coefficient for salts in this tissue. Although the situation is not so serious in carrot and beet, there are circumstances in which ion fluxes in these tissues are limited by diffusion even at rather high external salt concentrations (PITMAN, 1963).

2.3 Ageing of Tissue Slices

At the time of slicing, storage tissues are generally in a state of dormancy. The act of slicing initiates a sequence of developments (see Part B, *8*.3) in which various metabolic and transport activities increase progressively over a period of hours or days. In order to permit this development to take place, the disks of tissue are usually suspended after rinsing in 50–100 volumes of vigorously aerated distilled water, which is changed once or twice a day. At first the tissue loses ions and other material to the external solution, expecially from cells damaged in cutting. With time, this efflux ceases and is replaced by an increasingly vigorous capacity for accumulation.

Table *8*.2. Duration of washing required to develop maximal activity in slices of storage tissue

Tissue	Temperature of washing (°C)	Time to develop maximal rate of respiration (days)	Time to develop maximal rates of transport (days)
Red beet (*Beta vulgaris*)	25	3[a]	2–3[b,c] (cation) 6–14[d,e,f] (anion, or anion-limited)
	10	>10[d]	2[g] (K^+) 8[g] (Na^+)
Sugar beet (*Beta vulgaris*)	25	3[a]	3[b]
	10	20[a]	
Swede (*Brassica napus*)	25	1[a]	2–5[b,c]
	7	8[a]	
Carrot (*Daucus carota*)	25	1[a]	1–4[b,h,i]
	12	3–7[j]	
Potato (*Solanum tuberosum*)	23–25	2–4[a,k]	1[l]
	12	12[a]	

[a] MACDONALD and DECOCK (1958); [b] MACDONALD et al. (1960); [c] VAN STEVENINCK (1961); [d] R.J. POOLE (unpublished); [e] SUTCLIFFE (1952); [f] SUTCLIFFE (1957); [g] POOLE (1971b); [h] BIRT and HIRD (1956); [i] CRAM (1973); [j] STILES and DENT (1947); [k] STEWARD and PRESTON (1940); [l] STEWARD and HARRISON (1939).

Table *8*.2 shows the time required for the development of maximal activity in various storage tissues. The times are approximate, since they show some seasonal variation. In all cases, the rate of development is determined by the temperature during the washing period. In addition, it is seen that some tissues develop more rapidly than others, and that in any one tissue different activities may develop at different times. In red beet, for instance, the rates of respiration and of potassium transport peak after 2 or 3 days at 25° C, whereas the greatest capacity for chloride uptake is attained only after about a week of washing. These differences in the development of transport mechanisms for specific ions are discussed below for individual tissues.

2.4 Sterility

It should be noted that in most experiments with storage tissue, no special precautions are taken to ensure sterility, and quite large populations of bacteria (10^7 to 10^9 organisms per g of tissue) may be present. These bacteria may affect the rate of development of storage tissue aged in a limited volume of solution (MACDONALD, 1967), but they have little effect on ion fluxes, except when extremely small amounts of material are being measured (PALMER, 1970). If necessary, bacteria may be excluded by sterile procedures during cutting and washing the tissue.

3. Red Beet (*Beta vulgaris* L.)

3.1 Development of Transport Systems

If beet disks are transferred to a given salt solution at various times after slicing, there is a progressive increase in the initial net rate of ion uptake observed. This increase in uptake rate is observed over a wide range of external concentrations (OSMOND and LATIES, 1968; POOLE, 1971 b), and appears to indicate a progressive synthesis or activation of membrane transport systems.

It is significant that the capacity for rapid ion accumulation develops at different times for different ionic species (Table 8.2). Cl^- uptake develops only slowly (VAN STEVENINCK, 1964), and reaches a maximal rate of about 4 µmol g_{FW}^{-1} h^{-1} (PITMAN, 1964) after a weak at 25° C. K^+ uptake from KCl solutions develops parallel with Cl^- uptake. However, if measured under conditions where it is not dependent on the uptake of Cl^-, K^+ uptake is quite rapid even in freshly cut slices (VAN STEVENINCK, 1961; 1964), and reaches its maximal rate after only 2–3 days.

Differences in the development of uptake have also been observed between different cations and between different anions. Thus uptake of Cl^- develops later than the uptake of organic acids (DALE and SUTCLIFFE, 1959), and uptake of Na^+ develops later than the uptake of K^+ (SUTCLIFFE, 1957; POOLE, 1971 b). The development of Na^+ transport is promoted by a lower temperature of washing.

Changes in uptake of Na^+ and K^+ with ageing have also been observed in slices of bean stem (RAINS, 1969; RAINS and FLOYD, 1970). These observations differ from those in beet in several respects: the specificity is reversed in that K^+ uptake develops later than Na^+ uptake, the system is very sensitive to the presence of Ca^{2+}, and the changes are more pronounced for uptake from low external concentrations. Nevertheless, it seems likely that, in both bean stem and red beet, the observed changes in uptake reflect the independent development of a number of distinct transport mechanisms or channels.

3.2 Net Uptake Patterns

3.2.1 Anion-Dependent Uptake

The net uptake of one ionic species must, of course, be electrically balanced by the transport of other ions. As indicated above, red beet cells are capable of rapid cation uptake. However, in unbuffered external solutions at pH 6.5 or below, their capacity for net excretion of H^+ is limited. Consequently, cation uptake is conspicuously dependent on anion uptake. Thus potassium uptake is promoted by various anions in the following order: $KNO_3 > KCl > K_2HPO_4 > K_2SO_4$ (SUTCLIFFE, 1952), and the development of cation transport with ageing appears to follow the development of anion transport (cf. DALE and SUTCLIFFE, 1959).

3.2.2 The Effect of Calcium

In roots, the presence of Ca^{2+} in the external solution appears to be important both in maintaining healthy growth and in determining Na^+/K^+ selectivity (EPSTEIN, 1961). These effects are not observed in red beet (POOLE, 1971 b), perhaps because sufficient Ca^{2+} is normally present in the cell walls (BRIGGS et al., 1958 a). Addition of Ca^{2+} to the external solution does, however, promote anion uptake in both beet (PITMAN, 1964) and excised roots (ELGABALY, 1962). In beet, since cation uptake is closely dependent on anion uptake, the effect of Ca^{2+} is to increase both K^+ and Cl^- uptake at non-saturating external concentrations of KCl. At low K^+ concentrations, anion uptake may be increased by the addition of $CaCl_2$ to the point where KCl uptake becomes independent of Cl^- concentration and instead is dependent on K^+ concentration (PITMAN, 1964). Thus, in red beet, the rate of cation uptake may be determined by the rate of anion uptake, or *vice versa*. In either case, since H^+ efflux is slow, net cation uptake is approximately equal to net anion uptake. Calcium itself is not taken up to measurable extent by beet cells (PITMAN, 1964), and its effect may be due to a change in surface charge on the cell membrane. Other effects of Ca^{2+} in beet include an inhibition of net K^+-H^+ exchange at high pH, and an apparent decrease in permeability of the plasma membrane to K^+ (e.g. VAN STEVENINCK, 1965).

3.2.3 The Effect of pH

Fig. *8*.2 shows the remarkable effect of pH on cation uptake by red beet, first observed by HURD and SUTCLIFFE (1957), using solutions of KCl buffered with *tris*. At pH 6.5 or below, K^+ uptake is approximately equal to Cl^- uptake (see

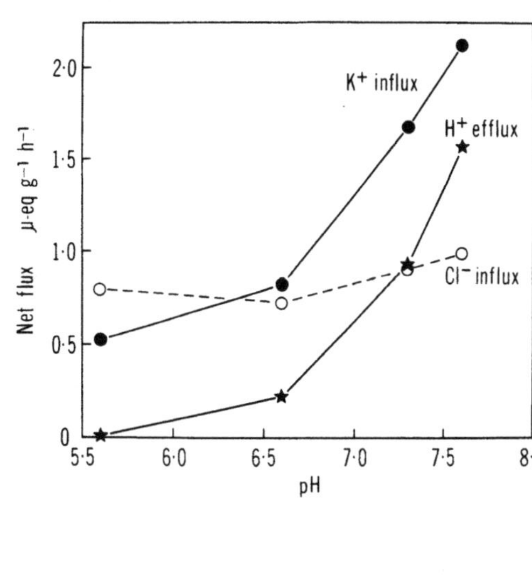

Fig. *8*.2. Effect of external pH on net fluxes in red beet tissue over a 24 h period at 25° C, after washing 4 days at 12° C. The external solution contained 2 mM KCl + 20 mM *tris* (hydroxymethyl) aminomethane adjusted to the appropriate pH with H_2SO_4. The H^+ flux was calculated from the additional amount of base (*tris*) needed to maintain a constant external pH during the experimental period (POOLE, 1973)

above). As the pH is increased from pH 6.5 up to pH 8.5, there is a progressive increase in K^+ uptake, while Cl^- uptake is virtually unchanged. The excess cation uptake is accompanied by the synthesis of an equivalent amount of organic acid (POOLE and POEL, 1965). HURD (1958) showed that label from $H^{14}CO_3^-$ was incorporated into organic acids, and considered that the K^+ flux at high pH was electrically balanced by HCO_3^- uptake. However, the results may equally be attributed to an exchange of K^+ for H^+, with a secondary fixation of HCO_3^- or CO_2^-. The latter interpretation is supported by the demonstration (POOLE and POEL, 1965; POOLE, 1974) that the rate of cation uptake is directly dependent on external pH rather than on HCO_3^- concentration. An apparent net efflux of H^+, equivalent to the excess cation uptake, is indicated by a steady decrease in pH of the external solution (Fig. *8*.2).

The increase of K^+-H^+ exchange and organic acid synthesis at high external pH values effectively releases K^+ uptake from dependence on Cl^- uptake. One consequence of this is to permit rapid cation uptake after only short periods of washing, before the tissue has acquired the ability to take up Cl^- (VAN STEVENINCK, 1961; 1964). Thus, *tris* buffer at high pH shortens or eliminates the "lag period" between cutting the tissue and the commencement of cation uptake. VAN STEVENINCK (1961) compared the ability of a number of different buffers to shorten the lag period in red beet. Since it turns out that the effectiveness of these substances at pH 8 correlates rather closely with their effectiveness as buffers at this pH (R.F.M. VAN STEVENINCK, personal communication) and does not require their penetration into the cells (POLYA, 1968), it seems likely that their mode of action is simply to control the pH at the cell membrane in the face of the outward flux of H^+.

HURD (1959) compared the effect of pH on uptake in a number of different storage tissues. The magnitude of the effect followed the order: parsnip > red beet and sugar beet > carrot xylem and phloem > artichoke > potato and swede = 0. No pH effect was found in potato and swede, and it was noted that these were tissues in which chloride uptake was large enough to balance the maximum rates of cation uptake. In beet also, when Cl^- uptake is high, there is little net H^+ efflux at high pH, and thus a smaller pH effect on cation uptake. (See also *11*.3.2, Chap. *12*, *13*.2.2, *13*.2.5, Part B, *3*.*3*.1, *3*.*3*.2, and *3*.*3*.3.)

3.2.4 The Preferential Uptake of Sodium

The net uptake of Na^+ by aged beet tissue shows all of the characteristics described above for net K^+ uptake. In chloride solutions, it is dependent on the rate of Cl^- uptake. Thus the rates of uptake of both cation and anion from NaCl solutions are approximately the same as from KCl solutions of the same concentration (Table *8*.3), and show the same responses to Ca^{2+} (PITMAN, 1964), and to pH (POOLE, 1971b). The net uptake of Na^+ in aged tissue is thus quantitatively and qualitatively similar to that of K^+. However, if the two cations are presented together, there is a strong preference for uptake of Na^+ (Table *8*.3). Since the total cation uptake rate is usually determined either by the rate of anion uptake or by pH, the uptake of Na^+ usually brings about a corresponding reduction in the uptake of K^+. In fact, addition of an equal quantity of Na^+ is often sufficient to reduce K^+ uptake to zero (BRIGGS et al., 1958b, POOLE, 1971a; 1971b), and excess Na^+ may induce K^+ efflux. Na^+ uptake and K^+ loss may then continue for several days until

Table 8.3. Net uptake of Na^+ and K^+ by red beet tissue. Average over 10 h at 25° C (Adapted from SUTCLIFFE, 1957)

Solution	Na^+ ($\mu mol\ g_{FW}^{-1}\ h^{-1}$)	K^+ ($\mu mol\ g_{FW}^{-1}\ h^{-1}$)	Total ($\mu mol\ g_{FW}^{-1}\ h^{-1}$)
10 mM KCl	—	3.28	3.28
10 mM NaCl	3.43	—	3.43
10 mM KCl + 10 mM NaCl	4.15	1.01	5.16
20 mM KCl	—	5.27	5.27
20 mM NaCl	5.51	—	5.51

the tissue contains about 250 mM Na^+ and less than 10 mM K^+ (MACDONALD et al., 1960). It is noteworthy, however, that in tissue which has not yet developed the ability to transport Na^+, even a 10-fold excess of Na^+ effects only a slight reduction of K^+ uptake (POOLE, 1971 b).

3.3 Fluxes during Net Uptake

PITMAN (1963) succeeded in estimating individual fluxes at the plasma membrane and tonoplast during steady-state net uptake in cells of red beet. Although the data are not readily confirmed by direct sampling of cytoplasm and vacuole as in the giant-celled algae, a number of lines of evidence support the reliability of the method. The rates of isotope uptake and loss with time are consistent with a model based on two intracellular compartments in series, and give plausible estimates of fluxes and of ion content in the cytoplasm, the latter being 2–6% of the tissue content for K^+. (See Chap. 5, Fig. 5.10.)

Independent evidence for a series arrangement of the two compartments was obtained in two ways. First, a long exposure to ^{22}Na was followed by a wash and a shorter exposure to ^{24}Na so as to label the two compartments with different isotopes. The ratio of labels in the subsequent efflux from the rapidly-exchanging compartment indicated a considerable exchange of isotopes between the compartments. Secondly, it was shown that, after labeling with ^{42}K, the total label eluted from the cytoplasm in an external solution of KCl was less than that eluted in NaCl. This suggests a variable proportion of label moves from the cytoplasm to the vacuole during elution, and that Na^+ competes with K^+ for transfer to the vacuole. Exposure to Na^+ for 2–3 h was required for maximum effect on ^{42}K efflux: in contrast, Na^+ uptake brings about an immediate inhibition of K^+ influx at the plasma membrane (POOLE, 1971 a). This dual competitive effect of Na^+ and K^+ at both plasma membrane and tonoplast accounts for the preferential uptake of Na^+ while maintaining a normal level of K^+ in the cytoplasm.

Fig. 8.3 shows the effect of external KCl concentration on fluxes and cytoplasmic content of K^+. The influx at both membranes is much larger than the net uptake. If the efflux were passive, the permeability coefficient is calculated to be $1\text{–}6 \cdot 10^{-10}\ m\ s^{-1}$ (Table 8.5). Since the plasma membrane fluxes increase progressively with increasing concentration, it seems likely that the saturation of net uptake at 20–40 mM KCl is an effect at the tonoplast. However, the possibility of exchange

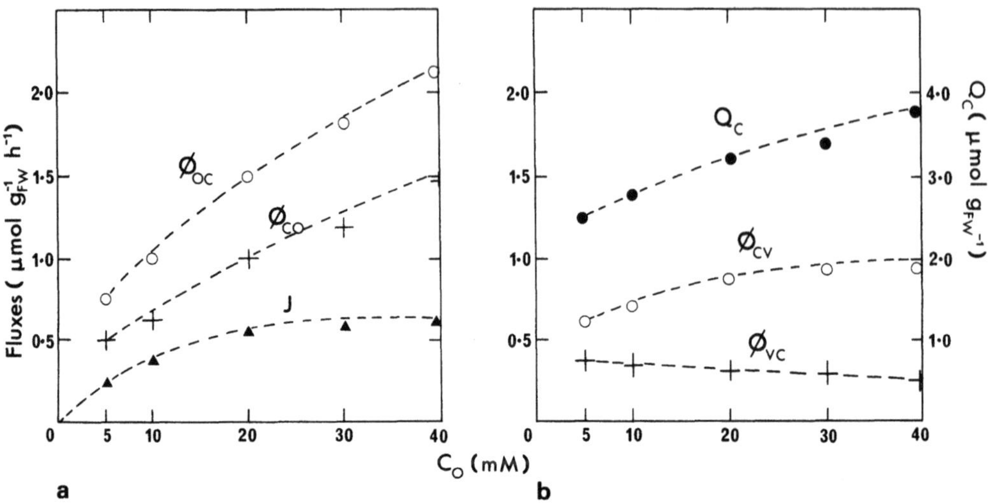

Fig. 8.3a and b. Effect of potassium chloride concentration (C_o) on potassium fluxes into and out of the cytoplasmic phase in red beet, and on Q_c, the content of the cytoplasmic phase. Temperature 2° C. (a) Net uptake rate (▲) and the fluxes across the outer boundary, ϕ_{oc} (○) and ϕ_{co} (+). (b) Content of the cytoplasmic phase (●), and the fluxes at the inner boundary, ϕ_{cv} (○) and ϕ_{vc} (+) (PITMAN, 1963)

diffusion at the plasma membrane (POOLE, 1969) may obscure a role of this membrane in control of net uptake.

Further flux studies by PITMAN showed that cessation of net uptake at high internal salt concentrations was attributable, in the case of K^+, to an increased efflux at both membranes, while, in the case of Br^-, there appeared to be an increased efflux from the vacuole but an inhibition of influx at the plasma membrane. Mechanisms of regulation of ion content are discussed in Chap. *11*.

Some of the patterns of net uptake discussed above have been illuminated by other flux studies. Thus the influx of both Na^+ (POOLE, 1971a) and Cl^- (VAN STEVENINCK, 1964) is very small during the "lag phase" after cutting the tissue, whereas the influx and efflux of K^+ during this period are both large (VAN STEVENINCK, 1964). This supports the earlier conclusions regarding the development of independent transport systems for the various ions, and also invites comparison of the general pattern of cation and anion fluxes in newly-cut tissue with that in salt-saturated material. Flux measurements by PITMAN (1964) also supported the idea that the primary effect of Ca^{2+} is on anion influx: increase in Ca^{2+} has the same general effect on plasma membrane fluxes, tonoplast fluxes and cytoplasmic content of Br^-, as does an increase in external Br^- concentration.

The flux analysis of PITMAN is applicable to steady-state uptake conditions. An attempt by OSMOND and LATIES (1968) to estimate fluxes during the initial approach to a steady state produced a paradoxical result. Pre-loading the cytoplasmic compartment with isotope increased the subsequent uptake of label from the external solution, whereas pre-loading with unlabeled ions decreased it. Since this effect was found only at low external concentrations, where fluxes may be limited by diffusion, it may reflect changes of specific activity in the intercellular space resulting from exchange of ions across the plasma membrane.

3.4 Membrane Potentials and Ion Pumps

Because of the technical difficulties involved in measuring membrane potentials in storage tissues, there is, as yet, insufficient data to answer the many questions about ion gradients and transport mechanisms which arise from the foregoing discussions. However, some potential measurements have been made in red beet by POOLE (1966; 1974). In unbuffered solutions of KCl or K_2SO_4, the potential is usually fairly close to the estimated value for K^+ diffusion equilibrium. On the addition of bicarbonate (Table *8*.4) or *tris* buffer at pH 7.8 (POLYA, 1968), both of which elicit rapid H^+ efflux, the membrane potential becomes much more negative. During ageing of the tissue, the degree of hyperpolarization correlates with the rate of K^+-H^+ exchange. Preliminary results (POOLE, 1974) also indicated that the hyperpolarization is rapidly and reversibly abolished by lowering the temperature to 6° C. It was concluded that H^+ efflux in beet is an active transport process at the plasma membrane, and that it becomes electrogenic at high external pH (see also *6.6*, *7.2.1*, Part B, *3.3.1* and *3.3.2*).

Table *8*.4. Effect of bicarbonate (pH) on membrane potential

External solution				Mean potential (mV)
K^+ (mM)	HCO_3^- (mM)	SO_4^{2-} (mM)	pH	
0.6	0	0.30	5.5	– 159 (±5)
0.6	0.1	0.25	6.8	– 182 (±6)
0.6	0.2	0.20	7.2	– 209 (±5)
				Estimated ψ_K: – 139 mV

Influx of Cl^- or Br^- in beet appears to be an active process at the plasma membrane, unless virtually all the cytoplasmic halide ions are sequestered in some way, which seems unlikely (cf. GERSON and POOLE, 1972). Cl^- uptake also appears to be unaffected by the large changes of membrane potential with external pH.

The mode of transport of K^+ is not so clear, since it appears generally to be close to equilibrium, or to show a net flux down a gradient. However, measured K^+ flux ratios in beet (POOLE, 1969) are not consistent with free diffusion across the plasma membrane, and the highest measured accumulation ratios (VAN STEVENINCK, 1961) would require a potential of −330 mV or more for uptake by diffusion. The independent development of transport systems with ageing of the tissue suggests the possibility of independent systems exchanging K^+ for H^+, Na^+ for H^+, Cl^- for OH^-, etc. In this case, interactions between cation and anion uptake, and between K^+ and Na^+ uptake, might be mediated by changes in cytoplasmic pH. It is apparent that more work needs to be done on the mechanisms of ion transport and their regulation.

3.5 Relation to Metabolism

Various metabolic inhibitors have been used in attempts to elucidate the biochemical mechanism and the energy source for ion accumulation. Cryptopleurine, which was found to inhibit protein synthesis without altering ATP levels, inhibited cation uptake in red beet (POLYA, 1968). However, the effect of this inhibitor on transport was not as rapid as its effect on protein synthesis. It was suggested that protein synthesis is not required for the transport process *per se*, but for the maintenance of membrane integrity. Cycloheximide had an effect on transport similar to that of cryptopleurine.

POLYA and ATKINSON (1969) found that red beet has the ability to maintain a constant level of ATP in the tissue when transferred to anaerobic conditions. Removal of O_2 nevertheless inhibits transport in this tissue. While this observation does not rule out the participation of ATP in ion transport, it does indicate some other link between ion transport and oxidative metabolism.

DNP and CCCP, which are well known as uncouplers of oxidative phosphorylation, also inhibit ion transport in red beet. As expected, this inhibition of transport was shown to be accompanied by a fall in ATP concentration. However, L-ethionine was found to depress the ATP concentration to an even greater extent, while its effect on transport was much less severe (POLYA and ATKINSON, 1969). Thus the inhibitory effects of DNP and CCCP on ion transport cannot readily be attributed to their effects on the cellular ATP concentration, but may indicate a direct action of these compounds as proton carriers at the cell membrane.

Until the physiology of transport in storage tissues is better understood, it is difficult to correlate the transport properties of membranes with the properties of isolated ATPase or other putative components of transport mechanisms. KYLIN (1973) has reported that ATPases isolated from various strains of sugar beet with characteristic differences in Na^+/K^+ selectivity show corresponding differences in activation by Na^+ and K^+. Since there is no evidence for active Na^+ efflux in beet, it is not yet clear what relation these enzymes might have to the Na-K ATPases of animal cells, nor whether they are derived from the plasma membrane or from the tonoplast. An interesting ATPase activity has also been found in turnip (*Brassica rapa* L.) by RUNGIE and WISKICH (1973). This showed specificity for anions, especially organic anions, and was tentatively implicated in tonoplast transport. (See also *10*.1, *10*.4.2.2, and *10*.4.3.2.)

4. Carrot (*Daucus carota* L.)

The pattern of ion transport in carrot is comparable in many respects to that in beet. There is a development of transport with time after cutting the tissue (Table *8*.2), a promotion of K^+-H^+ exchange at high pH (HURD, 1959), a dependence of $(Na^+ + K^+)$ uptake on anion uptake, and a consequent interaction between K^+ and Na^+ uptake (CRAM, 1968a). Here we will consider some characteristic aspects of transport in carrot which are different from beet, or which have received special experimental attention in this material.

4.1 Cation Selectivity

MACDONALD et al. (1960) found that carrot tissue, unlike aged beet tissue, will absorb K^+ in the presence of a 10-fold excess of Na^+. The relative affinities of different storage tissues for Na^+ rather than K^+ were: red beet > sugar beet > carrot > swede and potato. Carrot also differed from beet in absorbing a significant amount of Ca^{2+}. CRAM (1968a) showed that, with equal quantities of NaCl, KCl, and $CaCl_2$ in the external solution, carrot slices absorbed 3–4 times as much K^+ as Na^+. CRAM also demonstrated a ouabain-sensitive Na^+ efflux in carrot. Nevertheless, in single-salt solutions, Na^+ is readily accumulated to high concentrations in the vacuole (CRAM, 1973). Presumably there is a Na^+ pump from cytoplasm to vacuole as well as from cytoplasm to external solution, but there is no clear evidence for the nature of Na^+ influx at the plasma membrane, nor for the manner of regulation of the cation fluxes.

4.2 Chloride Fluxes

In carrot, as in red beet, the time course of isotope efflux gives evidence of intracellular compartmentation, both for ions (CRAM, 1968b) and for the non-metabolized, passively-distributed compound thiourea (GLINKA, 1974). CRAM (1968b) investigated Cl^- fluxes in this tissue and provided further evidence for the validity of the compartmental analysis developed by PITMAN (1963). Cl^- efflux was found to depend on the external salt concentration, and this property was used to confirm the series arrangement of the "cytoplasmic" and "vacuolar" components of ^{36}Cl efflux. If the tissue was transferred to water for an hour or two, so as to inhibit efflux,

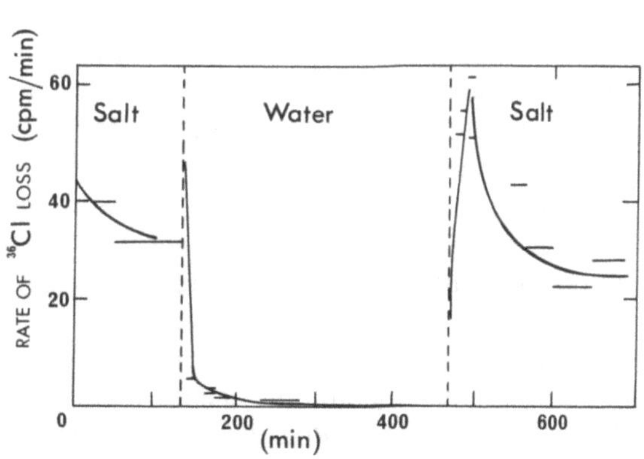

Fig. 8.4. Transient changes in rates of $^{36}Cl^-$ loss from carrot tissue after transfer from salt to water and water to salt. Tissue was loaded with $^{36}Cl^-$ and washed in inactive solution overnight. During this time the cytoplasmic component of the efflux had fallen to zero. The tissue was then washed in several changes of inactive solution, distilled water, and inactive solution, as indicated; the rate of $^{36}Cl^-$ loss over successive periods was measured, and plotted against time. The fall during the first 2 h in inactive solution is an after-effect of being in a relatively small volume of solution overnight. The rate of loss of $^{36}Cl^-$ to inactive solutions in controls (not shown) remained steady after 2 h. Electrical PD in salt = −40 mV; in water after salt = −95 mV. Temp. 20° C (CRAM, 1968b)

and then returned to a salt solution, there was a transient burst in the rate of loss of isotope. This burst showed similar kinetics to the cytoplasmic component of efflux, and indicated a transfer of isotope from vacuole to cytoplasm during the time that the tissue was held in water (Fig. *8*.4).

Net influx of Cl^- in this tissue was found to saturate at an external concentration of 8 mM, but Cl^- influx and efflux at the plasma membrane increased in parallel up to an external concentration of at least 95 mM. The increase in efflux could not be accounted for by changes in potential or cytoplasmic Cl^- concentration, and was ascribed to Cl^- exchange on an anion "carrier". If such a carrier can export other ions than Cl^-, it could function in net uptake as well as exchange.

The electrochemical gradients in this study indicated that Cl^- was actively transported at both the plasma membrane and the tonoplast. The regulation of this transport in carrot and other material is discussed in Chap. *11*.

4.3 Salt Respiration and the Energy Source for Transport

The rate of O_2 uptake by carrot root tissue may be approximately doubled on addition of salt to the external solution. This additional "salt respiration", which has been observed in many plant tissues, has been investigated extensively in carrot (BRIGGS et al., 1961), in the hope that it would provide a clue to the energy linkage of ion transport. It was found that salt respiration is particularly sensitive to inhibitors of cytochrome oxidase, and that, during rapid salt uptake, the net number of ions accumulated is approximately equal to the (additional) number of H atoms transferred to O_2.

ADAMS and ROWAN (1972) found an increase of ADP in the tissue within 30 s of the addition of salt, followed by an increase in O_2 uptake within 4 min. In addition, there was evidence that the ADP level regulates the rate of respiration in aged slices. The increase in ADP supports the idea that salt respiration results from the energy demands of ion transport, while the short times involved in this investigation point to transport at the plasma membrane. On the other hand, since the high rate of respiration may continue unabated when salt is removed, salt respiration cannot always be correlated with net uptake or with influx at the plasma membrane. LÜTTGE et al. (1971) did, however, find some evidence for correlation between salt respiration and transport at the tonoplast. At present, one might conclude that salt respiration, while of interest in itself, has not yet thrown much light on the processes of ion transport.

In carrot, as in beet, it has proven difficult to account for the effects of anaerobic conditions and uncoupling agents in terms of ATP depletion (ATKINSON et al., 1966; ATKINSON and POLYA, 1968). CRAM (1969) has shown that removal of O_2 inhibits Cl^- influx at the plasma membrane, but that this flux is not affected by CCCP or oligomycin. By contrast, the other Cl^- fluxes are sensitive to all three treatments. Although the significance of these results is not yet clear, they offer encouragement for further work on the differential inhibition of the various ion fluxes.

4.4 Amino Acids and Sugars

BIRT and HIRD (1958a; b) have shown that carrot slices accumulate both D- and L-isomers of amino acids against a concentration gradient, by a process which is inhibited by respiratory inhibitors. The uptake mechanism is saturated at high external concentrations, and shows competition between different amino acids. It further resembles ion transport in showing a development of activity with washing of the tissue slices, and a stimulation of respiration by the uptake process, even when the amino acid is not metabolized. Since slices may contain up to 70 mM of free amino acids (DALGARNO and HIRD, 1960), amino acids, like absorbed salts, are evidently accumulated in the cell vacuoles.

Sugars are likewise taken up against a concentration gradient by an energy-requiring, saturable process (GRANT and BEEVERS, 1964; REINHOLD and ESHHAR, 1968), and, like absorbed ions and organic acids (MACLENNAN et al., 1963), they are restricted in their movements between different cell compartments. (See also 7.7; 7.8.)

5. Potato (*Solanum tuberosum* L.)

A pioneering series of investigations by Steward and his co-workers on ion transport in potato slices laid particular emphasis on the active nature of accumulation, and its dependence on metabolism and on the metabolic development of the tissue after slicing. For a summary of this work, see STEWARD and MILLAR (1954). Further studies on the metabolic development of potato tissue, and the influence of the slice thickness, have been referred to above (LATIES, 1962; MACDONALD, 1968).

5.1 Kinetics of Chloride Uptake

LATIES (1959) described two interesting features of ^{36}Cl uptake by potato tissues at 0° C. First, if the tissue is transferred to water at 30° C for a short time, and then replaced in salt solution, there is a transient increase in the rate of uptake of ^{36}Cl. The additional amount of Cl^- absorbed is approximately 0.5 μmol g_{FW}^{-1}, and it is suggested (MACDONALD and LATIES, 1963) that this additional uptake represents the filling of the cytoplasm, which is presumed to transfer its ^{36}Cl into the vacuole during treatment in water at 30° C. The additional uptake appears to be sensitive to DNP applied during the 30° C pretreatment, and to cyanide applied during the subsequent uptake, but not *vice versa*. These effects may be compared with the inhibitor effects on specific fluxes in carrot tissue (CRAM, 1969). However, any conclusions about Cl^- fluxes in potato can only be tentative in the absence of a more detailed flux analysis.

The other observation of LATIES (1959) concerned the relation between ^{36}Cl uptake and external salt concentration. In fresh potato tissue, or in aged tissue at 0° C, Cl^- uptake increased more than proportionally with increasing external KCl concentration. This was shown to be due to the change in K^+ concentration (LATIES et al., 1964), and it was suggested that K^+ exerted its effect on Cl^- uptake *via* the membrane potential. MACKLON and MACDONALD (1966) confirmed that the potential does in fact change with KCl concentration in the required manner. While these results are consistent with Cl^- uptake by passive diffusion across the plasma membrane, it may be noted that any uptake mechanism in which anion uptake is dependent on cation uptake is likely to produce similar kinetics. More information is required about Cl^- gradients and fluxes in this tissue. Table *8.5* shows the permeability coefficients which would be required to account for various ion fluxes in storage tissues, if these fluxes were entirely passive. Since the involvement of carriers cannot be ruled out, these values can only be regarded as maximal estimates of membrane permeability.

Table *8.5*. Estimates of ion permeability in storage tissues. All of these estimates assume no carrier-mediated component in the flux used for calculation. (Estimates based on efflux also assume high ionic activity coefficients in cytoplasm)

Tissue	Ion	"Passive" flux	Permeability coefficient ($m\,s^{-1} \times 10^{10}$)
Beet (aged)[a]	K^+	ϕ_{co}	0.8–5.9
Beet (aged)[b]	K^+	Φ_{in}	4.0–15.0
Beet (aged)[c]	Br^-	ϕ_{co}	0.04–0.4
Carrot (aged)[d]	Cl^-	ϕ_{co}	0.08
Potato (fresh)[e]	Cl^-	Φ_{in}	0.24
Potato (aged)[e]	Cl^-	Φ_{in}	up to 2.5

[a] Based on PITMAN (1963) assuming ϕ_{co} passive, but see POOLE (1969); [b] Based on POOLE (1966) assuming Φ_{in} passive, but see POOLE (1969); [c] Based on PITMAN (1963); [d] Based on CRAM (1968b), using the lowest external concentration in the hope of avoiding exchange diffusion; [e] LATIES et al. (1964) and MACKLON and MACDONALD (1966).

5.2 Sugar Uptake

Both metabolized and non-metabolized sugars are taken up against a gradient in potato, and as in carrot, there is evidence of compartmentation of sugars within the cell (LATIES, 1964; LINASK and LATIES, 1973). The uptake kinetics of glucose or 3-0-methyl glucose (LINASK and LATIES, 1973) suggest a carrier which changes its properties at certain critical external concentrations, as proposed for ion transport by NISSEN (e.g., 1971, Part B, *3.2*.4.2.1).

6. Conclusion

Because of a lack of information on certain basic parameters, such as cytoplasmic electrochemical activities, the storage tissues of higher plants do not yet provide a comprehensive picture of transport at the cellular level. They do, however, serve to illustrate some features which appear to be of general significance. It is now clear that the cytoplasm and vacuole can be recognized as distinct compartments in transport studies in higher plants, as in the giant-celled algae. The plasma membrane and the tonoplast are both actively involved in the transport of Cl^-, and in Na^+/K^+ discrimination, and these membranes are also seen to react differently to metabolic inhibitors. Transport systems for different ions develop at different times, and transport in the higher plant cell is thus shown to be an adaptive and versatile developmental system (cf. Part B, Chap. *8*). Interrelationships between cation and anion transport are clearly seen in storage tissues, although more evidence is required on the question of whether these interactions are primarily electrical (LATIES et al., 1964) or chemical (POOLE, 1974). Another prominent feature of storage tissues, active proton efflux, is increasingly recognized as a basic function of the plant cell (Chap. *12*).

With the tools at hand, further progress can be expected in all these areas, both in storage tissues and in other simple cellular systems, such as cell suspension cultures. Meanwhile, despite the gaps in knowledge, the phenomena observed in storage tissue cells can form a basis from which to proceed to a study of transport in more complex tissues and organs.

References

ADAMS, P.B., ROWAN, K.S.: Regulation of salt respiration in carrot root slices. Plant Physiol. **50**, 682–686 (1972).

ATKINSON, M.R., ECKERMANN, G., GRANT, M., ROBERTSON, R.N.: Salt accumulation and adenosine triphosphate in carrot xylem tissue. Proc. Natl. Acad. Sci. U.S. **55**, 560–564 (1966).

ATKINSON, M.R., POLYA, G.M.: Effects of L-ethionine on adenosine triphosphate levels, respiration, and salt accumulation in carrot xylem tissue. Australian J. Biol. Sci. **21**, 409–420 (1968).

BENTRUP, F.W., PFRÜNER, H., WAGNER, G.: Evidence for differential action of indoleacetic acid upon ion fluxes in single cells of *Petroselinum sativum*. Planta **110**, 369–372 (1973).

BIRT, L.M., HIRD, F.J.R.: The uptake of amino acids by carrot slices. Biochem. J. **64**, 305–311 (1956).

BIRT, L.M., HIRD, F.J.R.: The uptake and metabolism of amino acids by slices of carrot. Biochem. J. **70**, 277–286 (1958a).

BIRT, L.M., HIRD, F.J.R.: Kinetic aspects of the uptake of amino acids by carrot tissue. Biochem. J. **70**, 286–292 (1958b).

BRIGGS, G.E., HOPE, A.B., PITMAN, M.G.: Exchangeable ions in beet disks at low temperature. J. Exptl. Bot. **9**, 128–141 (1958a).

BRIGGS, G.E., HOPE, A.B., PITMAN, M.G.: Measurement of ionic fluxes in red beet tissue using radioisotopes. In: Radioisotopes in scientific research. Proc. UNESCO Internal. Conf. vol. 4, p. 391–400 (1958b).

BRIGGS, G.E., HOPE, A.B., ROBERTSON, R.N.: Electrolytes and plant cells. Oxford: Blackwell 1961.

BRIGGS, G.E., ROBERTSON, R.N.: Diffusion and absorption in disks of plant tissue. New Phytologist **47**, 265–283 (1948).

CRAM, W.J.: The effects of ouabain on sodium and potassium fluxes in excised root tissue of carrot. J. Exptl. Bot. **19**, 611–616 (1968a).

CRAM, W.J.: Compartmentation and exchange of chloride in carrot root tissue. Biochim. Biophys. Acta **163**, 339–353 (1968b).

CRAM, W.J.: Respiration and energy-dependent movements of chloride at plasmalemma and tonoplast of carrot root cells. Biochim. Biophys. Acta **173**, 213–222 (1969).

CRAM, W.J.: Internal factors regulating nitrate and chloride influx in plant cells. J. Exptl. Bot. **24**, 328–341 (1973).

DALE, J.E., SUTCLIFFE, J.F.: The effects of aqueous extracts of red beet root on salt accumulation and respiration of discs of red beet root. Ann. Bot. (London), N.S. **23**, 1–21 (1959).
DALGARNO, L., HIRD, F.J.R.: Increase in the process of accumulation of amino acids in carrot slices with prolonged aerobic washing. Biochem. J. **76**, 209–215 (1960).
ELGABALY, M.M.: On the mechanism of anion uptake by plant roots II. Effect of valence of associated cation on Cl^- uptake by excised barley roots. Plant Soil **16**, 148–156 (1962).
EPSTEIN, E.: The essential role of calcium in selective cation transport by plant cells. Plant Physiol. **36**, 437–444 (1961).
FILNER, P.: Control of nutrient assimilation, a growth-regulating mechanism in cultured plant cells. In: Develop. Biol. Supplement **3**, 206–226 (1969).
GERSON, D.F., POOLE, R.J.: Chloride accumulation by mung bean root tips. A low-affinity active transport system at the plasmalemma. Plant Physiol. **50**, 603–607 (1972).
GLINKA, Z.: Fluxes of a nonelectrolyte and compartmentation in cells of carrot root tissue. Plant Physiol. **53**, 307–311 (1974).
GRANT, B.R., BEEVERS, H.: Absorption of sugars by plant tissues. Plant Physiol. **39**, 78–85 (1964).
HART, J.W., FILNER, P.: Regulation of sulphate uptake by amino acids in cultured tobacco cells. Plant Physiol. **44**, 1253–1259 (1969).
HEIMER, Y.M., FILNER, P.: Regulation of the nitrate assimilation pathway in cultured tobacco cells. III. The nitrate uptake system. Biochim. Biophys. Acta **230**, 361–372 (1971).
HELLER, R., GRIGNON, C., SCHEIDECKER, D.: Study of the efflux and the influx of potassium in cell suspensions of *Acer pseudoplatanus* and leaf fragments of *Hedera canariensis*. In: Ion transport in plants (W.P. ANDERSON, ed.), p. 337–356. London-New York: Academic Press 1973.
HURD, R.G.: The effect of pH and bicarbonate ions on the uptake of salts by disks of red beet. J. Exptl. Bot. **9**, 159–174 (1958).
HURD, R.G.: An effect of pH and bicarbonate on salt accumulation by disks of storage tissue. J. Exptl. Botany **10**, 345–358 (1959).
HURD, R.G., SUTCLIFFE, J.F.: An effect of pH on the uptake of salt by plant cells. Nature **180**, 233–235 (1957).
KYLIN, A.: Adenosine triphosphatases stimulated by (sodium+potassium): biochemistry and possible significance for salt resistance. In: Ion transport in plants (W.P. ANDERSON, ed.), p. 369–377. London-New York: Academic Press 1973.
LATIES, G.G.: The generation of latent-ion-transport capacity. Proc. Natl. Acad. Sci. U.S. **45**, 163–172 (1959).
LATIES, G.G.: Controlling influence of thickness on development and type of respiratory activity in potato slices. Plant Physiol. **37**, 679–690 (1962).
LATIES, G.G.: The relation of glucose absorption to respiration in potato slices. Plant Physiol. **39**, 391–397 (1964).
LATIES, G.G., MACDONALD, I.R., DAINTY, J.: Influence of the counter-ion on the absorption isotherm for chloride at low temperature. Plant Physiol. **39**, 254–262 (1964).
LINASK, J., LATIES, G.G.: Multiphasic absorption of glucose and 3-0-methyl glucose by aged potato slices. Plant Physiol. **51**, 289–294 (1973).
LÜTTGE, U., CRAM, W.J., LATIES, G.G.: The relationship of salt-stimulated respiration to localised ion transport in carrot tissue. Z. Pflanzenphysiol. **64**, 418–426 (1971).
MACDONALD, I.R.: Bacterial infection and ion absorption acapacity in beet disks. Ann. Bot. (London), N.S. **31**, 163–172 (1967).
MACDONALD, I.R.: Further evidence of oxygen diffusion as the determining factor in the relation between disk thickness and respiration of potato tissue. Plant Physiol. **43**, 274–280 (1968).
MACDONALD, I.R., DEKOCK, P.C.: Temperature control and metabolic drifts in ageing disks of storage tissue. Ann. Bot. (London), N.S. **22**, 429–448 (1958).
MACDONALD, I.R., DEKOCK, P.C., KNIGHT, A.H.: Variations in the mineral content of storage tissue disks maintained in tap water. Physiol. Plantarum **13**, 76–89 (1960).
MACDONALD, I.R., LATIES, G.G.: Kinetic studies of anion absorption by potato slices at 0 °C. Plant Physiol. **38**, 38–44 (1963).
MACKLON, A.E.S., MACDONALD, I.R.: The role of transmembrane electrical potential in determining the absorption isotherm for chloride in potato. J. Exptl. Bot. **17**, 703–717 (1966).
MACLENNAN, D.H., BEEVERS, H., HARLEY, J.L.: "Compartmentation" of acids in plant tissues. Biochem. J. **89**, 316–327 (1963).

MOTT, R.L., STEWARD, F.C.: Solute accumulation in plant cells. V. An aspect of nutrition and development. Ann. Bot. (London) **36**, 915–937 (1972).

NISSEN, P.: Uptake of sulfate by roots and leaf slices of barley: mediated by single, multiphasic mechanisms. Physiol. Plantarum **24**, 315–324 (1971).

OSMOND, C.B., LATIES, G.G.: Interpretation of the dual isotherm for ion absorption in beet tissue. Plant Physiol. **43**, 747–755 (1968).

PALMER, J.M.: The influence of microbial contamination of fresh and washed beetroot disks on their capacity to absorb phosphate. Planta **93**, 48–52 (1970).

PITMAN, M.G.: The determination of the salt relations of the cytoplasmic phase in cells of beetroot tissue. Australian J. Biol. Sci. **16**, 647–668 (1963).

PITMAN, M.G.: The effect of divalent cations on the uptake of salt by beetroot tissue. J. Exptl. Bot. **15**, 444–456 (1964).

POLYA, G.M.: Inhibition of protein synthesis and cation uptake in beetroot tissue by cycloheximide and cryptopleurine. Australian J. Biol. Sci. **21**, 1107–1118 (1968).

POLYA, G.M., ATKINSON, M.R.: Evidence for a direct involvement of electron transport in the high-affinity ion accumulation system of aged beet parenchyma. Australian J. Biol. Sci. **22**, 573–584 (1969).

POOLE, R.J.: The influence of the intracellular potential on potassium uptake by beetroot tissue. J. Gen. Physiol. **49**, 551–563 (1966).

POOLE, R.J.: Carrier-mediated potassium efflux across the cell membrane of red beet. Plant Physiol. **44**, 485–490 (1969).

POOLE, R.J.: Effect of sodium on potassium fluxes at the cell membrane and vacuole membrane of red beet. Plant Physiol. **47**, 731–734 (1971 a).

POOLE, R.J.: Development and characteristics of sodium-selective transport in red beet. Plant Physiol. **47**, 735–739 (1971 b).

POOLE, R.J.: The H^+ pump in red beet. In: Ion transport in plants (W.P. ANDERSON ed.), p. 129–134. London-New York: Academic Press 1973.

POOLE, R.J.: Ion transport and electrogenic pumps in storage tissue cells. Canad. J. Bot. **52**, 1023–1028 (1974).

POOLE, R.J., POEL, L.W.: Carbon dioxide and pH in relation to salt uptake by beetroot tissue. J. Exptl. Bot. **16**, 453–461 (1965).

RAINS, D.W.: Sodium and potassium absorption by bean stem tissue. Plant Physiol. **44**, 547–554 (1969).

RAINS, D.W., FLOYD, R.A.: Influence of calcium on sodium and potassium absorption by fresh and aged bean stem slices. Plant Physiol. **46**, 93–98 (1970).

REINHOLD, L., ESHHAR, Z.: Transport of 3-0-methylglucose into and out of storage cells of *Daucus carota*. Plant Physiol. **43**, 1023–1030 (1968).

RUNGIE, J.M., WISKICH, J.T.: Salt-stimulated adenosine triphosphatase from smooth microsomes of turnip. Plant Physiol. **51**, 1064–1068 (1973).

SKELDING, A.D., REES, W.J.: An inhibitor of salt absorption in the root tissues of red beet. Ann. Bot. (London), N.S. **16**, 513–529 (1952).

STEVENINCK, R.F.M. VAN: The "lag-phase" in salt uptake of storage tissue. Nature **190**, 1072–1075 (1961).

STEVENINCK, R.F.M. VAN: A comparison of chloride and potassium fluxes in red beet tissue. Physiol. Plantarum **17**, 757–770 (1964).

STEVENINCK, R.F.M. VAN: The effects of calcium and tris (hydroxymethyl)- aminomethane on potassium uptake during and after the lag phase in red beet tissue. Australian J. Biol. Sci. **18**, 227–233 (1965).

STEVENINCK, R.F.M. VAN: Abscissic acid stimulation of ion transport and alteration in K^+/Na^+ selectivity. Z. Pflanzenphysiol. **67**, 282–286 (1972).

STEWARD, F.C.: The absorption and accumulation of solutes by living plant cells V. Observations upon the effects of time, oxygen and salt concentration upon absorption and respiration by storage tissue. Protoplasma **18**, 208–242 (1933).

STEWARD, F.C., HARRISON, J.A.: The absorption and accumulation of salts by living plant cells IX. The absorption of rubidium bromide by potato discs. Ann. Bot. (London) N.S. **3**, 427–453 (1939).

STEWARD, F.C., MILLAR, F.K.: Salt accumulation in plants: a reconsideration of the role of growth and metabolism. In: Active transport and secretion. Symp. Soc. Exptl. Biol. vol. VIII, p. 367–406. Cambridge University Press 1954.

STEWARD, F.C., PRESTON, G.: Metabolic processes of potato discs under conditions conducive to salt accumulation. Plant Physiol. **15**, 23–61 (1940).

STEWARD, F.C., WRIGHT, R., BERRY, W.E.: The absorption and accumulation of solutes by living plant cells III. The respiration of cut discs of potato tuber in air and immersed in water, with observations upon surface: volume effects and salt accumulation. Protoplasma **16**, 576–611 (1932).

STILES, W., DENT, K.W.: Researches on plant respiration VI. The respiration in air and in nitrogen of thin slices of storage tissues. Ann. Bot. (London), N.S. **11**, 1–34 (1974).

SUTCLIFFE, J.F.: The influence of internal ion concentration on potassium accumulation and salt respiration of red beet root tissue. J. Exptl. Bot. **3**, 59–76 (1952).

SUTCLIFFE, J.F.: The selective uptake of alkali cations by red beet root tissue. J. Exptl. Bot. **8**, 36–49 (1957).

THOIRON, A., THOIRON, B., LE GUILE, J., GUERN, J., THELLIER, M.: A shock effect on the permeability to sulphate of *Acer pseudoplatanus* cell-suspension cultures. In: Membrane transport in plants (U. ZIMMERMANN, J. DAINTY, eds.), p. 234–238. Berlin-Heidelberg-New York: Springer 1974.

TURNER, J.S.: The respiratory metabolism of carrot tissue I. Material and methods. New Phytologist **37**, 232–253 (1938).

III. Regulation, Metabolism and Transport

9. Transport and Energy

U. LÜTTGE and M.G. PITMAN

1. Introduction

The general theme of this Section (III) is the control of the various transport processes in the cell. It should be clear from samples given in earlier Sections that the activity of transport processes is modified to meet different conditions of growth or availability of substrates. A problem in this type of study is to distinguish between correlations between, say, uptake and growth and the operation of specific control systems.

It can be seen intuitively that the rate of transport may be limited by the availability of metabolic energy. For example, inhibitors of energy metabolism reduce transport because of their effect on available energy for the process. The question that needs to be examined though, is whether either the energy supply may be varied or the activity of the transport sites changed, by some direct or indirect effect of the accumulation of material. Does the level of an ion or, say, of sugar, send information back to the component processes of transport so that the level can be regulated in the cell? Does the *amount of energy* required for transport determine the interaction of transport and metabolism, or is regulation achieved in a more subtle way by a signalling effect of the general *energy status* of the cell on transport? How can such control systems be recognized?

There is a large body of knowledge about control systems, derived from study of electronics, of information theory, and its application to many problems. Plant physiologists should borrow some of these ideas to apply to the search for regulation in plants. In Chap. *11*, CRAM discusses some simple aspects of negative feedback loops, and takes examples from balance of ions in cells and control of osmotic pressure.

At the cellular level there are many processes which interact and which are affected by transport processes. For example, the electrical potential in the cell, the pH of the cytoplasm, the synthesis of organic acids balancing uptake of metabolized ions (NO_3^-, phosphate, SO_4^{2-}) are all aspects of the internal environment of the cell affected by ion transport. Regulation of pH is the main topic of Chap. *12* and the connections between transport and organic acid synthesis are discussed in Chap. *13*.

The biochemical mechanisms of control are in some way less of an immediate problem than recognition of the presence of control. Feedback from the products or from cell pH to the sites of transport is an obvious means, but by analogy with animal systems one might expect plant hormones to affect transport. Part B, Chap. *7* shows that there are many ways in which transport can be affected by chemical "messengers" and it is interesting to find that some of these compounds are phytotoxins produced by pathological fungi.

In addition to biochemical mechanisms affecting sites of transport, the number and types of sites, and the controlled levels, appear to be determined by the plants genotype. Comparisons of plants during development (Part B, Chap. *8*) or growing in different locations (Part B, Chap. *9*) shows that the response of the plant varies. Some responses of plants to different environment (e.g. saline or high-Ca^{2+} soils) reflect different genetic content, but the responses may also be mimicked phenotypically by other species. Again, plant hormones may be involved in these environmental responses to ion transport, as they are in regulation of water potential in the plant.

Recognition of such kind of regulation in plants has been hampered for a long while because much of the response of the plant has been thought of as "growth", as if it were inevitable, like the diffusion of a gas into a vacuum. However, for example, study of control of water loss in the plant and its relation to abscisic acid or cytokinins shows that the plant has control systems operating at the whole-plant as well as at the cellular level. Evidence is accumulating—as discussed in Part B, Chap. *10*—that supply of nutrients and transport from root to shoot is closely regulated, and not the haphazard consequence of growth. Though lacking nervous systems as found in higher animals, chemical co-ordination seems well-developed in plants.

Understanding how processes of transport interact and are controlled requires too an understanding of the mechanisms coupling transport and metabolism. These mechanisms can be considered from various points of view, e.g. utilization of metabolic energy by transport, transport steps as parts of biochemical reaction sequences and cycles, feedback between transport mechanisms and metabolism in co-operating differentiated organs, and correlated control and regulation of transport and metabolism. Some information, which is necessary as a basis for such considerations is contained in Part B, Chap. *10* and in another volume of this series (see Vol. 3). The remainder of this Chap. (*9*) examines in particular the role of biochemical mechanisms coupling transport and energy metabolism.

2. Particular Energy Source for Particular Transport Mechanism

Studies of glycolysis, respiration and photosynthesis show that the overall process can often be regarded as a series of partial reactions characterized by different response to inhibitors, gas mixtures, light etc. The demonstration that transport too shows a differential response to inhibitors leads to the idea that the transport process may be linked to a particular partial reaction, such as oxidative phosphorylation or PS I.

An early example was the study of salt respiration, which is a cyanide-sensitive increase of respiratory O_2 uptake following addition of mineral ions. It was inferred that ion transport was driven by respiratory electron flow involving (CN-sensitive) cytochromes. However, what was not clear was whether respiratory electron flow was the immediate driving force or whether increased respiration reflected increased ATP-ADP turnover by ATP-utilizing ion uptake (LUNDEGÅRDH, 1939, 1950, 1955; ROBERTSON, 1968; LÜTTGE et al., 1971; cf. SUTCLIFFE, 1962; cf. LÜTTGE, 1973).

Similar problems are raised by the study of energy supply for ion transport in green tissues. MACROBBIE (1965) showed there was a different response of Cl^-

and K^+ uptake to far-red light and various inhibitors in *Nitella translucens* cells. She suggested that light-dependent Cl^- uptake was driven by non-cyclic photosynthetic electron flow, whereas K^+ influx was powered by ATP.

Considerable ingenuity has been employed in combining inhibitors and varying external conditions in efforts to demonstrate the closeness of correlation of ion transport and partial reactions of energy metabolism. This approach has produced evidence of linkage between transport and specific-energy metabolism in a few cases for eucaryotic cells (e.g. in *Nitella translucens* and *Hydrodictyon africanum,* Chap. *6*; for hexose uptake by *Chlorella vulgaris,* KOMOR and TANNER, 1974; and ion uptake by C_4 leaves, Part B, Chap. *4.2*). In many other cases investigated, the transport did not seem to be linked to one particular energy source but could be powered by alternatives when these were available (e.g. by either respiration or photosynthesis in certain leaf preparations, JOHANSEN and LÜTTGE, 1974; Part B, Chap. *4.2*). In this case transport stops only when all energy sources are inhibited.

This general observation is not surprising when one thinks of the reactions in various systems of energy metabolism as a *network* rather than a series of *linearly* arranged processes. It emphasizes, though, that there is a need for some coupling between electron flow and transport in many (if not all) cases.

3. Coupling between Sources of Metabolic Energy and Transport Mechanisms

Coupling between transport and sources of energy needs to be treated as a problem involving gross and cellular organization as well as molecular interactions. Often there is clear spatial separation as well as functional separation between energy source and transport.

3.1 Coupling at the Anatomical Level

Good examples of energetic coupling when spatial separation is a problem are multicellular systems, where energy is provided in one type of cells (or one tissue) and transport occurs in another type of cells (or in another tissue).

For example, such co-operation between differentiated parts of an organ is shown by leaf salt-gland (or salt-hair) systems, where light-dependent reactions in photosynthetically active leaf mesophyll cells drive ion excretion (or accumulation) in photosynthetically inactive gland cells (or salt-hairs). (Part B, Chap. *5.2*; OSMOND et al., 1969; LÜTTGE and OSMOND, 1970; HILL and HILL, 1973; LÜTTGE, 1971a, 1974.)

One solution to such problems of coupling at the anatomical level is the transport of energy-rich metabolites within the system. Another is the transfer of information as hormonal or electrical signals (Part B, Chaps. *5.2*; *7*; LÜTTGE, 1974). The problem may also be solved structurally by symplasmic linkages, as found in gland systems, across the bundle sheath of leaf vascular strands or across the root into the xylem (Part B, Chaps. *2*, *3.4*). In this case though, the site of energy consumption and transport are not necessarily separate, but can both be situated at the input to the symplasm, and coupling then occurs at the cellular level.

3.2 Coupling at the Cellular Level

This discussion is restricted to the situation in eucaryotic plant cells, where mitochondria and chloroplasts provide electron flow and produce ATP and reducing equivalents. Transport occurs at the plasmalemma and tonoplast, spatially separated from the energy source.

ATP and reducing equivalents can be coupled with transport by movement of chemical compounds carrying ~P and [H] from the mitochondria or chloroplasts respectively to the cell membranes. This has been referred to as "chemical coupling" (LÜTTGE, 1971b).

The discovery of multiple compartmentation of living cells has been a decisive step towards an understanding of the simultaneous operation of assimilative and dissimilative reaction sequences in cellular metabolism. The intracellular regulation of compartmented reaction sequences between glycolysis, respiration, photosynthesis and photorespiration, oxidative pentose-phosphate cycle, amino acid and protein metabolism, fatty acid synthesis and degradation, etc., depends on the shuttling of intermediates (transport-metabolites) between compartments. These transport steps become integral parts of the biochemical reaction sequences and cycles. The attempts to understand the regulation of intermediary metabolism by transport between compartments has developed into the flourishing field of "intracellular transport" which is treated in a separate volume (see Vol. 3). Transport across a membrane is also frequently a prerequisite of polymer synthesis. The membrane involved often is an organelle membrane within the cell, but it can also be a membrane separating "inside" and "outside" of the cell. The basic principle is that monomers are synthesized on one side of the membrane, transported across the membrane and polymerized on the other side of the membrane. A good illustration of this is the uptake of precursors and the synthesis of polysaccharides in the cisternae and vesicles of the Golgi apparatus (Part B, *5.3.1.3*; and Vol. 3). But there are many more examples, e.g. cellulose synthesis, polymerization of lignin (Part B, Chap. *1*), etc.

3.3 Coupling at the Membrane or Molecular Level

Much use has been made of irreversible thermodynamics in defining active transport, with the aim of investigating molecular coupling between transport and energy use. In this sense, active transport is a movement of particles across an asymmetric membrane, and which is energetically coupled with a chemical reaction within the membrane and does not utilize the transported particles to form other products from them (cf. KEDEM, 1961). This approach has been applied to a number of model systems, one of which is shown in Fig. *9.1*.

In living cells a considerable number of cases have been found where chemical reactions in the membrane drive active transport across it. For example, certain transport mechanisms in bacteria (e.g. sugar and amino-acids) depend on transphosphorylation or redox reactions in the plasma membrane (Part B, *5.3.1.1*; KABACK, 1970; BARNES and KABACK, 1970; BARNES, 1972). In *Chlorella vulgaris* the uptake of electrically neutral glucose molecules seems to be driven by an H^+ electrochemical gradient at the plasmalemma (Part B, *5.3.1.1.2.3*; KOMOR and TANNER, 1974). This seems to be an example of functional coupling between H^+ or OH^- fluxes and fluxes of electrically neutral solute particles in which there is a stoichiometric 1:1 hexose:H^+ "symport" or 1:1 hexose:OH^- "antiport" (WEST and MITCHELL, 1972, 1973; KASHKET and WILSON, 1973; KOMOR, 1973).

Fluxes of H^+ ions are associated with many ion-transport processes and appear to have a special role in both ion transport and energy metabolism. According

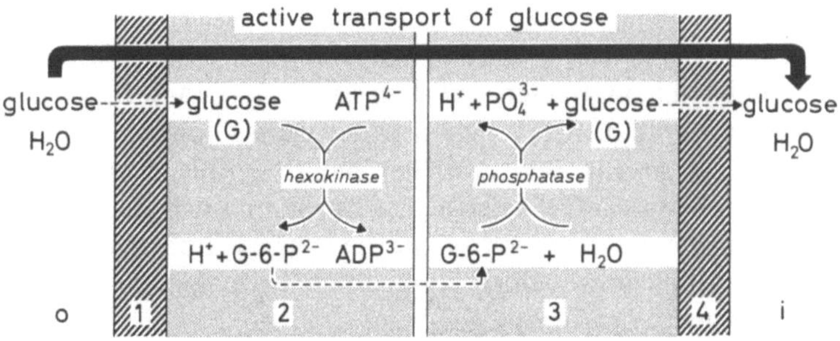

Fig. *9.*1. Model of active transport in an asymmetric artificial membrane. The membrane consists of 4 layers. Layer 1 and 4 are cation exchange membranes which are impermeable to the anions involved in the system ($G\text{-}6\text{-}P^{2-}$, ATP^{4-}, ADP^{3-}, PO_4^{3-}). Layer 2 is a hydrophilic polymer gel phase containing hexokinase, layer 3 is a similar gel containing phosphatase. Both phase 2 and phase 3 contain ATP which cannot leave the sandwich membrane because of the ion exchange properties of layers 1 and 4. Glucose can penetrate these layers and can enter from the outer compartment (o) across layer 1 into layer 2. By phosphorylation of glucose to glucose-6-phosphate ($G\text{-}6\text{-}P^{2-}$), a concentration gradient of glucose between o and layer 2 is maintained. G-6-P can diffuse into layer 3. By hydrolysis of the phosphate ester in layer 3 a concentration gradient of G-6-P is maintained between layers 2 and 3; and glucose can diffuse into the inner compartment (i). As long as the ATP is not exhausted, glucose can move actively from o to i driven by concentration gradients established by the two enzyme reactions. It is important to note that the total amount of glucose present in the system is not affected by these reactions, i.e. the chemical reactions involving glucose in order to transport glucose do not utilize glucose to form other products from it. Glucose concentration is always smaller in 2 than in o, glucose concentration in i may be smaller or larger than in o (Redrawn after NÉEL, 1974)

to MITCHELL's hypothesis (see *12*.4.) charge separation (or H^+—OH^- separation) is seen as a consequence of operation of an ATPase in the membrane, producing a vectorial biochemical reaction.

The H^+ gradient so set up may be coupled electrogenically with a K^+ influx; in this case the cell potential responds to inhibition of energy metabolism (see Table *4*.2 and discussion in *4*.2.6). A particularly good example is the PD in *Neurospora* which is highly correlated with the ATP-level in the cell (SLAYMAN et al., 1970; Fig. *4*.7, *7*.2.1).

Involvement of H^+ (or OH^-) gradients in both energy metabolism and in transport processes provides an integrative principle that suggests how many processes can be coupled with energy. Vectorial reactions may set up gradients of H^+ and OH^- at the membrane, but these gradients may in turn affect the rates of reaction of other enzymes (e.g. through pH-sensitivity). The importance of H^+ fluxes and their regulation is extensively discussed in Chap. *12*. There are other systems though, that may not operate in this way, and an alternative view has been that the carrier operates by consumption of ATP. For example, the K^+/Na^+ antiport system of algal and of animal cell membranes has been considered to be an ATPase in with both K^+ and Na^+ are transported (*6*.6.3.3). In *Neurospora*, Na^+ may replace H^+ to a certain extent (*7*.2) and in barley roots Na^+ efflux develops as H^+ efflux decreases (Part B *3*.*3*.1). Certain ATPase's isolated from root cell membranes are considered to be carriers of K^+ (or Na^+) (Chap. *10*).

4. Energy Requirement for Ion Transport

Rates at which energy is made available from photosynthesis and respiration can be calculated from free-energy changes involved, and from the free energy available from hydrolysis of ATP. The energy required for ion transport can be calculated from the active ion flux and the change in free energy involved. In the steady state the change in free energy due to the pump is dissipated by passive diffusion of the ions in the direction of the free-energy gradient. HOPE and WALKER (1975) have used this approach to compare energy available and used for ion transport in *Chara* and *Griffithsia*.

Energy from photosynthesis was calculated from the flux of CO_2 taken up and the molar free energy change of glucose oxidation divided by 6.

$$\frac{1}{A} \cdot \frac{dG}{dt} = J_{CO_2} \times 2.89 \times 10^6/6 \text{ W m}^{-2} \tag{9.1}$$

For respiration, the rate of hydrolysis of ATP was used as a basis, together with ATP production relative to O_2 uptake.

$$\frac{1}{A} \cdot \frac{dG}{dt} = J_{O_2} \times 33 \times 10^3 \times 38/6 \text{ W m}^{-2} \tag{9.2}$$

(Note that 33 kJ mol^{-1} is used for the free energy of hydrolysis, but this does not take into account the actual concentration of ions in the cytoplasm. WALKER and SMITH (1975) suggest 54 kJ mol^{-1} is more accurate for *Chara corallina*. In this case the energy available from respiration would have been correspondingly higher.)

Ion transport can be approximated from the change in electro-chemical potential.

$$\frac{1}{A} \cdot \frac{dG}{dt} = \phi_{\text{active}} \Delta \bar{\mu} = \phi_{\text{active}} \left(RT \ln \frac{c_1}{c_2} + z_j F \psi_{12} \right) \tag{9.3}$$

These Equations apply to fluxes expressed relative to surface area; other units would be needed for fluxes expressed relative to fresh weight, for example.

For *Chara* HOPE and WALKER calculated that photosynthesis produced 190 mW m^{-2} and respiration 29 mW m^{-2} (or 47 mW m^{-2} if ATP gives 54 kJ mol^{-1}) while ion transport consumed approximately 7.7 mW m^{-2} in the light and 1.8 mW m^{-2} in the dark. For *Griffithsia* photosynthesis was 170 mW m^{-2}, respiration was 26 mW m^{-2} and Cl^- transport consumed 250 μW m^{-2} in the light and 55 μW m^{-2} in the dark. For barley roots we have calculated using the above approach that respiration provides 870 μW g_{FW}^{-1} and ion fluxes at the steady state consume about 50 μW g_{FW}^{-1}. In all these cases the consumption of energy in fluxes (based on free-energy change) is a small proportion of that available from respiration or photosynthesis.

An alternative approach is to calculate the number of ATP molecules needed, making an assumption about the stoichiometry. The conclusions are essentially the same; thus for *Chara* HOPE and WALKER (1975) calculated a requirement of 8.0 mW m^{-2} assuming there was a stoichiometry of 1 ATP:1 Cl^- and 1 ATP:1 Na^+ for fluxes at the plasmalemma. The increased energy was because $\Delta\mu_{Cl}$ was about 23 kJ mol^{-1} whereas hydrolysis of ATP was assumed to be 33 kJ mol^{-1}.

If the energy of hydrolysis of ATP is taken at 54 kJ mol^{-1} then there is enough energy to transfer two ions per ATP. Substitution of values for concentrations and potentials in Eq. (*9*.3) gives estimates of $\Delta\mu$ of about 25 kJ mol^{-1} for most ions. In a study of H^+ fluxes and pH of the cytoplasm of *Chara corallina*, WALKER and SMITH estimated that $\Delta\mu_H$ was 27 kJ mol^{-1} and showed that there was extremely close agreement between observed and predicted ψ_{co} if it were assumed there were $2H^+$:1 ATP and the potential was set up by H^+ extrusion by an ATPase. Similar proposals were made by SLAYMAN et al. (1973).

Examples have been reported, where by comparison of the properties of electrogenic pumps and of energy transfer rates in the cells, a requirement of 30–40% of the total energy by the pumps is calculated (active H^+ flux in Characean cells, *12*.3.1, electrogenic ion pump in *Neurospora crassa*, SLAYMAN et al., 1973). However, in these cases either properties of the electrogenic pump are not unequivocally established (*12*.3.1) or the energy turnover was determined by a doubtful method. SLAYMAN et al. (1973) determine the ATP-turnover by the initial rate of the decrease of the ATP-levels of their cells upon addition of a high concentration of the uncoupler CCCP. Application of this method is based on the assumption that the uncoupler affects phosphorylation immediately while ~P consuming reactions continue unaltered for some while. This assumption is not justified because CCCP acts unspecifically on membranes, and it may immediately affect not only energy-providing but also energy-consuming reactions at membranes (*viz*. energy consumption by the pump itself; activities of intracellular transport metabolite-shuttle systems; synthetic activities of the ER membrane system, e.g. protein synthesis; etc.). But even if under certain conditions, high energy requirements of transport appear to be given, an important feature of active negative feedback (Fig. *11*.2) must be borne in mind: the power consumption is minimal at equilibrium in a steady state, it may be high in transient situations arising in response to perturbation (*11*.2.2).

Since consumption of energy in fluxes is only a small proportion of that available from respiration or photosynthesis, suggestions have been made that the "energy state" of the cell may have some signalling effect on transport. Even with inhibited energy metabolism and reduced bulk energy available, there should be more than enough energy to supply the modest needs of transport. This never is the case, transport is closely correlated to the general energy state of a cell. Conversely, fluxes of ions in giant algal cells react to inhibitors and to light and dark, when there is little change in ATP-level or in "energy charge" (ATKINSON, 1969; LILLEY and HOPE, 1971; see also *6*.6.2.5 and *6*.6.3.3). This re-emphasizes that the "energy state" relevant for transport is not solely determined by high-energy phosphate but alternatively or additionally by electron flow (*9*.2). What the "signalling" effect of the changed energy state may be is obscure unless some other intermediate couples energy source and transport at the membrane level.

The discrepancies between models of clear stoichiometric coupling of energy and transport (e.g. Fig. *9*.1) and the more general interdependence between metabolic energy and transport as obvious in complex biological systems, have often led to unnecessary controversies between thermodynamicists and biologists about usage of the term active transport. The use of the term "metabolic transport" has therefore been suggested in biological systems, whenever the rigorous application of thermodynamic definitions is impossible (e.g. LÜTTGE, 1969, 1973).

References

ATKINSON, E.D.: Adenine nucleotides as universal stoichiometric coupling agents. Advan. Enzyme Regulation **9**, 207–219 (1970).

BARNES, E.M.: Respiration-coupled glucose transport in membrane vesicles from *Azotobacter vinelandii*. Arch. Biochem. Biophys. **152**, 795–799 (1972).

BARNES, E.M., KABACK, H.R.: β-galactoside transport in bacterial membrane preparations: energy coupling *via* membrane-bound D-lactic dehydrogenase. Proc. Natl. Acad. Sci. U.S. **66**, 1190–1198 (1970).

HILL, B.S., HILL, A.E.: ATP-driven chloride pumping and ATPase activity in the *Limonium* salt gland. J. Membrane Biol. **12**, 145–158 (1973).

HOPE, A.B., WALKER, N.A.: Physiology of giant algal cells. Cambridge: Cambridge University Press 1975.

JOHANSEN, C., LÜTTGE, U.: Respiration and photosynthesis as alternative energy sources for chloride uptake by *Tradescantia albiflora* leaf cells. Z. Pflanzenphysiol. **71**, 189–199 (1974).

KABACK, H.R.: Transport. Ann. Rev. Biochem. **39**, 561–598 (1970).

KASHKET, E.R., WILSON, T.H.: Proton-coupled accumulation of galactoside in *Streptococcus lactis* 7962. Proc. Natl. Acad. Sci. U.S. **70**, 2866–2869 (1973).

KEDEM, O.: Criteria of active transport. In: Membrane transport and metabolism. London-New York: Academic Press 1961.

KOMOR, E.: Proton-coupled hexose transport in *Chlorella vulgaris*. F.E.B.S. Letters **38**, 16–18 (1973).

KOMOR, E., TANNER, W.: The nature of the energy metabolite responsible for sugar accumulation in *Chlorella vulgaris*. Z. Pflanzenphysiol. **71**, 115–128 (1974).

LILLEY, R. M.C., HOPE, A.B.: Adenine nucleotide levels in cells of the marine alga, *Griffithsia*. Australian J. Biol. Sci. **24**, 1351–1354 (1971).

LUNDEGÅRDH, H.: An electrochemical theory of salt absorption and respiration. Nature **143**, 203–204 (1939).

LUNDEGÅRDH, H.: The translocation of salts and water through wheat roots. Physiol. Plantarum **3**, 103–151 (1950).

LUNDEGÅRDH, H.: Mechanisms of absorption, transport, accumulation and secretion of ions. Ann. Rev. Plant Physiol. **6**, 1–24 (1955).

LÜTTGE, U.: Aktiver Transport (Kurzstreckentransport bei Pflanzen). Protoplasmatologia, Vol. VIII/7b. Wien-New York: Springer 1969.

LÜTTGE, U.: Structure and function of plant glands. Ann. Rev. Plant Physiol. **22**, 23–44 (1971a).

LÜTTGE, U.: Localized ion transport in complex systems of higher plants as related to respiration and photosynthesis. Proc. 1st European Biophys. Congr. Baden. (E. BRODA, A. LOCKER and H. SPRINGER-LEDERER, eds.), vol. III, p. 353–362, Membranes, transport, p. 119–123. Wien: Verlag der Wiener Medizinischen Akademie 1971b.

LÜTTGE, U.: Stofftransport der Pflanzen. Berlin-Heidelberg-New York: Springer 1973.

LÜTTGE, U.: Co-operation of organs in intact higher plants: A review. In: Membrane transport in plants (U. ZIMMERMANN, J. DAINTY, eds.) p. 353–362. Berlin-Heidelberg-New York: Springer 1974.

LÜTTGE, U., CRAM, W.J., LATIES, G.G.: The relationship of salt stimulated respiration to localized ion transport in carrot tissue. Z. Pflanzenphysiol. **64**, 418–426 (1971).

LÜTTGE, U., OSMOND, C.B.: Ion absorption in *Atriplex* leaf tissue. III. Site of metabolic control of light-dependent chloride secretion to epidermal bladders. Australian J. Biol. Sci. **23**, 17–25 (1970).

MACROBBIE, E.A.C.: The nature of coupling between light energy and active ion transport in *Nitella translucens*. Biochim. Biophys. Acta **94**, 64–73 (1965).

NÉEL, J.: Les membranes artificielles. La Recherche **5**, 33–43 (1974).

OSMOND, C.B., LÜTTGE, U., WEST, K.R., PALLAGHY, C.K., SCHACHER-HILL, B.: Ion absorption in *Atriplex* leaf tissue. II. Secretion of ions to epidermal bladders. Australian J. Biol. Sci. **22**, 797–814 (1969).

ROBERTSON, R.N.: Protons, electrons, phosphorylation and active transport. Cambridge: Cambridge University Press 1968.

SLAYMAN, C.L., LONG, H.S., LU, C.Y.-H.: The relationship between ATP and an electrogenic pump in the plasma membrane of *Neurospora crassa*. J. Membrane Biol. **14**, 305–338 (1973).

SLAYMAN, C.L., LU, C.Y.-H., SHANE, L.: Correlated changes in membrane potential and ATP concentrations in *Neurospora*. Nature **226**, 274–276 (1970).

SUTCLIFFE, J.F.: Mineral salts absorption in plants. Oxford-London-New York-Paris: Pergamon Press 1962.

WALKER, N.A., SMITH, F.A.: Intracellular pH in *Chara corallina* measured by DMO distribution. Plant Sci. Letters **4**, 125–132 (1975).

WEST, I.C., MITCHELL, P.: Proton-coupled β-galactoside translocation in non-metabolizing *Escherichia coli*. J. Bioenergetics **3**, 445–462 (1972).

WEST, I.C., MITCHELL, P.: Stoichiometry of lactose—H^+ symport across the plasma membrane of *Escherichia coli*. Biochem. J. **132**, 587–592 (1973).

10. ATPases Associated with Membranes of Plant Cells

T.K. HODGES

1. Introduction

Adenosine triphosphatases (ATPases) have been implicated as the energy-transfer agents for the transport of inorganic ions in animal and bacterial cells. Animal cells have a $(Na^{+}+K^{+})$-ATPase on the plasma membrane (SKOU, 1965; HOKIN and DAHL, 1972), a (Ca^{2+})-ATPase on the sarcoplasmic reticulum of muscle cells (MARTONOSI, 1972), and an ion translocating ATPase on the inner mitochondrial membrane (CEREIJO-SANTALÓ, 1970). Bacterial cells have a (K^{+})-ATPase on their plasma membrane (ABRAMS et al., 1972; ROTHSTEIN, 1972). The evidence is now convincing that each of these ATPases functions as an energy transducer in the transport of ions across membranes.

Are ATPases involved in ion transport in plant cells? This question may seem presumptuous since it has only been demonstrated for a few systems that ATP can serve as the energy source for ion transport in plant cells (see Chaps. *4, 6, 7* and *9*). Thus, it could be argued that a search for a transport-ATPase in plant cells is like putting the cart before the horse. However, it can also be argued that the strongest evidence that could be obtained for implicating ATP in ion transport would be to demonstrate that the ion transport system itself can use ATP. Thus, an affirmative answer to the above question would not only be relevant to the identity of the energy source for ion transport, but it would also provide information about the mechanism of ion transport. It is in this spirit that several efforts have been made to determine whether ATPases exist in plant cells and whether these enzyme activities have any relationship to ion transport.

Plant cells do contain several ATPases (POUX, 1967; HALL, 1971b; LEONARD et al., 1973), and it now appears that some of these enzymes are involved in ion transport. The ATPase of mitochondria clearly participates in the transport of ions into these organelles (HODGES and HANSON, 1965; ELZAM and HODGES, 1968). Another ATPase is associated with the plasma membrane[1] (HODGES et al., 1972; WILLIAMSON and WYN JONES, 1972), and there is substantial evidence that this enzyme mediates cation transport across the plasma membrane (HODGES, 1973). A third ATPase, which is especially sensitive to anions, may be associated with the tonoplast (RUNGIE and WISKICH, 1973; BALKE et al., 1974). Although there is no direct evidence that links the latter ATPase to ion transport, the unique sensitivity of the enzyme to anions suggests it may also have a transport function.

This Chapter will consider investigations dealing with ATPases that might play a role in ion transport across the plasma membrane or tonoplast of plant cells. The role of the mitochondrial ATPase in ion transport (HANSON and HODGES, 1967) will be considered in volume 3.

---

[1] "Plasma membrane" is used commonly instead of "plasmalemma" in studies of ATPases in plant cells, and is used in this Chapter with that meaning.

2. Difficulties in Establishing a Role for ATPases in Ion Transport

It is difficult to establish that an enzyme or enzyme complex has a specific and vital role in any physiological process. The primary reason for this is that physiological processes are studied in cells, tissues, organs, or organisms, whereas enzyme studies usually require cell-free extracts. To extrapolate conclusions based on *in vitro* results to the intact system is risky, and it must be done with caution. This is one of the major difficulties in establishing that an ATPase is involved in ion absorption. However, the problem is not insurmountable. Other physiological processes (photosynthesis and respiration are classic examples) have been elucidated most clearly by studies that employed subcellular fractions.

There are other problems in showing that an ATPase is involved in ion absorption. For example, it is not obvious how to identify a transport-ATPase of plants. In animal cells, the transport ATPase of the plasma membrane was identified initially by its being stimulated synergistically by Na^+ and K^+ and by its inhibition with ouabain (SKOU, 1957). This followed logically because it was known that K^+ influx was coupled to Na^+ efflux (K^+/Na^+ exchange) and that ouabain inhibited the coupled transport of these ions (SCHATZMANN, 1953). In plant cells, however, transport is more complex since both the plasma membrane and tonoplast are involved and several ions appear to be actively pumped (HIGINBOTHAM, 1973). For example, specific transport processes appear to be involved for K^+ and Cl^- transport at the plasma membrane (Chaps. *6*, Part B, Chap. *3.3*), H^+ efflux (Chap. *12*), divalent cation fluxes (Part B, Chap. *3.3*) as well as reduction sites for converting Fe^{3+} to Fe^{2+} (Part B, Chap. *3.3*). It is also probable that specific bidirectional exchanges occur such as K^+/Na^+, K^+/H^+, Cl^-/OH^-, Cl^-/HCO_3^-, etc. (Chaps. *6, 7, 12;* Part B, Chap. *3.3*). Furthermore, there is no known inhibitor which will specifically block ion transport in plant cells i.e. one that does not affect energy conservation or protein synthesis. Without knowing these aspects of ion transport a definitive test for a transport-ATPase simply does not exist.

In attempts to find a transport-ATPase in plant cells, two criteria have been used to test for the ATPase. Most investigators have assumed that a sensitivity (either a stimulation or inhibition) to single salts of monovalent ions would be an indication of a potential transport function (DODDS and ELLIS, 1966; GRUENER and NEUMANN, 1966; FISHER and HODGES, 1969; RUNGIE and WISKICH, 1973). Others have assumed that a transport-ATPase of plant cells should be similar to the transport-ATPase of mammalian cells, and therefore the plant transport-ATPase should be stimulated synergistically by Na^+ and K^+ (HANSSON and KYLIN, 1969; KYLIN and GEE, 1970). Both criteria could be justly criticized, however, considering our lack of information about the transport processes, and the absence of a specific inhibitor of ion transport, both of these initial criteria represented reasonable assumptions.

Another big problem in the investigations of ATPases of plants is that plant cells contain many enzymes capable of hydrolyzing ATP. Furthermore, these ATPases seem to be ubiquitous (LEONARD et al., 1973). They are present on nearly all membranes (POUX, 1967; HALL, 1971b; LEONARD et al., 1973) as well as in the soluble or supernatant fraction of cell extracts (ATKINSON and POLYA, 1967; FISHER and HODGES, 1969; GRUENER and NEUMANN, 1966). Thus, most measurements of ATPases represent the activities of several different ATP hydrolyzing enzymes. Furthermore, different experimental techniques, such as homogenizing procedures, centrifugation schedules, etc. undoubtedly give rise to subcellular fractions enriched in particular ATPases. The presence of different membranes in a fraction must be responsible for many of the divergent reports on the characteristics of plant ATPases.

3. ATPase Activity of Soluble Fractions

The supernatant or soluble fraction that remained after plant extracts were centrifuged at high gravity forces had the capacity for hydrolyzing ATP (BROWN and ALTSCHUL, 1964; GRUENER and NEUMANN, 1966; FISHER and HODGES, 1969; KASAMO and YAMAKI, 1974). This fraction of soluble enzymes contained much greater ATP hydrolyzing activity than the membranes pelleted by the centrifugation (GRUENER and NEUMANN, 1966; FISHER and HODGES, 1969). The ATPase activities of the soluble fraction have been reported to possess both acidic and alkaline pH optima and to be either stimulated or inhibited by inorganic salts.

An acidic pH optimum has been reported for the ATPase activity of soluble fractions from several different plants (e.g. peanut seedlings, BROWN and ALTSCHUL, 1964, BROWN et al., 1965, carrots, beets, and *Chara,* ATKINSON and POLYA, 1967, and oat roots FISHER and HODGES, 1969). The ATP hydrolyzing activity of soluble fractions of carrots, beets, *Chara,* and oats, was inhibited by inorganic salts. On the other hand, the ATPase activity of a similar fraction from peanut seedlings was stimulated by Mg^{2+} as well as by NaCl or KCl. In the latter study, however, the centrifugation forces were low (20,000 g for 20 min), and it was probable that the supernatant fraction still contained several membrane fragments (BROWN and ALTSCHUL, 1964). KASAMO and YAMAKI (1974) have separated the soluble fraction of mung bean hypocotyls into 3 different phosphatases. Two of these enzymes were inhibited while one of them was stimulated by Mg^{2+}. The investigations by ATKINSON and POLYA (1967) represent the most thorough characterization of the soluble ATPase activity that has the acidic pH optimum. They showed by gel electrophoresis that the enzyme primarily responsible for hydrolyzing ATP was a non-specific acid-phosphatase. This also appeared to be true for the soluble ATPase activity of oat roots because several phosphate esters, including nucleoside diphosphates, were hydrolyzed (FISHER et al., 1970).

Neutral to alkaline pH optima were found for the soluble ATPase activity in other plants. This was true for roots of bean (GRUENER and NEUMANN, 1966; HOROVITZ and WAISEL, 1970), barley (HALL, 1971a) maize, *Atriplex,* and *Suaeda* (HOROVITZ and WAISEL, 1970). Several phosphate esters including ATP, GTP, ITP, and ADP, but not AMP and p-nitrophenyl phosphate, were hydrolyzed by the soluble enzyme(s) of bean roots (GRUENER and NEUMANN, 1966). In beans, barley, and maize, the soluble ATPase activity was stimulated by inorganic salts including $MgCl_2$, $CaCl_2$, NaCl, and KCl. However, in the two halophytes, *Atriplex* and *Suaeda,* NaCl inhibited the soluble ATPase activity (HOROVITZ and WAISEL, 1970). From the latter study it was inferred that the opposite responses to NaCl by the ATPases obtained from halophytic and glycophytic plants might be related to different ion transport properties of these two groups of plants. This is an interesting possibility, but much additional information is needed about the enzymes involved and the ion transport properties of these plants. It is probable that most of the soluble ATPase activity that exhibits the high pH optimum is due to an apyrase enzyme (GRUENER and NEUMANN, 1966) or to a non-specific alkaline-phosphatase.

There is insufficient information about the soluble ATPases to assess their involvement in ion transport. One reason for the different responses to pH and inorganic ions is that genuine differences probably exist in the soluble enzymes of different plant species. For example, different plants surely contain different amounts of acid- and alkaline-phosphatases. In addition, some of the supernatant phosphatase activity may be due to ATPases of membrane origin. As pointed out in *10*.4 ATPases associated with membranes also exhibit different pH optima and sensitivity to inorganic ions. These enzymes could contaminate the supernatant by either being stripped off the membranes during homogenization and centrifugation, or they could exist on small membrane fragments that did not sediment at the centrifugation forces employed. Regardless of the basis for the contrasting results, two points are obvious: (1) the supernatant fraction remaining after high-speed centrifugation of plant extracts does exhibit considerable ATP hydrolyzing activity, which is probably due to the presence of several enzymes, and (2) there is no evidence to date that these enzymes function in ion absorption.

4. ATPase Activity of Membrane Fractions

4.1 General Features

If ATPases are involved in ion transport across the plasma membrane and tonoplast, they are probably integral parts of these membranes. However, until recently (HODGES et al., 1972; WILLIAMSON and WYN JONES, 1972), it was not possible to isolate either of these membranes from root cells. Thus, the experimental approach taken initially was to determine whether any membrane preparation exhibited ATPase activity.

All membrane preparations obtained by differential centrifugation of cell-free extracts of plant tissue did contain ATPase activity (BROWN et al., 1965; GRUENER and NEUMANN, 1966; FISHER and HODGES, 1969; HANSSON and KYLIN, 1969). The membrane fractions examined included those obtained at low gravitational forces (frequently referred to as a cell-wall fraction), intermediate gravitational forces (consisting mainly of mitochondria), and high gravitational forces (generally called a microsomal fraction). All of these fractions were heterogeneous, i.e. they contained several different kinds of membranes. Nevertheless, the initial findings of ATPase activity in these crude membrane fractions provided encouragement that the plasma membrane and/or tonoplast might contain specific ATPases.

4.2 Non-Purified Membrane Preparations

4.2.1 Effects of Divalent Cations

The ATPase activity of all membrane preparations was generally stimulated by certain divalent cations (Table *10*.1). The most effective cations were Mg^{2+}, Ca^{2+}, and Mn^{2+}. Most often, Mg^{2+} stimulated the ATPase activity more than the other divalent cations, but Ca^{2+} stimulated more than Mg^{2+} the ATPase activity of a cell wall preparation of barley roots (HALL and BUTT, 1969) and a microsomal preparation of wheat roots (KYLIN and KÄHR, 1973).

The pH optimum for ATPase activity of membranes in the presence of divalent cations depended on the particular membrane fraction examined, the salt status of the tissue, and the plant species. Thus, the large differences reported (Table *10*.1) may be explained by the different conditions used to grow the plants and to genuine differences between species. In addition, the diverse results may have resulted from the use of membrane preparations that were heterogeneous, but enriched in a particular type of membrane. For example, oat roots were found to have at least five different membranes that exhibited ATPase activity, and these ATPases had different pH optima and sensitivity to inorganic ions (LEONARD et al., 1973; HODGES and LEONARD, 1974; see *10*.4.3). The various membrane-associated ATPases were distinguished and partially purified by both differential and density gradient centrifugation. It was impossible to obtain a membrane preparation by differential centrifugation that contained fewer than two different kinds of membranes. Thus, some of the differences noted above in the ATPase activity of membranes prepared by differential centrifugation must have been because different enzymes were being measured.

4.2.2 Effects of Monovalent Ions

If ATPases are involved in ion transport, the enzymes should be capable of recognizing and responding to the ions that are transported. The response of an ATPase to such ions would presumably be a stimulation in the activity of the enzyme. This occurs for the transport-ATPase of animal (SKOU, 1965) and bacterial (ABRAMS et al., 1972) cells.

ATPase activity of crude membrane preparations from plant roots and other plant tissues was generally found to be stimulated by salts of monovalent ions (Table *10*.1). The monovalent ion stimulated ATPase activity occurred only in the presence of divalent cations; Mg^{2+} and Mn^{2+} were the most effective (DODDS and ELLIS, 1966; FISHER and HODGES, 1969). Ca^{2+} did not substitute for the Mg^{2+} or Mn^{2+} requirement (FISHER and HODGES, 1969), except possibly at very high pH (HALL and BUTT, 1969).

Many salts of monovalent ions were capable of stimulating the ATPase activity of membranes. Examples were KCl, KBr, KNO_3, KNO_2, NaCl, RbCl, LiCl, NH_4Cl, CsCl, *tris*-Cl, choline-Cl, tetramethylammoniumchloride, ethanolamine-Cl, K-acetate, K_2SO_4, NaBr, Na-succinate, Na-malate and Na-oxaloacetate (see Table *10*.1). Few attempts have been made to determine whether the stimulation of ATPase activity by monovalent salts was caused by the cation or the anion: DODDS and ELLIS (1966) and FISHER and HODGES (1969) concluded that cations were responsible

Table *10*.1. ATPase activity of non-purified membrane fractions of plants

Plant, common and latin name	Plant, organ and salt status[a]	Membrane fraction (centrifuge)	Effects of divalent cations[a]	Effects of monovalent ions or salts[a]	pH	Ref.
Oats (*Avena sativa*)	roots low salt	1,500 g 12,000 g 100,000 g	(+)Mg=Mn>Ca (+)Mg (+)Mg	(+)K>Rb>Na>Cs>NH_4>Li (+)KCl (+)KCl	8–8.5 – –	FISHER and HODGES (1969), FISHER et al. (1970)
	roots low salt	1,500 g 12,000 g 100,000 g	(+)Mg	(+)KCl>*tris*-Cl>choline Cl>ethanolamine Cl>NaCl	8.2	RATNER and JACOBY (1973)
	roots low salt	10,000 g to 30,000 g	(+)Mg>Ca	–	5.2–7.7	KYLIN and KÄHR (1973)
Wheat (*Triticum vulgare*)	roots high salt	10,000 g to 30,000 g	(+)Ca>Mg	–	5.2–8.3	KYLIN and KÄHR (1973)
(*Triticum aestivum*)	roots low salt	10,000 g to 30,000 g	(+)Ca>Mg	–	5.2–8.3	KYLIN and KÄHR (1973)
	roots low salt	1,500 g 12,000 g 100,000 g	(+)Mg (+)Mg (+)Mg	(+)Rb (+)Rb (+)Rb	7.2 7.2 7.2	FISHER et al. (1970)
Barley (*Hordeum vulgare*)	roots –	cell walls	(+)Ca>Mg	(+)NaCl>KCl (in presence of Ca)	7–9	HALL and BUTT (1969)
	roots	cell walls	(+)Ca	Na>Li>K>Rb	9.0	HALL (1971a)
	roots low salt	1,000 g 12,000 g 100,000 g	(+)Mg (+)Mg (+)Mg	(+)Rb (+)Rb (+)Rb	7.2 7.2 7.2	FISHER et al. (1970)
Turnip (*Brassica rapa*)	roots	27,000 g to 78,000 g	(+)Mg	(+)KCl, NaCl, *tris*-Cl, choline Cl, CsCl, RbCl, LiCl, NH_4Cl, NaBr, $NaHCO_3$, K—CH_3COO, NaF, Na-malate, Na-succinate, Na-fumarate Na-tartrate, Na-aspartate, Na-oxalate (—)KI, K_2SO_4, KNO_3, Na—$C_6H_5SO_3$	7.5–9.0	RUNGIE and WISKICH (1973)

[a] A (+) or (−) indicates a stimulation or inhibition, respectively.

Table *10*.1 (continued)

Plant, common and latin name	Plant, organ and salt status	Membrane fraction (centrifuge)	Effects of divalent cations[a]	Effects of monovalent ions or salts	pH	Ref.
Beans (*Phaseolus vulgaris*)	roots	20,000 g	+Mg	(−)NaCl	9.6	GRUENER and NEUMANN (1966)
		120,000 g	+Mg	(+)NaCl	9.6	
	cotyledons	unidentified membrane	(+)Mg	(+)K, Na	8.0	LAI and THOMPSON (1971, 1972)
(*Phaseolus vulgaris*)	roots	1,500 g	(+)Mg	(+)KCl	7.0	KUIPER (1972)
		12,000 g	(+)Mg	(+)KCl		
		100,000 g	(+)Mg	(+)KCl		
Mung Beans (*Phaseolus mungo*)	hypocotyls	1,500 g	(+)Mg	–	6.5	KASAMO and YAMAKI (1974)
		20,000 g	(+)Mg	–		
		105,000 g	(+)Mg	–		
Maize (*Zea mays*)	roots	1,500 g	(+)Mg	(+)choline-Cl > KCl > *tris*-Cl > ethanolamine-Cl > NaCl	8.2	RATNER and JACOBY (1973)
		12,000 g	(+)Mg	(+) same as above		
		100,000 g	(+)Mg	(+) ethanolamine-Cl > NaCl > *tris*-Cl		
	root	1,500 g	(+)Mg	(+)KCl	8.0	LEONARD and HANSON (1972)
		12,000 g	(+)Mg	(+)KCl	8.0	
		80,000 g	(+)Mg	(+)KCl	6.8	
	root low salt	cell walls	(−)Mg, Ca	–	7.0	EDWARDS and HALL (1973)
		nuclei	(−)Mg, Ca	–	6.2	
		mito-chondria	(+)Mg, Ca	–	8.5–9.0	
	root	500 to 100,000 g	(+)Mg	(+)K	6.5	WILLIAMSON and WYN JONES (1971)
	root low salt	1,500 g	(+)Mg	(+)Rb	7.2	FISHER et al. (1970)
		12,000 g	(+)Mg	(+)Rb		
		100,000 g	(+)Mg	(+)Rb		

Table *10*.1 (continued)

Plant, common and latin name	Plant, organ and salt status	Membrane fraction (centrifuge)	Effects of divalent cations[a]	Effects of monovalent ions or salts	pH	Ref.
Sugar Beet (*Beta vulgaris*)	roots	2,500 g to 20,000 g	(+)Mg	(+)KCl > NaCl (+)(Na+K)	5.75 to 6.6	HANSSON and KYLIN (1969), KYLIN et al. (1972)
	conducting bundles of leaf	dialyzed homog.	(−)Mg (+)Ca	+KCl, NaCl	5.4	BOWLING et al. (1972)
		1,000 g	(+)Mg	+KCl, NaCl		
		20,000 g	(+)Mg	+KCl, NaCl		
		201,000 g	(+)Mg	+KCl, NaCl		
	root	4,000 g to 20,000 g	(+)Mg	(+)K+Na	6.75	HANSSON et al. (1973)
Gourd (*Cucurbita* sp.)	roots	1,000 g	(+)Mg	(+)NaCl, KCl	7.2	KRASAVINA and VYSKREBENTSEVA (1972)
		4,000 g	(+)Mg	(+)NaCl, KCl		
		20,000 g	(+)Mg	(+)NaCl, KCl		
		105,000 g	(+)Mg	(+)NaCl, KCl		
Cow parsnip (*Heracleum sosnovskyi*)	phloem of leaf petiole	dialyzed homog.	(−)Mg, (+)Ca	(+)KCl, NaCl	5.4	BOWLING, et al. (1972)
Peas (*Pisum sativum*)	roots	homog.	(+)Mg=Ca	(−)Na	7.0	SEXTON and SUTCLIFFE (1969)
	roots	cell walls	(+)Mg, Ca, Mn, Fe	(+)K, Rb, Li, NH_4	4.5	DODDS and ELLIS (1966)
Sea Lettuce (*Ulva lactuca*)	—	homog.	(+)Mg	No effect of K and Na	7.5	BONTING and CARAVAGGIO (1966)

for stimulating the ATPase activity of cell wall preparations from pea and oat roots, respectively. However, recent reports suggested that anions were responsible for stimulating the ATPase activity of membrane preparations from leaves of *Limonium* (HILL and HILL, 1973), turnip roots (RUNGIE and WISKICH, 1973) and oat roots (BALKE et al., 1974). It appears from these results that both cation-sensitive and anion-sensitive ATPases are present on plant cell membranes and that measurement of ATPase activity of heterogeneous membrane preparations includes both of these enzymes.

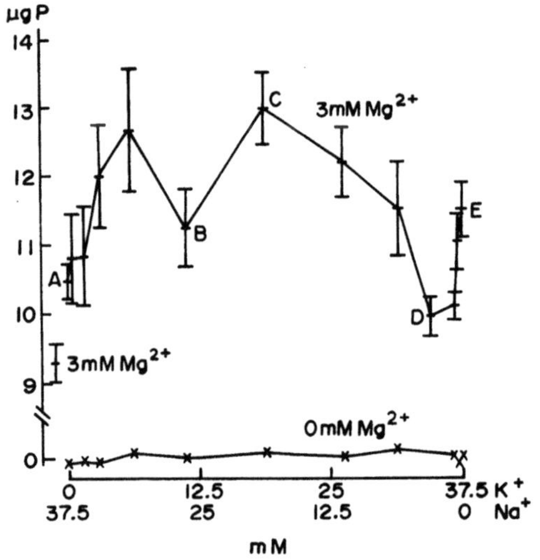

Fig. *10*.1. The effect of various K^+ and Na^+ concentrations on the ATPase activity of a membrane preparation from sugar beet roots. *A*, *B*, *C*, *D*, and *E* correspond to mixtures of K^+ and Na^+ that result in either a minimum or maximum in ATPase activity. (From HANSSON and KYLIN, 1969)

In nearly all investigations, individual salts of monovalent ions were as effective in stimulating ATPase activity as were various combinations of salts. However, KCl and NaCl stimulated synergistically the ATPase activity of membrane preparations from beet roots (HANSSON and KYLIN, 1969), mangrove leaves (KYLIN and GEE, 1970), conducting bundles of sugar beet petioles, phloem of cow parsnip (BOWLING et al., 1972), and roots of gourd (KRASAVINA and VYSKREBENTSEVA, 1972). In all of these studies, however, the ATPase activity in the presence of Mg^{2+} + NaCl + KCl was only slightly, if any, greater than in the presence of Mg^{2+} + NaCl, or Mg^{2+} + KCl. However, significant stimulations by NaCl and KCl together were found by KYLIN and associates (Fig. *10*.1). Two different combinations of NaCl and KCl gave greater ATPase activity than either NaCl or KCl alone in membranes isolated from sugar beets (HANSSON and KYLIN, 1969). These synergistic responses to NaCl and KCl were induced by pretreating the membrane preparations with 0.1% deoxycholate for 1 h. Subsequently (KYLIN et al., 1972), it was shown that deoxycholate extracted over 50% of the membrane lipids. Thus, this treatment may have exposed sites on the ATPase(s) that were originally "buried" within the membrane. Although the findings of a $(Na^+ + K^+)$-ATPase in crude membrane preparations of plants is of interest, a more thorough characterization of this enzyme activity is needed in order to assess whether it plays any role in ion transport.

Both acid and alkaline pH optima have been reported for the monovalent ion stimulation of membrane-bound ATPases (Table *10*.1). For example, pH values from 4–7 gave the greatest stimulation of ATPase activity in a cell-wall fraction of peas and carrots (DODDS and ELLIS, 1966), in dialyzed crude homogenates of phloem tissue of cow parsnips and homogenates of conducting bundles of sugar beets (BOWLING et al., 1972), and in a microsome fraction of oat roots (LEONARD et al., 1973). On the other hand, pH values above 7 resulted in the greatest stimulations by monovalent ions of cell wall fractions from oat roots (FISHER and HODGES, 1969) and barley roots (HALL and BUTT, 1969), and in microsome fractions from corn roots (RATNER and JACOBY, 1973) and turnip roots (RUNGIE and WISKICH, 1973). The stimulation of ATPase activity by KCl at both low and high pH is illustrated in Fig. *10*.2.

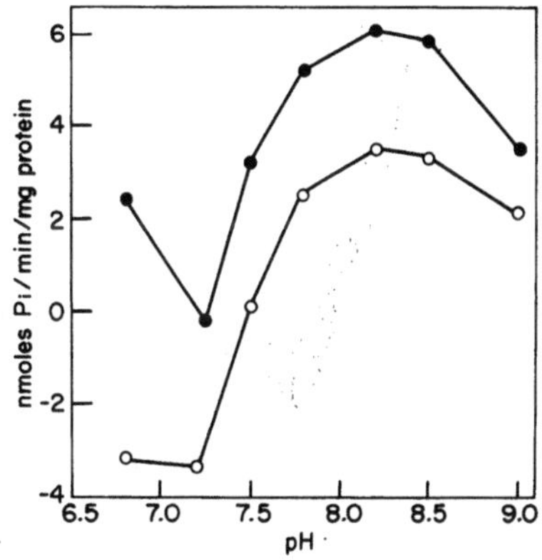

Fig. *10*.2. The effect of pH on the ATPase activity of a membrane preparation of turnip roots. Open symbols represent the activities due to 2 mM $MgSO_4$ and the closed symbols represent the activities due to 2 mM $MgSO_4$ and 50 mM KCl. (From RUNGIE and WISKICH, 1973)

As already pointed out, crude membrane preparations were used for all of the above studies and some of the fractions were probably enriched in a particular ATPase while other fractions were undoubtedly enriched in a different ATPase. Recent studies using more purified membrane fractions support this interpretation.

4.3 Semi-Purified Membrane Preparations

Several different membranes of oat and corn roots have been partially purified and they all appear to possess ATPase activity (HODGES et al., 1972; WILLIAMSON and WYN JONES, 1972; LEONARD et al., 1973). The procedures employed for the purification of oat root membranes involved both continuous and discontinuous sucrose gradients; the latter is illustrated in Fig. *10*.3. With the discontinuous gradient shown in Fig. *10*.3, a crude microsomal fraction can be separated into 6 different membrane fractions. Each of these membrane fractions exhibited ATPase activity (Table *10*.2). Based on the ATPase activity at different pH's and the sensitivity to ions, membrane density, chemical composition, staining properties of the membranes, and the association of other enzyme activities, the various membrane fractions have been concluded to be enriched in specific membranes (Table *10*.3). None of the membrane fractions contain a single type of membrane and further purification studies are needed. However, fractions E and F were found to consist of more

Table 10.2. ATPase activity at pH 6.0 and 9.0 in the presence and absence of 50 mM KCl for various membrane fractions of oat roots. (Adapted from HODGES et al., 1972; LEONARD et al., 1973)

	ATPase activity							
	pH 6.0				pH 9.0			
	−KCl	+KCl	ΔKCl	Enrichment	−KCl	+KCl	ΔKCl	Enrichment
	(μmol mg protein^{-1} h^{-1})				(μmol mg protein^{-1} h^{-1})			
Homogenate	36.65	40.24	4.59	–	21.42	31.02	9.62	–
13K Pellet	5.29	14.62	9.34	2.0	35.33	62.08	26.75	2.8
13–80K Pellet	11.19	26.45	15.26	3.3	14.17	23.08	8.91	0.9
(13–80K overlay) *Gradient fractions*								
A	4.68	7.38	2.70	0.6	5.52	7.59	2.07	0.2
B	7.96	14.88	6.92	1.5	10.55	17.19	6.64	0.7
C	8.96	18.15	9.19	2.0	21.14	30.28	9.18	1.0
D	10.32	30.25	19.93	4.3	24.39	38.06	13.67	1.4
E	14.19	46.30	32.11	7.0	17.11	26.96	9.85	1.0
F	13.98	49.10	35.12	7.7	13.11	20.50	7.39	0.8

Table 10.3. Properties and probable composition of membrane fractions of oat roots separated on discontinuous sucrose gradients

Membrane fraction[a]	Approximate membrane density[b]	Sterol/phospholipid (molar basis)[c]	Major enzyme enrichments	Probable membrane distribution: Major	Probable membrane distribution: Minor
A	1.08	0.308	NADPH-cytochrome-c-reductase[d]	–	endoplasmic reticulum
B	1.10	0.226	NADH-cytochrome-c-reductase[d]	endoplasmic reticulum	tonoplast Golgi membranes
C	1.13	0.537	IDPase[d], pyrophosphatase[d], Cl-ATPase[e]	Golgi membranes	endoplasmic reticulum, tonoplast
D	1.15	0.837	High pH-ATPase[d]	–	Golgi membranes plasma membrane tonoplast
E	1.17	1.074	K^+-ATPase[c,d], glucan synthetase[f], malic dehydrogenase[g]	plasma membrane	mitochondrial fragments
F	1.20	1.213	K^+-ATPase[c,d], glucan synthetase[f], malic dehydrogenase[g]	plasma membrane	mitochondrial fragments

[a] The membrane fractions correspond to the fractions shown on the discontinuous sucrose gradient of Fig. 3. – [b] From LEONARD et al. (1973) and HODGES and LEONARD (1974). – [c] From HODGES et al. (1972). – [d] From LEONARD et al. (1973). – [e] From BALKE et al. (1974). – [f] From HODGES and LEONARD (1974). – [g] From D.L. HENDRIX and T.K. HODGES, unpublished.

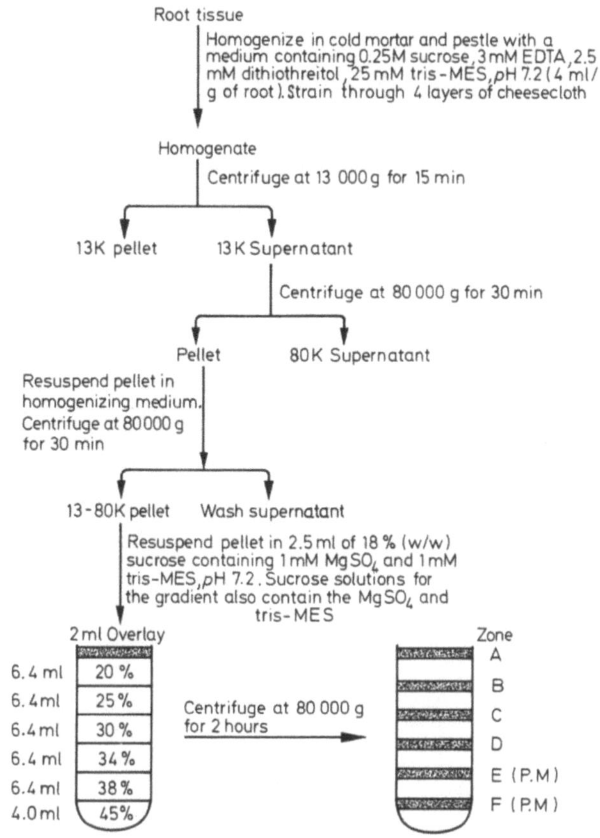

Fig. *10*.3. Flow sheet describing the procedure for isolating plasma membrane vesicles from oat roots

than 75% plasma membrane vesicles (HODGES et al., 1972). The identity of the other membranes in this fraction is unknown but one very minor contaminant is inner mitochondrial membranes (D.L. HENDRIX and T.K. HODGES, unpublished). The ATPase of the inner mitochondrial membrane does not contribute significantly when measuring the plasma membrane ATPase because the pH optima of the two enzymes are very different (D.L. HENDRIX and T.K. HODGES, unpublished).

4.3.1 Plasma Membrane

The membrane fractions of oat roots that were enriched in KCl-ATPase, pH 6.0, (Table *10*.2) were identified as plasma membrane vesicles by their associated glucan synthetase activity, their high sterol: phospholipid ratio, and by their ability to stain with phosphotungstic acid-chromic acid (HODGES et al., 1972). The latter stain is specific for the plant plasma membrane (ROLAND, 1969; ROLAND et al., 1972). The plasma membrane vesicles are of various sizes with an average diameter of about 0.3 µm (Fig. *10*.4). The size of the vesicles indicates that the plasma membrane broke into many small pieces during the isolation procedure.

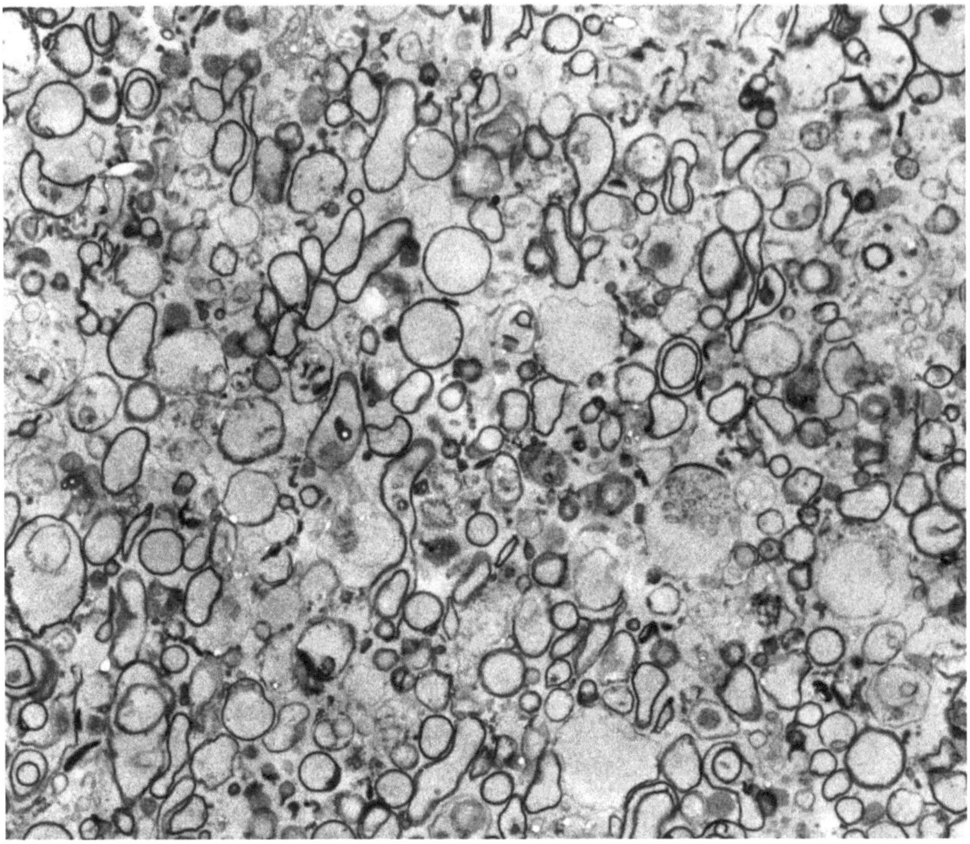

Fig. *10*.4. Electron micrograph of plasma membrane vesicles isolated from oat roots. Membranes were stained with a mixture of phosphotungstic acid and chromic acid. The magnification is 23,000 ×. (From HODGES et al., 1972)

The ATPase on the plasma membrane of oat roots had a pH optimum of 7.5 in the presence of Mg^{2+} and 6.5 in the presence of Mg^{2+} and KCl (LEONARD and HODGES, 1973, see Fig. *10*.5). A slightly lower pH optimum was found for the ATPase of plasma membranes of maize roots (WILLIAMSON and WYN JONES, 1972). Other divalent cations stimulated the plasma membrane ATPase of oat roots with the order of effectiveness being $Mg^{2+} = Mn^{2+} > Zn^{2+} > Fe^{2+} > Ca^{2+}$. Calcium activation was negligible suggesting that the stimulation of ATPase activity by Ca^{2+} in crude membrane preparations (see *10*.4.2.1) was caused by ATPase on other membranes (FISHER and HODGES, 1969; HALL and BUTT, 1969; KYLIN and KÄHR, 1973). In support of this interpretation, the ATPases of a mitochondrial fraction and of fraction D (Fig. *10*.3) were stimulated by Ca^{2+} (LEONARD and HODGES, 1973). The ATPase of the plasma membrane was actually inhibited by Ca^{2+} (BALKE et al., 1974). This inhibition was thought to be the result of the vesicles being inside-out i.e. the cytoplasmic side of the membrane being exposed directly to the experimental solution.

The Mg^{2+} requirement by the ATPase on the plasma membrane was partially explained by Mg-ATP being the substrate for the enzyme (Fig. *10*.6). The K_m values

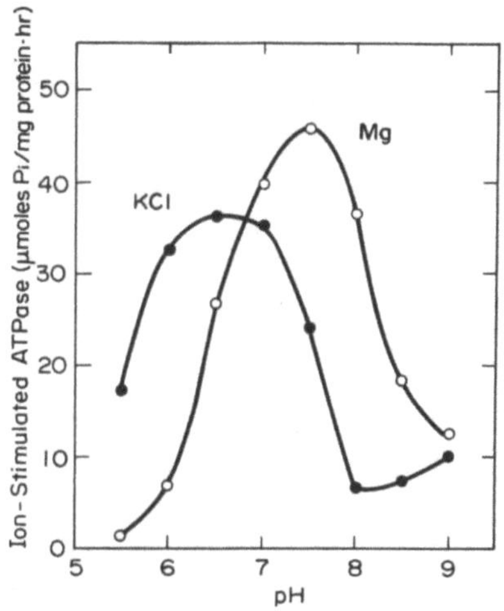

Fig. *10*.5. The effect of pH on the Mg^{2+}- and KCl-stimulated components of the ATPase that is associated with the plasma membrane of oat roots. (From LEONARD and HODGES, 1973)

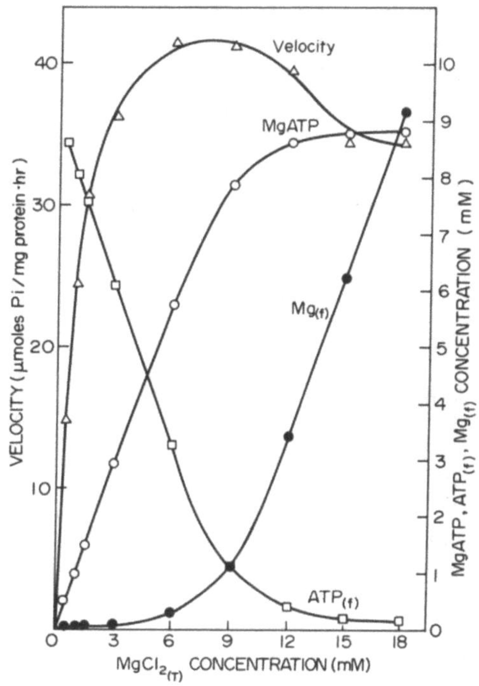

Fig. *10*.6. ATPase activity of the plasma membrane of oat roots as a function of total concentration of $MgCl_{2(T)}$. The concentrations of free Mg^{2+} and free $ATP_{(f)}$ as well as complexed Mg·ATP are shown. (From BALKE and HODGES, 1975)

for Mg^{2+} and ATP were 0.84 and 0.38 mM, respectively (LEONARD and HODGES, 1973); however, these values depended on the experimental conditions. The reason for this was because free ATP at low concentrations, as well as Mg-ATP and $MgCl_2$ at high concentrations, inhibited the ATPase, and the relative amounts of these components altered enzyme activity (BALKE and HODGES, 1975). As a result, the K_m values for Mg-ATP, under different experimental conditions, varied from 0.68 mM to 1.24 mM. Other nucleoside triphosphates were not hydrolyzed appreciably by this enzyme (Table *10*.4; HODGES et al., 1972; HODGES and LEONARD, 1974). Table *10*.4 also shows that the plasma membrane has negligible phosphatase, cytochrome-c-oxidase, NADH- and NADPH-cytochrome-c-reductase activities.

Monovalent cations specifically stimulated the plasma membrane ATPase (BALKE et al., 1974); anions, other than toxic ones like F^- or I^-, had little effect (LEONARD and HODGES, 1973). This ATPase was not stimulated synergistically by Na^+ and K^+ (Fig. *10*.7), but rather showed a definite order of specificity toward the alkali cations (Fig. *10*.8). These responses to the alkali cations and the lack of a ($Na^+ + K^+$) stimulation, illustrate that the plasma membrane ATPase of oat roots is qualitatively different from the plasma membrane ATPase of mammalian cells. The basis for these differences is as yet unknown but may be the result of their different ionic environments during evolution.

The kinetics of the alkali cation stimulation of the plasma membrane ATPase has been described as negatively cooperative (LEONARD and HODGES, 1973). This suggested that the ATPase consists of subunits that interact when cation binding

Table 10.4. Substrate specificity of the plasma membrane ATPase of oat roots. (Adapted from HODGES et al., 1972; HODGES and LEONARD, 1974)

Substrate	Enzyme activity (μmoles substrate utilized/$mg_{protein}$/h)	
	Homogenate	Plasma membrane
ATP[a]	4.59	36.91
ITP[a]	6.13	1.43
UTP[a]	9.19	1.42
GTP[a]	2.84	1.29
CTP[a]	0.66	0.43
ADP	30.30	0.18
IDP	29.04	1.38
UDP	28.62	0.66
CDP	29.76	1.02
IDP	29.52	0.84
AMP	2.28	0.24
IMP	2.28	0.18
UMP	2.28	0.30
CMP	2.82	0.36
IMP	3.12	0.60
Glucose-6-Phosphate	8.28	0.60
pyrophosphate	46.68	5.82
p-nitrophenylphosphate (pH 5.0)	39.66	1.98
p-nitrophenylphosphate (pH 9.0)	3.78	0.42
Glucan Synthetase	0.175[b]	0.696
Cytochrome Oxidase	19.68	6.06
NADH[c]	7.08	2.16
NADPH[c]	2.16	0.84

[a] The nucleoside triphosphatase activity is that due to the stimulation by 50 mM KCl e.g. the activity in the presence of 3 mM $MgSO_4$ was subtracted.
[b] Homogenate data for glucan synthetase was for a total membrane fraction which sedimented at 80,000 g for 30 min.
[c] Cytochrome c served as the electron acceptor, thus the enzymes being measured were NADH- and NADPH-cytochrome-c-reductase.

occurs (HODGES, 1973) as described in detail for ligand binding to many other enzymes (KOSHLAND, 1970). The complex nature of the kinetics of K^+ stimulation of the ATPase is illustrated by the non-linear nature of a Lineweaver-Burk (Fig. 10.9A) and an Eadie, Hofstee (Fig. 10.9B) plot of the data. ^{42}K influx into oat roots exhibited similar kinetics (Fig. 10.9A and 10.9B) which supports the concept that the ATPase is involved in cation transport.

Further studies are needed to determine whether the plasma membrane ATPase is indeed involved in ion transport. However, with the technique for isolating this membrane now available, it is possible to test this hypothesis critically as well as to evaluate and characterize the passive and possibly the "active" fluxes of ions across the isolated plasma membrane vesicles.

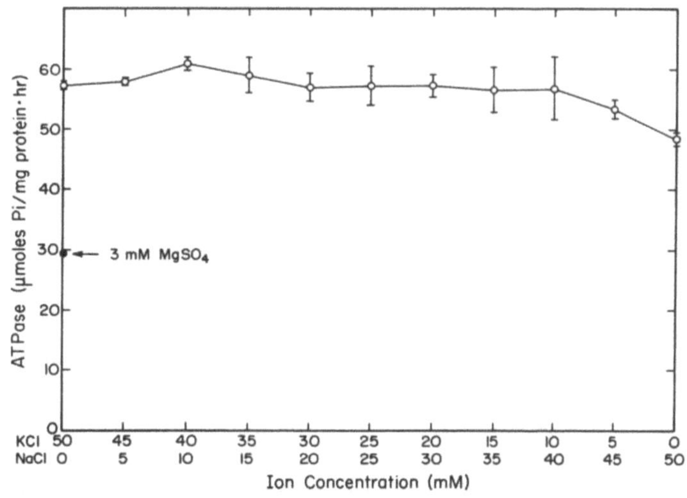

Fig. *10*.7. The effect of various K^+ and Na^+ concentrations on the ATPase activity of plasma membrane vesicles of oat roots. The ATPase activity in the presence of $MgSO_4$ is also shown. (From LEONARD and HODGES, 1973)

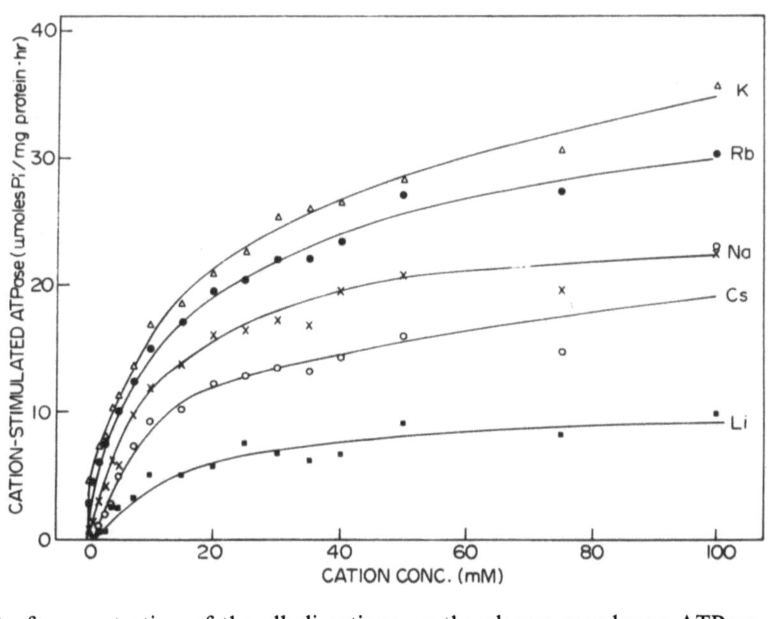

Fig. *10*.8. The effect of concentration of the alkali cations on the plasma membrane ATPase of oat roots. (H. SZE and T.K. HODGES, unpublished)

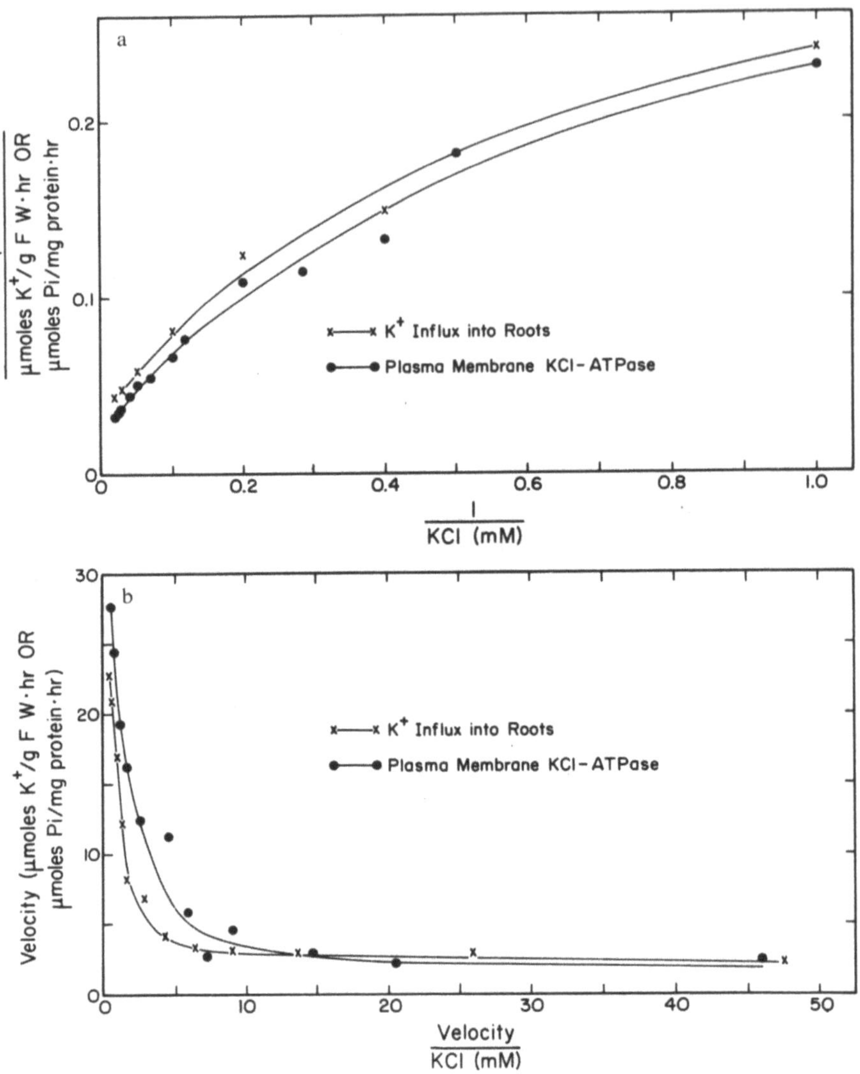

Fig. *10.9*. (a) Lineweaver-Burk plot of K^+-stimulated ATPase of plasma membranes of oat roots and $^{42}K^+$ influx into excised oat roots. (b) Eadie, Hofstee plot of K^+-stimulated ATPase of plasma membranes of oat roots and $^{42}K^+$ influx into excised oat roots. (Adapted from LEONARD and HODGES, 1973)

4.3.2 Tonoplast

The tonoplast of higher plant cells has not been isolated. However, it has been suggested that the ATPase that was sensitive to anions (see *10*.4.2.2) may be located on the tonoplast (RUNGIE and WISKICH, 1973; BALKE et al., 1974). This ATPase had a high pH optimum—probably about 8 (see Fig. *10*.2), and it was stimulated by organic acids such as malic and oxalacetic as well as by inorganic ions such as Cl^- (RUNGIE and WISKICH, 1973; BALKE et al., 1974). Because organic acids accumulate in vacuoles (TORII and LATIES, 1966), it is reasonable that the tonoplast would have a transport system for organic acids. Another very interesting property of this ATPase was its inhibition by NO_3^-. This is also a feature which might have been predicted for a tonoplast transport system of root cells since most of the NO_3^- absorbed by roots moves through the root and up the xylem to the leaves where it is reduced and assimilated (BEEVERS and HAGEMAN, 1969, see *13*.2.3).

The membrane fraction of oats that was most enriched in anion-sensitive ATPase was fraction C (Table *10*.3). This fraction was very heterogeneous, based on the presence of enzymes believed to be associated with Golgi membranes and endoplasmic reticulum (LEONARD et al., 1973). Thus, it is possible that it was one of these membranes, rather than tonoplast, that contained the anion-sensitive ATPase. Nevertheless, further studies on the purification and identification of the membrane containing the anion-sensitive ATPase could be of great importance in further elucidating the role of ATPases in ion transport.

4.4 Effects of Inhibitors

There has been little success thus far in finding a specific inhibitor of the ATPase activity of plant membranes. Ouabain has been occasionally reported as slightly inhibiting plant ATPases (BROWN and ALTSCHUL, 1964; HALL, 1971a; BOWLING et al., 1972; KRASAVINA and VYSKREBENTSEVA, 1972). Other studies have found ouabain to have no effect (DODDS and ELLIS, 1966; BONTING and CARAVAGGIO, 1966; GRUENER and NEUMANN, 1966; HALL and BUTT, 1969; FISHER and HODGES, 1969; RUNGIE and WISKICH, 1973; LEONARD and HODGES, 1973; see Table *10*.5). Oligomycin inhibited the ATPase activity of a microsomal fraction of oat roots, but this may have been an effect of oligomycin on the contaminating mitochondrial ATPase (FISHER and HODGES, 1969). The ATPase of the plasma membrane was not inhibited by oligomycin (LEONARD and HODGES, 1973). The anion-sensitive ATPase of turnip roots was also insensitive to oligomycin (RUNGIE and WISKICH, 1973). DCCD was a potent inhibitor of the plasma membrane-ATPase and of K^+ transport (Table *10*.5). This finding was of interest because the K^+-ATPase activity on the plasma membrane of *Streptococcus faecalis* was also inhibited by

Table *10*.5. Effect of ions and other solutes on ATPase activity of plasma membrane from oat roots. Reaction mixture contained 3 mM ATP (pH 6.0), 33 mM *tris*-MES (pH 6.0), and other additions at concentrations indicated. (From LEONARD and HODGES, 1973)

Additions	ATPase activity (μmoles P_i/mg $_{protein}$/h)
None	3.53
50 mM KCl	4.82
1.5 mM $MgSO_4$	9.46
1.5 mM $MgSO_4$+50 mM KCl	45.06
1.5 mM $MgSO_4$+50 mM KCl+0.01 mM DCCD	12.20
1.5 mM $MgSO_4$+50 mM NaCl	40.08
1.5 mM $MgSO_4$+25 mM KCl+25 mM NaCl	44.82
1.5 mM $MgSO_4$+25 mM KCl+25 mM NaCl+0.01 mM Ouabain	39.46

DCCD (ABRAMS et al., 1972). However, it has not been determined whether the inhibitory effect of DCCD on K^+ absorption by roots is due to its effect on the plasma membrane ATPase or to some other effect. Mercury, silver, PCMB, and PCMBS inhibit the plasma membrane ATPase (D.L. HENDRIX and T.K. HODGES, unpublished) which indicates that the enzyme contains sulfhydryl groups. The anion-sensitive ATPase was also found to be inhibited by mersalyl (RUNGIE and WISKICH, 1973) which is a sulfhydryl inhibitor that interferes with transport processes in mitochondria (BERTAGNOLLI and HANSON, 1973). A toxin produced by *Helminthosporium maydis*-race T has been found to inhibit the ATPase activity that is stimulated by KCl in a microsomal fraction of maize roots (TIPTON et al., 1973). The toxin only inhibited the ATPase activity of roots of a maize variety that was susceptible to the pathogen; no effect was observed with a variety of maize resistant to the pathogen. This result is particularly interesting because the toxin has been identified as a terpenoid glycoside having a structure similar to ouabain (GARY STROBEL, personal communication). Further evaluations of this toxin on ATPase activity and on ion transport might be very rewarding.

5. Evidence for ATPase Involvement in Cation Absorption by Roots

There is considerable indirect evidence that the ATPase stimulated by monovalent cations is involved in cation absorption by roots. The first experimental support for this came from studies that correlated the absorption of K^+ or Rb^+ by roots of oats, wheat, barley, and maize with the KCl- or RbCl-stimulated ATPase activity of membrane preparations from these roots (FISHER and HODGES, 1969; FISHER et al., 1970; Fig. *10*.10). A high positive correlation ($r=0.94$) existed between the

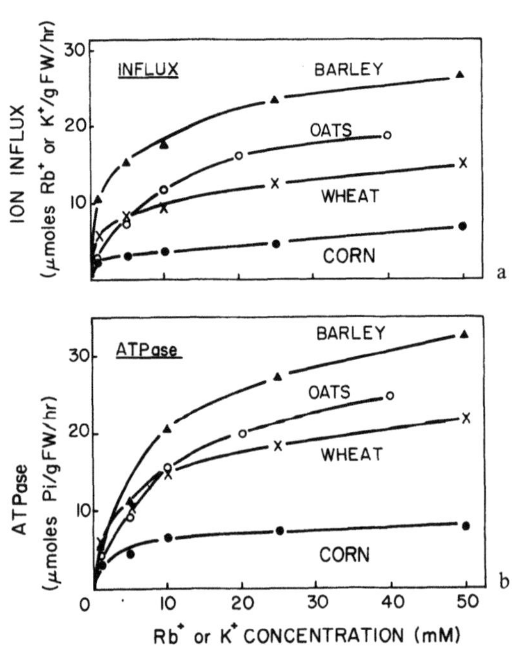

Fig. *10*.10. (a) The effect of KCl or RbCl concentrations on the influx of K^+ into oat roots and the influx of Rb^+ into roots of barley, corn, and wheat. (b) The effect of KCl or RbCl concentrations on the ATPase activity that is stimulated by these ions (the activity in the presence of Mg^{2+} was subtracted) in membranes isolated from roots of the four plant species. (From FISHER et al., 1970)

ATPase activity and ion transport. When these experiments were conducted, less was known about plant ATPases than is known today. Because of this, and also because of the potential importance of the high correlation, a reappraisal of these investigations is needed.

Three experimental conditions employed in the correlation studies are in need of reappraisal. First, the temperature of the ion absorption experiments was 30° C, whereas the temperature of the ATPase experiments was 37–38° C. Second, ion absorption periods were for 30 min which would probably represent a measure of transport across both the plasma membrane and tonoplast (CRAM and LATIES, 1971; *5*.4.4). In subsequent experiments we have found that the temperature optimum was about 42° C for both K^+ absorption by oat roots (in 10-min experiments which represents a better estimate of plasma membrane influx, CRAM and LATIES, 1971) and for the K^+-ATPase of the plasma membrane of oat roots (H. SZE and T.K. HODGES, unpublished). When expressed on a 1 h basis, K^+ transport at 30° C for 30 min was about 10% lower than K^+ transport at 38° C for 10 min. When this correction was made for the transport data for oat roots the correlation between transport and ATPase activity was virtually unaffected. However, the ratios of K^+ transported per K^+-stimulated ATPase activity in oat roots was shifted from an average value of 0.76 (FISHER and HODGES, 1969) to 0.84. Similar corrections for the other 3 plant species would presumably be of a similar magnitude. Another experimental condition that could have influenced the results was the pH of the ATPase assays. This is much more difficult to analyze. The pH employed for the ATPase assays was 7.5 for membranes of oat roots (FISHER and HODGES, 1969) and 7.2 for the membranes of maize, wheat, and barley roots (FISHER et al., 1970). We now know that the ATPase of the plasma membrane of oat roots is stimulated by cations and has a pH optimum of 6.5 to 7.0 (see Fig. *10*.5) and that an anion-stimulated ATPase exists on another membrane and has a pH optimum of about 8.0–8.5. The consequences of the presence of these 2 enzymes, with regard to the correlation study, is that the plasma membrane ATPase would have been somewhat underestimated, which would have caused the measured values to be low. In contrast, the ATPase that is stimulated by anions would have contributed to the measured ATPase activity because of the Cl^-, and this would have caused the measured values to be too high. The net result is that the two effects appear nearly to off-set each other (at least in oat roots) and the measured ATPase values appear to be reasonably accurate measurements of the plasma membrane ATPase. In summary, these correlation studies could now be performed with greater insight, but the basic result would probably be the same—namely, K^+ or Rb^+ transport across the plasma membrane and K^+ or Rb^+ stimulated ATPase activity of the plasma membrane are highly correlated.

The significance of the correlation between ion transport and ATPase activity (FISHER et al., 1970) has been questioned recently (RATNER and JACOBY, 1973) because of apparent differences in selectivity of transport and specificity of the ATPase. RATNER and JACOBY (1973) found that substances such as *tris*-Cl, choline-Cl, and ethanolamine-Cl would stimulate the ATPase activity of microsome preparations from roots of maize, oats, and barley, but that choline and ethanolamine were absorbed very slowly by the roots. However, they did not determine whether the cation or the anion was responsible for the stimulation of the ATPase activity. Since the experiments were conducted at an alkaline pH (8.2 in most experiments) and the activating solutes were always added as the Cl^- salt, the measured ATPase activities must have been largely the result of the anion-sensitive ATPase (see *10*.4.3.2). Certainly at pH 8.2 the plasma membrane ATPase that is specifically stimulated by cations would have contributed very little to the measured ATPase activity (see the KCl stimulation in Fig. *10*.5). Thus, the reservations of RATNER and JACOBY (1973) about the correlations between ATPase activity and transport are understandable, but not justified.

Additional evidence that an ATPase may be involved in ion transport was obtained from studies that involved the induction or development of an increased ion transport capacity in various tissues. For example, a 3-h treatment of maize roots in $CaCl_2$ caused about a 300% increase in K^+ absorption rates and a 20–30% increase in ATPase activity (LEONARD and HANSON, 1972). The accelerated rates of both ion transport and ATPase activity were prevented from developing by cold temperatures, cycloheximide, and 6-methylpurine. A similar response to a washing treatment was obtained with slices of turnip roots (RUNGIE and WISKICH, 1973). In this tissue, 60 h of washing caused an 8-fold increase in accumulation of KCl and about a 2-fold increase in ATPase activity. In both corn roots and slices of turnip, the induced ATPase activity was small, and the large discrepancy in the relative changes in ATPase activity and ion transport makes the significance of the results uncertain. However, an apparently closer correspondence was found in leaves of *Limonium* between the ion transport capacity and ATPase activity induced by an overnight wash with 100 mM NaCl (HILL and HILL, 1973; see Part B. Chap. *9*). In this tissue, which contains salt glands, the salt-loading treatment resulted in a full induction in ion transport capacity and a 2- to 4-fold increase in ATPase activity, however the levels of ATPase activity were very low. Nevertheless, it was found that puromycin prevented the development of the ATPase activity as well as the ion transport potential. Thus, all these studies involving washing in dilute or concentrated salt solutions are consistent with the view that ATPases are involved in ion transport, but a direct relationship between the two phenomena remains to be established.

The finding that isolated plasma membrane vesicles contained an ATPase (HODGES et al., 1972; see *10*.4.3.1) provided strong support for the view that ATP is the energy source for ion transport across the plasma membrane (HODGES, 1973). The specific activation of the plasma membrane ATPase by monovalent cations (BALKE et al., 1974) suggests that the enzyme is probably more directly involved in cation transport than in anion transport across the plasma membrane (HODGES, 1973). Further support for plasma membrane ATPase involvement in cation absorption came from the studies that showed a striking similarity in the kinetics of K^+ activation of the ATPase and K^+ influx into roots (see Fig. *10*.9). Finally, the specificity of the plasma membrane ATPase toward the alkali cations is $K^+ > Rb^+ > Na^+ > Cs^+ > Li^+$ for the concentration range of 0.5 to 50 mM (Fig. *10*.8) and this resembles the order of specificity for influx of these ions into oat roots (H. SZE and T.K. HODGES, unpublished).

On the basis of the correlations cited above between ATPase activity and ion influx and the presence of an ATPase on the plasma membrane that is uniquely stimulated by alkali cations, it is probable that the primary energy transduction involved in cation influx across the plasma membrane of root cells is mediated by this plasma membrane ATPase.

6. Summary

Plant cells contain several enzymes that are capable of hydrolyzing ATP. Soluble fractions of plant extracts seem to be rich in either acid- or alkaline-phosphatases, and there is no evidence that these enzymes play a role in ion transport. Several membranes of plant cells also contain ATPase activity, and it is quite probable that some of these enzymes are involved in ion transport.

A general feature of membrane-bound ATPase is their sensitivity to monovalent ions. ATPase activity in heterogeneous membrane preparations is generally stimulated by monovalent ions, but occasionally it is inhibited. In certain instances, the ATPase activity is stimulated synergistically by Na^+ and K^+, but the activity does not seem to be inhibited by ouabain.

Two membrane-bound ATPases from oat roots are of particular interest. One of these ATPases was present on isolated vesicles of plasma membranes. The plasma membrane ATPase had a pH optimum of 6.5 to 7.0, required Mg^{2+}, and was stimulated by monovalent cations. The other ATPase was on a different membrane, it had an alkaline pH optimum, required Mg^{2+}, and was sensitive to anions. The particular characteristics of the anion-sensitive ATPase suggested that it may be bound to the tonoplast.

The plasma membrane ATPase of roots may play an important role in cation transport across the plasma membrane of these cells. In addition to the enzyme being stimulated by monovalent cations, there was a high, positive correlation between the K^+- or Rb^+-stimulated ATPase activity and K^+ or Rb^+ absorption by roots. In addition, both the kinetics and the specificity of the plasma membrane ATPase toward the alkali cations was similar to the kinetics and specificity of alkali cation transport into root cells.

Additional evidence that an ATPase may be involved in ion transport has come from induction-type studies. Various pretreatments of tissues cause an increased ion transport capacity and an increased ATPase activity. The magnitude of the increase in ATPase activity was, however, less than the increase in ion transport capacity. The increased rates of both the transport and ATPase activity required protein synthesis.

References

ABRAMS, A., SMITH, J.B., BARON, C.: Carbodiimide-resistant membrane adenosine triphosphatase in mutants of *Streptococcus faecalis*. I. Studies of the mechanism of resistance. J. Biol. Chem. **247**, 1484–1488 (1972).

ATKINSON, M.R., POLYA, G.M.: Salt-stimulated adenosine triphosphatases from carrot, beet, and *Chara australis*. Australian J. Biol. Sci. **20**, 1069–1086 (1967).

BALKE, N.E., HODGES, T.K.: Plasma membrane adenosine triphosphatase of oat roots: Activation and inhibition by Mg^{2+} and ATP. Plant Physiol. **55**, 83–86 (1975).

BALKE, N.E., SZE, H., LEONARD, R.T., HODGES, T.K.: Cation sensitivity of the plasma membrane ATPase of oat roots. In: Membrane transport in plants (U. ZIMMERMANN, J. DAINTY, eds.). Berlin-Heidelberg-New York: Springer 1974.

BEEVERS, L., HAGEMAN, R.H.: Nitrate reduction in higher plants. Ann. Rev. Plant Physiol. **20**, 495–522 (1969).

BERTAGNOLLI, B.L., HANSON, J.B.: Functioning of the adenine nucleotide transporter in the arsenate uncoupling of corn mitochondria. Plant Physiol. **52**, 431–435 (1973).

BONTING, S.L., CARAVAGGIO, L.L.: Studies on Na^+-K^+-activated adenosine triphosphatase

XVI. Its absence from the cation transport system of *Ulva lactuca*. Biochim. Biophys. Acta **112**, 519–523 (1966).

BOWLING, D.J.F., TURKINA, M.V., KRASAVINA, M.S., KRYUCHESHNIKOVA, A.L.: Na^+-, K^+-activated ATPase of conducting tissue. Fiziol. Rast. **19**, 968–977 (1972).

BROWN, H.D., ALTSCHUL, A.M.: Glycoside-sensitive ATPase from *Arachis hypogaea*. Biochem. Biophys. Res. Comm. **15**, 479–483 (1964).

BROWN, H.D., NEUCERE, N.J., ALTSCHUL, A.M., EVANS, W.J.: Activity patterns of purified ATPase from *Arachis hypogaea*. Life Sci. **4**, 1439–1447 (1965).

CEREIJO-SANTALÓ, R.: Ion movements in mitochondria. In: Membranes and ion transport, (E.E. BITTER, ed.), vol. 2, p. 229–258. New York: Wiley-Interscience 1970.

CRAM, W.J., LATIES, G.G.: The use of short-term and quasi-steady influx in estimating plasmalemma and tonoplast influx in barley root cells at various external and internal chloride concentrations. Australian J. Biol. Sci. **24**, 633–646 (1971).

DODDS, J.J.A., ELLIS, R.J.: Cation-stimulated adenosine triphosphatase activity in plant cell walls. Biochem. J. **101**, 31 (1966).

EDWARDS, M.L., HALL, J.L.: Intracellular localization of the multiple forms of ATPase activity in maize root tips. Protoplasma **78**, 321–338 (1973).

ELZAM, O.E., HODGES, T.K.: Characterization of energy-dependent Ca^{2+} transport in maize mitochondria. Plant Physiol. **43**, 1108–1114 (1968).

FISHER, J.D., HANSEN, D., HODGES, T.K.: Correlation between ion fluxes and ion-stimulated adenosine triphosphatase activity of plant roots. Plant Physiol. **46**, 812–814 (1970).

FISHER, J.D., HODGES, T.K.: Monovalent ion stimulated adenosine triphosphatase from oat roots. Plant Physiol. **44**, 385–395 (1969).

GRUENER, N., NEUMANN, J.: An ion-stimulated adenosine triphosphatase from bean roots. Physiol. Plantarum **19**, 678–682 (1966).

HALL, J.L.: Further properties of adenosine triphosphatase and β-glycerophosphatase from barley roots. J. Exptl. Bot. **22**, 800–808 (1971a).

HALL, J.L.: Cytochemical localization of ATP-ase activity in plant root cells. J. Microscopy **93**, 219–225 (1971b).

HALL, J.L., BUTT, V.S.: Adenosine triphosphatase activity in cell-wall preparations and excised roots of barley. J. Exptl. Bot. **20**, 751–762 (1969).

HANSON, J.B., HODGES, T.K.: Energy-linked reactions of plant mitochondria. In: Current topics in bioenergetics (D.R. SANADI, ed.), vol. 2, p. 65–98. New York: Academic Press 1967.

HANSSON, G., KUIPER, P.J.C., KYLIN, A.: Effect of preparation method on the induction of (sodium + potassium)-activated adenosine triphosphatase from sugar beet root and its lipid composition. Physiol. Plantarum **28**, 430–435 (1973).

HANSSON, G., KYLIN, A.: ATPase activities in homogenates from sugarbeet roots, relation to Mg^{2+} and $(Na^+ + K^+)$-stimulation. Z. Pflanzenphysiol. **60**, 270–275 (1969).

HIGINBOTHAM, N.: Electropotentials of plant cells. Ann. Rev. Plant Physiol. **24**, 25–46 (1973).

HILL, B.S., HILL, A.E.: ATP-driven chloride pumping and ATPase activity in the *Limonium* salt gland. J. Membrane Biol. **12**, 145–148 (1973).

HODGES, T.K.: Ion absorption by plant roots. Advan. Agron. **25**, 163–207 (1973).

HODGES, T.K., HANSON, J.B.: Calcium accumulation by maize mitochondria. Plant Physiol. **40**, 101–109 (1965).

HODGES, T.K., LEONARD, R.T.: Purification of a plasma membrane-bound adenosine triphosphatase from plant roots. In: Methods in enzymology (S.P. COLOWICK, N.O. KAPLAN, eds.), vol. 32B, p. 392–406. New York: Academic Press 1974.

HODGES, T.K., LEONARD, R.T., BRACKER, C.E., KEENAN, T.W.: Purification of an ion-stimulated adenosine triphosphatase from plant roots: Association with plasma membranes. Proc. Natl. Acad. Sci. U.S. **69**, 3307–3311 (1972).

HOKIN, L.E., DAHL, J.L.: The sodium-potassium adenosinetriphosphatase. In: Metabolic pathways, (L.E. HOKIN, ed.), vol. 6, p. 269–315. New York: Academic Press 1972.

HOROVITZ, C.T., WAISEL, Y.: Different ATPase systems in glycophytic and halophytic plant species. Experientia **26**, 941–942 (1970).

KASAMO, K., YAMAKI, T.: Mg^{++}-activated and -inhibited ATPases from mung bean hypocotyls. Plant Cell Physiol. **15**, 507–516 (1974).

KOSHLAND, D.E., JR.: The molecular basis for enzyme regulation. In: The enzymes (P.O. BOYER, ed.), vol. I, ed. 3, p. 342–396. New York: Academic Press 1970.

KRASAVINA, M.S., VYSKREBENTSEVA, E.I.: ATPase activity and transport of potassium and sodium in root tissues. Fiziol. Rast. **19**, 978–983 (1972).

KUIPER, P.J.C.: Temperature response of adenosine triphosphatase of bean roots as related to growth temperature and to lipid requirement of the adenosine triphosphatase. Physiol. Plantarum **26**, 200–205 (1972).

KYLIN, A., GEE, R.: Adenosine triphosphatase activities in leaves of the mangrove *Avicennia nitida* Jacq. Influence of sodium to potassium ratios and salt concentration. Plant Physiol. **45**, 169–172 (1970).

KYLIN, A., KÄHR, M.: The effect of magnesium and calcium ions on adenosine triphosphatases from wheat and oat roots at different pH. Physiol. Plantarum **28**, 452–457 (1973).

KYLIN, A., KUIPER, P.J.C., HANSSON, G.: Lipids from sugar beet in relation to the preparation and properties of (sodium+potassium)-activated adenosine triphosphatases. Physiol. Plantarum **26**, 271–278 (1972).

LAI, Y.F., THOMPSON, J.E.: The preparation and properties of an isolated plant membrane fraction enriched in (Na^+-K^+)-stimulated ATPase. Biochim. Biophys. Acta **233**, 84–90 (1971).

LAI, Y.F., THOMPSON, J.E.: Effects of germination on Na^+-K^+-stimulated adenosine 5′-triphosphatase and ATP-dependent ion transport of isolated membranes from cotyledons. Plant Physiol. **50**, 452–457 (1972).

LEONARD, R.T., HANSEN, D., HODGES, T.K.: Membrane-bound adenosine triphosphatase activities of oat roots. Plant Physiol. **51**, 749–754 (1973).

LEONARD, R.T., HANSON, J.B.: Increased membrane-bound adenosine triphosphatase activity accompanying development of enhanced solute uptake in washed corn root tissue. Plant Physiol. **49**, 436–440 (1972).

LEONARD, R.T., HODGES, T.K.: Characterization of plasma membrane-associated adenosine triphosphatase activity in oat roots. Plant Physiol. **52**, 6–12 (1973).

MARTONOSI, A.: Biochemical and clinical aspects of sarcoplasmic reticulum function. In: Current topics in membranes and transport (F. BRONNER, A. KLEINZELLER, eds.), vol. 3, p. 83–197. New York: Academic Press 1972.

POUX, N.: Localisation d'activités enzymatiques dans les cellules du méristème radiculaire de *Cucumis sativus* L. I. Activités phosphatasiques neutres dans les cellules du protoderme. J. Microscopie **6**, 1043–1058 (1967).

RATNER, A., JACOBY, B.: Non-specificity of salt effects on Mg^{2+}-dependent ATPase from grass roots. J. Expt. Bot. **24**, 231–238 (1973).

ROLAND, J.C.: Mise en évidence sur coupes ultrafines de formations polysaccharidiques directement associées au plasmalemme. Compt. Rend. **269**, 939–942 (1969).

ROLAND, J.C., LEMBI, C.A., MORRE, D.J.: Phosphotungstic acid-chromic acid as a selective electron-dense stain for plasma membranes of plant cells. Stain Technol. **47**, 195–200 (1972).

ROTHSTEIN, A.: Ion transport in microorganisms. In: Metabolic pathways (L.E. HOKIN, ed.), vol. 6, p. 17–39. New York: Academic Press 1972.

RUNGIE, J.M., WISKICH, J.T.: Salt-stimulated adenosine triphosphatase from smooth microsomes of turnip. Plant Physiol. **51**, 1064–1068 (1973).

SCHATZMANN, J.H.: Herzglykoside als Hemmstoffe für den aktiven Kalium- und Natriumtransport durch die Erythrocytenmembran. Helv. Physiol. Pharmacol. Acta **11**, 346 (1953).

SEXTON, R., SUTCLIFFE, J.F.: Some observations on the characteristics and distribution of adenosine triphosphatases in young roots of *Pisum sativum*, cultivar Alaska. Ann. Bot. (London) **33**, 638–694 (1969).

SKOU, J.C.: The influence of some cations on an adenosine triphosphatase from peripheral nerves. Biochim. Biophys. Acta **23**, 394–401 (1957).

SKOU, J.C.: Enzymatic basis for active transport of Na^+ and K^+ across cell membrane. Physiol. Rev. **45**, 596–617 (1965).

TIPTON, C.L., MONDAL, M.H., UHLIG, J.: Inhibition of the K^+-stimulated ATPase of maize root microsomes by *Helminthosporium maydis*-Race T pathotoxin. Biochem. Biophys. Res. Commun. **51**, 725–728 (1973).

TORII, K., LATIES, G.G.: Organic acid synthesis in response to excess cation absorption in vacuolate and non-vacuolate sections of corn and barley roots. Plant Cell Physiol. **7**, 395–403 (1966).

WILLIAMSON, F.A., WYN JONES, R.G.: The isolation of plant root protoplasts and their use in the separation of subcellular components. Biochem. J. **123**, 4P-5P (1971).

WILLIAMSON, F.A., WYN JONES, R.G.: On the role of plasma membrane and tonoplast adenosine triphosphatases in ion uptake. IAEH/FAO Use of isotopes in soil-plant relations. 69. Intern. Atomic Energy Agency, Vienna (1972).

11. Negative Feedback Regulation of Transport in Cells. The Maintenance of Turgor, Volume and Nutrient Supply

W.J. CRAM

1. Introduction

Regulation concerns the flow of information. It can primarily be distinguished from the flow of energy. Changes in the rate of a process can take place in response to an inflow of either, but it seems likely that in biological systems the inflow of information would be specific and integrative, while the inflow of energy would simply impose a more general limitation under certain conditions.

The distinction between information and energy transfer in biological systems is quantitative rather than qualitative. Information transfer naturally dissipates energy, but at a much lower rate than does the controlled process. Even in the extreme of no power consumption by the information flow, the information is still itself indistinguishable from energy. The relationship between free energy, entropy, probability and hence information is well known (e.g. MOROWITZ, 1970). The idea that information transfer can intrinsically involve no energy dissipation is the fallacy on which Maxwell's demon is based.

The relation between energy and information flow is further considered, after the properties of negative feedback systems have been outlined, in *11*.2.2.

In this Chapter, the processes to be considered are transport across cell membranes and, to a lesser extent, biochemical conversions. In considering their regulation certain terms will be used. The result of the activity of such a process will be called its *"output"*; this may be the quantity or concentration of a substance in the cell, or it may be the generalized result of several such processes, for instance osmotic presure. A process, its output, and their relationship to each other will be called a *"system"*. Energy and information flows into the system will be called the *"inputs"*.

The effect of information input is often referred to as "control" or "regulation", but common usage of these terms is somewhat vague. Based on their etymology, "control" will be used in the sense of "exercise of restraint or direction upon", and "regulation" in the sense of "adjustment or control (*sic*) with regard to some law, standard or reference". It is not uncommon to read that an output (concentration, *etc.*) has been found to be "regulated". This usage of the term, while valid, conceals the fact that the primary effect of an inflow of information to a system is always on a process, not its output. In this Chapter, therefore, "regulation" will be confined to the immediate effect on a process. Terms such as "osmoregulation" will not be used for this and other reasons (*11*.3.1).

The simplest type of regulation is one in which the information input is from sources outside the system, as in hormonally and genetically directed changes in rates of transport into cells during development (Part B, Chaps. *7, 8, 9*). This gives a coarse direction (*Steuerung*—RASCHKE, 1965) to the course of events.

The other type of regulation is one in which the source of information is within the system itself, as when information flows from the output (concentration *etc.*)

to the controlled transport process, i.e. feeds back. This is a fine control, operating to maintain an output at a "desired" or reference level (*Regelung*—RASCHKE, 1965) in the face of naturally and inevitably fluctuating conditions.

The first type of regulation is also often called "open-loop control" and the second "closed-loop control". The terms "coarse" and "fine control" are also sometimes used, in a sense corresponding roughly to open-loop control and closed-loop control. However, these terms, although graphic, have no sufficiently defined meaning and are best not used in conjunction with the more precise terms developed in *11*.2.

Conceptually the simplest "desired" level is one which is constant with time; for instance, a constant turgor during cell expansion growth or fluctuating external osmotic pressure, or constant intracellular concentration of a nutrient during changes in the rate of its incorporation into organic form. Such constancy would tend to be achieved if a decrease in internal concentration stimulated inwards transport. The increased inflow would then change the internal concentration back towards its initial value. This type of control system has been recognized in biology since the 19th century (BERNARD, 1859), and has been termed "homeostatic" or "regulatory". The inflow of information from the output (concentration) to the controlled process (transport) is known as "negative feedback", since a decrease in the output value leads to an increase in the rate of the process, and *vice versa*.

In addition to systems for maintaining constancy, the plant needs systems for controlling change. Negative feedback with a variable "desired" level provides a mechanism for a system to respond to one signal while at the same time minimizing the effects of other influences on the system. An automatic tracking system is the obvious example, and "osmotic adaptation" of plants, where internal osmotic pressure changes somewhat in parallel with changes in external water potential (*11*.3.1.3.2), superficially appears to be another. These have been called "servo" or, perhaps more graphically, "follow-up" systems. ("Servo system" has also been used as synonymous with all types of negative feedback system on occasion.)

One set of problems about regulation concerns mechanisms of information flow and action. As will appear, very little is yet known about this aspect. An independent and equally important set of problems concerns the quantitative description of the information flow. This aspect will be discussed in *11*.2, where general properties of negative feedback systems will be summarized. The operation of such control systems in plant cells is summarized and discussed in the rest of the Chapter.

2. Some Elementary Properties of Control Systems

Only some fairly bald, qualitative statements about control systems will be made, since space precludes an outline of the development of the concepts, and their quantitative application demands more detailed information than is available for any of the systems to be discussed. The treatment is similar to that of GRODINS (1963) and TOATES (1975). Development of concepts, with various emphases, may also be found in BAYLISS (1966), MILSUM (1966), KALMUS (1966), HASSENSTEIN (1971), and others.

2.1 Negative Feedback

One can first consider a cell in which the concentration is built up to a particular value by an active transport inwards together with a passive leak outwards, which will be represented for simplicity as proportional to P times the internal concentration (Fig. *11*.1). On switching on the system (for instance by transferring from water to salt solution) the internal concentration will rise until efflux equals influx. Such a "pump and leak" system operates, for instance, in red blood cells (POST et al., 1967). In this system there is nothing that would be called "regulation". Nevertheless, it can be represented as a negative feedback loop, but the feedback is a purely passive phenomenon.

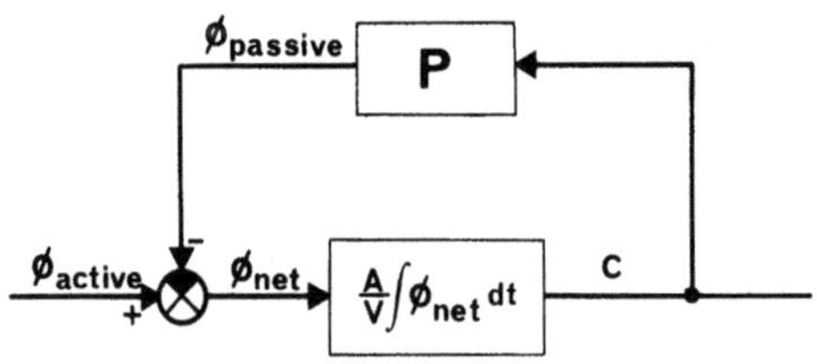

Fig. *11*.1. Block diagram of a "pump and leak" system. Active influx (ϕ_a) is constant. Passive efflux increases with internal concentration (c), being shown as $P \times c$ for simplicity. The passive efflux can be subtracted from the active influx to give the net flux, which, over time t, leads to accumulation of the transported substance to a concentration c in a cell of volume V and surface area A. The upper limb represents the passive feedback of internal concentration on net flux

The Equation relating internal concentration (c) to net flux in Fig. *11*.1 is

$$\frac{dc}{dt} = \frac{A}{V} \cdot \phi_a - P \cdot \frac{A}{V} \cdot c \tag{11.1}$$

where A = surface area of the cell; V = volume of the cell; ϕ_a = active influx; and P = "permeability".

Hence

$$c = \frac{\phi_a}{P}\left[1 - \exp\left(-P \cdot \frac{A}{V} \cdot t\right)\right] + c_{t=0} \cdot \exp\left(-P \cdot \frac{A}{V} \cdot t\right) \tag{11.2}$$

and at a steady state ($t \to \infty$)

$$c_\infty = \frac{\phi_a}{P}. \tag{11.3}$$

Two properties of this system should be noted: (1) the internal concentration rises as an exponential function to its final steady state; and (2) any perturbation causing ϕ_a or P to change will lead to a *proportional* change in the final internal concentration.

Secondly one can consider a cell in which, in addition, active transport is related to the internal concentration *via* negative feedback. In all negative feedback the primary signal is the output of the system (in this case c), and the actuating signal is the difference between the actual output and a desired or reference value (in

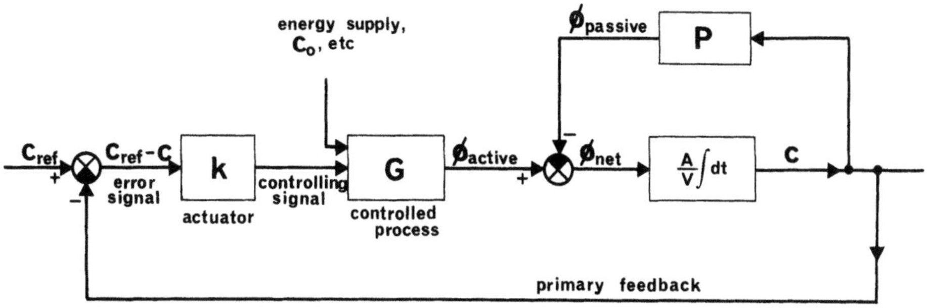

Fig. *11*.2. Block diagram of a system similar to that of Fig. *11*.1, but including regulated rather than constant active transport. Here the internal concentration also feeds back a signal to alter active influx. This is not a model of a plant cell. The lower loop shows the main conceptual components of negative feedback. The primary signal (the value of c) is subtracted from a reference value (c_{ref}) to give the deviation of the output from the desired value—the error signal (c_{ref}–c). This signal is modified and transduced into a form in which it can act on active transport. The instantaneous value of active transport is also determined by other factors, of which external concentration (c_o) and energy supply are shown. The final steady value of the internal concentration is primarily dependent on the value of c_{ref}, rather than on the value of P or the factors influencing the active flux

this case c_{ref}, the input, which may be zero). This difference is known as the "error". In Fig. *11*.2 the error or actuating signal is shown as being transduced into a "controlling" signal (error × k), which in turn can act on the controlled process (active transport).

The Equation relating the internal concentration (c) to net flux in Fig. *11*.2 is

$$\frac{dc}{dt}=k\cdot G\cdot\frac{A}{V}(c_{ref}-c)-P\cdot\frac{A}{V}\cdot c \qquad (11.4)$$

where k is the transfer function or gain of the actuator, relating the controlling signal to the error signal; and G is the transfer function of the controlled process.

Hence

$$c=\frac{kG}{kG+P}\cdot c_{ref}\left[1-\exp\left(-(kG+P)\frac{A}{V}\cdot t\right)\right]+c_{t=0}\cdot\exp\left(-(kG+P)\frac{A}{V}\cdot t\right) \qquad (11.5)$$

and at a steady state

$$c_\infty=\frac{kG}{kG+P}\cdot c_{ref}. \qquad (11.6)$$

The system of Fig. *11*.2 contains the minimum components (an active and a passive flux) needed to describe the ion relations of a plant cell, but it is not presented as a model of the ion relations of a cell. The reader will readily see that Fig. *11*.2 can be modified to correspond to an actual system, particularly by including an electrical term in the output of the system and its effect on active and passive fluxes; and that a one-to-one correspondence between the elements of Fig. *11*.2 and those of a living system need to obtain. Analytical solutions to equations describing other systems rapidly become more complex, but the basic features remain the same as those illustrated by the simpler equations describing Fig. *11*.2. The latter are therefore an adequate basis for discussion.

In comparison with the system of Fig. *11*.1 the system of Fig. *11*.2 has the properties that (1) everything else remaining constant, the rise towards a new steady level of c is faster; and (2) external perturbations (such as a change in external concentration or power input (see below) altering G or k and hence ϕ_a), and perturbations within the system (such as alterations in P) will give rise to *less than proportional changes* in internal concentration. (A change in G or P alters the fraction $(kG/kG+P)$ less than proportionally.) In addition, the system of Fig. *11*.2 has the important property that the output (c) is related to the input (c_{ref}). If c_{ref} is constant, the system is said to be operating in the *homeostatic* mode: if c_{ref} varies, the system is said to be operating in the *servo* or *follow-up* mode.

Thus the system of Fig. *11*.2 rapidly adjusts internal concentration towards a desired or reference value while minimizing the effects on c of changes in the rate of active transport or of P due to external and internal perturbations.

The system of negative feedback represented in the lower loop of Fig. *11*.2 differs from the passive upper loop and the system of Fig. *11*.1 in that the feedback in the former involves an act of measurement of the state of the system and the generation of a signal dependent on this. It may hence be called active feedback in contrast to the passive response represented in the upper loop of Fig. *11*.2 and Fig. *11*.1.

Two further features of the system of Fig. *11*.2 are the steady-state error and the possibility of oscillations. In Fig. *11*.2 the final steady concentration is less than the reference value, the difference being the steady-state error. This error is less when the value of k (the "gain") is larger. (At the same time a larger gain would decrease the effects of perturbations on the system.) The steady-state error would also be less if other types of control operated. The system of Fig. *11*.2 is shown with proportional control (error $\times$ k). Alternatively, or in addition, the rate of change of the error signal could be transduced ("derivative" control), which contributes to speeding up the response of the system: or the integrated error signal could be transduced ("integral" control), which contributes to reducing the steady state error to zero.

Time lags in the flow of information in the system may under certain conditions cause transient overshoots and oscillations in the response to changes in input, or permanently maintained oscillations. These may be the basis of oscillations and endogenous rhythms in, for instance, electrical potential differences (SCOTT, 1967), stomata (COWAN, 1972), and Crassulacean acid metabolism (WILKINS, 1969). See also COWAN's (1972) discussion of the physiological significance of oscillations. It should be noted that negative feedback, even with time lags, only leads to oscillations under certain conditions.

2.2 Further Considerations of Energy Flow

Fig. *11*.2 represents the flow of information in the system, and energetic considerations are not included. As pointed out earlier (*11*.1), information transfer will be accompanied by a relatively small dissipation of energy compared with that in the controlled process. As discussed further by MILSUM (1966), in many control systems the rate of the controlled process is determined by the information input, and power consumption is dependent on the rate of the process. The energy input would then be a dependent variable which need not be explicitly included in the description of the dynamic properties of the system. Energy could then be shown as a dependent input to the actuator (box k), raising the low-energy error signal to a higher-energy controlling signal. In many systems it may in fact be appropriate to consider the controlling element of a negative feedback system as an amplifier.

Energy is treated in this way in Fig. *11*.2, the energy input simply representing any independent variation in energy supply. This can be compensated for in reaching the steady state internal concentration in the same way as any other perturbation (cf. JESCHKE, 1974).

In certain systems where some function of energy is an independent input signal to the controlled process (ATKINSON, 1972) this would be shown as an influence on the value of c_{ref}.

Transient power requirements of a system in response to perturbation are of some interest. In the negative feedback system of Fig. *11*.2 the power consumption is not constant, but decreases to a minimum as the steady state is reached. In the system of Fig. *11*.1 on the other hand, the power consumption remains the same at the steady state. In these examples when P is the same, the more rapidly adjusting system with negative feedback has a temporarily larger power intake. Alternatively, if P were smaller in the system of Fig. *11*.2 it might adjust no faster than the system of Fig. *11*.1, but its power consumption at the steady state would be less. Thus minimum power consumption may be an important, but not the only, performance criterion (cf. MILSUM, 1966).

3. Experimental Observations

3.1 Control of the Total Number of Osmotically Active Particles in Cells

The walled plant cell under normal conditions has a higher osmotic pressure inside the cell than outside. The consequent tendency for fluid to flow into the cell and increase its volume is opposed by the internal hydrostatic pressure (turgor) generated by the stretched cell wall. The elasticity of cell walls varies, so the extension needed to generate the normal turgor may range from a few percent (*Valonia*) to several tens percent (some red algae). (See *2*.3).

In wall-less cells (which are characteristic of the reproductive stage of the life cycle of plant species as high in the evolutionary scale as *Cycas*, and of the whole of the life-span of certain unicellular algae; cf. Chap. *6*) no internal hydrostatic pressure can be sustained. The internal osmotic pressure equals the external osmotic pressure in some marine algae, but in fresh water algae the internal osmotic pressure is higher than the external, and the inflow of fluid is balanced by the secretion from contractile vacuoles of what must be a dilute fluid (Fig. *11*.3).

If external osmotic pressure is suddenly increased around a walled cell fluid flows out, the volume decreases, and the wall shrinks until the decreased pressure it generates equals the reduced osmotic pressure difference between inside and outside the cell. It has been found that some cells treated in this way subsequently recover their initial turgor. This is achieved by increasing internal osmotic pressure. Osmotica are accumulated by either transport or biochemical interconversions. To give the exact recovery, there is most probably a negative feedback signal from turgor to the accumulatory process, and hence turgor must be regarded as the output of the accumulatory control system. This type of self-adjustment is often termed "osmoregulation", but since osmotic pressure may not be the output of the control system involved the term may be misleading and it will not be used.

If the external osmotic pressure is suddenly increased round a wall-less cell it shrinks like an osmometer, internal osmotic pressure increasing until it equals the external osmotic pressure or until the fluid flow inwards again equals the outwards secretion of fluid by the contractile vacuoles. Again, it has been found that some cells treated in this way subsequently recover their initial volume. This is achieved by accumulation of solutes in the cell, accompanied, not by a rise in internal osmotic pressure, but by an inflow of water and an increase in cell volume at a more or less *constant* osmotic pressure. To give the exact recovery observed there is most probably a negative feedback signal from volume to the accumulatory process, and volume must be regarded as the output of the accumulatory control system. This type of self-adjustment could clearly *not* be termed "osmoregulatory".

The main components of osmotic pressure in plant cells under natural conditions will first be considered, since it would seem probable that the process of accumulation of a single major component is regulated.

3.1.1 The Main Components of Osmotic Pressure in Plant Cell Sap

The results gathered in Tables *11*.1 to *11*.3 correspond to vacuolar, rather than cytoplasmic, concentrations.

In several marine and fresh water algae the osmotic pressure (π) is totally accounted for by inorganic salts, often mainly KCl (Table *11*.1; Chap. *6*, Tables *6*.3–*6*.5, see also *6*.3). Organic acids have rarely been found in algae, in contrast to higher plants (Table *11*.3). Significant, but not large, amounts of neutral organic compounds occur widely (Tables *11*.1 and *11*.4), mannitol being characteristic of brown algae and floridoside of red algae (FOGG, 1953; MEEUSE, 1962; BONEY, 1965).

In many higher plants the main osmoticum is again K^+, most often associated with organic acids, though the absolute amounts vary with conditions (Table *11*.2). There is a complete range of Na^+ concentrations, increasing in the more halophylic species (COLLANDER, 1941). Cl^- is generally a minor component in glycophytes except in water plants, where it may be the major anion (Tables *11*.2 and *11*.3). Divalent cations, nitrate, proline, and various neutral organic substances also occur in certain species (STEINER and ESCHRICH, 1958; WRIGHT and DAVISON, 1964; STEWART and LEE, 1974). There is no indication that sugars are an appreciable component of osmotic pressure in shoots of full-grown cultivated plants (Table *11*.2), though under natural conditions the sugar concentrations in glycophytes may be considerable (Table *11*.3). This may be related to the relative supply of nutrients (see PITMAN and CRAM, 1973).

In halophytes and halophylic xerophytes the osmotic pressure is higher, and this is associated largely with a higher NaCl concentration. Fluctuations in osmotic pressure in halophytes during growth or in response to external conditions are generally *via* changes in NaCl concentration (see STEINER and ESCHRICH, 1958; *11*.3.1.3.2). Decreasing succulence has been correlated with increasing concentrations of SO_4^{2-} (WALTER, 1971).

Osmotic pressure and its components may vary in different parts of the plant (e.g. BERNSTEIN, 1963); during development (e.g. STEINER and ESCHRICH, 1958; ÖNAL, 1966, MOTT and STEWARD, 1972b); and in different external media (*11*.3.1.4).

3.1.2 Control of the Total Number of Osmotically Active Particles by Biochemical Interconversions

Lower plants in which a change in the concentration of an organic substance contributes to volume or turgor maintenance are listed, together with the compound involved and brief notes, in Table *11*.4.

Ochromonas malhamensis, Fig. *11*.3, the most intensively studied, is a wall-less, unicellular alga. In fresh water, inwards osmosis is balanced by secretion of water *via* the contractile vacuole. When the external osmotic pressure is raised, the cells shrink rapidly, and then slowly recover their original volume (but with a slightly

Table *11*.1. Principle components of osmotic pressure in algae

Species	K^+ (mM)	Na^+ (mM)	Cl^- (mM)	other (mM)	Turgor (100 kPa)	π (% accounted for)	Notes
Chaetomorpha linum	697	68	754	Mg^{2+}:54; SO_4^{2-}:41	12[a]	100	Marine[d]. *C. aerea* similar[d]
Valonia ventricosa	550	75	625	–	2[a, c]	98	Marine[e, f, g, h, i]. *V. macrophysa*[h, j] and *V. utricularis*[c, k] similar
Codium fragile	16	475	495	SO_4^{2-}:116	2a	100	Marine[d, l]. *C. tomentosum* similar[d]
Halicystis osterhoutii	6	557	603	–	0.3[a]	100	Marine[m, n]
Enteromorpha intestinalis	450	260	370	++	(25[b])	(50)	Marine[o]. Turgor value is for *E. clathrata*[p]
Acetabularia mediterranea	355	65	480	oxalate:55 Mg^{2+}:50	v.low	90	Marine[q]
Ceramium rubrum	515	31	573	Mg^{2+}:75; SO_4^{2-}:44	5[a]	100	Marine[d]. Mannoglycerate may be accumulated[r]
Griffithsia pulvinata	550	60	650	+	(15)	(65)	Marine[s]. Turgor value from[t]
Porphyra perforata	480	51	81	+	low?	(50)	Marine. Water relations are complex[u, v]
Laminariales	542	45	(400)	mannitol:250	0–5[a, b]	80	Marine[l, w, x, y]
Chara ceratophylla	88	142	225	$Ca^{2+}+Mg^{2+}$:20 SO_4^{2-}:4	8[b]	100	Brackish water[z]
Chlorella pyrenoidosa	110	1	1	+	?	40?	Fresh water[aa]. Small vacuole
Nitella clavata	54	10	91	$Ca^{2+}+Mg^{2+}$:28	5[a]	85	Fresh water[bb]
Nitella flexilis	75	11	134	Ca^{2+}:9	6[b]	90	Fresh water[cc, dd]
Hydrodictyon patenaeforme	75	3	60	$Ca^{2+}+Mg^{2+}$:4 SO_4^{2-}:8	2[b]	100	Fresh water[ee]

[a] Turgor measured as the difference between freezing point depressions of vacuolar and external solutions. [b] Turgor measured by incipient plasmolysis. [c] Turgor measured directly. [d] KESSELER (1965). [e] AIKMAN and DAINTY (1966). [f] GUTKNECHT (1966). [g] VILLEGAS (1967). [h] JACQUES (1938a). [i] GESSNER (1969). [j] HASTINGS and GUTKNECHT (1974). [k] ZIMMERMANN and STEUDLE (1974). [l] BINET (1956). [m] JACQUES (1938b). [n] BLINKS and JACQUES (1929). [o] BLACK and WEEKS (1972). [p] BIEBL (1956). [q] SADDLER (1970). [r] FOGG (1953). [s] FINDLAY et al. (1969). [t] TRAMÈR (1957). [u] GUTKNECHT and DAINTY (1968). [v] EPPLEY and CYRUS (1960). [w] EPPLEY and BOVELL (1958). [x] BONEY (1965). [y] ALLEN et al. (1972). [z] COLLANDER (1930). [aa] BARBER (1968). [bb] HOAGLAND and DAVIS (1923). [cc] TAZAWA (1964). [dd] KISHIMOTO and TAZAWA (1965). [ee] BLINKS and NIELSON (1939).

Table *11*.2. Principle components of the osmotic pressure in leaves and shoots of higher plants

Species	K^+ (mM)	Na^+ (mM)	Ca^{2+} (mM)	Mg^{2+} (mM)	Cl^- (mM)	SO_4^{2-} (mM)	Organic acids (μeq g^{-1})	Sugars (mM)	Osmolarity (osmol m^{-3})	Osmolarity accounted for (%)
Sorghum vulgare[g]	179	1	32	37	26	11	–	–	495	100[a]
Medicago sativa[g,h]	265	25	54	31	25	19	200	–	660	80–90
Gossypium hirsutum[g,i]	228	18	126	54	18	92	–	–	620	100
	77	8	99	23	11	36	250[b]	10	–	
Hordeum vulgare[j] (cf.[k])	228	22	18	18	72	8	–	15	450	100[a]
Beta vulgaris[g,l]	167	229	1	51	44	20	–	–	610	100
	200	68	22	57	23	10	274[c]	17	–	
Lycopersicum esculentum[g,m]	161	6	58	42	25	75	–	–	425	100
	58	19	40	15	12	11	180[d]	–	–	
Phaseolus vulgaris[n]	170	0	45	20	5	–	–	–	360	100[a]
Capsicum frutescens[n]	190	13	2	30	5	–	–	–	360	100[a]
Potamogeton schweinfurthii[o]	223	37	73	37	133	12	–	–	550	83
Chenopodiaceae[p,q] (9 semi-arid spp)	75	370	38	34	50	–	140[e]	–	–	–
Salt marsh plants[r,v] (6 spp)	80	500	–	–	480	18	–	–	1250	87
Dune plants[r] (11 spp)	110	96	–	–	166	9	–	–	760	50
Daucus carota[s] tissue culture	113	3	–	–	10	–	105[f]	270	430	100
Helianthus annuus[u] expanding hypocotyl tissue	30	3	<2	–	5	1	–	100	190	90[a]

[a] Assuming the presence of divalent organic ions sufficient to balance the inequality between inorganic ion equivalents. [b] Mainly citrate and malate. [c] 240 oxalate. [d] 130 malate. [e] oxalate. [f] Probably mainly malate[t]. [g] EATON (1942). [h] PIERCE and APPLEMAN (1943). [i] ERGLE and EATON (1949). [j] GAUCH and EATON (1942). [k] PITMAN and CRAM (1973). [l] BRETELER (1973). [m] KIRKBY and MENGEL (1967). [n] BERNSTEIN (1963). [o] DENNY and WEEKS (1968). [p] OSMOND (1963). [q] OSMOND (1968). [r] ÖNAL (1966). [s] CRAVEN et al. (1972). [t] CRAM (1973a). [u] MCNEIL (1975). [v] STEWART and LEE (1974).

This table was compiled from data for plants grown under very different conditions, and comparisons between species must take this into account. The range of variation is indicated by the values for *Gossypium*, *Beta* and *Lycopersicum*. Data for sugar and nitrate levels under comparable conditions are scanty.

Table *11*.3. Contributions of sugars and Cl^- to osmotic pressure in higher plants in natural habitats

Plant type	No. of species	Osmotic pressure (100 kPa) (Range)	Cl^- conc. (mM)	Cl^- (as NaCl) % of osmotic pressure)	Sugar (mM)
Mangroves [a]	15	32 (23–46)	475	66±8	80 (10–180)
Halophytes [a]	16	33 (14–44)	475	65±11	–
Salt marsh plants [b]	18	30 (15–40)	338	50±22	76 (10–180)
Glycophytes [a]	30	14 (9–20)	67	23	–
Glycophytes [b]	28	17 (10–25)	–	–	300 (90–600)
Glycophytes [b]	13	13 (9–22)	7	3 (0–8)	130 (50–250)
Cultivated plants [a]	20	16 (10–23)	69	18	–
Water plants [a]	14	15	143	44 (29–65)	–

[a] From ARNOLD (1955). [b] from STEINER (1939). See also STEINER and ESCHRICH (1958).

different shape) within 30 min. During shrinkage the internal osmotic pressure must rise, and recovery must be caused by an increase in the number of osmotically active particles in the cell at constant osmotic presure. The main osmoticum to change during recovery appears to be α-galactosyl glycerol (isofloridoside), whose intracellular concentration after recovery of the original volume equals 80% or more of the increase in external osmotic pressure (KAUSS, 1967; SCHOBERT et al., 1972; KAUSS, 1973, 1974). A concomitant decrease in contractile vacuole activity (ETTL, 1961; STONER and DUNHAM, 1970) would account for a smaller increase in internal than in external osmotic pressure.

The volume maintenance system therefore appears to involve the regulation of the reversible formation of isofloridoside from the related polymer, chrysolaminarin. After increasing the external osmotic pressure, both the rate of formation of isofloridoside and the activity of one of the enzymes involved increase; and this occurs within 2 min of increasing the external osmotic pressure, and is not affected by actinomycin D or CHM (KAUSS, 1969; KAUSS and SCHOBERT, 1971; SCHOBERT et al., 1972; KAUSS, 1973). It would therefore appear that a pre-existing enzyme is activated, and that a change in external osmotic pressure is sensed not by the enzyme but by some property of the intact cell, such as tension in the plasmalemma or microtubules (BOUCK and BROWN, 1973).

Other changes apparently involved in volume maintenance are a decrease in the rate of isofloridoside formation during the volume recovery phase, and a sudden increase in the rate of disappearance of isofloridoside on transfer back to a low external osmotic pressure (KAUSS, 1973). These changes may all possibly result from non-linear relationships between rates of formation and removal of isofloridoside and cell volume or shape, but they cannot be explained by linear relationships (i.e. by simple proportional control) without postulating more than one primary signal.

In other cells a range of organic substances has been found to respond to changes in external osmotic pressure (Table *11*.4). Except for cyclohexanetetrol these are all fairly simple molecules not far removed from the central biochemical pathways. The primary signal for these adaptive responses must be a function of cell volume in wall-less cells, while it is most probably a function of turgor in walled cells.

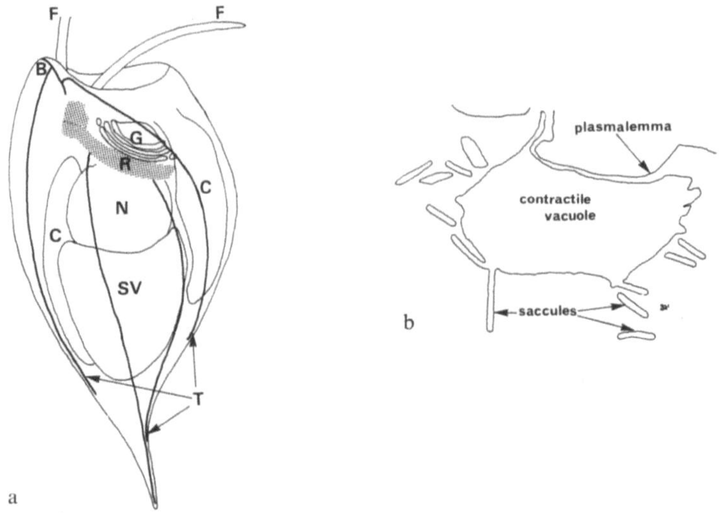

Fig. *11*.3. (a) Details of the structure of *Ochromonas danica*. *B* beak; *C* chloroplast; *F* flagellum; *G* Golgi apparatus; *N* nucleus; *R* rhizoplast; *SV* storage vacuole; *T* microtubules. There are two sets of microtubules maintaining the cell's asymmetric shape, radiating from the beak and the rhizoplast. Two only of each set are shown. Details of the eyespot, contractile vacuole, mitochondria and flagellar bases at the apical end of the cell are not illustrated. Redrawn from BOUCK and BROWN (1973). The storage vacuole contains chrysolaminarin. The interconversion of this polymer and the related monomer is the main process in volume maintenance of the cell. (b) Details of the fine structure of the contractile vacuole of *Ochromonas malhamensis*. Note the absence of any cell wall, and the unusual saccules whose internal contents appear to be continuous with that of the contractile vacuole. The diameter of the contractile vacuole is approximately 50 μm. Redrawn from AARONSON and BEHRENS (1974)

In some cases the response is sufficient to account for volume or turgor maintenance, but in others it is too small. In yet other cases the response is large, but the amounts involved are insufficient to play any part in volume or turgor maintenance.

In terrestrial higher plants starch-sugar interconversions have long been thought to be involved in the generation or maintenance of hydrostatic pressure in stomatal guard cells and wilting cells (MOTHES, 1956; VAADIA et al., 1961; HSIAO, 1973; Part B, Chap. *4.3*). Similarly, increases in the concentrations of unknown, probably organic substances have been found to account for the increase in osmotic pressure in leaves in response to increases in osmotic pressure of the growth medium (SLATYER, 1961; JARVIS and JARVIS, 1963; JANES, 1966; LAWLOR, 1969; MEYER and BOYER, 1972), and the accumulation of large amounts of proline in some salt marsh plants has been observed (STEWART and LEE, 1974). The quantitative significance of such changes has not yet been established, but the production of the organic compound may be signalled by a decrease in hydrostatic pressure and thus may be part of a system maintaining hydrostatic pressure.

The control of the volume of the cytoplasm is a problem analogous to the control of the volume of a wall-less cell, since the tonoplast has no structural rigidity. Transport of salt from the cytoplasm to the vacuole seems the most likely process preventing Donnan swelling of the cytoplasm (cf. DAINTY, 1968; MCNEIL, 1975). Cytoplasmic volume could be maintained constant by a signal from cytoplasmic volume to transport or to the production of an organic substance (e.g. proline—cf. STEWART and LEE, 1974; CHU et al., 1975).

Table *11*.4. Algae and fungi in which the concentration of an organic compound(s) increases in response to increasing external osmotic pressure

Species	Compound	Response	Notes
Algae			
Ochromonas malhamensis (fresh water)	isofloridoside (α-galactosyl glycerol, IF)	$\Delta c_i \geqq 80\% \; \Delta \pi_o$	Volume recovers [a]
Scenedesmus obliquus (fresh water)	soluble carbohydrates	$\Delta c_i < \Delta \pi_o$	Recovers from plasmolysis. Turgor constant? Light-dependent. Quantities uncertain [b]
Tetrahymenia pyriformis (fresh water animal)	amino acids	$\Delta c_i = \Delta \pi_i$	Volume recovers. Contractule vacuole activity decreases at higher π_o [c]
Nitella flexilis (fresh water)	unknown	$\Delta \pi_i < \Delta \pi_o$	[d,e]
Platymonas suecica (marine)	mannitol	$\Delta c_i = \Delta \pi_o$	Walled. Turgor recovers? [f]
Monochrysis lutheri (marine)	cyclohexanetetrol	$\Delta c_i = \Delta \pi_o$	Volume recovers. Light-dependent [g]
Dunaliella parva *D. tertiolecta* *D. viridis* (marine and halophylic)	glycerol	$\Delta c_i < \Delta \pi_o$	Volume recovers. In light or dark [h, i, j, k]
Phaeodactylum tricornutum (marine)	amino acids (mainly proline)	several-fold	Absolute amounts are osmotically insignificant [l]
Ulva fasciata (marine)	amino acids and sugars	several-fold	Absolute amounts are osmotically insignificant [m]
Iridophycus flaccidum *Porphyra perforata*	floridoside and isofloridoside	absolute amounts uncertain	[n] Widespread in red algae [o]
Lichen			
Lichinea pygmaea	mannoside-mannitol	location and quantities uncertain	[p]
Fungi			
Dendryphiella salina	mannitol and arabitol	$\Delta c_i = \Delta \pi_i$	[q] Vegetative stage tolerates sea water. See also [r]
Saccharomyces rouxii	arabitol		[s] Not accumulated in sugar-intolerant yeasts

[a] KAUSS (1973, 1974). [b] WETHERELL (1963). [c] STONER and DUNHAM (1970). [d] KAMIYA and KURODA (1956). [e] TAZAWA (1961). [f] HELLEBUST (1973). [g] CRAIGIE (1969). [h] BEN-AMOTZ and AVRON (1973). [i] BOROWITZKA and BROWN (1974). [j] WEGMANN (1971). [k] BEN-AMOTZ (1974). [l] BESNIER et al. (1969). [m] MOHSEN et al. (1972). [n] KAUSS (1968). [o] MEEUSE (1962). [p] FEIGE (1973). [q] JENNINGS (1973). [r] JENNINGS and AUSTIN (1973). [s] BROWN and SIMPSON (1972).

3.1.3 Control of the Total Number of Osmotically Active Particles by Transport

3.1.3.1 Giant Algal Cells

Several cases in which transport responds to changes in internal osmotic or hydrostatic pressure, and is involved in homeostasis of internal osmotic or hydrostatic pressure, have been found in giant algal cells whose vacuolar contents can be artificially altered.

Valonia (see Fig. *6*.3) is a genus of giant marine algae which grow and maintain turgor constant in a range of salinities (HASTINGS and GUTKNECHT, 1974; ZIMMERMANN and STEUDLE, 1974; see also HASTINGS and GUTKNECHT, 1975). In the laboratory, turgor recovers after a change in external osmotic pressure within 10–30 h (ZIMMERMANN and STEUDLE, 1974; GESSNER, 1969; see also HASTINGS and GUTKNECHT, 1975). The principle component of its internal solutes is KCl (Table *11*.1). Active K^+ influx (GUTKNECHT, 1966, 1967; see also HASTINGS and GUTKNECHT, 1975) must be the main process generating the internal osmotic pressure and hence turgor, and it is therefore significant that K^+ influx varies with turgor.

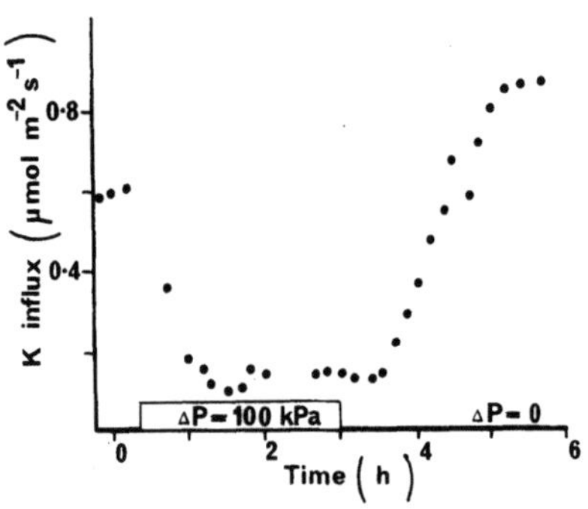

Fig. *11*.4. The effect of changing intracellular hydrostatic pressure on the active K^+ influx to *Valonia ventricosa*. The intracellular hydrostatic pressure is raised to 100 kPa (1 bar) above the extracellular hydrostatic pressure over the period indicated. Redrawn from GUTKNECHT (1968)

K^+ influx increases 4-fold or more after normal turgor is reduced to zero by inserting capillaries (JACQUES, 1938a; GUTKNECHT, 1968), and can be reversibly reduced to its normal value by applying an internal hydrostatic pressure (Fig. *11*.4; GUTKNECHT, 1968; HASTINGS and GUTKNECHT, 1974). K^+ efflux and urea influx and efflux (GUTKNECHT, 1968) are not significantly affected by turgor changes, so that the effect on K^+ influx appears to be fairly specific. On the other hand, in *V. utricularis* the membrane conductance (half of which is carried by K^+ in normally turgid *V. ventricosa*—GUTKNECHT, 1967) decreases as turgor rises from 0 to 100 kPa, but rises again at higher turgor (ZIMMERMANN and STEUDLE, 1974). It was suggested that increasing turgor stimulates K^+ efflux as well as inhibiting (and possibly reversing) active K^+ influx, but the nature of the ions carrying the current remains to be established.

A hydrostatic pressure difference of the magnitude found in *Valonia* does not constitute a significant driving force on ion movements, and therefore the change in turgor must be transduced and amplified before it takes effect, i.e. it must be a signal. Even the most pressure-sensitive biological process would respond by only about 0.3% to a 100 kPa (1 bar) isotropic pressure change (JOHNSON and EYRING, 1970), so it is as expected that a general increase in pressure within and without the cell has no effect on K^+ influx (HASTINGS and GUTKNECHT, 1974). The first response is therefore probably the 0.4% per 100 kPa anisotropic change in the area of the plasmalemma (VILLEGAS, 1967), which could significantly alter rates of processes fixed in the membrane. But see also COSTER et al. (1976).

The slowness of the response (GUTKNECHT, 1968; HASTINGS and GUTKNECHT, 1974) suggests that the intermediate steps are chemical rather than physical. The quantitative relationship between K^+ influx and the error signal (deviation from normal turgor) has yet to be investigated. Since turgor does not vary with changes in external K^+ concentration (ZIMMERMANN and STEUDLE, 1974, but cf. JACQUES and OSTERHOUT, 1931) the turgor signal must override the potentially perturbing effect of external K^+ concentration on K^+ influx at the steady state.

The effect of turgor changes on K^+ influx can qualitatively be understood as a negative feedback signal involved in turgor homeostasis in a fluctuating environment and during growth. It probably serves not only to keep the cell turgid but also to prevent turgor from rising excessively, since the large size of the cells makes them more susceptible than smaller cells to bursting (cf. GUTKNECHT and DAINTY, 1968).

A similar turgor-dependent transport system occurs in *Halicystis*, and in *Codium* (BISSON and GUTKNECHT, 1975), where active Cl^- influx increases 10-fold in response to a fall in turgor of only 30 kPa (0.3 bar) (JACQUES, 1938b).

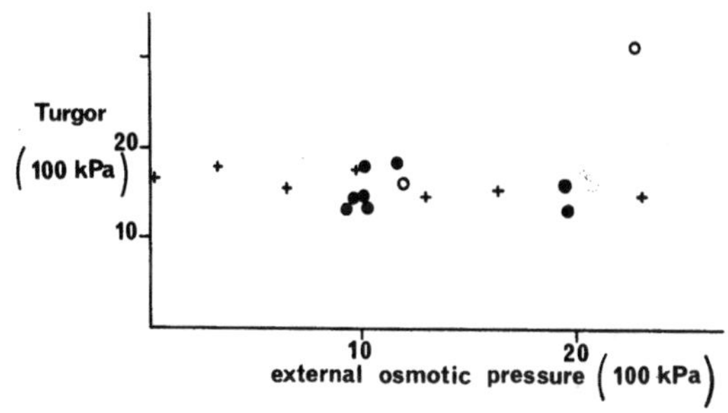

Fig. *11*.5. Turgor constancy in *Chaetomorpha linum* adapted to a range of dilutions or concentrations of sea water. Natural habitats: ● (KESSELER, 1958, 1965). In the laboratory: + (KESSELER, 1959), ○ (ZIMMERMANN and STEUDLE, 1971)

Chaetomorpha linum, a multicellular temperate marine alga, is very like *Valonia* in having KCl as its main internal solute and in maintaining turgor constant in a range of salinities (Table *11*.1, Fig. *11*.5 KESSELER, 1959, 1964; STEUDLE and ZIMMERMANN, 1971; ZIMMERMANN and STEUDLE, 1971). K^+ fluxes have not been measured in *C. linum*, but K^+, rather than Cl^-, is actively transported inwards in *C. darwinii* (Fig. *6*.3) (FINDLAY et al., 1971). *Chaetomorpha* differs from *Valonia* in maintaining a higher turgor (1,500–3,000 kPa) over a wider range of salinities both in its natural habitats and in artificially diluted sea water (Fig. *11*.5). It also differs in that its turgor depends on (rather than being independent of) external K^+ concentration. Turgor changes to a limited extent when external K^+ concentration changes at constant external osmotic pressure, as well as remaining constant when external osmotic pressure changes at constant K^+ concentration (STEUDLE and ZIMMERMANN, 1971; ZIMMERMANN and STEUDLE, 1971). The change in internal solute concentration in diluted sea water may therefore be partly a response to decreasing external K concentration, as well as to increasing turgor as in *Valonia*.

Nitella is a temperate fresh-water algal genus with giant internodal cells whose vacuolar contents (mainly KCl—Table *11*.1) can be altered by perfusion (TAZAWA, 1964) or by transcellular osmosis (KAMIYA and KURODA, 1956). The latter procedure sweeps solutes from one end of the cell to the other, leaving more concentrated (H) and less concentrated (L) ends which can be tied off and separated. (H) cells lose KCl and (L) cells gain KCl over 7 days until the original vacuolar K^+ concentration and hence also turgor are restored (KAMIYA and KURODA, 1956; TAZAWA, 1961; TAZAWA and NAGAI, 1960, 1966). The recovery of internal osmotic pressure in (L) cells is determined by the photosynthetically driven, temperature-sensitive, active influx of Cl^- with coupled flows of K^+ and Na^+ (TAZAWA, 1961; KISHIMOTO and TAZAWA, 1965; TAZAWA and NAGAI, 1966; NAKAGAWA et al., 1974). Recovery of internal osmotic pressure in (H) cells by loss of salts differs from recovery in (L) cells in light and temperature sensitivity and in cation selectivity (TAZAWA, 1961; NAKAGAWA et al., 1974). This, together with the observation of recovery of (L) cells in moist air (Table *11*.4, KAMIYA and KURODA, 1956; TAZAWA, 1961) suggests that more than one process may respond to the same primary signal.

The initial net flux of KCl in (L) or (H) cells is proportional to the difference between the initial osmotic pressure in the cell and the normal value (NAKAGAWA et al., 1974), suggesting a proportional negative feedback to net KCl flux from an internal parameter, which might be turgor, osmotic pressure, K^+ concentration, or Cl^- concentration.

If turgor were the signal in *Nitella*, as in *Valonia* and *Chaetomorpha*, then lowering turgor by raising the external osmotic pressure should increase net KCl influx. However, 250 mM sucrose in the external medium did not stimulate KCl influx in normal cells (TAZAWA, 1961) nor did it decrease KCl loss in (H) cells. Indeed, loss of KCl from (H) cells in 340 mM sucrose continued to such an extent that the cells plasmolyzed themselves (KAMIYA and KURODA, 1956). Thus turgor is not the regulatory signal for KCl influx, nor is turgor homeostatically controlled in *Nitella*.

Vacuolar perfusion allows the internal osmotic pressure and K^+ concentration to be varied independently. When osmotic presure is lower but K^+ concentration is higher than normal, so that possible tendencies for the two to return to normal oppose, it is found that osmotic pressure returns to normal while KCl concentration increases (NAKAGAWA et al., 1974). This, and several other experiments, show that the main signal to KCl transport is more closely related to osmotic pressure than to internal K^+ concentration. The relationship to internal Cl^- concentration remains to be examined.

The osmotic pressure of fresh water is so much less than that of the vacuole of *Nitella* that the largest natural fluctuations would have no significant effect on turgor. Hence a mechanism for maintaining constant internal osmotic pressure would also ensure constancy of turgor. The *Nitella* system may therefore still be seen as a turgor homeostat, its natural function being to maintain turgor during growth, but the question then remains as to why internal osmotic pressure, rather than turgor, should have been selected as the primary control signal.

The primary detector of internal osmotic pressure may be the volume of an organelle (cf. *Ochromonas*). It will be interesting to examine the situation in one of the brackish water Characeae.

3.1.3.2 Other Algae and Higher Plants

In several other algae, turgor also remains constant in a range of salinities in different habitats or in diurnally varying salinities in the same habitat (*Chaetomorpha aerea, Bryopsis plumosa, B. hypnoides, Ceramium rubrum*—KESSELER, 1965; *Enteromorpha clathrata* and other species referred to by BIEBL, 1962; GUILLARD, 1962; GUTKNECHT and DAINTY, 1968). This suggests the operation of turgor-dependent regulation of the accumulation of at least the principle osmoticum, which in some cases is known (Table *11*.1). However, when the external fluctuations are slow compared with the rate of growth the contribution of turgor-dependent expansion growth (*11*.3.1.4) to the constancy of internal turgor needs to be established.

In contrast to *Valonia* and some other algae previously discussed, higher plant cells may have a somewhat variable turgor (during growth—see *11*.3.1.2; and in active, nongrowing cells—ARISZ, 1947; SUTCLIFFE, 1954a; ORDIN et al., 1956; MOTT and STEWARD, 1972a; MOWAT, 1973; CRAM, 1974a).

The influence of turgor on transport processes in several higher plant tissues has been examined. The results of experiments in which turgor is reduced by raising the external osmotic pressure with a non-penetrating osmoticum such as mannitol (but cf. TRIP et al., 1964) are shown in Table *11*.5. In most cases decreasing turgor does not increase the influx of KCl, which is perhaps not surprising, since K-organate, not KCl, is the principle component of osmotic pressure in most glycophytes (Table *11*.2). In a few cases turgor has been found to have a distinct effect on transport.

In *Beta vulgaris* (perhaps only coincidentally a halophyte) K^+ influx is stimulated by reducing turgor. In washed root tissue in 20 mM KCl, turgor is about 1,000 kPa and net K^+ influx is 7 μmol g_{FW}^{-1} h^{-1}. In KCl loaded tissue, turgor is about 1,900 kPa and net K^+ influx is 0.4 μmol g_{FW}^{-1} h^{-1}, but if the external osmotic pressure is raised by 1,000 kPa, so reducing turgor to 900 kPa, net K^+ influx increases to 2.4 μmol g_{FW}^{-1} h^{-1} (SUTCLIFFE, 1954a).

Bryophyllum daigremontianum exhibits typical Crassulacean acid metabolism, with a diurnal fluctuation of malic acid concentration in the leaves which is maintained for some time in continuous light or darkness. During the dark phase malate is transported (possibly actively) from the cytoplasm to the vacuole and vacuolar malate concentration and osmotic pressure increase. The efflux of malate and protons from slices of such acidified leaves is high, but the efflux is reduced 5 to 10 times when turgor is reduced from 650 kPa to 150 kPa (LÜTTGE and BALL, 1974a). This effect is specific since K^+ and Cl^- influxes are not affected by the same change in turgor (LÜTTGE and BALL, 1974b). The flux(es) varying with turgor have not yet been defined in detail.

The operation of turgor-dependent transport processes in terrestrial plants is also suggested by the observation of "osmotic adaptation", superficially similar to the responses of algae to varying salinities discussed above. In many plants the osmotic pressure of leaf extracts increases when the external osmotic pressure is raised (references in MAGISTAD, 1945; HAYWARD and WADLEIGH, 1949; BERNSTEIN and HAYWARD, 1958; SLATYER, 1963, 1969; WAISEL, 1972; CRAM, 1974a; CHU et al., 1975). The increase in osmotic pressure in the leaf extract is often approximately equal to, but may also be greater or less than, that in the external solution. In NaCl the internal osmoticum which changes is generally NaCl, but K^+ and organic acids may also change (BERNSTEIN, 1963; MEIRI et al., 1970), and in non-penetrating osmotica organic substances may increase (references in previous section). In a few cases water potential and osmotic pressures have been measured simultaneously and hence turgor has been calculated and found to increase to about its original value during the recovery phase (LAWLOR, 1969; GREACEN and OH, 1972; MEYER and BOYER, 1972; CHU et al., 1975). Growth recovery after increasing the external

Table 11.5. The effect of increased external osmotic pressure on K^+, Cl^- and organic acid fluxes in plant cells and tissues

Species and tissue	Measured flux	External osmotic pressure increase ($osmol\ m^{-3}$)	Effect	Notes
Valonia ventricosa	K^+ influx	90	×1.5	Effect underestimated [a]
Nitella flexilis	Net K^+	250	none	[b]
Citrus sinensis leaf slices	Cl^- influx	600	none	[c]
Pisum sativum leaf fragments	light-dependent K^+ influx	260	none	[d]
	dark K^+ loss	260	decreased	[d]
Bryophyllum daigremontianum leaf slices	H^+ efflux	200	decreased	[e] Note the specificity of the effect
	malate efflux	200	(×0.1–0.2)	
	K^+ and Cl^- influx	200	none	[f]
Avena sativa coleoptiles	Cl^- influx	300	×1.3	[g] Less stimulation with IAA present
Zea mays root freshly excised	Cl^- influx	200	none	[h]
	net Cl^-	150	none	
	net K^+	150	none	
KCl loaded	Cl^- influx	200	none	[h]
	net Cl^-	150	none	
	net K^+	150	none	
Hordeum vulgare root KNO_3 loaded	Br^- influx	450	none	[i] No correlation of Cl^- influx with turgor [a]
Daucus carota fresh cut storage tissue	Cl^- influx	200	none	[j]
	net Cl^-	200	none	
washed storage tissue	Cl^- influx	200	none	[k, l]
	net Cl^-	200	none	
Beta vulgaris washed storage tissue	net K^+	220	×1.5–6	[m] Greater stimulation in more turgid tissue

[a] CRAM (1973a). [b] TAZAWA (1961). [c] ROBINSON and SMITH (1970). [d] NOBEL (1969). [e] LÜTTGE and BALL (1974a). [f] LÜTTGE and BALL (1974b). [g] RUBINSTEIN and LIGHT (1973). [h] CRAM (1973b). [i] GREENWAY et al., (1968). [j] CRAM (1972). [k] PITMAN and CRAM (1973). [l] BRIGGS (1971). [m] SUTCLIFFE (1954a).

osmotic pressure is not correlated with the ability of the leaf to adjust its osmotic pressure (e.g. HAYWARD and WADLEIGH, 1949; BERNSTEIN and HAYWARD, 1958; WAISEL, 1972; GREENWAY, 1973; CHU et al., 1975).

There are two points to be made about these observations. First, a decrease in turgor would decrease cell expansion growth (11.3.1.4), and this would automatically lead to a build-up of turgor if transport continued unchecked. This in turn would tend to restore turgor and turgor-dependent expansion growth. Thus the system is inherently at least partly self-regulating in a passive sense (cf. Fig. *11*.1) as pointed out by GREENWAY and THOMAS (1965), GREENWAY (1968), and OERTLI (1971).

Secondly, a transient change in growth rate with constant flux cannot account for all the change in internal osmotic pressure observed, and this suggests that some of the responses may also involve active negative feedback regulation (in the sense of *11*.2.1) of the transport of the major osmotica, similar to that in *Valonia*. The simultaneous measurements of flux and turgor necessary to examine this possibility in a range of ecological types have yet to be made. Flux measurement *in vivo* appears a difficult problem, and one must not overlook the influence on the flux of a changing concentration in the extracellular spaces (cf. OERTLI, 1968).

It may also be dangerous to consider the leaf cells in isolation, since the net flux into the leaf is regulated by uptake into the root and/or phloem unloading, as well as (or instead of) uptake by the leaf cells (PITMAN and CRAM, 1973; PITMAN et al., 1974; MCNEIL, 1975).

3.1.4 Turgor and Volume Maintenance in a Fluctuating Environment and during Growth

PFEFFER (1900) pointed out that "all cells which can accommodate themselves to concentrated nutrient solutions must be able to increase the amount of osmotic substances which they contain. This is attained either by a direct absorption of the salts present in the external medium, or by a corresponding increased production of osmotically active substances". This we can amplify in two ways. Some cells maintain volume rather than turgor, though in the case of some red algae with elastic walls (EPPLEY and CYRUS, 1960), volume and turgor maintenance are not distinct: and the accommodation is achieved by a negative feedback signal from turgor or volume to the transport or biochemical process of accumulation, and perhaps partly in some cases by a growth response. We do not yet know the quantitative behaviour of such feedback systems, their mechanism, or the variants in different cell types.

Other cases of osmotic responses to *external* perturbations are the rapid volume changes associated with K-organate movements in walled cells such as stomatal guard cells (Part B, Chap. *4.3*), *Mimosa pudica* pulvini (ALLEN, 1969), and *Albizzia* leaf pinnules (SATTER and GALSTON, 1971).

The regulation of transport to maintain turgor and volume in the face of *internal* perturbing factors has received less attention, although it has long been realized that "since during the growth and stretching of the cell wall, turgidity remains fairly constant, a regulatory and correlating adjustment must continually go on" (PFEFFER, 1900).

The rate of cell expansion, dV/dt, is proportional to turgor (P) in excess of a critical value (P_c): –

$$dV/dt = \varepsilon(P - P_c) \qquad (11.7)$$

where ε is a coefficient describing cell-wall extensibility. This equation is obeyed by coleoptiles (CLELAND, 1967; RAY, 1969), roots (GREACEN and OH, 1972), leaves (LAWLOR, 1969; ACEVEDO et al., 1971; MEYER and BOYER, 1972), cotyledons (KIRKHAM et al., 1972), and *Nitella* (GREEN et al., 1971).

Initiation of growth is associated most often with a slight fall in turgor (URSPRUNG and BLUM, 1924; ORDIN et al., 1956; BURSTRÖM, 1961; MOTT and STEWARD, 1972b; MCNEIL, 1975), though not always (NEEB, 1952). During expansion growth (BURSTRÖM, 1961; GREEN et al., 1971; CRAVEN et al., 1972; MCNEIL, 1975) and at the end of growth (LOCKHART, 1960; MCNEIL, 1975) turgor remains fairly constant. (In some experiments on tissues *in vivo* only the overall osmotic pressure was measured, and turgor may not have been so constant.) Thus changes in wall extensibility (ε and P_c) appear to be the factors initiating, maintaining, and ending growth, although the situation is complicated by the fact that wall synthesis is itself turgor-dependent (ORDIN, 1960; SCHRÖTER and SIEVERS, 1971; LOESCHER and NEVINS, 1973). A mechanism that could maintain turgor in a fluctuating external osmotic pressure—a negative feedback regulation of accumulatory processes—could also serve to keep internal osmotic pressure from falling excessively while the cell expands, and prevent its rising excessively when growth stops.

3.2 Control of the Uptake and Accumulation of Specific Substances in Plant Cells

3.2.1 Halide and Alkali Cations

The first suggestion of regulation of transport in higher plant cells came from HOAGLAND and BROYER'S observation (1936) of a decrease in net Br^- influx in roots which had accumulated salt. Their introduction of the term "high salt" tissue was perhaps unfortunate, as it obscured the fact that the accumulation of different salts has different effects on the influx of an ion (HUMPHRIES, 1951; CRAM, 1973a).

This study was followed by other workers in two ways (see also CRAM, 1973a). First, the question as to whether the decreased net influx in salt-loaded tissue was due to decreased influx or to increased efflux was examined in several tissues, although not always with the distinction between "pump and leak" and pump with negative feedback explicitly stated. The general conclusion was that "influx" did decrease, but, owing to the complexity of the quasi-steady flux measured, many of these results are not fully conclusive and some are contradictory (CRAM, 1968). Secondly, the question as to whether this decreased influx was due to a relationship of influx to the internal concentration of the substance transported, to turgor, or to sugar concentration, was investigated.

More recently it has been shown that during the accumulation of KCl there is a reduction in the influx of Cl^- across the tonoplast in carrot, maize, and barley, and also in Cl^- influx across the plasmalemma in barley and sometimes in carrot (CRAM, 1968; CRAM and LATIES, 1971; CRAM, 1973a, 1973b; cf. discussion in PITMAN, 1969).

The fall in influx of Cl^- to carrot and barley root tissue is related to the rise in the internal concentration of Cl^- plus NO_3^- (CRAM, 1973a). After loading with various salts, all the variation in Cl^- influx to the vacuole (Φ_{ov}) was accounted for by a relationship of the form

$$\Phi'_{ov} = \Phi^{o}_{ov} - A \cdot \log [Cl^- + NO_3^-]_v \qquad (11.8)$$

where Φ^{o}_{ov} is the influx in non-loaded tissues, A is a constant for each tissue, and $[Cl^- + NO_3^-]_v$ is the vacuolar concentration. It is difficult to recast this into a form where an error signal is apparent. Other cases of decreased KCl influx in KCl- or KNO_3-loaded tissues are listed in Table *11*.6. In most cases the location and specificity of the effects have not been established, but it will be assumed that they are all cases of the same negative feedback phenomenon.

The absence of a relation between Cl^- influx and reducing sugar concentration in barley and carrot tissues (HOAGLAND and BROYER, 1936; CRAM, 1973a) and the fact that respiration rate changes very little during accumulation in barley and beet tissues (SUTCLIFFE, 1952; JACKSON and EDWARDS, 1966) shows that the reduced influx in KCl-loaded tissues is not due to respiratory substrate limitation of energy supply to active transport. On the other hand, a relationship of KCl influx to the level of reducing sugar and respiration rate was observed over the period of loading of barley roots with KCl (PITMAN et al., 1971), and ringing or darkening the shoots of plants decreases the flow of sugar to the roots and also the influx of KCl (PITMAN, 1972; HATRICK and BOWLING, 1973). Hence, since it does not invariably accompany the reduction in Cl^- influx during the accumulation of KCl or KNO_3, a fall in hexose concentration cannot be the essential mechanism of the feedback system; but some indirect relationship of KCl (and KNO_3) transport to intracellular hexose concentration would not be unexpected.

KCl is accumulated to a more or less constant internal concentration independent of the external concentration, despite the fact that the latter alters KCl influx in the non-loaded root (PITMAN, 1969; Part B, Chap. *3.3*; but see also NEIRINCKX and BANGE, 1971). The negative feedback system may therefore be part of a homeostat for internal KCl concentration, or at least contribute to KCl concentration maintenance (possibly in conjunction with changes in membrane permeability—CRAM, 1973b). KCl accumulation cannot then at the same time be a turgor maintenance system (CRAM, 1974a); and the fact that KCl influx is not sensitive to turgor (CRAM, 1973a, 1973b; *11*.3.1.3.2; Table *11*.5) is as expected.

The function of such a regulatory system for Cl^- transport is not apparent, since Cl^- has no known function at high concentration, and, in the shoots of glycophytes at least, is not normally accumulated to these levels. It is not a question of Cl^- transport in a system which normally transports NO_3^-, since Cl^- and NO_3^- are transported independently (CRAM, 1973a; JACKSON et al., 1973). The non-specific effects on NO_3^-, malate, and SO_4^{2-} (Table *11*.6) suggest that the function is related to some more general aspect of inorganic ion accumulation.

In KCl loaded roots there is, in addition to the decreased Cl^- influx, an increased selectivity for K^+ and also a decreased H^+ efflux (PITMAN et al., 1968; Part B, Fig. *3*.15; these aspects are discussed in Part B, *3.3*.1). Thus linkage between flows of ions, as well as absolute rates of inflow, are controlled in relation to internal contents (see also *11*.5), but this does not appear to be a negative feedback regulation.

Table *11*.6. Apparently specific negative feedback effects of internal concentration on influx of univalent inorganic ions in higher plant cells

Tissue	External conc. of transported ion (mM)	Non-loaded		Salt-loaded		Inhibited rate	Notes
		Tissue conc. μmol g_{FW}^{-1}	Influx μmol g_{FW}^{-1} h^{-1}	Tissue conc. μmol g_{FW}^{-1}	Influx μmol g_{FW}^{-1} h^{-1}	(%)	
Beta vulgaris storage root	20 K^+	60	3.5	150	0.04	1	[a] Fluxes calculated approximately. Probable homeostatic action[b]. Internal NaCl also feeds back[c]. See also Ref.[d]
	20 Br^-	5	2.8	100	0.03	1	
Daucus carota storage root	1.5 Cl^-	19	1.7	102	0.15	9	[e] Main effect at tonoplast[f]. Probable homeostatic action[g]. NO_3^- also feeds back[e].
	5 NO_3^-	18 Cl^-	1.1	67 Cl^-	0.3	27	[e] Net flux measured
Hordeum vulgare root	1 Cl^-	12	5.8	68	0.8	14	[e] Effect at both plasmalemma and tonoplast[e, h].
	10 Cl^-	5	8	65	2.5	31	[i] NO_3^- also feeds back[e, k, l]. Probable homeostatic action[h, m]
	0.05 K^+	14	5.3	52	0.06	1	[i] Probable homeostatic action[n]
	10 K^+	15	13.5	95	5	37	[j] See also[o, p]
	10 Na^+	23	8.2	75	4.5	55	[q] Also Na^+ and K^+ interactions with ($Na^+ + K^+$) content[q] See also[r]
Zea mays Root cortex	0.1 Cl^-	5	1.0	60	0.7	70	[s] Main effect at tonoplast.
	50 Cl^-	5	6.4	84	1.6	25	Probable homeostatic action[s, t]
Root segment	0.1 K^+	30	4	70	0.4	8	[u] K/Na selectivity also changes
	30 K^+	30	7	70	4.5	65	[u]
	0.1 Rb^+	44	2.2	100	1.8	82	[v]
Triticum vulgare root	10 Br^-	(10)	5	(80)	2	40	[w] NO_3^- also feeds back
	10 K^+	(30)	8	(100)	3	38	
Allium cepa	2 Cl^-	–	4	–	0.4	10	[x] Tissue pretreated in $CaSO_4$ or KCl
Lemna minor	K^+	40	1.12	140	0.13	12	[y] Fluxes in μmol $m^{-2} s^{-1}$. See also[z]

[a] SUTCLIFFE (1954b). [b] SUTCLIFFE (1952). [c] SUTCLIFFE (1954c). [d] PITMAN (1963). [e] CRAM (1973a). [f] CRAM (1968). [g] MOTT and STEWARD (1972a). [h] CRAM and LATIES (1971). [i] A.D.M. GLASS, personal communication. [j] PITMAN et al. (1971). [k] SMITH (1973). [l] HOAGLAND and BROYER (1936). [m] JACKSON and EDWARDS (1966). [n] ASHER and OZANNE (1967). [o] JOHANSEN et al. (1970). [p] MARSCHNER and OSSENBERG-NEUHAUS (1970). [q] PITMAN et al. (1968). [r] HUMPHRIES (1956). [s] CRAM (1973b). [t] MEIRI (1973). [u] LEIGH and WYN JONES (1973). [v] VENRICK and SMITH (1967). [w] CSEH et al. (1970). [x] HODGES and VAADIA (1964). [y] YOUNG et al. (1970). [z] YOUNG and SIMS (1970).

Values of quasi-steady influx (Φ_{ov}) in tissue that has been variously loaded are shown. The loading conditions, temperatures, growth conditions, and external solutions vary greatly, so that quantitative comparisons cannot be drawn from the values in the Table. The order of magnitude of the effects (which were probably not maximal in some cases) can be seen in the right-hand column of figures (% inhibition).

Table *11.7*. Apparently specific negative feedback effects of internal substances on influx of metabolized ions into plant cells

Species	Transported ion	Feedback substance	Notes
Hordeum vulgare root	SO_4^{2-}	methionine	[a] At least 80% inhibition by externally supplied methionine, but cf[b]
	NO_3^-	NO_3^-?	Increased net uptake in deficient plants (*e.g.*[c,d])
	NO_3^-	Cl^-	50% inhibition by internal Cl[e]
	$H_2PO_4^-$	?	[c,f]
Daucus carota excised storage root tissue	NO_3^-	Cl^-	[g] up to 75% inhibition by internal Cl^-
	malate	Cl^-	[h] up to 50% inhibition by internal Cl^-
	SO_4^{2-}	methionine, cysteine, SO_4^{2-}, Cl^-	[i] inhibition by externally supplied methionine, but cf. [b]
Nicotiana tabacum callus	SO_4^{2-}	L-methionine L-cyst(e)ine L-isoleucine	[j] 50–80% inhibition by externally supplied amino acids. No other amino acid inhibits SO_4^{2-} uptake
	NO_3^-	NH_4^+ amino acids	[k] >10-fold inhibition by externally supplied NH_4^+ or casein hydrolysate
Riccia fluitans	SO_4^{2-}	?	[l] 5–8-fold stimulation after S starvation
Penicillium chrysogenum *P. notatum* *Aspergillus nidulans*	SO_4^{2-}	SO_4^{2-}, S-amino acids	[m,n] 40-fold stimulation after S starvation. Many reduced metabolites are known to be similarly controlled in fungi[o]
Chlorella pyrenoidosa	SO_4^{2-}	not SO_4^{2-} or S-amino acids	[p] 4-fold stimulation in S-starved cells. A metabolite acts directly on the transport system
	$H_2PO_4^-$	?	[q,r] 6-fold stimulation in P-starved cells, *via* gene action
Scenedesmus sp.	$H_2PO_4^-$	polyphosphate?	[s,t] Slight stimulation in P-starved cells
Monochrysis lutheri	SO_4^{2-}	methionine cysteine	[u] 3-fold stimulation in S-starved cells largely reversed by externally supplied S-amino acids

[a] FERRARI and RENOSTO (1972). [b] DAVIES (1966). [c] HUMPHRIES (1951). [d] IVANKO and INGVERSEN (1971). [e] SMITH (1973). [f] BARBER (1972). [g] CRAM (1973a). [h] CRAM (1974b). [i] W.J. CRAM, unpublished. [j] HART and FILNER (1969). [k] HEIMER and FILNER (1971). [l] THOIRON et al. (1970). [m] YAMOMOTO and SEGEL (1966). [n] BRADFIELD et al. (1970). [o] JENNINGS (1975). [p] VALLÉE and JEANJEAN (1968). [q] JEANJEAN (1973). [r] FUHS et al. (1972). [s] RHEE (1973). [t] KYLIN (1964). [u] DEANE and O'BRIEN (1975).

3.2.2 Inorganic Metabolites (NO_3^-, $H_2PO_4^-$ and SO_4^{2-})

Table *11.7* lists various cases in which negative feedback from an ion or its metabolic product to the influx of the ion appears to occur. The influx of the three major inorganic metabolites appears to be regulated in this way in representative higher plants, unicellular algae and fungi. Bacteria have also been extensively investigated but will not be considered here.

Several less clear cases, where increased net uptake has been observed in nutrient-deficient plants, have not been recorded. The main difficulty has been in characterizing the transport system involved. An increased net influx in deficient plants may simply be due to a smaller efflux associated with a low internal concentration; and an increased tracer influx may be due to more rapid metabolism. In both cases influx may be unchanged.

Nitrate transport is quantitatively the most extensive in many plants (LATSHAW and MILLER, 1924; DIJKSHOORN, 1970; RAVEN, *6*.5.5.2 and *6*.6.4.3). The four aspects of the regulation of its transport which have appeared show what may be general features of the regulation of nutrient transport.

1. NO_3^- transport is inducible in tobacco tissue culture (HEIMER and FILNER, 1971) and maize seedlings (JACKSON et al., 1973). (See also Part B, Chap. *9*).

2. It is not known if internal NO_3^- feeds back on NO_3^- influx, in the way that Cl^- does on both Cl^- and NO_3^- influx (*11*.3.2.1), since only net NO_3^- influx has yet been measured. NO_3^- is accumulated to a constant internal concentration independent of external concentration in *Chenopodium* (AUSTENFELD, 1972), suggesting homeostasis.

3. Of the products of NO_3^- reduction—NH_4^+, amino acids, and organic acids—internal malate has little if any effect on NO_3^- influx in barley and carrot root tissue (CRAM, 1973a; SMITH, 1973), while NH_4^+ and some amino acids may inhibit NO_3^- uptake in tobacco-tissue culture (HEIMER and FILNER, 1971). A specific negative feedback.

4. In non-loaded tissue NO_3^- influx is coupled to the influx of K^+; but in tissue loaded with K-malate, NO_3^- influx is independent of external K^+ (barley roots—SMITH, 1973; carrot root tissue—W.J. CRAM, unpublished). In the latter state, NO_3^- presumably exchanges, probably indirectly, with internal organic acids (CRAM and LATIES, 1974; see also OSMOND, *13*.2.3). If this were a change in K^+/H^+ selectivity it would be a similar type of regulation to the change in K^+/Na^+ selectivity referred to in *11*.3.2.1 and Part B, *3*.3.1.

Negative feedback regulation by metabolic products may be general (Table *11*.7). Such systems would appear to function in matching supply to demand during growth without excessive accumulation. Where the nutrient is also stored (phosphate as polyphosphates in algae—O'KELLEY, 1968—or nitrate in vacuoles) the feedback signal(s) for regulation of transport may be more complex.

4. Mechanisms of Negative Feedback Regulation

As will be apparent from *11*.3, as yet very little is known or can be guessed about the physical nature or dynamic characteristics of information flow in the negative feedback systems that have been found in plant cells. Nevertheless, the possible nature of each of the components of a feedback system shown in Fig. *11*.2 will be briefly considered.

The build-up of an electrochemical gradient in opposition to active influx (as with Na^+ and K^+ transport in red blood cells—LEW et al., 1970) would constitute a passive feedback. (It would be represented as a slight modification of Fig. *11*.1). Such a system has no capacity for homeostasis, and is best regarded as a non-regulatory interaction. It is not the origin of the decrease in Cl^- influx in carrot tissue accumulating KCl (CRAM, 1975).

Most negative feedback systems have been postulated after finding a correlation between a flux and some parameter which cannot be a driving force on the flux and which therefore appears to be a signal. Such parameters (e.g. the value of turgor or osmotic pressure) if established as the *primary negative feedback signal* must be regarded as an output of the transport system. The first step in the flow of information within the cell must be the measurement of this output. Intracellular devices for sensing the concentration of a specific substance may consist of single molecules; but those for measuring osmotic pressure, hydrostatic pressure or volume would have to be more elaborate. Hydrostatic pressure changes appear likely to be detected by an anisotropic response such as plasmalemma stretching (*11*.3.1.3). Volume could be detected similarly or by other organelles such as microtubules (*11*.3.1.2) or may be detected by the concentration of a substance present exclusively in the cell in fixed amount at each stage of growth. Osmotic pressure might be detected by the volume of a semi-permeable organelle.

The *reference value* (e.g. 300 osmol m^{-3} internal osmotic pressure in *Nitella flexilis*) and *comparator* may not be distinct from the primary signal detector in biological systems (cf. MILSUM, 1966). Some hormonal and other effects may involve modification of the reference value.

The *actuator*, transcribing the error signal into a controlling signal, may similarly not be distinct from the other parts of the controlling system.

The *controlling signal* might take a physical form (such as membrane stretching) or a chemical form. In the latter case the intracellular messenger must operate analogously to cyclic AMP in response to hormones in animal cells.

The signal may take effect on transport (the *controlled process*) by repression of synthesis of one or more components of the process (e.g. as for phosphate transport in *Chlorella*—JEANJEAN, 1973) or by a direct, presumably allosteric, inhibition of the process (as for SO_4^{2-} transport in *Chlorella*—VALLÉE and JEANJEAN, 1968).

Time lags could occur at any of these stages.

5. Interrelations between Systems Regulating Transport

The integration within an organism must involve a multiplicity of interactions, many of which must involve more than simple mass-action effects. One may therefore expect that many more regulatory interactions will be discovered. Negative feedback is just one type of regulation, serving to stabilize certain parameters in the cell amongst fluctuations and developmental changes.

In addition to the simple negative feedback regulation discussed in this Chapter, other more integrative regulatory inputs from related metabolites (possibly including energy sources) or from pH (SMITH and RAVEN, Chap. *12*) may be expected to occur, as well as hormonally mediated regulation from without the cell (VAN STEVENINCK, Part B, Chap. 7). These may alter the reference value of the system, or may alter parameters of it (such as the values of k or G in Fig. *11*.2). Another type of input to the system is one which directly alters the output. The most important of these is expansion growth, which, by increasing the volume into which the substance is transported, alters the relationship between the rate of transport and the rate of change of the output and the value of the output at the steady state in the steadily expanding cell.

The systems considered in this Chapter fall naturally into two groups, involving feedback from osmotic pressure and related parameters, or from concentrations of specific substances. If the concentration of a particular substance is maintained constant, then it cannot at the same time be involved in maintaining turgor (cf. CRAM, 1974a). There are, on the other hand, aspects of the uptake of nutrients and of the generation of turgor which are not independent. In mature *Zea mays* plants, 447 μmol g_{FW}^{-1} N were found in the tops (LATSHAW and MILLER, 1924). If all had entered as KNO_3, and one organic acid anion produced per NO_3^- reduced, the osmotic pressure in the tops from this source alone would be about 900 osmol m^{-3}, or 2,000 kPa, which is more than twice what is frequently found (LAWLOR, 1969). In fact the average K^+ concentration is about 100 mM and other cations are present in much smaller quantities, so that NO_3^- must have entered in exchange for an internal, probably organic, anion. The maintenance of NO_3^--organic acid exchange rather than KNO_3 influx (SMITH, 1973; *11*.3.2.2) would therefore serve a dual purpose. It would make economical use of K^+ (as suggested by DIJKSHOORN, 1970; BEN ZIONI et al., 1971), and maintain turgor at a lower level than would otherwise be the case, though the latter may be an aspect of the former. An apparently specific regulation of the linkage between fluxes may therefore also be related to turgor maintenance.

References

AARONSON, S., BEHRENS, U.: Ultrastructure of an unusual contractile vacuole in several Chrysomonad phytoflagellates. J. Cell Sci. **14**, 1–9 (1974).

ACEVEDO, E., HSIAO, T.C., HENDERSON, D.W.: Immediate and subsequent growth responses of maize leaves to changes in water status. Plant Physiol. **48**, 631–636 (1971).

AIKMAN, D.P., DAINTY, J.: Ionic relations of *Valonia ventricosa*. In: Some contemporary studies in marine science (H. BARNES, ed.), p. 37–44. London: Allen and Unwin 1966.

ALLEN, R.D.: Mechanism of the seismonastic rection in *Mimosa pudica*. Plant Physiol. **44**, 1101–1107 (1969).

ALLEN, R.D., JACOBSEN, L., JOAQUIN, J., JAFFE, L.J.: Ionic concentrations in developing *Pelvetia* eggs. Develop. Biol. **27**, 538–545 (1972).

ARISZ, W.H.: Uptake and transport of chlorine by parenchymatic tissue of leaves of *Vallisneria spiralis*. 1. The active uptake of chlorine. Proc. Koninkl. Ned. Akad. Wetenschap. **50**, 1019–1032 (1947).

ARNOLD, A.: Die Bedeutung der Chlorionen für die Pflanze, insbesondere deren physiologische Wirksamkeit. Botan. Studien H.2 (1955).

ASHER, C.J., OZANNE, P.G.: Growth and potassium content of plants in solution cultures maintained at constant potassium concentrations. Soil Sci. **103**, 155–161 (1967).

ATKINSON, D.E.: The adenylate charge in metabolic regulation. In: Horizons of bioenergentics (A. SAN PIETRO, H. GEST, eds.), p. 83–96. New York: Academic Press 1972.

AUSTENFELD, F.A.: Untersuchungen zur Physiologie der Nitratspeicherung und Nitratassimilation von *Chenopodium album* L. Z. Pflanzenphysiol. **67**, 271–281 (1972).

BARBER, D.A.: "Dual isotherms" for the absorption of ions by plant tissues. New Phytologist **71**, 255–262 (1972).

BARBER, J.: Measurement of the membrane potential and evidence for active transport of ions in *Chlorella pyrenoidosa*. Biochim. Biophys. Acta **150**, 618–625 (1968).

BAYLISS, L.E.: Living control systems. London: English Universities Press 1966.

BEN-AMOTZ, A.: Osmoregulation mechanism in the halophilic alga *Dunaliella parva*. In: Membrane transport in plants (U. ZIMMERMANN, J. DAINTY, eds.), p. 95–100. Berlin-Heidelberg-New York: Springer 1974.

BEN-AMOTZ, A., AVRON, M.: The role of glycerol in the osmotic regulation of the halophylic alga *Dunaliella parva*. Plant Physiol. **51**, 875–878 (1973).

BEN ZIONI, A., VAADIA, Y., LIPS, S.H.: Nitrate uptake by roots as regulated by nitrate reduction products of the shoots. Physiol. Plantarum **24**, 288–290 (1971).

BERNARD, C.: Leçons sur les propriétés physiologiques et les alterations pathologiques des liquides de l'organisme. Paris: Bailliere 1859.

BERNSTEIN, L.: Osmotic adjustment of plants to saline media. II. Dynamic phase. Amer. J. Bot. **50**, 360–370 (1963).

BERNSTEIN, L., HAYWARD, H.E.: Physiology of salt tolerance. Ann. Rev. Plant Physiol. **9**, 25–46 (1958).

BESNIER, V., BAZIN, M., MARCHELIDON, J., GENEVOT, M.: Étude de la variation du pool intracellulaire des acides aminés libres d'une diatomée marine en fonction de la saltinité. Bull. Soc. Chim. Biol. **51**, 1255–1262 (1969).

BIEBL, R.: Zellphysiologisch-ökologische Untersuchungen an *Enteromorpha clathrata* (Roth) Greville. Ber. Deut. Botan. Ges. **69**, 75–86 (1956).

BIEBL, R.: Seaweeds. In: Physiology and biochemistry of algae (R.A. LEWIN ed.), p. 799–815. New York-London: Academic Press 1962.

BINET, P.: Revues de biologie végétale marine. 1. La pression osmotique des algues marines. Bull. Soc. Botan. France **103**, 376–400 (1956).

BISSON, M.A., GUTKNECHT, J.: Osmotic regulation in the marine algae, *Codium decorticatum*. I. Regulation of turgor pressure by control of ionic composition. J. Membrane Biol. in press. (1975).

BLACK, D.R., WEEKS, D.C.: Ionic relations of *Enteromorpha intestinalis*. New Phytologist **71**, 119–127 (1972).

BLINKS, L.R., JACQUES, A.G.: The cell sap of *Halicystis*. J. Gen. Physiol. **13**, 733–737 (1929).

BLINKS, L.R., NIELSEN, J.P.: The cell sap of *Hydrodictyon*. J. Gen. Physiol. **23**, 551–559 (1939).

BONEY, A.D.: Aspects of the biology of the seaweeds of economic importance. Advan. Marine Biol. **3**, 105–253 (1965).

BOROWITZKA, L.J., BROWN, A.D.: The salt relations of marine and halophilic species of the unicellular green alga, *Dunaliella*. The role of glycerol as a compatible solute. Arch. Mikrobiol. **96**, 37–52 (1974).

BOUCK, G.B., BROWN, D.L.: Microtubule biogenesis and cell shape in *Ochromonas*. I. The distribution of cytoplasmic and mitotic microtubules. J. Cell Biol. **56**, 340–359 (1973).

BRADFIELD, G., SOMERFIELD, P., MEYN, T., HOLBY, M., BABCOCK, D., BRADLEY, D., SEGEL, I.H.: Regulation of sulphate transport in filamentous fungi. Plant Physiol. **46**, 720–727 (1970).

BRETELER, H.: A comparison between ammonium and nitrate nutrition of young sugar beet plants grown in nutrient solutions at constant acidity. 1. Production of dry matter, ionic balance and chemical composition. Neth. J. Agr. Sci. **21**, 227–244 (1973).

BRIGGS, G.E.: Effect of cane sugar on the uptake of chloride by discs from carrot root. New Phytologist **70**, 403–407 (1971).

BROWN, A.D., SIMPSON, J.R.: Water relations of sugar-tolerant yeasts: the role of intracellular polyols. J. Gen. Microbiol. **72**, 589–591 (1972).

BURSTRÖM, H.: Physics of cell elongation. In: Encyclopedia of plant physiology (W. RUHLAND, ed.), vol. XVI, p. 285–310. Berlin-Heidelberg-New York: Springer 1961.

CHU, T.M., ASPINALL, D., PALEG, L.G.: Salinity and proline accumulation. Australian J. Plant Physiol. submitted for publication.

CLELAND, R.: A dual role of turgor pressure in auxin-induced cell elongation in *Avena* coleoptiles. Planta **77**, 182–191 (1967).

COLLANDER, R.: Permeabilitäts-studien an *Chara ceratophylla*. 1. Die normale Zusammensetzung des Zellsaftes. Acta Botan. Fenn. **6**, 1–20 (1930).

COLLANDER, R.: Selective absorption of cations by higher plants. Plant Physiol. **16**, 691–720 (1941).

COSTER, H.G.L., ZIMMERMANN, U., STEUDLE, E.: Turgor pressure sensing in plant cell membranes. Planta (1976) in press.
COWAN, I.R.: Oscillations in stomatal conductance and plant functioning associated with stomatal conductance: observations and a model. Planta **106**, 185–220 (1972).
CRAIGIE, J.S.: Some salinity-induced changes in growth, pigments and cyclohexanetetrol content of *Monochrysis lutheri*. J. Fisheries Res. Board Can. **26**, 2959–2967 (1969).
CRAM, W.J.: The control of cytoplasmic and vacuolar ion contents in higher plant cells. Abhandl. Deut. Akad. Wiss. Berl. 117–126 (1968).
CRAM, W.J.: The initiation of developmental drifts in excised plant tissues. Australian J. Biol. Sci. **25**, 855–859 (1972).
CRAM, W.J.: Internal factors regulating nitrate and chloride influx in plant cells. J. Exptl. Bot. **24**, 328–341 (1973a).
CRAM, W.J.: Chloride fluxes in cells of the isolated root cortex of *Zea mays*. Australian J. Biol. Sci. **26**, 757–779 (1973b).
CRAM, W.J.: The regulation of concentration and hydrostatic pressure in cells in relation to growth. Bull. Roy. Soc. New Zealand **12**, 183–189 (1974a).
CRAM, W.J.: Effects of Cl^- on HCO_3^- and malate fluxes and CO_2 fixation in carrot and barley root cells. J. Exptl. Bot. **25**, 253–268 (1974b).
CRAM, W.J.: Relationships between Cl transport and electrical potential differences in carrot root cells. Australian J. Plant Physiol. **2**, 301–310 (1975).
CRAM, W.J., LATIES, G.G.: The use of short-term and quasi-steady influx in estimating plasmalemma and tonoplast influx in barley root cells at various external and internal chloride concentrations. Australian J. Biol. Sci. **24**, 633–646 (1971).
CRAM, W.J., LATIES, G.G.: The kinetics of bicarbonate and malate exchange in carrot and barley root cells. J. Exptl. Bot. **25**, 11–27 (1974).
CRAVEN, G.H., MOTT, R.L., STEWARD, F.C.: Solute accumulation in plant cells. IV. Effects of ammonium ions on growth and solute content. Ann. Bot. (London) **36**, 897–914 (1972).
CSEH, E., BÖSZÖRMENYI, Z., MESZES, G.: Characterisation of some parameters of ion transport and translocation. II. The effect of the excision, pretreatment with nutrition elements on bromide and potassium transport and translocation. Acta Botan. Acad. Sci. Hung. **16**, 267–278 (1970).
DAINTY, J.: The structure and possible function of the vacuole. In: Plant cell organelles (J.B. PRIDHAM, ed.), p. 40–46. London: Academic Press 1968.
DAVIES, D.D.: The control of respiration of turnip disks by L-methionine. J. Exptl. Bot. **17**, 320–331 (1966).
DEANE, E.M., O'BRIEN, R.W.: Sulphate uptake and metabolism in the Chrysomonad *Monochrysis lutheri*. Arch. Mikrobiol. **105**, 295–301 (1975).
DENNY, P., WEEKS, D.C.: Electrochemical potential gradients of ions in an aquatic angiosperm, *Potamogeton schweinfurthii* (Benn). New Phytologist **67**, 875–882 (1968).
DIJKSHOORN, W.: Partition of ionic constituents between organs. Proc. 6th Internat. Colloq. Plant Analysis and Fertilizer Problems. Tel-Aviv, p. 447–476 (1970).
EATON, F.M.: Toxicity and accumulation of chloride and sulfate salts in plants. J. Agr. Res. **64**, 357–399 (1942).
EPPLEY, R.W., BOVELL, C.R.: Sulfuric acid in *Desmarestia*. Biol. Bull. **115**, 101–106 (1958).
EPPLEY, R.W., CYRUS, C.C.: Cation regulation and survival of the red alga, *Porphyra perforata*, in diluted and concentrated sea water. Biol. Bull. **118**, 55–65 (1960).
ERGLE, D.R., EATON, F.M.: Organic acids in the cotton plant. Plant Physiol. **24**, 373–388 (1949).
ETTL, H.: Über pulsierende Vacuolen bei Chlorophyceen. Flora (Jena) **151**, 88–98 (1961).
FEIGE, G.B.: Untersuchungen zur Ökologie und Physiologie der marinen Blaualgenflechte *Lichina pygmaea* Ag. II. Die Reversibilität der Osmoregulation. Z. Pflanzenphysiol. **68**, 415–421 (1973).
FERRARI, G., RENOSTO, F.: Regulation of sulfate uptake by excised barley roots in the presence of selenate. Plant Physiol. **49**, 114–116 (1972).
FINDLAY, G.P., HOPE, A.B., PITMAN, M.G., SMITH, F.A., WALKER, N.A.: Ionic relations of marine algae. III. *Chaetomorpha*: membrane electrical properties and chloride fluxes. Australian J. Biol. Sci. **24**, 731–745 (1971).
FINDLAY, G.P., HOPE, A.B., WILLIAMS, E.J.: Ionic relations of marine algae. I. *Griffithsia*: membrane electrical properties. Australian J. Biol. Sci. **22**, 1163–1178 (1969).

FOGG, G.E.: The metabolism of algae. London: Methuen 1953.

FUHS, G.W., DEMMERLE, S.D., CARELLI, E., CHEN, M.: Characterization of phosphorus-limited plankton algae (with reflections on the limiting nutrient concept). Limnol. Oceanogr. Special Symposia **1**, 113–133 (1972).

GAUCH, H.G., EATON, F.M.: Effect of saline substrate on hourly levels of carbohydrates and inorganic constituents of barley plants. Plant Physiol. **17**, 347–365 (1942).

GESSNER, F.: The osmotic regulations in *Valonia ventricosa*. A.J. Agardh. Intern. Rev. Ges. Hydrobiol. **54**, 529–532 (1969).

GREACEN, E.L., OH, J.S.: Physics of root growth. Nature New Biol. **235**, 24–25 (1972).

GREEN, P.B., ERICKSON, R.O., BUGGY, J.: Metabolic and physical control of cell elongation rate. *In vivo* studies in *Nitella*. Plant Physiol. **47**, 423–430 (1971).

GREENWAY, H.: Growth stimulation by high chloride concentrations in halophytes. Israel J. Botany **17**, 169–177 (1968).

GREENWAY, H.: Salinity, plant growth, and metabolism. J. Australian Inst. Agr. Sci. **39**, 24–34 (1973).

GREENWAY, H., KLEPPER, B., HUGHES, P.G.: Effects of low water potential on ion uptake and loss of excised roots. Planta **80**, 129–141 (1968).

GREENWAY, H., THOMAS, D.A.: Plant responses to saline substrates. V. Chloride regulation in the individual organs of *Hordeum vulgare* during treatment with sodium chloride. Australian J. Biol. Sci. **18**, 505–524 (1965).

GRODINS, F.S.: Control theory and biological systems. New York-London: Columbia University 1963.

GUILLARD, R.R.L.: Salt and osmotic balance. In: Physiology and biochemistry of algae (R.A. LEWIN, ed.), p. 529–540. New York: Academic 1962.

GUTKNECHT, J.: Sodium, potassium, and chloride transport and membrane potentials in *Valonia ventricosa*. Biol. Bull. **130**, 331–344 (1966).

GUTKNECHT, J.: Ion fluxes and short-circuit current in internally perfused cells of *Valonia ventricosa*. J. Gen. Physiol. **50**, 1821–1834 (1967).

GUTKNECHT, J.: Salt transport in *Valonia*: inhibition of potassium uptake by small hydrostatic pressures. Science **160**, 68–70 (1968).

GUTKNECHT, J., DAINTY, J.: Ionic relations of marine algae. Oceanogr. Marine Biol. Ann. Rev. **6**, 163–200 (1968).

HART, J.W., FILNER, P.: Regulation of sulfate uptake by amino acids in cultured tobacco cells. Plant Physiol. **44**, 1253–1259 (1969).

HASSENSTEIN, B.: Information and control in the living organism. London: Chapman and Hall 1971 (Revised from: Biologische Kybernetik. Heidelberg: Quelle and Meyer 1970).

HASTINGS, D.F., GUTKNECHT, J.: Turgor pressure regulation – modulation of active potassium transport by hydrostatic pressure gradients. In: Membrane transport in plants. (U. ZIMMERMANN and J. DAINTY, eds.), p. 79–83. Berlin-Heidelberg-New York: Springer 1974.

HASTINGS, D.F., GUTKNECHT, J.: Ionic relations and the regulation of turgor pressure in the marine alga, *Valonia macrophysa*. Submitted to J. gen. Physiol. (1975).

HATRICK, A.A., BOWLING, D.J.F.: A study of the relationship between root and shoot metabolism. J. Exptl. Bot. **24**, 607–613 (1973).

HAYWARD, H.E., WADLEIGH, C.H.: Plant growth on saline and alkali soils. Advan. Agron. **1**, 1–38 (1949).

HEIMER, Y.M., FILNER, P.: Regulation of nitrate assimilation pathway in cultured tobacco cells. III. The nitrate uptake system. Biochim. Biophys. Acta **230**, 362–372 (1971).

HELLEBUST, J.A.: Mannitol metabolism and osmoregulation in the green flagellate *Platymonas suecica*. Plant Physiol., Suppl. **51**, 20 (1973).

HOAGLAND, D.R., BROYER, T.C.: General nature of the process of salt accumulation by roots with description of experimental methods. Plant Physiol. **11**, 471–507 (1936).

HOAGLAND, D.R., DAVIS, A.R.: The composition of the cell sap of the plant in relation to the absorption of ions. J. Gen. Physiol. **5**, 629–646 (1923).

HODGES, T.K., VAADIA, Y.: Uptake and transport of radiochloride and tritiated water by various zones of onion roots of different chloride status. Plant Physiol. **39**, 104–108 (1964).

HSIAO, T.C.: Plant responses to water stress. Ann. Rev. Plant Physiol. **24**, 519–570 (1973).

HUMPHRIES, E.C.: The absorption of ions by excised root systems. II. Observations on roots of barley grown in solutions deficient in phosphorous, nitrogen, or potassium. J. Exptl. Bot. **2**, 344–379 (1951).

HUMPHRIES, E.C.: The relation between the rate of nutrient uptake by excised barley roots and their content of sucrose and reducing sugars. Ann. Bot. (London) N.S., **20**, 411–417 (1956).

IVANKO, S., INGVERSEN, J.: Investigation on the assimilation of nitrogen by maize roots and the transport of some major nitrogen compounds by xylem sap. 1. Nitrate and ammonia uptake and assimilation in the major nitrogen fractions of nitrogen-starved maize roots. Physiol. Plantarum **24**, 59–65 (1971).

JACKSON, P.C., EDWARDS, D.G.: Cation effects on chloride fluxes and accumulation levels in barley roots. J. Gen. Physiol. **50**, 224–241 (1966).

JACKSON, W.A., FLESHER, D., HAGEMAN, R.H.: Nitrate uptake by dark-grown corn seedlings. Some characteristics of apparent induction. Plant Physiol. **51**, 120–127 (1973).

JACQUES, A.G.: The kinetics of penetration. XV. The restriction of the cellulose wall. J. Gen. Physiol. **22**, 147–163 (1938a).

JACQUES, A.G.: The kinetics of penetration. XIX. Entrance of electrolytes and of water into impaled Halicystis. J. Gen. Physiol. **22**, 757–733 (1938b).

JACQUES, A.G., OSTERHOUT, W.J.V.: The accumulation of electrolytes. IV. Internal versus external concentrations of potassium. J. Gen. Physiol. **15**, 537–550 (1931).

JANES, B.E.: Adjustment mechanisms of plants subject to varied osmotic pressures of nutrient solution. Soil Sci. **101**, 180–188 (1966).

JARVIS, P.G., JARVIS, M.S.: Effects of several osmotic substrates on the growth of *Lupinus albus* seedlings. Physiol. Plantarum **16**, 485–500 (1963).

JEANJEAN, R.: The relationship between the rate of phosphate absorption and protein synthesis during phosphate starvation in *Chlorella pyrenoidosa*. F.E.B.S. Letters **32**, 149–151 (1973).

JENNINGS, D.H.: Cations and filamentous fungi: invasion of the sea and hyphal functioning. In: Ion transport in plants (W.P. ANDERSON, ed.), p. 323–335. London-New York: Academic Press 1973.

JENNINGS, D.H.: Transport and translocation in filamentous fungi. In: The filamentous fungi (J.E. SMITH, D.E. BERRY, eds.), vol. 2, p. 32–63. London: Arnold 1975.

JENNINGS, D.H., AUSTIN, S.: The stimulatory effect of the nonmetabolised sugar 3-0-methyl glucose on the conversion of mannitol and arabitol to polysaccharide and other insoluble compounds in the fungus *Dendryphiella salina*. J. Gen. Microbiol. **75**, 287–294 (1973).

JESCHKE, W.D.: The effect of inhibitors on the K^+-dependent Na^+ efflux and the K-Na selectivity of barley roots. In: Membrane transport in plants. (U. ZIMMERMANN, J. DAINTY, eds.), p. 397–405. Berlin-Heidelberg-New York: Springer 1974.

JOHANSEN, C., EDWARDS, D.G., LONERAGAN, J.F.: Potassium fluxes during potassium absorption by intact barley plants of increasing potassium content. Plant Physiol. **45**, 601–603 (1970).

JOHNSON, F.H., EYRING, H.: In: High pressure effects on cellular processes (A.M. ZIMMERMANN, ed.). New York-London: Academic Press 1970.

KALMUS, H. (ed.): Regulation and control in living systems. London-New York: Wiley 1966.

KAMIYA, N., KURODA, K.: Artificial modification of the osmotic pressure of the plant cell. Protoplasma **46**, 423–436 (1956).

KAUSS, H.: Isofloridosid und Osmoregulation bei *Ochromonas malhamensis*. Z. Pflanzenphysiol. **56**, 453–465 (1967).

KAUSS, H.: α-Galaktosylglyzeride und Osmoregulation in Rotalgen. Z. Pflanzenphysiol. **58**, 428–433 (1968).

KAUSS, H.: Osmoregulation mit α-Galaktosylglyzeriden bei *Ochromonas* und Rotalgen. Ber. Deut. Botan. Ges. **82**, 115–125 (1969).

KAUSS, H.: Turnover of galactosylglycerol and osmotic balance in *Ochromonas*. Plant Physiol. **52**, 613–615 (1973).

KAUSS, H.: Osmoregulation in *Ochromonas*. In: Membrane transport in plants (U. ZIMMERMANN, J. DAINTY, eds.), p. 90–94. Berlin-Heidelberg-New York: Springer 1974.

KAUSS, H., SCHOBERT, B.: First demonstration of UDP-GAL:*sn*-glycero-3-phosphoric acid 1-α-galactosyl-transferase and its possible role in osmoregulation. F.E.B.S. Letters. **19**, 131–135 (1971).

KESSELER, H.: Eine mikrokryoskopische Methode zur Bestimmung des Turgors von Meeresalgen. Kiel. Meeresforsch. **14**, 23–41 (1958).

KESSELER, H.: Mikrokryoskopische Untersuchungen zur Turgorregulation von *Chaetomorpha linum*. Kiel. Meeresforsch. **15**, 51–73 (1959).

KESSELER, H.: Die Bedeutung einiger anorganischer Komponenten des Seewassers für die Turgorregulation von *Chaetomorpha linum* (Cladophorales). Helgoländer Wiss. Meeresuntersuch. **10**, 73–90 (1964).

KESSELER, H.: Turgor, osmotisches Potential und ionale Zusammensetzung des Zellsaftes einiger Meeresalgen verschiedener Verbreitungsgebiete. In: Proc. 5th Mar. Biol. Symp. Botanica Gothoburgensia **3**, 103–111 (1965).

KIRKBY, E.A., MENGEL, K.: Ionic balance in different tissues of the tomato plant in relation to nitrate, urea, or ammonium nutrition. Plant Physiol. **42**, 6–14 (1967).

KIRKHAM, M.B., GARDNER, W.R., GERLOFF, G.C.: Regulation of cell division and cell enlargement by turgor pressure. Plant Physiol. **49**, 961–962 (1972).

KISHIMOTO, U., TAZAWA, M.: Ionic composition of the cytoplasm of *Nitella flexilis*. Plant Cell Physiol. (Tokyo) **6**, 507–518 (1965).

KYLIN, A.: The influence of phosphate nutrition on growth and sulphur metabolism of *Scenedesmus*. Physiol. Plantarum **17**, 384–402 (1964).

LATSHAW, W.L., MILLER, E.C.: Elemental composition of the corn plant. J. Agr. Res. **27**, 845–860 (1924).

LAWLOR, D.W.: Plant growth in polyethylene glycol solutions in relation to the osmotic potential of the root medium and the leaf water balance. J. Exptl. Bot. **20**, 895–911 (1969).

LEIGH, R.A., WYN JONES, R.G.: The effect of increased internal ion concentration upon the ion uptake isotherms of excised maize root segments. J. Exptl. Bot. **24**, 787–795 (1973).

LEW, V.L., GLYNN, I.M., ELLORY, J.C.: Net synthesis of ATP by reversal of the sodium pump. Nature **225**, 865–866 (1970).

LOCKHART, J.A.: Intracellular mechanism of growth inhibition by radiant energy. Plant Physiol. **35**, 129–135 (1960).

LOESCHER, W.H., NEVINS, D.J.: Turgor-dependent changes in *Avena coleoptile* cell wall composition. Plant Physiol. **52**, 248–251 (1973).

LÜTTGE, U., BALL, E.: Proton and malate fluxes in cells of *Bryophyllum diagremontianum* leaf slices in relation to potential osmotic pressure of the medium. Z. Pflanzenphysiol. **73**, 326–338 (1974a).

LÜTTGE, U., BALL, E.: Mineral ion fluxes in slices of acidified and de-acidified leaves of the CAM plant *Bryophyllum diagremontianum*. Z. Pflanzenphysiol. **73**, 339–348 (1974b).

MAGISTAD, O.C.: Plant growth relations on saline and alkali soils. Bot. Rev. **11**, 181–230 (1945).

MARSCHNER, H., OSSENBERG-NEUHAUS, H.: Bedeutung des Begleitanions bei den Wechselbeziehungen zwischen K^+ und Ca^{++} im Bereich hoher Außenkonzentrationen. Z. Pflanzenernähr. Düng. Bodenk. **126**, 217–228 (1970).

MCNEIL, D.L.: The basis of osmotic pressure maintenance during expansion growth in *Helianthus annuus* hypocotyls. Australian J. Plant Physiol. **2**, (1975).

MEEUSE, B.J.D.: Storage products. In: Physiology and biochemistry of algae (R.A. LEWIN, ed.), p. 289–313. New York: Academic Press 1962.

MEIRI, A.: Potassium and chloride accumulation and transport by excised maize roots of different salt status. In: Ion transport in plants (W.P. ANDERSON, ed.), p. 519–530. London-New York: Academic Press 1973.

MEIRI, A., MOR, E., POLYAKOFF-MAYBER, A.L.: Effect of time of exposure to salinity on growth, water status, and salt accumulation in bean plants. Ann. Bot. (London) N.S., **34**, 383–391 (1970).

MEYER, R.F., BOYER, J.S.: Sensitivity of cell division and cell elongation to low water potentials in soybean hypocotyls. Planta **108**, 77–87 (1972).

MILSUM, J.H.: Biological control systems analysis. New York: McGraw-Hill 1966.

MOHSEN, A.F., NASR, A.H., METWALLI, A.M.: Effect of different salinities on growth, reproduction, amino acid synthesis, fat and sugar content in *Ulva fasciata* Delile. Botan. Marina **15**, 177–181 (1972).

MOROWITZ, H.J.: Entropy for biologists. New York-London: Academic Press 1970.

MOTHES, K.: Der Einfluß des Wasserzustandes auf Fermentprozesse und Stoffumsatz. In: Encyclopedia of plant physiology (W. RUHLAND, ed.), vol. III, p. 656–664. Berlin-Heidelberg-New York: Springer 1956.

MOTT, R.L., STEWARD, F.C.: Solute accumulation in plant cells. 1. Reciprocal relations between electrolytes and non-electrolytes. Ann. Bot. (London) N.S., **36**, 621–639 (1972a).

MOTT, R.L., STEWARD, F.C.: Solute accumulation in plant cells. V. An aspect of nutrition and development. Ann. Bot. (London) N.S., **36**, 915–937 (1972b).
MOWAT, J.L.: Some relations between salt and sugar in *Hordeum vulgare*. Ph. D. Thesis, University of Sydney (1973).
NAKAGAWA, S., KATAOKA, H., TAZAWA, M.: Osmotic and ionic regulation in *Nitella*. Plant Cell Physiol. (Tokyo) **15**, 457–468 (1974).
NEEB, O.: Hydrodictyon als Object einer vergleichenden Untersuchung physiologischer Größen. Flora (Jena) **139**, 39–95 (1952).
NEIRINCKX, L.J.A., BANGE, G.G.J.: Irreversible equilibration of barley roots with Na^+ ions at different external Na concentrations. Acta Botan. Neerl. **20**, 481–488 (1971).
NOBEL, P.S.: Light-dependent potassium uptake by *Pisum sativum* leaf fragments. Plant Cell Physiol. (Tokyo) **10**, 597–605 (1969).
ÖNAL, M.: Vergleichende ökologische Untersuchungen bei Halophyten und Glycophyten in der Nähe von Neapel. Rev. Fac. Sci. Univ. Istanbul, Ser. B **31**, 209–248 (1966).
OERTLI, J.J.: Extracellular salt accumulation, a possible mechanism of salt injury in plants. Agrochimica **22**, 461–469 (1968).
OERTLI, J.J.: A whole-system approach to water physiology in plants. New Delhi 1971.
O'KELLEY, J.C.: Mineral nutrition of algae. Ann. Rev. Plant Physiol. **19**, 89–112 (1968).
ORDIN, L.: Effect of water stress on cell wall metabolism of *Avena* coleoptile tissue. Plant Physiol. **35**, 443–450 (1960).
ORDIN, L., APPLEWHITE, T.G., BONNER, J.: Auxin-induced water uptake by *Avena* coleoptile sections. Plant Physiol. **31**, 44–53 (1956).
OSMOND, C.B.: Oxalates and ionic equilibria in Australian saltbushes (*Atriplex*). Nature **198**, 503–504 (1963).
OSMOND, C.B.: Acid metabolism in *Atriplex*. 1. Regulation of oxalate synthesis by the apparent excess cation absorption in leaf tissue. Australian J. Biol. Sci. **20**, 575–578 (1968).
PFEFFER, W.: The physiology of plants, vol. 1. Oxford: Clarendon 1900 (Transl. from: Pflanzenphysiologie, 2nd Ed., 1897, by A.J. EWART).
PIERCE, E.C., APPLEMAN, C.O.: Role of ether soluble organic acids in the cation-anion balance in plants. Plant Physiol. **18**, 224–238 (1943).
PITMAN, M.G.: The determination of the salt relations of the cytoplasmic phase in cells of beetroot tissue. Australian J. Biol. Sci. **16**, 647–668 (1963).
PITMAN, M.G.: Simulation of Cl^- uptake by low-salt barley roots as a test of models of salt uptake. Plant Physiol. **44**, 1417–1427 (1969).
PITMAN, M.G.: Uptake and transport of ions in barley seedlings. III. Correlation between transport to the shoot and relative growth rate. Australian J. Biol. Sci. **25**, 905–919 (1972).
PITMAN, M.G., COURTICE, A.C., LEE, B.: Comparison of potassium and sodium uptake by barley roots at high and low salt status. Australian J. Biol. Sci. **21**, 871–881 (1968).
PITMAN, M.G., CRAM, W.J.: Regulation of inorganic ion transport in plants. In: Ion transport in plants (W.P. ANDERSON, ed.), p. 465–481. London-New York: Academic Press 1973.
PITMAN, M.G., LÜTTGE, U., LÄUCHLI, A., BALL, E.: Ion uptake to slices of barley leaves, and regulation of K content in cells of the leaves. Z. Pflanzenphysiol. **72**, 75–88 (1974).
PITMAN, M.G., MOWAT, J.L., NAIR, H.: Interactions of processes for accumulation of salt and sugar in barley plants. Australian J. Biol. Sci. **24**, 619–631 (1971).
POST, R.L., ALBRIGHT, C.D., DAYANI, K.: Resolution of pump and leak components of sodium and potassium ion transport in human erythrocytes. J. Gen. Physiol. **50**, 1201–1220 (1967).
RASCHKE, K.: Die Stomata als Glieder eines schwingungsfähigen CO_2-Regelsystems. Experimenteller Nachweis an *Zea mays* L. Z. Naturforsch. **20b**, 1261–1270 (1965).
RAY, P.M.: The action of auxin on cell enlargement in plants. Develop. Biol., Suppl. 3, 172–205 (1969).
RHEE, G.Y.: A continuous culture study of phosphate uptake, growth rate and polyphosphate in *Scenedesmus* sp. J. Phycol. **9**, 495–506 (1973).
ROBINSON, J.B., SMITH, F.A.: Chloride influx in citrus leaf slices. Australian J. Biol. Sci. **23**, 953–960 (1970).
RUBINSTEIN, B., LIGHT, E.N.: Indoleacetic-acid-enhanced chloride uptake into coleoptile cells. Planta **110**, 43–56 (1973).
SADDLER, H.D.W.: The ionic relations of *Acetabularea mediterranea*. J. Exptl. Bot. **21**, 345–359 (1970).

SATTER, R.L., GALSTON, A.W.: Potassium influx: a common feature of *Albizzia* leaflet movement controlled by phytochrome or endogenous rhythm. Science **174**, 518–530 (1971).

SCHOBERT, B., UNTNER, E., KAUSS, H.: Isofloridosid und die Osmoregulation bei *Ochromonas malhamensis*. Z. Pflanzenphysiol. **67**, 385–398 (1972).

SCHRÖTER, K., SIEVERS, A.: Wirkung der Turgorreduktion auf den Golgi-Apparat und die Bildung der Zellwand bei Wurzelhaaren. Protoplasma **72**, 203–211 (1971).

SCOTT, B.I.H.: Electric fields in plants. Ann. Rev. Plant Physiol. **18**, 409–418 (1967).

SLATYER, R.O.: Effects of several osmotic substrates on the water relationships of tomato. Australian J. Biol. Sci. **14**, 519–540 (1961).

SLATYER, R.O.: Climatic control of plant water relations. In: Environmental control of plant growth (L.T. EVANS, ed.), p. 33–54. New York-London: Academic Press 1963.

SLATYER, R.O.: Physiological significance of internal water relations to crop yield. In: Physiological aspects of crop yield (J.D. EASTIN, ed.), p. 53–83. Madison: Amer. Soc. Agron., Crop Sci. Soc. Amer. 1969.

SMITH, F.A.: The internal control of nitrate uptake into excised barley roots with differing salt contents. New Phytologist **72**, 769–782 (1973).

STEINER, M.: Die Zusammensetzung des Zellsaftes bei höheren Pflanzen in ihrer ökologischen Bedeutung. Ergeb. Biol. **17**, 151–254 (1939).

STEINER, M., ESCHRICH, W.: Die osmotische Bedeutung der Mineralstoffe. In: Encyclopedia of plant physiology (W. RUHLAND, ed.), vol. IV, p. 334–354. Berlin-Heidelberg-New York: Springer 1958.

STEUDLE, E., ZIMMERMANN, U.: Zellturgor und selektiver Ionentransport bei *Chaetomorpha linum*. Z. Naturforsch. **26b**, 1276–1282 (1971).

STEWART, G.R., LEE, J.A.: The role of proline accumulation in halophytes. Planta **120**, 279–289 (1974).

STONER, L.C., DUNHAM, P.B.: Regulation of cellular osmolarity and volume in *Tetrahymenia*. J. Exptl. Biol. **53**, 391–399 (1970).

SUTCLIFFE, J.F.: The influence of internal ion concentration on potassium accumulation and salt respiration of red beet root tissue. J. Exptl. Bot. **3**, 59–76 (1952).

SUTCLIFFE, J.F.: The absorption of potassium ions by plasmolysed cells. J. Exptl. Bot. **5**, 215–231 (1954a).

SUTCLIFFE, J.F.: The exchangeability of potassium and bromide ions in cells of red beetroot tissue. J. Exptl. Bot. **5**, 313–326 (1954b).

SUTCLIFFE, J.F.: Cation absorption by non-growing plant cells. Symp. Soc. Exptl. Biol. **8**, 325–342 (1954c).

TAZAWA, M.: Weitere Untersuchungen zur Osmoregulation der *Nitella*-Zelle. Protoplasma **53**, 227–258 (1961).

TAZAWA, M.: Studies on *Nitella* having artificial cell sap. I. Replacement of the cell sap with artificial solutions. Plant Cell Physiol. (Tokyo) **5**, 33–43 (1964).

TAZAWA, M., NAGAI, R.: Die Mitwirkung von Ionen bei der Osmoregulation der *Nitella*zelle. Plant Cell Physiol. (Tokyo) **1**, 255–267 (1960).

TAZAWA, M., NAGAI, R.: Studies on osmoregulation of *Nitella* internode with modified cell saps. Z. Pflanzenphysiol. **54**, 333–344 (1966).

THOIRON, A., THOIRON, B., THELLIER, M.: Absorption du sulfate par la *Riccia fluitans*: effet des conditions antérieures de nutrition en sulfate. Compt. Rend. **270**, 328–330 (1970).

TOATES, F.M.: Control theory in biology and experimental psychology. London: Hutchinson 1975.

TRAMÈR, P.: Zur Kenntnis der Saugkraft des Meerwassers und einiger Hydrophyten. Ber. Schweiz. Botan. Ges. **67**, 411–419 (1957).

TRIP, P., KROTKOV, G., NELSON, C.D.: Metabolism of mannitol in higher plants. Amer. J. Bot. **51**, 828–835 (1964).

URSPRUNG, A., BLUM, G.: Eine Methode zur Messung des Wand- und Turgor-Druckes der Zelle, nebst Anwendungen. Jahrb. Wiss. Botanik **63**, 1–110 (1924).

VAADIA, Y., RANEY, F.C., HAGAN, R.M.: Plant water deficits and physiological processes. Ann. Rev. Plant Physiol. **12**, 265–292 (1961).

VALLÉE, M., JEANJEAN, R.: Le système de transport de $SO_4^=$ chez *Chlorella pyrenoidosa* et sa régulation. II. Recherches sur la régulation de l'entrée. Biochim. Biophys. Acta **150**, 607–617 (1968).

VENRICK, D.M., SMITH, R.C.: The influence of initial salt status on absorption of rubidium by corn root segments of two stages of development. Bull. Torrey Botan. Club **94**, 501–510 (1967).

VILLEGAS, L.: Changes in volume and turgor presure in *Valonia* cells. Biochim. Biophys. Acta **136**, 590–593 (1967).

WAISEL, Y.: Biology of halophytes. New York-London: Academic Press 1972.

WALTER, H.: Ecology of tropical and subtropic vegetation. Edinburgh: Oliver and Boyd 1971.

WEGMANN, K.: Osmotic regulation of photosynthetic glycerol production in *Dunaliella*. Biochim. Biophys. Acta **234**, 317–323 (1971).

WETHERELL, D.F.: Osmotic equilibrium and growth of *Scenedesmus obliquus* in saline media. Physiol. Plantarum **16**, 82–91 (1963).

WILKINS, M.B.: Circadian rhythms in plants. In: Physiology of plant growth and development (M.B. WILKINS, ed.), p. 647–671. London: McGraw-Hill 1969.

WRIGHT, M.J., DAVISON, K.L.: Nitrate accumulation in crops and nitrate poisoning in animals. Advan. Agron. **16**, 197–247 (1964).

YAMOMOTO, L.A., SEGEL, I.H.: The inorganic sulfate transport system of *Penicillium chrysogenum*. Arch. Biochem. Biophys. **114**, 523–538 (1966).

YOUNG, G.M., JEFFERIES, R.L., SIMS, A.P.: The regulation of potassium uptake in *Lemna minor* L. Abhandl. Deut. Akad. Wiss. Berl. 67–82 (1970).

YOUNG, G.M., SIMS, A.P.: The potassium relations of *Lemna minor* L. I. Potassium uptake and plant growth. J. Exptl. Bot. **23**, 958–969 (1970).

ZIMMERMANN, U., STEUDLE, E.: Effects of potassium concentration and osmotic presure of sea water on the cell-turgor pressure of *Chaetomorpha linum*. Marine Biol. **11**, 132–137 (1971).

ZIMMERMANN, U., STEUDLE, E.: The pressure-dependence of the hydraulic conductivity, the membrane resistance and membrane potential during turgor pressure regulation in *Valonia utricularis*. J. Membrane Biol. **16**, 331–352 (1974).

12. H^+ Transport and Regulation of Cell pH

F.A. SMITH and J.A. RAVEN

1. Introduction

Exchange of H^+ or OH^- between plant cells and their surroundings is associated with a wide range of metabolic processes, and especially with solute transport. One major field of investigation has been the causes and effects of the large balancing fluxes of H^+ or OH^- which occur during excess absorption of cations (e.g. K^+ from K_2SO_4 solutions) or anions (e.g. Cl^- from $CaCl_2$ solutions). A second, biophysical approach, deals with the contribution of H^+ and OH^- fluxes to the electrical parameters of cell membranes. A third approach considers the separation of H^+ and OH^- across cell membranes as an energy source for other ion transport processes within the cell. These aspects have been discussed separately in other Chapters of this Volume. The aim of this Chapter is to present the main findings as part of a wider appraisal of the significance of H^+ and OH^- transport in cellular metabolism. The starting-point is simply that the intracellular distribution of H^+ (i.e. the intracellular pH), far from being a haphazard consequence of metabolism, is a closely controlled component of the ion balance within cells. Particular attention is thus paid to the pH-regulating role of H^+ and OH^- transport. Interactions with other ion transport processes and various morphogenetic phenomena are also considered in relation to the need for pH regulation. Some of the processes discussed in this Chapter (e.g. organic acid synthesis, and NO_3^- reduction) are also discussed in the next Chapter by OSMOND. However, the approach there is quite different, in that OSMOND assesses in detail the regulation of metabolism by ions other than H^+ or OH^-.

2. Effects of Metabolism on the pH of Cells and Their Surroundings

2.1 General Considerations

Plant cell metabolism involves many processes which result in intracellular production or consumption of H^+, or transfer of ions from one compartment to another. It is axiomatic that in any large compartment of the cell such as cytoplasm or vacuole there should be equality of positive and negative charges. (But see *3.5.2.* for discussion of charge differences associated with membrane potential.) To some extent this balance is achieved by diffusion in response to gradients of electrochemical activity, but in addition the cell possesses interdependent processes that assist in maintenance of charge balance, in many cases operating through the common currency of H^+ and OH^- ions.

For example the metabolic products of cell growth (containing C, H, O and N) are found to contain a net negative charge. This is due to an excess of free carboxyl over free amino groups, and is balanced by inorganic cations (see DIJKSHOORN, 1969; KIRBY, 1969; DAMADIAN, 1973). Depending on whether the carbon source is CO_2, hexose or HCO_3^- and the N-source NH_4^+ or NO_3^-, there will be need for either H^+ or OH^- removal from the cytoplasm, or formation of carboxylate anions to maintain charge balance (Fig. *12.*1a–c). A further example is the assimilation of SO_4^{2-} which, like NO_3^-, leads to OH^- production but to a much lesser extent compared with NO_3^- (DIJKSHOORN and VAN WIJK, 1967). A final example is the synthesis of large amounts of carboxylate anions used in turgor generation in many types of plant cells, leading to massive production of H^+ within the cell. This is in fact a special case of the situation shown in Fig. *12.*1a (excluding the nitrogen assimilation), and will be considered in detail below.

Disposal of metabolically-produced H^+ or OH^- to the bathing medium is well established, and can result in extremely large external pH changes. Yeast cells can reduce the external pH below 2 (CONWAY and O'MALLEY, 1946), while "low salt" barley roots placed in K_2SO_4 solution reduce the pH below 4 (HOAGLAND and BROYER, 1940). Conversely, some aquatic plants assimilating HCO_3^- in the light raise the pH above 11 (SCHUTOW, 1926; RUTTNER, 1947). The effect

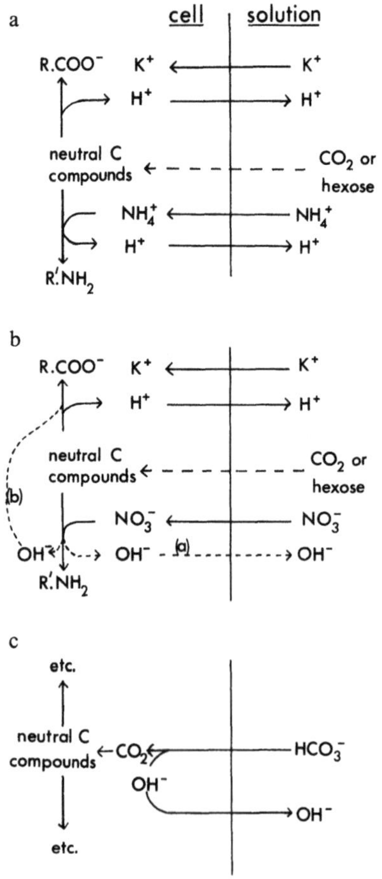

Fig. *12.*1a–c. Major metabolic processes involving membrane transport of H^+ or OH^-. (a) General case of cells using neutral carbon source (CO_2 in photosynthetic growth or hexose in heterotrophic growth) and NH_4^+ as an N-source. In this case there is net uptake of positive charge and potentially an increase in cytoplasmic H^+. Maintenance of cytoplasmic pH near neutrality (pH 7) thus requires removal of H^+ from the cell. (b) With neutral carbon source (as in a) but NO_3^- as the N-source there is net uptake of negative charge and potentially an increase in cytoplasmic OH^-. In this case there is a need for removal of OH^- from the cell (a), or synthesis of carboxylate ions (b), or both. (c) During photosynthetic growth using HCO_3^- as the carbon source, excess OH^- needs to be lost to the bathing medium to balance internal pH of the form of nitrogen nutrition on the pH of nutrient solutions is shown in Fig. *12.2*.

It is important to stress that not all external pH changes result from H^+ or OH^- transport across the plasmalemma. There may in addition be considerable production or consumption of H^+ or OH^- within the bathing medium, resulting from movement of unionized forms of weak electrolytes into or out of the cell.

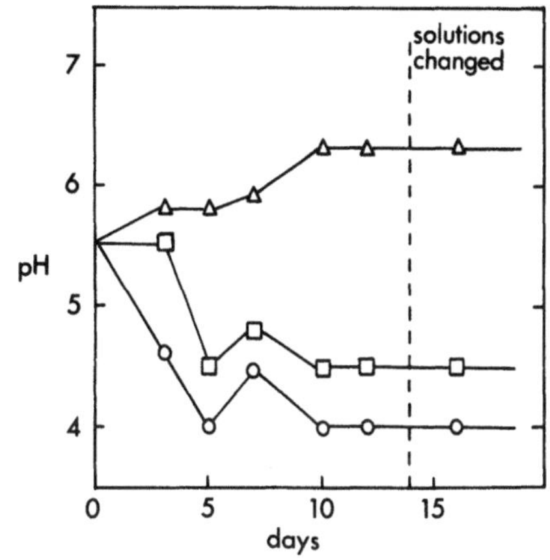

Fig. *12.2*. Effects of nitrate, urea and ammonium nutrition on pH of nutrient solutions bathing tomato plants (results of KIRKBY and MENGEL, 1967, redrawn). The change in pH is plotted, as pH was adjusted back to 5.5 after every determination (△) NO_3^-, (□) urea, (○) NH_4^+; all at 5 mM

Table *12*.1. Summary of major processes leading to production or consumption of H^+ (or OH^-) inside or outside the cell, with resulting external pH changes (RAVEN and SMITH, 1974)

Metabolic process	Site of H^+ (or OH^-) production or consumption	Direction of external pH change
Photosynthetic CO_2 fixation: CO_2 entering cell	Bathing solution	Increase
Photosynthetic CO_2 fixation: HCO_3^- entering cell	Cell	Increase
Nitrogen assimilation: NH_4^+ entering cell	Cell	Decrease
Nitrogen assimilation: NO_3^- entering cell	Cell	Increase
Nitrogen assimilation: NH_4OH entering cell	Bathing solution	Decrease
Excess cation influx (e.g. K^+ carboxylate accumulation)	Cell	Decrease
Excess anion influx (e.g. Cl^-/HCO_3^- exchange)	Cell	Increase
Organic acid assimilation (undissociated form entering)	Bathing solution	Increase
Excretion of organic acids (in undissociated form)	Bathing solution	Decrease
Rapid hormonal responses (H^+ secretion)	Cell	Decrease

External reequilibration of ionized and unionized forms can result in an increased external pH (e.g. during photosynthesis with CO_2, when solutions contain CO_2 and HCO_3^-) or a decrease (e.g. during respiration, or NH_3 assimilation). These changes can obscure those due to H^+ or OH^- movement across the plasmalemma. The problem of interpretation is at its most acute in experiments with green cells in the light, when pH changes may be due to photosynthetic uptake of CO_2 or HCO_3^-, or associated with imbalance in transport of cations (K^+) and anions (Cl^-, NO_3^- etc). Table *12.*1 summarizes the major processes which affect external pH by production or consumption of H^+ inside or outside the cell, including those shown in Fig. *12.*1.

In drawing up a balance sheet of movements of ions between cells and their surroundings it may seem immaterial whether an external pH change results from H^+ exchange, or whether OH^- is the ion species moving across the membrane. In terms of the mechanisms of ion transport, however, the distinction will be important. For example, in Figure *12.*1b, NO_3^- influx is balanced by OH^- efflux. It would be equally plausible to regard NO_3^- influx as coupled to H^+ influx, yet at the molecular level the transport processes, and their control, would be quite different. This point will be discussed in more detail below (*12.*3).

2.2 Intracellular pH Measurements and Their Significance

Measurements with plant material grown under a wide range of nutritional conditions suggest that large variations in intracellular pH do not normally occur. Most of the measurements (e.g. with cell-sap extracts or using indicator dyes) refer to vacuolar pH (HURD-KARRER, 1939; SMALL, 1946; DRAWERT, 1955). Accurate estimations of vacuolar pH are possible with giant algal cells such as *Nitella* (HOAGLAND and DAVIS, 1923). In general, the vacuolar pH values fall within the range 5–6, indicating that organic acids, where present, are dissociated and form salts with inorganic cations. However, some exceptionally low values have been recorded. *Begonia* and *Oxalis* tissues have saps of pH 1–2, due to the presence of free oxalic acid (Table *12.*2; see also HURD-KARRER, 1939; RANSON, 1965; *12.*2.3). The marine alga *Desmarestia* contains sap with pH less than 1, due to the presence of free sulfuric acid (EPPLEY and BOVELL, 1958; KESSELER, 1964).

Table *12.*2. Titratable acidity (normality) and pH values of expressed saps. (Results of BROWN, quoted by THOMAS, 1951. See also RANSON, 1965)

Organ from which sap was expressed	Titratable acidity (normality) of sap	pH of sap	Major organic acids present
Lemon fruit	0.95	2.4	citric
Blackberry fruits	0.23	2.7	isocitric, oxalic
Rhubarb petioles	0.22	3.2	malic, oxalic
Unripe grapes	0.21	3.0	malic, tartaric
Oxalis leaves	0.16	2.3	oxalic
Begonia leaves	0.11	2.2	oxalic
Ripe tomato fruit	0.063	4.4	citric, malic
Ripe apple fruit	0.045	3.9	malic

During excess absorption of cations or anions, the vacuolar pH changes are quite small (ULRICH, 1941, 1942). HIATT (1967a) found that after excised barley roots were placed for 2 h in K_2SO_4 solutions the expressed sap was about 0.2 pH units higher than the sap from roots placed in KCl solutions. The sap from roots placed in $CaCl_2$ was about 0.25 units lower than the sap from the roots in KCl. Actual values ranged from 5.07 to 5.56. KIRKBY and MENGEL (1967) found that leaf tissue extracts from tomato plants grown on NO_3^- had a pH of 5.50, compared with 5.00 for plants grown on NH_4^+. The differences for roots were somewhat greater: 5.60 compared with 4.70 for NO_3^--grown and NH_4^+-grown plants respectively. It should be noted that many early workers found that there was relatively little change in intracellular pH when plant material was grown under a wide range of external pH conditions (HOAGLAND and DAVIS, 1923; see also references in HURD-KARRER, 1939).

Interactions between metabolism and intracellular pH must largely be mediated by cytoplasmic and not by vacuolar pH, but measurement presents formidable difficulties. In non-vacuolate cells (animal cells and prokaryotes) the cytoplasmic pH is usually near 7 (e.g. KASHKET and WONG, 1969; HAROLD et al., 1970; PAILLARD, 1972). Values obtained with plant cells having only small vacuoles (e.g. fungi, root tips, unicellular algae) range from 6.0 to 7.5 (CONWAY and DOWNEY, 1950; SHIEH and BARBER, 1971; VOTTA et al., 1971; GERSEN, 1973). Using microelectrodes, it is possible to distinguish between the pH of the cytoplasm and vacuole, as shown in Fig. *12*.3a. Effects of light (photosynthetic metabolism) are clearly shown. Estimates of cytoplasmic pH also have been obtained from DMO distribution for *Chara* cells (Fig. *12*.3b). However, the significance and interpretation of cytoplasmic pH measurements is open to question, in that the cytoplasm consists of several distinct compartments. Apart from the major membrane-bound organelles (mitochondria and chloroplasts), the endoplasmic reticulum forms a discrete phase within the "ground cytoplasm", and appears to have ion-transporting prop-

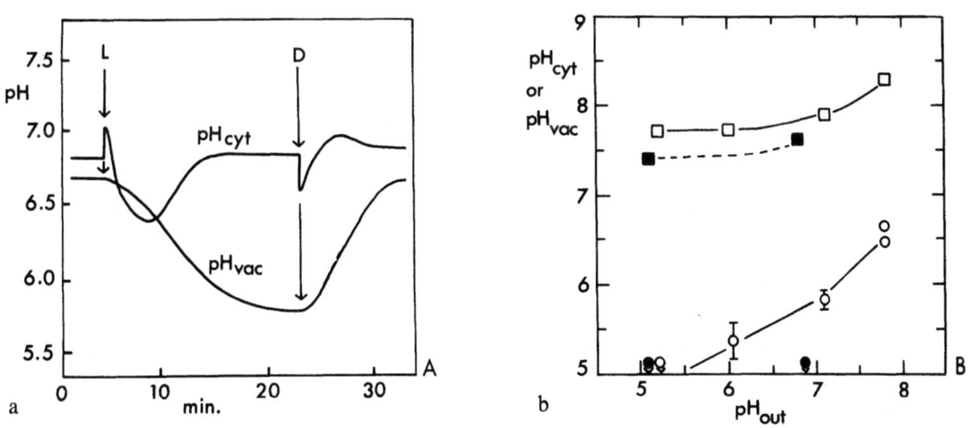

Fig. *12*.3 (a) Measurements of the intracellular pH of *Phaeoceros laevis* (a hornwort), using antimony-covered microelectrodes. Effects of light (*L*) and darkness (*D*) are shown. Results of DAVIS (1974), redrawn. (b). Determination of cytoplasmic and vacuolar pH of *Chara corallina*, using DMO (WALKER and SMITH, 1975). Cytoplasmic pH in light (□) and dark (■). Vacuolar pH in light (○) and dark (●). Arrowed symbols are values too low to measure (i.e. less than about 5.5)

erties (RUNGIE and WISKICH, 1973). There is likely to be considerable variation in pH between the different compartments, especially those primarily concerned with energy-transfer. Nevertheless, the available evidence supports the view that cytoplasmic pH remains relatively constant, despite large-scale metabolic production of H^+ or OH^- (*12*.2.1).

2.3 Internal Regulation of Intracellular pH

In accounting for the fate of excess H^+ or OH^- produced during metabolism, it is necessary to consider the possible significance of internal disposal mechanisms, as opposed to loss to the bathing medium. It must be stressed that in most cases the buffering capacity of the cell is inadequate to neutralize H^+ produced during growth. For example, the mixture of organic (carboxylic) acids and their salts in many vacuolate plant cells will buffer against the addition of H^+ or OH^-. However, the accumulation of carboxylate salts itself involves the prior production and disposal of H^+. Likewise, although cytoplasmic proteins may contribute to the buffering power of the cytoplasm, the nitrogen assimilation involved in their synthesis will involve production of H^+ or OH^- (Fig. *12*.1; Table *12*.1). It has been suggested that changes in enzymic activity resulting from pH-induced feedback can play a major part in regulating cytoplasmic pH. DAVIES (1973a, 1973b) has proposed a biochemical "pH stat", resulting from synthesis of malate at high pH, using phosphoenolpyruvate carboxylase (PEP carboxylase) and breakdown at low pH, using malic enzyme. The pH of the cytoplasm would be maintained between the pH optima of the two enzymes (at about pH 7). The scheme may be summarized as follows:

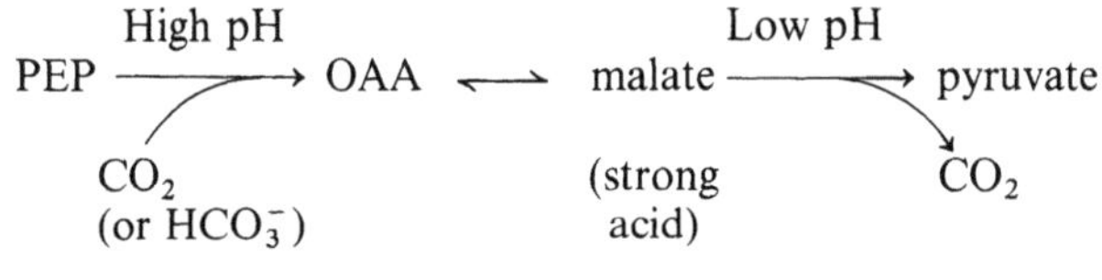

The scheme again cannot account for internal disposal of H^+ over the long term, due to the production of H^+ during synthesis of malate. Nevertheless, it has important implications with respect to disposal of OH^- (e.g. resulting from assimilation of HCO_3^- or NO_3^-) and will be discussed in more detail below (*12*.5). It is also of interest in considering the rapid changes in the organic acid content of cells in response to excess uptake of K^+ or Cl^- (*12*.5).

In some exceptional cases, the vacuole can be considered as a significant internal sink for H^+. These include those cells with very acid sap (mentioned above). The pH of the sap does not itself indicate the concentration (normality) of the acids present (i.e. their titratable acidity). The weaker the acid, the higher will be the normality for a given pH. This is shown in Table *12*.2 where, for example, the pH of the sap from lemon fruit is slightly higher than that from *Begonia* leaves, even though the normality of the acids in the lemon is higher. This is because the lemon contains mainly citric acid, a weaker acid than the oxalic acid in *Begonia* leaves.

It is possible that although storage of weak acids (as opposed to their salts) in the vacuole will dispose of cytoplasmic H^+, this may be a severe drain on the carbon metabolism of the cell. Green cells with Crassulacean acid metabolism are a special case in which free acids formed in the dark are stored temporarily, prior to their decarboxylation in the light (see RANSON, 1965; also *13*.2.1).

RAVEN and SMITH (1974) have considered briefly as an alternative internal "pH stat" the production and breakdown of bases, such as polyamines (suggested by T.A. SMITH, 1971). These could be produced from ammonia (a weak base) and carbohydrate (neutral), but the buffer range of the amines and the pH optima of the enzymes may not be suitable for a pH-regulating role. Nevertheless, it is possible that accumulation of amino acids such as proline, or amino acid derivatives such as putrescine may be a consequence of the necessity for the removal of H^+ under conditions of nutrient stress (see T.A. SMITH, 1971; *12*.5.2.4), apart from its possible role in osmoregulation (*11*.3.1.2; *13*.5).

2.4 Conclusions: pH Regulation by H^+ or OH^- Transport

In *12*.2 we have shown that many metabolic reactions lead to the production of H^+ or OH^- within the cytoplasm. In most cases the internal pH changes cannot be prevented by other internal metabolic reactions, and transport of H^+ or OH^- to the bathing medium or (in some cases) the vacuole takes place. We have suggested previously (RAVEN and SMITH, 1973) that H^+ or OH^- transport is in itself a primary mechanism for intracellular pH regulation. Besides responding to changes in intracellular pH (e.g. produced by carbon or nitrogen metabolism), we believe that changes in H^+ or OH^- transport in response to metabolic signals can initiate changes in metabolic pathways. The sensitivity of enzymic activity to pH changes is in fact so great that control of intracellular pH within well-defined limits must be of fundamental importance in the regulation of metabolism. As pointed out by LEHNINGER (1970) the optimum pH of an enzyme is not necessarily the same as the pH of its intracellular surroundings. Since the hundreds of enzymes in cells respond in many different ways to pH, small changes in intracellular pH will have profound effects on the operation of metabolic pathways. This theme is developed in more detail in *12*.5 and *12*.6. In *12*.3 and *12*.4 we discuss the possible nature of the mechanisms for membrane transport of H^+ and OH^-, and their interactions with other ion transport processes.

3. Biophysical Implications of H^+ Transport

3.1 The Necessity for Active Transport

To determine whether ion movements across membranes are passive or active, measurements of the chemical and electrical driving forces are required (Chaps. *3* and *4*). Even in the absence of accurate measurements of cytoplasmic pH, it is possible to make some deductions about the nature of the H^+ and OH^- fluxes considered above (*12*.2). The cytoplasm of plant cells has an electrical potential of -40 to -200 mV with respect to the bathing solution (SCOTT, 1967; GUTKNECHT and DAINTY, 1968; HIGINBOTHAM, 1973; Chaps. *3* and *4*; Chap. *6*: Table *6*.3;

Part B, Chaps. *3.3*, *4.2*: Table *4*.1). If the H^+ in the cytoplasm is in passive equilibrium with H^+ in the bathing solution, then the familiar Nernst equation tells us that the cytoplasm must be 0.7 to 3.5 pH units more acid than the bathing solution. (An electrical PD of 58 mV will balance a difference of 1 pH unit.) The pH of the cytoplasm could be maintained at pH 6.5–7 by passive diffusion of H^+ only if the bathing solution had a constant pH within the range 7.5–10.5. With the exception of the sea, bathing media do not normally meet this specification, and the cytoplasm can be maintained near pH 7 only by transport of H^+ or OH^- against an energy-gradient.

Where the bathing solution has a low pH there must be active extrusion of H^+, as has been demonstrated in a number of specific cases, such as fungi (SLAYMAN, 1970), algal coenocytes (KITASATO, 1968) and higher plant cells (POOLE, 1966, 1974; PITMAN, 1970; HIGINBOTHAM and ANDERSON, 1974). Rates of H^+ extrusion are greatest when there is massive internal synthesis of organic acids, as mentioned in *12*.2. This is discussed with respect to storage tissues in *8*.3.2.3, and low-salt roots in Part B, Chap. *3.3*.

In contrast to the situation at the plasmalemma, the electrical PD across the tonoplast of most plant cells is relatively small (SCOTT, 1967; MACROBBIE, 1970; GUTKNECHT and DAINTY, 1968; Chap. *3* and *4*), and is normally such that the vacuole is positive with respect to the cytoplasm. Passive distribution of H^+ would lead to vacuolar H^+ concentrations about twice those in the cytoplasm. Measurement of actual pH values (*12*.2.2) suggest that the vacuolar H^+ concentrations are at least 10 times higher than in the cytoplasm, indicating that there is active transport of H^+ across the tonoplast into the vacuole. This must certainly be the case in marine algae such as *Valonia* and *Chaetomorpha* in which the vacuoles are at least 40 mV positive with respect to the cytoplasm (GUTKNECHT and DAINTY, 1968). H^+ transport must also be involved in maintaining the large pH gradients across the tonoplast of cells with very acid sap. Where plant cells accumulate organic acids, it is possible that these are transported across the tonoplast in the undissociated form, essentially as a neutral H^+ pump, depending on the dissociation constant of the particular acids.

In the case of plant cells which can tolerate a range of external pH values, the thermodynamic necessity for H^+ extrusion will vary. For example, the freshwater alga *Chara corallina* (Fig. *6.3*a) will grow in solutions with pH up to about 9.5, at which pH it does not seem necessary to postulate active H^+ transport. The situation in *Chara* is, however, complex. In the light, large areas of the surface of these giant cells are maintained at about pH 5.5, even when the pH of the bulk solution is increased by the addition of buffers at low concentrations. It is postulated that this acidification reflects light-promoted H^+ extrusion (LUCAS and SMITH, 1973). Other areas are maintained at pH 9.5–10.0 even when the bulk solution is at lower pH. This latter effect is due to photosynthetic HCO_3^- assimilation (see *12*.5.2.1). The extent of this external alkalinization may again give an indication of the cytoplasmic pH. Assuming an electrical PD of −150 mV (as in the normal unbuffered solutions: see Table *6*.3), then a cytoplasmic pH of greater than 7.5 would be necessary for net passive OH^- efflux into the external alkaline zone (pH 10.0).

The current emphasis on H^+ fluxes in giant algal cells (such as those of the Characeae) results from the increasing conviction that the electrical properties of the cell membranes cannot be explained satisfactorily in terms of passive or active movements of other ion species such as K^+ or Cl^- (KITASATO, 1968; SPANSWICK et al., 1967; SPANSWICK, 1970, 1973; *4*.2.6). Some electrical studies have indicated that the permeability to H^+ in these cells is extremely high (KITASATO, 1968; RENT et al., 1972). This would indicate a very large energy expenditure involved in pumping H^+ from the cells. If the active H^+ efflux from Characean cells is 400 nmol m^{-2} s^{-1} (KITASATO, 1968) and the rate of O_2 uptake during respiration about 50 nmol m^{-2} s^{-1} (SMITH, 1968), then about 40% of the total free energy of respiration would be used solely in maintaining the H^+ pump. (The calculation assumes that the total electrochemical difference for H^+ between cytoplasm and bathing solution is −240 mV.) This would seem an impossibly wasteful use of respiratory energy, especially as it does not take into account energy expended in pumping H^+ into the vacuole. High rates of H^+ extrusion from Characean cells in the light are more credible as the energy contribution from photosynthesis might be ten times that available from respiration (SMITH, 1967). By analogy with other ion pumps in the Characeae, light stimulation of H^+ transport is to be expected, and light effects on the membrane potential of *Nitella translucens* have been described by SPANSWICK (1972, 1974), VREDENBERG (1973) and VREDENBERG and TONK (1973). (See also *6*.6.6.2.)

Internal generation of OH^- during assimilation of HCO_3^- or NO_3^- would in general appear to remove the requirement for H^+ extrusion. Under these conditions net OH^- efflux (or H^+ influx) and organic acid synthesis can be regarded as alternative OH^- balancing processes (Fig. *12*.1 b). Nevertheless, if the permeability of the plasmalemma to H^+ were so high that H^+ entered cells faster than required to neutralize internal OH^-, extrusion of H^+ would again be necessary to maintain the cytoplasm at or near pH 7.

Finally, it should be noted that net OH^- loss into very alkaline solutions (e.g. during HCO_3^- assimilation) is almost certainly due to active OH^- extrusion. The presence or absence of an OH^- pump may be indicated by the upper pH limit tolerated by plant cells (RAVEN, 1970).

3.2 The Electrogenic H^+ Pump

The effects of external pH changes on the electrical potential (using well buffered solutions) support the view that in the Characeae, as in other plant cells, the H^+ pump is electrogenic and plays a major part in determining the PD across the plasmalemma. These pH effects vary in magnitude in different species and are further complicated by the interactions with light. In *Nitella flexilis* the maximum hyperpolarization with increasing external pH is about 80 mV at pH 8–9, as shown in Fig. *12*.4. In *Chara corallina* the membrane hyperpolarizes by about 100–120 mV at pH 9, compared with the value at pH 5.5 (SMITH and LUCAS, 1973). In *Nitella translucens* there is a hyperpolarization of about 60 mV (SPANSWICK, 1972). In all three species these very large changes are temporary, but in *N. flexilis* and *C. corallina* the hyperpolarized state is maintained at pH 7.3–7.5 (Fig. *12*.4; see also RICHARDS and HOPE, 1974). In *N. translucens* the hyperpolarization is not maintained above pH 6 (SPANSWICK, 1972).

SPANSWICK (1972, 1973), has proposed that rate of H^+ transport is a function of the membrane electrical potential. He has further postulated that the electrical potential is affected by other variables, including the free-energy change for the chemical (i.e. non-transported) components of the system, and the pH gradient across the plasmalemma. The latter suggestion has important implications with respect to pH regulation, and will be discussed briefly below (*12*.4.4). It still remains

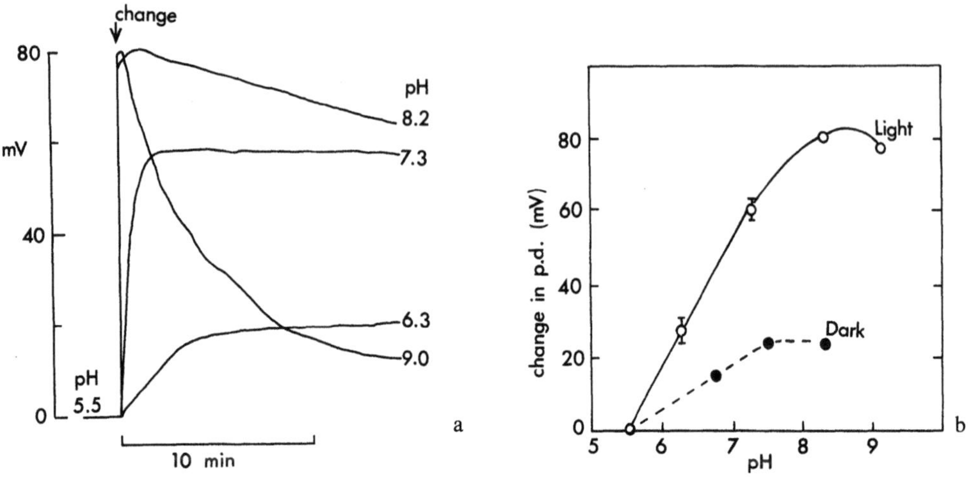

Fig. *12.*4a and b. Effects of external pH on the membrane potential of *Nitella flexilis*. (a) Time-courses resulting from external pH changes. (b) The magnitude of the maximum hyperpolarizations obtained at various pH values. Vertical bars indicate ± standard error of the mean of 4 cells, shown only where this is larger than the symbols. Tricine buffer containing 0.1 mM KCl plus 1.0 mM NaCl. Results of SAITO and SENDA (1973), redrawn

possible that in some Characean species HCO_3^- transport from solutions with pH greater than 6 may have a role in determining membrane electrical parameters (see LUCAS and SMITH, 1973; VOLKOV, 1973). The wider significance of electrogenic ion transport is assessed in *3.*5.4.2; *4.*2.6, *6.*6.6 and *7.*2.

4. Models for H^+ Transport

4.1 H^+ Transport and "Charge Separation"

Most of the proposed ion transport mechanisms have been developed specifically to explain membrane transport of inorganic ions such as K^+ or Cl^-, and movements of H^+ or OH^- (where included) have generally been regarded as secondary or balancing movements. Attempts in the 1920's and 1930's to explain KCl accumulation in terms of physical ion exchange (K^+ for H^+ and Cl^- for OH^-) were unsatisfactory in that the necessity for input of metabolic energy was not in general appreciated. Little attention was paid to mechanisms for separating H^+ and OH^-, without which ion-exchange systems could not function (BRIGGS, 1930, 1932; OSTERHOUT, 1931, 1936). This criticism does not apply to the hypothesis developed by LUNDEGÅRDH (1939, 1954). This considers anion accumulation as specifically linked to electron flow through the cytochrome system, with H^+ liberated by respiratory dehydrogenases exchanging with external cations. The scheme is summarized in Fig. *12.*5a. The scheme developed by ROBERTSON (1960, 1968) differs from that of LUNDEGÅRDH in that the anion carrier is an (unspecified) molecule, which is reduced at the inside of the plasmalemma, and diffuses in unionized form to the outside. There it is oxidized by losing an electron, picks up a Cl^-

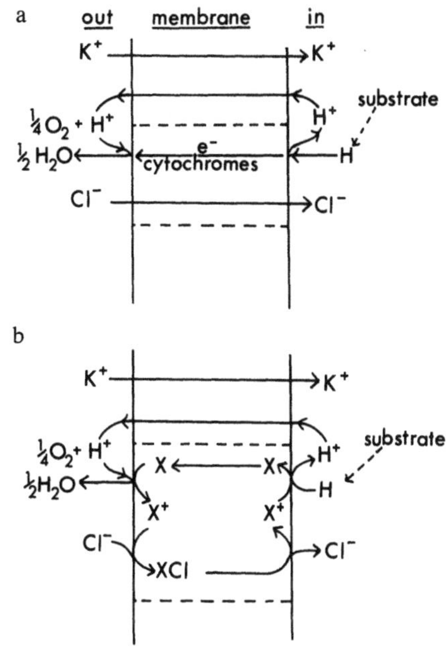

Fig. *12*.5. (a) Model for salt accumulation, as proposed by LUNDEGÅRDH (1939, 1954). (b) Model for salt accumulation, as proposed by ROBERTSON (1960, 1968)

and diffuses back to the inside (Fig. *12*.5b). It is again postulated that H^+ exchanges for K^+ by a separate system. Both of these schemes suffer from the disadvantage that as far as H^+ transport is concerned they are the reverse of what is required; H^+ is released inside the cell, which becomes positively charged. This situation arises as a result of considering Cl^- influx (effectively Cl^-/OH^- exchange) as the primary process, rather than K^+/H^+ exchange, and applies equally to the proposal (ROBERTSON, 1960, 1968) that salt accumulation in plants can be directly powered by charge-separation across mitochondrial membranes.

The direct implication of redox reactions in ion transport is also a feature of CONWAY's scheme for H^+ extrusion from yeast (see 7.2.2.3 and CONWAY, 1954). According to CHANEY et al. (1972), redox reactions at the plasmalemma are involved in Fe^{3+} reduction prior to Fe^{2+} absorption by roots.

4.2 H^+ Transport and ATP

An alternative energy source for transport of H^+ or OH^- would be ATP, with the possibility that K^+/H^+ exchange is analogous to K^+/Na^+ exchange in animal cells. In general, although the involvement of ATP would be assumed by many workers, there is little direct evidence to distinguish between the alternatives of redox energy versus ATP. SLAYMAN et al. (1970) and SLAYMAN et al. (1973) have provided strong evidence that ATP from oxidative phosphorylation is the energy source in *Neurospora* (Fig. *4*.7). The light-stimulated H^+ pump in *Atriplex* leaves appears to be coupled to the redox reactions involved in non-cyclic photosynthetic electron flow (LÜTTGE et al., 1970) while in *Nitella translucens* ATP from cyclic photophosphorylation has been implicated (SPANSWICK, 1974; VREDENBERG and TONK, 1973). The possible significance of ion-transporting ATPases is discussed in detail in Chap. *10*.

4.3 H^+ Transport and the Chemiosmotic Hypothesis

The chemiosmotic hypothesis of energy-coupling, developed by MITCHELL (1961, 1970) provides a much more comprehensive basis for models of H^+ transport than any of the above considerations. According to this hypothesis, electron transport in respiration or photosynthesis creates a free-energy gradient of H^+ (i.e. a pH gradient plus an electrical potential) across the membrane containing the electron carriers. This energy gradient can then be used to drive an H^+ transporting ATPase in the direction of ATP synthesis (see Fig. 12.6). These reactions occur

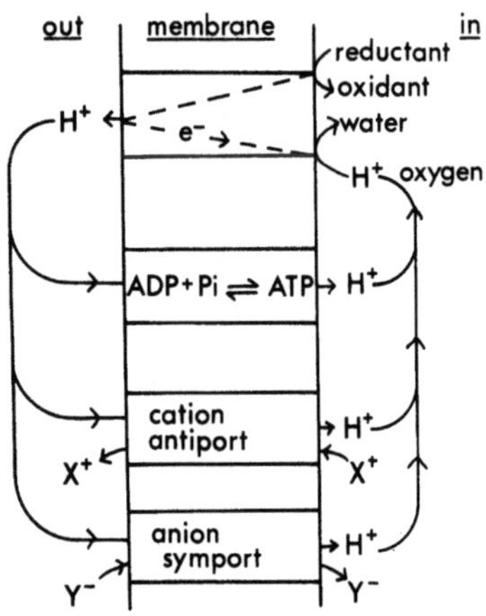

Fig. 12.6. Composite H^+ circuit diagram, illustrating the coupling between metabolism and transport in mitochondria and prokaryotes (see MITCHELL, 1970)

in the outer membranes of prokaryotes, the cristae membranes of mitochondria and the thylakoid membranes of chloroplasts. When ATP synthesis in these "coupling membranes" does not proceed (e.g. in the absence of ADP) the H^+ energy gradient can be expended in the transport of other ions (Fig. 12.6). H^+ recirculation during ATP synthesis is an essential feature of this hypothesis. By contrast, according to the chemical intermediate hypothesis for energy-coupling, and also the conformational hypothesis, H^+ transport is an alternative to ATP synthesis. It should be noted that all three hypotheses require that the passive permeability of the coupling membrane to H^+ should be low, otherwise the H^+ energy gradient is wasted (the system becomes uncoupled). A much more detailed treatment of MITCHELL's hypothesis is given in Vol. 3; only aspects relevant to ionic relations of cells and tissues are discussed here.

In the present context, the important features of the MITCHELL hypothesis are those which deal with ion transport. In coupling membranes active H^+ transport is brought about during the electron transport reactions, or by ATP hydrolysis. There are then various ways in which fluxes of H^+ or OH^- can be linked to other solute fluxes (MITCHELL, 1970). Some of these may be summarized as follows.

1. If active H^+ transport is electrogenic, then the electrical PD is used to move cations in the opposite direction to the H^+ pump (electrogenic antiport) or anions in the same direction (electrogenic symport).

2. The H^+ pump may be coupled chemically to either cation antiport or anion symport. The system is then electrically neutral.

3. The pH gradient generated by the H^+ pump is used in active transport. In this case the downhill flux of H^+ may be coupled chemically to cation antiport, anion symport (Fig. *12*.6) or non-electrolyte symport (see Part B, *5.3*.1.1.2.3). Alternatively, downhill flux of OH^- may be coupled to anion antiport or cation symport.

4. The pH gradient will determine the distribution of weak acids and bases between the solution and the cytoplasm, and has been implicated in transport of IAA (RUBERY and SHELDRAKE, 1974; RAVEN, 1975; *12*.6).

It is important to note that mechanism 1 can bring about only downhill ion transport, i.e. the metabolic energy supplied to the H^+ pump generates an increased electrical component of the passive driving forces. In mechanism 2 part of the chemical energy supplied to the H^+ pump is used directly in the transport of some other ion, which may move actively, i.e. against its own free-energy gradient. In mechanism 3 transport of other ions can also be active, but in this case the energy comes from the downhill flux of H^+ or OH^-.

There is evidence for involvement of H^+ in transport of solutes in both eukaryotes and prokaryotes (lactose in *Escherichia coli,* WEST and MITCHELL, 1973; hexose transport in *Chlorella,* KOMOR, 1973; general transport in prokaryotes, HAROLD, 1972; see also Part B, *5.3*.1.1.2.3).

Evidence for the operation of these mechanisms in ion transport is reviewed by MITCHELL (1970) and SKULACHEV (1971) for prokaryotes, mitochondria and chloroplasts.

Mechanisms based on MITCHELL's proposals have been suggested as the basis for the accumulation of inorganic salts and salts of organic acids (SMITH, 1970, 1972, 1973; RAVEN and SMITH, 1973). The original scheme (Fig. *12*.7) was specifically designed to account for the loose coupling between cation influx and Cl^- influx in Characean cells. It allows variation between cation and Cl^- influx at the expense of changes in external and internal pH, the extent of the latter depending on the internal buffering capacity (particularly carboxylate levels).

Experimental work with *Chara corallina* does not allow distinction between the possibilities that the H^+ extrusion mechanism is a MITCHELL-type ATPase, or alternatively a redox system, and the nature of the coupling between H^+ extrusion and cation influx was not specified. The model (Fig. *12*.7) did not specify the nature of ion transport systems at the tonoplast. Coupling between active influx of H^+ and Cl^- to the vacuole could be achieved by mechanism 2, above. Active cation influx into the vacuole (if necessary) could be linked to H^+ efflux from the vacuole by mechanism 3.

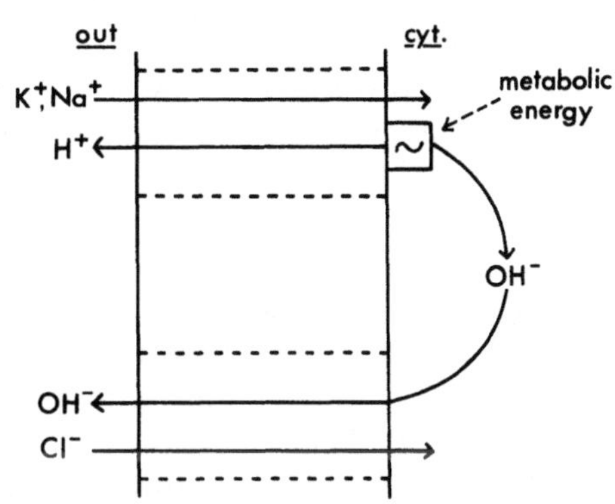

Fig. *12*.7. Model for ion transport coupled to charge-separation at the plasmalemma (SMITH, 1970)

The most controversial feature was the suggestion that the energy for Cl^- influx could be provided by Cl^-/OH^- exchange (mechanism 3), and that direct links with metabolic energy were not required. An alternative but similar suggestion (SPEAR et al., 1969) was that uptake of molecular HCl took place (again *via* mechanism 3).

The main evidence for this type of model is that Cl^- influx in *Chara* cells is greatly reduced at high pH, and can be stimulated (at pH 5–6) by pre-treatment at pH 9.0–9.5 (SMITH, 1970, 1972). The model has been criticized (MACROBBIE, 1970) on the grounds that if Cl^- transport is mediated by Cl^-/OH^- exchange, it should be even more sensitive to external pH changes, than is in fact the case. In particular, an unreasonably high cytoplasmic pH might be necessary to permit net Cl^- influx with an external pH greater than 7. (For example, with one-for-one Cl^-/OH^- exchange, a ten-fold accumulation of Cl^- from a solution of pH 8 would require a cytoplasmic pH of at least 9.) The validity of this criticism depends on the magnitude of the total (electrochemical) H^+ energy gradient across the plasmalemma, its response to external pH changes, and the stoichiometry of Cl^-/OH^- exchange. It is possible that the latter could vary with changing pH.

With respect to plant cells in general, it is important to consider the possibility that at high pH, H^+ extrusion may be replaced by OH^- extrusion. It is therefore possible that there may be direct input of metabolic energy (i.e. ATP or redox energy) to both the K^+/H^+ and Cl^-/OH^- exchange systems, but that this varies with changes in the H^+ free energy gradient. Perhaps the most important feature of this type of mechanism is that by involving pH regulation of Cl^- influx it suggests that this process can be considered as analogous to accumulation of organic anions, both being dependent on primary H^+ extrusion (see also HODGES, 1973; RAVEN and SMITH, 1973, 1974; SMITH, 1973).

4.4 Regulation of the H^+ Pump

The main factors which will contribute to the control of the rate of H^+ transport may be summarized as follows:

a) Metabolic Energy. Whether provided as redox energy or ATP, the energy supply for H^+ transport can be regarded as a potentially limiting factor. Light effects on membrane electrical parameters (see *4*.2.6) or intracellular pH (Fig. *12*.3a) may reflect changes in H^+ transport due to availability of photosynthetic energy.

b) The H^+ Free Energy Gradient. This may be regarded as the product of the H^+ pump. It is convenient (but hazardous) to consider the electrical and chemical (pH) components separately. SPANSWICK (1972, 1973) has suggested that H^+ transport is subject to negative feedback from the electrical PD across the membrane. However, in terms of pH regulation, feedback from external and cytoplasmic pH must be of paramount importance. Thus with increasing external pH the chemical concentration gradient for H^+ decreases, and the activity of the H^+ pump results in an increased electrical PD (hyperpolarization). If the internal pH increases, then the rate of pumping ought to decrease, resulting in depolarization. Interactions with other ion movements will also affect the H^+ free energy gradient. Increasing inward diffusion of K^+ would tend to decrease the electrical PD, but not the pH gradient across the plasmalemma. Conversely,

Cl^-/OH^- exchange would tend to lower cytoplasmic pH, without directly affecting the electrical PD. The major internal processes determining cytoplasmic pH would be the synthesis and dissociation of organic acids, and ion transport to the vacuole. One result of these interactions would be that the apparent "electrogenicity" of the H^+ pump could change under different conditions, as proposed by POOLE (1973) for the H^+ pump in beet cells (see also *8*.3.2.3).

The overall situation is best summarized by MITCHELL's model, according to which the rate of H^+ transport would be greatest when the H^+ free energy gradient was being dissipated by ion transport and other metabolic processes (as shown in Fig. *12*.6).

c) Other Metabolic Triggers. The processes mentioned above will result in the control of cytoplasmic pH by H^+ transport. We believe that other metabolic signals may in some circumstances over-ride this control so that well-defined changes in cytoplasmic pH may in turn control various metabolic processes. Some specific examples are discussed in *12*.5, below, and the role of plant hormones as metabolic triggers affecting H^+ transport is assessed in *12*.6.

5. H^+ Fluxes and the Regulation of Solute Accumulation

5.1 General Principles

The evidence that accumulation of salts of organic acids involves both internal disposal of OH^- or HCO_3^- and extrusion of H^+ has been summarized in *12*.2. Fig. *13*.4a, shows the situation in which there is extrusion of H^+, coupled to cation influx. We believe that the H^+ pump has a primary role so that as well as being subject to feedback from internal pH changes, it can respond to other signals which might regulate the total ion content of the cell (*12*.4.4; see also Chaps. *16* and *13*). Such signals (e.g. hormones, fusicoccin-like compounds, Part B, Chap. *7*) would over-ride the internal pH stat discussed in *12*.2.3. Thus when H^+ extrusion is stimulated by a requirement for additional turgor, cytoplasmic pH will start to increase, as will cytoplasmic K^+ content. Increasing pH will stimulate organic acid synthesis, as proposed by HIATT (1967a, 1967b) and HIATT and HENDRICKS (1967). The pH stat will not shut down organic acid synthesis as long as K^+/H^+ exchange is the pacemaker, or as long as the product is removed from the site of synthesis, i.e. to the vacuole (JACOBY and LATIES, 1971; *13*.2.5). Where the external pH is high, influx of HCO_3^- (plus K^+) will maintain increased internal pH, and hence organic acid synthesis. An additional K^+/H^+ exchange is necessary in both situations to create the double K^+ salt of malate. Synthesis of salts of tricarboxylic acids will require a further K^+/H^+ exchange. It is suggested that the response of the sequence of reactions to pH changes is such that the pH of the cytoplasm will only fluctuate within narrow limits (e.g. 0.5 pH units; see DAVIES, 1973a). Whether the substrate for carboxylation is CO_2 or HCO_3^- (see WAYGOOD et al., 1969; COOPER and WOOD, 1971) is of secondary importance in the present context. It may be noted that with increasing cytoplasmic pH, carboxylation would be favored if HCO_3^- were the substrate.

When plant cells accumulate inorganic ions (e.g. K^+ plus Cl^-), it is proposed that increased cytoplasmic OH^- availability resulting from K^+/H^+ exchange favors Cl^- transport. This effect need not necessarily be regarded as the result of an increased pH gradient on a simple Cl^-/OH^- exchange system (Fig. *12*.7). It is possible that a transport ATPase (e.g. Cl^-/OH^-) could be subject to feedback control from cytoplasmic pH (*12*.4.4).

Where plant cells can accumulate both inorganic and organic anions, there must be a delicate balance, which may change during cell development (MOTT and STEWARD, 1972a, 1972b). One possibility is that the cytoplasmic pH maintained by the balance between K^+/H^+ exchange and Cl^-/OH^- exchange is somewhat lower than that required for rapid carboxylation. LEONARD and HODGES (1973) found that the KCl-stimulated plasma membrane ATPase from oat roots has a pH optimum of 6.5 (*10*.4.3.1). PEP carboxylases isolated from a range of tissues have been shown to have pH optima greater than 7.0 (OSMOND and GREENWAY, 1972).

Effects of pH on interactions between accumulation of inorganic anions and carboxylates are also indicated by the work of VAN STEVENINCK (1965, 1966). It was shown that in the presence of *tris* buffer (pH 8) Cl^- accumulation by some storage tissues is replaced by rapid synthesis of malate and citrate. VAN STEVENINCK suggested that this may be due to increasing K^+/H^+ exchange resulting from efficient removal of H^+ from the cell surface. It is also possible that penetration of *tris* into the cytoplasm directly increases cytoplasmic pH. Whatever the true explanation of this effect, the results show that it is an over-simplification to regard the Cl^- pump as necessarily dominant over the carboxylation system. In the early stages of KCl absorption by low-salt barley roots it has been found that net K^+ influx initially exceeds net Cl^- influx (see PITMAN et al., 1971). This suggests that there is an essential minimum requirement for accumulation of organic anions. This may also be the explanation of lags in NO_3^- absorption (JACKSON et al., 1972, JACKSON et al., 1973) rather than induction of the NO_3^- pump (see also Part B, Chap. *9*).

Barley roots and carrot slices which have accumulated organic anions such as malate will subsequently absorb Cl^- and NO_3^- at rapid rates (CRAM, 1973; SMITH, 1973). This process must involve decarboxylation and is apparently independent of net K^+ influx. It can be regarded as a special case of excess anion absorption, and is shown in Fig. *13*.4b. The end result of the processes shown in Figs. *13*.4a and *13*.4b is accumulation of KCl or KNO_3, but in this case, influx of cations and inorganic anions is completely separated in time. It is again tempting to implicate control by cytoplasmic pH in this replacement of internal carboxylate anions by inorganic anions. Although this effect provides evidence for independent cation and anion exchange processes, it seems unlikely that anion exchange could operate in the absence of K^+/H^+ exchange without direct input of metabolic energy at the plasmalemma (cf. Fig. *12*.7).

5.2 H^+ Fluxes in Photosynthetic Tissues

In assessing the involvement of H^+ fluxes in salt accumulation by photosynthesizing cells it is necessary to distinguish pH changes resulting from uptake and assimilation of CO_2 or HCO_3^- from those associated with the ion transport processes discussed above (*12*.5.1). This distinction may, however, be somewhat artificial in the sense that photosynthetic carbon fixation may result in the formation of neutral carbon compounds (e.g. carbohydrate) or ionized compounds (e.g. carboxylates). The proportions may vary, not only in different plant species, but also in the same species under different conditions. As an example of the latter effect, OUELLET and BENSON (1952) showed that in short-term photosynthetic ^{14}C fixation by *Scenedesmus* the proportion of ^{14}C fixed into malic and aspartic acids increased markedly with increasing external pH.

5.2.1 External pH Changes Associated with Uptake of CO_2 or HCO_3^-

There is no doubt that large external pH increases (up to about pH 8.5) can result from photosynthetic uptake of CO_2 by aquatic plants in closed containers (STEEMANN NIELSEN, 1960). NEUMANN and LEVINE (1971) showed that this effect was responsible for "apparent H^+ fluxes" in *Chlamydomonas*, rather than the direct reflection of pH shifts across the chloroplast envelope, as suggested previously (SCHULDINER and OHAD, 1969; see also BEN-AMOTZ and GINZBURG, 1969). Even larger external pH increases result from HCO_3^- uptake by some aquatic plants (Fig. *12*.1c; Part B, *4*.2.3.5), and some of the biophysical implications of this have been discussed above.

HCO_3^- assimilation is frequently associated with calcification at the cell surface as shown in Fig. *12*.8 (see also RAVEN, 1970, and references therein). Submerged leaves of some aquatic angiosperms absorb HCO_3^- through both surfaces, but

Fig. *12*.8. (A) Calcification at the surface of *Chara corallina* often shows bands associated with different degrees of H^+ and OH^- transport (×2; photograph by J. FAIRBURN). (B) Calcification of algae such as this *Lithophyllum* sp. contributes considerably to the growth of coral reefs. The physiological basis of the pH changes associated with calcification in marine algae is obscure (×0.4; photograph by B. LESTER and M. BOROWITZKA)

OH^- is released only through the upper (adaxial) surface. OH^- release is accompanied by transport of inorganic cations from the lower to the upper surface (ARENS, 1936a, 1936b; GESSNER, 1937; STEEMANN NIELSEN, 1947; LOWENHAUPT, 1956; Part B, *4.2.3.2.2*). STEEMANN NIELSEN (1947) proposed that OH^- is extruded by a polarized OH^- pump at the upper leaf surface, with HCO_3^- and cation fluxes occurring as secondary processes. LOWENHAUPT (1956) placed more emphasis on active cation transport as the primary process. Rapid influx of HCO_3^- by diffusion seems unlikely (RAVEN, 1970) even though photosynthesis would keep the internal concentration of this ion low.

Detailed measurements of effects of cations on HCO_3^- transport into Characean cells have not yet been made. However, comparison of rates of HCO_3^- assimilation with cation influx (SMITH, 1968; FINDLAY et al., 1969; LUCAS, 1975) shows that HCO_3^-, assimilation does not involve influx of cations plus HCO_3^-, followed by efflux of cations plus OH^-.

Under natural conditions, with large volumes of bathing solution, large external pH changes at the surface of algal cells or aquatic angiosperms can be to some extent dissipated by diffusion into the bulk medium. In comparison, in intact leaves of other angiosperms such dissipation would not be possible.

JONES and OSMOND (1973) showed that cotton leaf slices did not assimilate external HCO_3^-, and pointed out that this is in accord with the evidence that CO_2 is the substrate for the carboxylating enzyme (ribulose diphosphate carboxylase) in plants having the C_3 pathway of photosynthesis. They suggested that C_4 plants, in which PEP carboxylase is the initial carboxylating enzyme, should be able to absorb HCO_3^-. In such plants the initial carboxylation products (the C_4 acids) would temporarily dispose of OH^-, but this would be regenerated during subsequent conversion to carbohydrate, which involves decarboxylation and internal refixation of CO_2. The involvement of external HCO_3^- in C_4 photosynthesis has yet to be demonstrated, but although it is attractive to associate automatically the ability of cells to absorb HCO_3^- with the use of this ion as a substrate for carboxylation, this temptation should be avoided. There is little doubt that HCO_3^- uptake involves a specific membrane carrier (RAVEN, 1970), while the internal HCO_3^-/CO_2 ratio will be affected by internal pH regulation. Within the intact leaf of a C_4 plant charge balance must be maintained without the involvement of membrane transport processes resulting in large-scale net transport of H^+ or OH^-.

5.2.2 The Regulation of Solute Accumulation in Photosynthetic Tissues

The evidence for H^+ transport in Characean cells, derived from electrical studies, has been summarized in *12.3*. Similar evidence has been obtained with other green plant cells (Part B, *4.2.3.5*), including *Vallisneria* (BENTRUP et al., 1973), *Elodea* (JESCHKE, 1970; SPANSWICK, 1973) and *Phaeoceros* (DAVIS, 1974). DAVIS (1974) has combined electrical studies with measurements of intracellular pH; the latter are shown in Fig. *12.4*. HOPE et al. (1972) showed that there is light-promoted net H^+ influx into illuminated *Elodea* leaves, which is associated with O_2 evolution, but under some circumstances apparently independent of CO_2 fixation (Fig. *12.9*). It is not clear how this can be related to the H^+ *extrusion* postulated by SPANSWICK (1973), or the polar transport in *Elodea* as discussed above (*12.5.2.1*).

In many vacuolate algae, including the Characeae, organic acids are apparently confined to the cytoplasm, but may still be of significance in the pH control of Cl^- uptake and HCO_3^- assimilation. In other green cells, and particularly cells of higher plants, synthesis and accumulation of organic anions may be even

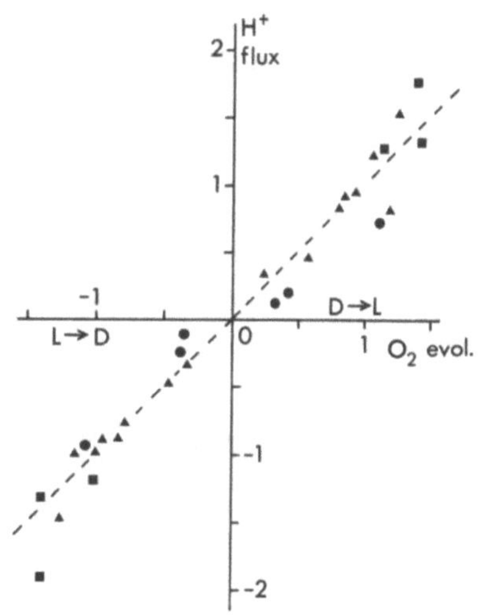

Fig. *12.9*. Relationship between apparent H^+ fluxes and O_2 exchange in *Elodea* leaves. Rates are in μmol g_{FW}^{-1} min^{-1}. Positive and negative values for apparent H^+ flux represent net influx and efflux respectively. Different symbols represent results from three batches of cells. The dashed line has unit slope. Results of HOPE et al. (1972), redrawn

more important (relative to inorganic anions) than in non-green cells such as root cells. NOBEL (1969) showed that chopped pea leaves bathed in KCl showed light-promoted net K^+ influx well in excess of Cl^- influx, and largely balanced by H^+ efflux. KHOLDEBARIN and OERTLI (1970a) likewise showed that barley leaf slices in KCl absorbed K^+ in excess of Cl^-, with synthesis of organic acids, in both light and darkness. *Tris* buffer reduced Cl^- influx, and stimulated organic acid synthesis (KHOLDEBARIN and OERTLI (1970b). LÜTTGE et al. (1970) found that there was net H^+ efflux from slices of *Atriplex spongiosa* leaves in light and a small net influx in darkness. It was suggested that the H^+ efflux was linked to photosynthetic redox reactions. Light-promoted transient changes in the electrical potential of *A. spongiosa* leaf cells were investigated by PALLAGHY and LÜTTGE (1970) who correlated them with changes in H^+ fluxes. It was proposed that the latter directly reflected H^+ movement across the thylakoid membranes. This is a problematical explanation, as it implies that energy conversion within the chloroplast is inefficient and that the chloroplast envelope and plasmalemma play no role in pH regulation.

LÜTTGE (1973) and LÜTTGE et al. (1974) measured external pH changes when slices from etiolated barley and maize leaves (at various stages of greening) were

Table *12.3*. Ratios of changes in external H^+ concentrations to O_2 evolution ($\Delta[H^+]:O_2$) and of CO_2 fixation to O_2 evolution ($CO_2:O_2$) found with barley and maize leaf slices. Results of LÜTTGE et al. (1974)

	Short greening times		Long greening times	
	$\Delta[H^+]:O_2$	$CO_2:O_2$	$\Delta[H^+]:O_2$	$CO_2:O_2$
Barley	1.47±0.06 (20)[a]	0.15±0.02 (20)[a]	1.10±0.03 (14)[c]	1.22±0.21 (4)[d]
Maize	0.70±0.07 (3)[b]	0.25±0.05 (6)[b]	0.67±0.08 (3)[e]	0.45±0.06 (6)[e]

[a] Plants were greened for 30–120 min; [b] 3 h; [c] 3–24 h; [d] 24–30 h; [e] 24 h. H^+ and O_2 concentrations were measured with a pH combination electrode in an oxygen electrode assembly. CO_2 fixation was measured with $^{14}CO_2$ added as $NaH^{14}CO_3$.

bathed in solutions containing KCl plus $NaHCO_3$. O_2 evolution, CO_2 fixation and Cl^- influx were also measured. During the early stages of greening the external pH increase was accompanied by O_2 evolution. Rates of CO_2 fixation (measured with $^{14}CO_2$) were lower than rates of O_2 evolution, and were inadequate to explain the increase in pH (Table *12*.3). The results certainly suggest that under these conditions there is net H^+ influx linked to photosynthetic redox reactions but the problem is to account for the fate of the H^+ apparently absorbed by the cells and the excess reducing power resulting from O_2 evolution. Rates of Cl^- uptake were insufficient to account for the pH rise in terms of Cl^-/HCO_3^- exchange (LÜTTGE, 1973).

Although leaf slices are proving to be very useful experimental material, enabling comparisons to be made with excised roots and other non-green tissues, it is yet to be determined how relevant the experimental conditions are to the situation in the intact leaf. It is important to remember that the solution bathing the cells of intact leaves, and of other aerial plant parts, is of very small volume. Consequently, net H^+ fluxes resulting in large extracellular pH changes could have dire effects on membrane structure and permeability. The buffering power of the cell wall may be important in this respect. H^+ fluxes are also presumably kept within manageable limits by control of the ions entering the xylem stream, and by internal disposal of H^+ or OH^-. Perhaps, within the leaf, processes resulting in H^+ production can be balanced by those producing OH^- (Table *12*.1). The most obvious example of the latter is NO_3^- reduction.

Export of organic compounds *via* the phloem could also have a pH-regulating role. BEN-ZIONI et al. (1971) have proposed that malate produced during NO_3^- reduction in the shoot moves down to the root, where it is decarboxylated and NO_3^-/HCO_3^- exchange takes place (Fig. *13*.4b). The whole system acts as a NO_3^-/malate shuttle, with K^+ as balancing cation, resulting in excess NO_3^- uptake by the plant (*13*.2.3).

While it is possible to account for the control of accumulation of ions in green plant cells in terms of pH-controlled mechanisms similar to those suggested for non-green cells, there is one major difference. In illuminated green cells endogenous CO_2 levels must be low, due to removal of CO_2 by photosynthesis. In effect there is competition between PEP-carboxylase and ribulose-1,5-diphosphate carboxylase. The control of enzymic activity by pH may be an important factor in this competition (*12*.2.3). Furthermore, the HCO_3^-/CO_2 balance (also controlled by pH) may be involved. If the substrate for PEP carboxylase is HCO_3^- rather than CO_2, then this enzyme will be favored if the pH rises high enough to reduce CO_2 to very low levels. (Even at pH 7.4, there is a 10 to 1 ratio of HCO_3^- to CO_2.)

The balance of HCO_3^- to CO_2 in the cytoplasm will be dependent on cytoplasmic pH and could have important implications in plants having the C_4 pathway of photosynthesis and those with Crassulacean acid metabolism (CAM). For example, photosynthesis in C_4 plants would be promoted by a higher pH in the mesophyll cells (the site of PEP carboxylase) than the bundle sheath (the site of decarboxylation and subsequent refixation of CO_2). In CAM plants, prolonged operation of PEP carboxylase might depend on relatively high cytoplasmic pH, i.e. removal of organic acids to the vacuole. Daytime CO_2 fixation would again be promoted by a relatively low pH, giving high CO_2 levels for the efficient operation of RUDP carboxylase.

These speculations about the role of cytoplasmic pH emphasize how important it is to have data for variation of cytoplasmic pH between different plants and under different conditions (see Fig. *12*.3a, b).

5.2.3 H^+ Fluxes and Stomatal Physiology

The physiology of stomatal guard cells has been discussed in Part B, Chap. *4.3* by HSIAO. The aim here is to summarize those aspects which can be related to the pH-regulating systems which have been described in this Chapter.

Firstly it should be noted that regulation by pH was an integral part of the starch-sugar hypothesis. It has been suggested (see LEVITT, 1967) that the pH changes could be brought about by carboxylation or decarboxylation of organic acids, depending on the ambient CO_2 level. Attempts to verify this hypothesis by measurements of internal pH are unfortunately of doubtful validity, as the indicator dyes used for this purpose would only give a measure of vacuolar pH. Nevertheless stomatal opening does seem to be associated with an increase in pH (LEVITT, 1967), though the starch-sugar hypothesis is now in disfavor.

There is at present considerable interest in K^+ uptake as a turgor-generating mechanism in stomatal guard cells. In some cases K^+ uptake is accompanied by synthesis of organic acids, and must be balanced by H^+ extrusion (see Part B, Chap. *4.3*). This situation is directly analogous to that shown in Fig. *12*.8, and raises the possibility that H^+ extrusion has a primary role in pH regulation and turgor generation in guard cells. In maize, K^+ uptake is balanced in part by Cl^- uptake (RASCHKE and FELLOWS, 1971), a situation analogous to that shown in Fig. *12*.7. It was suggested (ZELITCH and WALKER, 1964; ZELITCH, 1969) that stomatal opening in light is due to an increased supply of ATP to the K "pump" (strictly, a K^+/H^+ exchange system), but this does not account for stomatal opening in the dark under some conditions, e.g. in CO_2-free air. Opening in CO_2-free air in both light and dark has been used as evidence against organic acid synthesis as a turgor-generating system (LEVITT, 1967; see also ZELITCH, 1969), on the grounds that removal of CO_2 might be expected to remove the substrate for carboxylation. Nevertheless, carboxylation could proceed if, even under "CO_2-free" conditions, the cytoplasm could maintain an endogenous supply of HCO_3^- from glycolysis for use by PEP carboxylase. Synthesis of malate in beet tissue under "CO_2-free" conditions has been shown by SPLITTSTOESSER and BEEVERS (1964), and was attributed to refixation of endogenous CO_2 or (more likely) HCO_3^-. Similar reactions could account for excess K^+ influx into *Citrus* leaf slices in CO_2-free air (SMITH and RAVEN, 1974).

It is more difficult to account for stomatal closing (or prevention of opening) when the ambient CO_2 concentration is increased. Increased CO_2 would not in itself be sufficient to reduce cytoplasmic pH, but reduction might be achieved by an effect of CO_2 on membrane permeability to H^+ (RASCHKE, 1972). Decreasing pH would swing the pH stat system over towards net decarboxylation, and hence loss of osmotic potential. To account for rapid turgor decreases in guard cells which accumulate KCl, it is similarly necessary to postulate a large increase in membrane permeability (on receipt of the "closing signal"). Changes in fluxes of H^+ in response to other (ill-defined) triggers could account for stomatal changes inexplicable by conventional mechanisms involving CO_2 supply, photosynthesis, etc. (MEIDNER and MANSFIELD, 1965).

5.2.4 Conclusions; Further Aspects of pH Regulation in Photosynthetic Cells

This part of the Chapter (*12*.5.2) has been necessarily speculative, because of the relative lack of information concerning H^+ transport in green plant cells when compared with non-green cells. We have virtually ignored the possibility of control of pH by H^+ transport between chloroplasts and cytoplasm. H^+ transport is an important part of the energy-transfer reaction within chloroplasts, and is implicated in the transport of inorganic ions into isolated chloroplasts (see Vol. 3). However, the significance of H^+ or OH^- transport from the intact chloroplast within the intact cell is as yet unknown.

Finally, much could be said about the possible consequences of nutrient imbalance on the pH-regulating mechanisms within the intact leaf. Such imbalance could be specific (e.g. deficiency of K^+ or NO_3^-) or general, as in the absence of nutrient input during water stress. The result would be to prevent some H^+-disposal mechanisms, and produce intracellular acidity. If this is so, then production of basic or neutral amino acids from acidic amino acids could be regarded as an H^+ disposal mechanism under stress. Amino acid derivatives could serve a similar purpose. Excellent examples would be the accumulation of proline (see SINGH et al., 1973) and putrescine (T.A. SMITH, 1971). That this is not just speculation is suggested by the promotion of proline synthesis, in intact plants and excised tissues under water stress, by abscisic acid (ASPINALL et al., 1973). This hormone is known to affect a range of membrane transport processes (see Part B, *7*.5) including specifically potassium uptake into stomatal guard cells (MANSFIELD and JONES, 1971; see also KRIEDEMANN et al., 1972). However, it would be premature to add these hormonal responses to the list of those definitely known to involve H^+ transport (see *12*.6).

6. H^+ Transport in Morphogenesis

There is good evidence that a wide range of morphogenetic phenomena are associated with changes in membrane permeability and ion transport (Part B, Chap. *7*). Increasing H^+ extrusion is an early response to the action of indoleacetic acid (auxin), and has been invoked as an important part of the wall-loosening mechanism in coleoptiles (CLELAND, 1971; HAGER et al., 1971; RAYLE and CLELAND, 1972). RAYLE and JOHNSON (1973) have shown that effects of auxin on cell extension in *Avena* coleoptiles can be simulated by reducing external pH. Auxin-induced H^+ extrusion is currently being investigated in detail (see the review by P.J. DAVIES, 1973). RAYLE (1973) and RAYLE and JOHNSON (1973) have shown that abscisic acid inhibits auxin-induced H^+ secretion. A link between these effects and the biophysical approach to H^+ transport (*12*.3) is provided by the work of ETHERTON (1970), who showed that auxin increased the (negative) electrical potential of coleoptile cells in proportion to the growth response. Little is known about the intracellular biochemical events accompanying auxin-induced H^+ extrusion. However there is evidence that this process is associated with K^+ uptake, carboxylation and turgor generation (LÜTTGE et al., 1972; HASCHKE and LÜTTGE, 1974).

RAVEN (1975) has suggested that the distribution of IAA is determined by passive entry of undissociated IAA, and passive efflux of both IAA and IAA^-. Polar transport at the cell level could be accounted for if P_{IAA^-} were higher at the apical than the basal end, and this could be extended to symplastic transport. If IAA itself stimulates H^+ transport, then it could have an autocatalytic effect on its own transport *via* increases in both the pH and electrical potential gradients across the membrane.

There is increasing support for the theory that phytochrome acts *via* changes in membrane transport or permeability properties. Conversion of phytochrome to the far-red form increases H^+ efflux from bean roots (JAFFE, 1970; YUNGHANS and JAFFE, 1972) and electrical effects of this process are being further investigated (RACUSEN, 1973). MARMÉ et al. (1973) and QUAIL et al. (1973) have shown that the reversible membrane binding of phytochrome *in vitro* is pH-sensitive, with greatly increased binding below pH 7.

Permeability changes were implicated in phytochrome-mediated rapid nyctinastic movements of excised pinnae of *Albizzia* (JAFFE and GALSTON, 1967). Thigmotropic stimulation of pea tendrils again increases H^+ extrusion (JAFFE and GALSTON, 1968). Implication of H^+ transport in this wide range of developmental processes brings us back to the possibility, or even likelihood, that hormonal control is involved in those aspects of H^+ transport considered above. Of especial significance would be the overall regulation by pH of enzyme pathways (*12*.2.4). A further important example would be the changes in the pattern of solute accumulation as cells develop (STEWARD and MOTT, 1970).

7. Conclusions: the Evolution of H^+ Transport

It has been suggested previously (RAVEN and SMITH, 1973) that transport of H^+ and OH^- evolved as a means of regulating intracellular pH, and we believe that this is still its primary function. If metabolic processes lead to net H^+ production, then H^+ is actively transported from the cell in an exchange process involving cations such as K^+ or (to a lesser extent) Na^+. If there is net production of OH^-, then this ion is likewise lost from the cell. OH^- efflux can be passive (probably coupled to active uptake of anions such as Cl^-), or possibly active, again probably coupled to influx of inorganic anions. It is suggested that these transport systems are subject to control by feedback from changes in cytoplasmic pH. Internal regulation of cytoplasmic pH takes place also, by means of the biochemical pH stat, but this is more efficient at disposing of OH^- than H^+. The balance between these two control mechanisms varies in different cell types, and may depend on the stage of development of any specific cell.

Transport of H^+ or OH^- can be powered by redox reactions or by ATP, and it was also suggested previously (RAVEN and SMITH, 1973) that a chemiosmotic coupling resulting in ATP synthesis could have arisen if a redox-driven H^+ pump and an ATP-driven H^+ pump occurred in the same membrane. However, competition between the use of H^+ transport in redox-coupled ATP synthesis and in solute transport across the same membrane could be disadvantageous. It is tempting to consider in this way the low P/O ratios measured (or estimated) in bacteria (STOUTHAMER and BETTENHAUSEN, 1973). Also, the intolerance of the Cyanophyceae to low external pH (BROCK, 1973) may be due to limitations in pH-regulating mechanisms in these cells.

In the present-day eukaryotes, coupling between redox reactions and ATP synthesis is confined to the mitochondria and chloroplasts. These organelles can also carry out H^+-linked solute transport (MITCHELL, 1970). In the outer membrane, energy from H^+ transport is dissipated in other solute transport systems. This appears a more efficient partitioning of metabolic energy flow, whether it evolved by division of labor (compartmentation within the cell) or by endosymbiosis.

The three components of energy-transfer which are associated with H^+ transport, namely electron transport, ATPase activity and ion transport, also occur at other cellular membranes. The tonoplast contains ion transport systems possibly linked *via* H^+ transport to electron transport or ATP hydrolysis. Microsomal fractions derived from the endoplasmic reticulum have non-phosphorylating electron transport systems, and anion-stimulated ATPase activity has also been demonstrated (RUNGIE and WISKICH, 1972, 1973; *10*.4.3.2).

The involvement of H^+ and OH^- transport in accumulation of solutes above the levels needed biochemically in intermediary metabolism can be regarded as a further evolutionary development. In some cells the solutes accumulated are largely non-electrolytes. Otherwise, there is accumulation of inorganic anions in exchange for OH^-, or synthesis and accumulation of organic anions. These can be regarded as alternatives, both normally dependent on initial K^+/H^+ exchange. Of the two alternatives, Cl^-/OH^- exchange or OH^- extrusion would be about ten times less energetically expensive than malate synthesis as a means of countering internal alkalinity. This calculation is based on a maximum of 1 ATP used for each Cl^- accumulated, or OH^- extruded, with an effective cost of 10 ATP per carboxyl group synthesized (RAVEN and SMITH, 1974; ATKINSON, 1970). Nevertheless, accumulation of inorganic anions may be disadvantageous in some circumstances, possibly due to specific ion effects within the cytoplasm. Excess anion uptake in exchange for OH^- will lead to external pH changes which might be undesirable, as at the surface of cells of aerial parts of plants.

Ion transport at the tonoplast is the final major factor in the regulation of the ionic relations of plant cells, and is especially important in determining the balance between vacuolar accumulation of inorganic and organic anions. It should be pointed out that the nature of the signals which determine the final osmotic potential of plant cells is unknown (CRAM, 1973; also Chap. *11*) but it is essential that besides the control exerted by internal pH, H^+ transport must respond to environmental and developmental triggers, including hormone effects.

We started this Chapter by discussing the regulation of cell pH by metabolism (including transport). We continued by discussing the regulation of metabolism (including transport) by pH. We wish to conclude by reiterating that the various roles of H^+ and OH^- transport which we have suggested must be mutually compatible. Thus, if H^+ or OH^- fluxes (occurring during salt accumulation or as a response to hormonal activity) vary cytoplasmic pH, such variation must be within the relatively narrow pH range in which metabolism can operate.

References

ARENS, K.: Physiologisch polarisierter Massenaustausch und Photosynthese bei submersen Wasserpflanzen. II. Die $Ca(HCO_3)_2$-Assimilation. Jb. wiss. Bot. **83**, 513–560 (1936a).

ARENS, K.: Photosynthese von Wasserpflanzen in Kaliumbikarbonatlösungen. Jb. wiss. Bot. **83**, 561–566 (1936b).

ASPINALL, D., SINGH, T.N., PALEG, L.G.: Stress metabolism. V. Abscisic acid and nitrogen metabolism in barley and *Lolium temulentum* L. Australian J. Biol. Sci. **26**, 319–327 (1973).

ATKINSON, D.E.: Adenine nucleotides as universal stoichiometric coupling agents. Advan. Enzyme Regulation **9**, 207–219 (1970).

BEN-AMOTZ, A., GINZBURG, B.Z.: Light-induced proton uptake in whole cells of *Dunaliella parva*. Biochim. Biophys. Acta **183**, 144–154 (1969).

BENTRUP, F.W., GRATZ, H.J., UNBEHAUEN, H.: The membrane potential of *Vallisneria* leaf cells: evidence for light-dependent proton permeability changes. In: Ion transport in plants (W.P. ANDERSON, ed.), p. 171–182. London-New York: Academic Press 1973.

BEN-ZIONI, A., VAADIA, Y., LIPS, W.: Nitrate uptake by roots as regulated by nitrate reduction products of the shoot. Physiol. Plantarum **24**, 288–290 (1971).

BRIGGS, G.E.: The accumulation of ions in plant cells—a suggested mechanism. Proc. Roy. Soc. B. **107**, 248–269 (1930).

BRIGGS, G.E.: The absorption of salts by plant tissues, considered as ionic interchange. Ann. Bot. (London) **46**, 301–322 (1932).

BROCK, T.D.: Lower pH limit for the existence of blue-green algae: evolutionary and ecological implications. Science **179**, 480–483 (1973).

CHANEY, R.L., BROWN, J.C., TIFFIN, L.O.: Obligatory reduction of ferric chelates in iron uptake by soybeans. Plant Physiol. Lancaster **50**, 208–213 (1972).

CLELAND, R.: Cell wall extension. Ann. Rev. Plant Physiol. **22**, 197–222 (1971).

CONWAY, E.J.: Some aspects of ion transport through membranes. Symp. Soc. Exp. Biol. **8**, 297–324 (1954).

CONWAY, E.J., DOWNEY, M.: pH values of the yeast cell. Biochem. J. **47**, 355–360 (1950).

CONWAY, E.J., O'MALLEY, E.: The nature of the cation exchanges during yeast fermentation with formation of 0.02 N-H ion. Biochem. J. **40**, 59–67 (1946).

COOPER, T.G., WOOD, H.G.: The carboxylation of phosphoenolpyruvate and pyruvate. II. The active species of "CO_2" utilized by phosphoenolpyruvate carboxylase and pyruvate carboxylase. J. Biol. Chem. **246**, 5488–5490 (1971).

CRAM, W.J.: Internal factors regulating nitrate and chloride influx in plant cells. J. Exptl. Botany **24**, 328–342 (1973).

DAMADIAN, R.: Cation transport in bacteria. CRC Critical Rev. Microbiology **2**, 377–422 (1973).

DAVIES, D.D.: Control of and by pH. Symp. Soc. Exp. Biol. **27**, 513–529 (1973a).

DAVIES, D.D.: Metabolic control in higher plants. In: Biosynthesis and its control in plants (B.V. MILBORROW, ed.), p. 1–20. London-New York: Academic Press 1973b.

DAVIES, P.J.: Current theories on the mode of action of auxin. Botan. Rev. **39**, 139–171 (1973).

DAVIS, R.F.: Photoinduced changes in electrical potentials and H^+ activities of the chloroplast, cytoplasm and vacuole of *Phaeoceros laevis* In: Membrane transport in plants (U. ZIMMERMANN, J. DAINTY, eds.), p. 197–201. Berlin-Heidelberg-New York: Springer 1974.

DIJKSHOORN, W.: The relation of growth to the chief ionic constituents of the plant. In: Ecological aspects of the mineral nutrition of plants (I.H. RORISON, ed.), p. 201–213. Oxford-Edinburgh: Blackwell Scientific Publications 1969.

DIJKSHOORN, W., WIJK, A.L. VAN: The sulphur requirements of plants as evidenced by the sulphur: nitrogen ratio in the organic matter. A review of published data. Plant Soil **26**, 129–157 (1967).

DRAWERT, H.: Der pH-Wert des Zellsaftes. In: Encyclopedia of plant physiology (W. RUHLAND, ed.), vol. 1, p. 627–648. Berlin-Göttingen-Heidelberg: Springer 1955.

EPPLEY, R.W., BOVELL, C.R.: Sulfuric acid in *Desmaresia*. Biol. Bull. **115**, 101–106 (1968).

ETHERTON, B.: Effect of indole-3-acetic acid on membrane potentials of oat coleoptile cells. Plant Physiol. Lancaster, **45**, 527–528 (1970).

FINDLAY, G.P., HOPE, A.B., PITMAN, M.G., SMITH, F.A., WALKER, N.A.: Ionic fluxes in cells of *Chara corallina*. Biochim. Biophys. Acta **183**, 565–576 (1969).

FISCHER, R.A.: Role of potassium in stomatal opening in the leaf of *Vicia faba*. Plant Physiol. Lancaster **47**, 555–558 (1971).

GERSEN, D.F.: Intracellular pH measurements in roots of *Phaseolus aureus*. Plant Physiol. Lancaster **51** (Supplement), 44 (1973).

GESSNER, F.: Untersuchungen über Assimilation und Atmung submerser Wasserpflanzen. Jb. wiss. Bot. **85**, 267–326 (1937).

GUTKNECHT, J., DAINTY, J.: Ionic relations of Marine algae. Oceanogr. mar. Biol. Ann. Rev. **6**, 163–200 (1968).
HAGER, A., MENZEL, H., KRAUSS, A.: Versuche und Hypothese zur Primärwirkung des Auxins beim Streckungswachstum. Planta **100**, 47–75 (1971).
HAROLD, F.M.: Conservation and transformation of energy by bacterial membranes. Bacteriol. Rev. **36**, 172–230 (1972).
HAROLD, F.M., PAVLASOVA, E., BAARDA, J.R.: A transmembrane pH gradient in *Streptococcus faecalis:* origin and dissipation by proton conductors and N,N′-dicyclohexylcarbodiimide. Biochim. Biophys. Acta **196**, 235–244 (1970).
HASCHKE, H.-P., LÜTTGE, U.: β-Indolylessigsäure (-IES)-abhängiger K^+-H^+-Austauschmechanismus und Streckungswachstum bei *Avena*-Koleoptilen. Z. Naturforsch. **28C**, 555–558 (1973).
HIATT, A.J.: Relation of cell sap pH to organic acid changes during ion uptake. Plant Physiol. Lancaster **42**, 294–298 (1967a).
HIATT, A.J.: Reactions *in vitro* of enzymes involved in CO_2 fixation accompanying salt uptake by barley roots. Z. Pflanzenphysiol. **56**, 233–245 (1967b).
HIATT, A.J., HENDRICKS, S.B.: The role of CO_2 fixation in accumulation of ions by barley roots. Z. Pflanzenphysiol. **56**, 220–232 (1967).
HIGINBOTHAM, N.: Electropotentials of plant cells. Ann. Rev. Plant Physiol. **24**, 25–46 (1973).
HIGINBOTHAM, N., ANDERSON, W.P.: Electrogenic pumps in higher plant cells. Can. J. Botany **52**, 1011–1021 (1974).
HOAGLAND, D.R., BROYER, T.C.: Hydrogen-ion effects and the accumulation of salt by barley roots as influenced by metabolism. Am. J. Bot. **27**, 173–185 (1940).
HOAGLAND, D.R., DAVIS, A.R.: The composition of the cell sap of the plant in relation to the absorption of ions. J. Gen. Physiol. **5**, 629–646 (1923).
HODGES, T.K.: Ion absorption by plant roots. Advan. Agron. **25**, 163–207 (1973).
HOPE, A.B., LÜTTGE, U., BALL, E.: Photosynthesis and apparent proton fluxes in *Elodea canadensis.* Z. Pflanzenphysiol. **68**, 73–81 (1972).
HURD-KARRER, A.M.: Hydrogen-ion concentration of leaf-juice in relation to environment and plant species. Am. J. Bot. **26**, 834–846 (1939).
JACKSON, W.A., FLESHER, D., HAGEMAN, R.H.: Nitrate uptake by darkgrown corn seedlings. Some characteristics of apparent induction. Plant Physiol. Lancaster **51**, 120–127 (1973).
JACKSON, W.A., VOLK, R.J., TUCKER, T.C.: Apparent induction of nitrate uptake in nitrate-depleted plants. Agron. J. **64**, 518–521 (1972).
JACOBY, B., LATIES, G.G.: Bicarbonate fixation and malate compartmentation in relation to salt-induced stoichiometric synthesis of organic acid. Plant Physiol. Lancaster **47**, 525–531 (1971).
JAFFE, M.J.: Evidence for the regulation of phytochrome-mediated processes in bean roots by the neurohumor, acetylcholine. Plant Physiol. Lancaster, **46**, 768–777 (1970).
JAFFE, M.J., GALSTON, A.W.: Phytochrome control of rapid nyctanistic movements and membrane permeability in *Albizzia julibrissin.* Planta **77**, 135–141 (1967).
JAFFE, M.J., GALSTON, A.W.: Physiological studies on pea tendrils. V. Membrane changes and water movement associated with contact coiling. Plant Physiol. Lancaster **43**, 537–542 (1968).
JESCHKE, W.D.: Lichtabhängige Veränderungen des Membranpotentials bei Blattzellen von *Elodea densa.* Z. Pflanzenphysiol. **62**, 158–172 (1970).
JONES, H.G., OSMOND, C.B.: Photosynthesis by thin leaf slices in solution. I. Properties of leaf slices and comparison with whole leaves. Australian J. Biol. Sci. **26**, 15–24 (1973).
KASHKET, E.R., WONG, P.T.S.: The intracellular pH of *Escherichia coli.* Biochim. Biophys. Acta **193**, 212–214 (1969).
KESSELER, H.: Collection of cell sap, apparent free space and vacuole concentration of the osmotically most important mineral components of some Helgoland marine algae. Helgolaender Wiss. Meeresuntersuch. **11**, 258–269 (1964).
KHOLDEBARIN, B., OERTLI, J.J.: Changes of organic acids during salt uptake by barley leaf tissues under light and dark conditions. Z. Pflanzenphysiol. **62**, 237–244 (1970a).
KHOLDEBARIN, B., OERTLI, J.J.: The effect of Tris-buffer on salt uptake and organic acid synthesis by leaf tissues under light and dark conditions. Z. Pflanzenphysiol. **62**, 231–236 (1970b).

KIRKBY, E.A.: Ion uptake and ionic balance in plants in relation to the form of nitrogen nutrition. In: Ecological aspects of the mineral nutrition of plants (I.H. RORISON, ed.), p. 215–235. Oxford-Edinburgh: Blackwell Scientific Publications 1969.

KIRKBY, E.A., MENGEL, K.: Ionic balance in different tissues of the tomato plant in relation to nitrate, urea, or ammonium nutrition. Plant Physiol. Lancaster, **42**, 6–14 (1967).

KITASATO, H.: The influence of H^+ on the membrane potential and ion fluxes of *Nitella*. J. Gen. Physiol. **52**, 60–87 (1968).

KOMOR, E.: Proton-coupled hexose transport in *Chlorella vulgaris*. F.E.B.S. Lett. **38**, 16–18 (1973).

KRIEDEMANN, P.E., LOVEYS, B.R., FULLER, G.L., LEOPOLD, A.C.: Abscisic acid and stomatal regulation. Plant Physiol. Lancaster **49**, 842–847 (1972).

LEHNINGER, A.L.: Biochemistry. New York: Worth 1970.

LEONARD, R.T., HODGES, T.K.: Characterization of plasma membrane-associated adenosine triphosphatase activity of oat roots. Plant Physiol. Lancaster **52**, 6–12 (1973).

LEVITT, J.: The mechanism of stomatal action. Planta **74**, 101–118 (1967).

LOWENHAUPT, B.: The transport of calcium and other cations in submerged aquatic plants. Biol. Rev. Cambridge Phil. Soc. **31**, 371–395 (1956).

LUCAS, W.J.: Photosynthetic fixation of ^{14}Carbon by internodal cells of *Chara corallina*. J. Exp. Bot. **26**, 331–346 (1975).

LUCAS, W.J., SMITH, F.A.: The formation of alkaline and acid regions at the surface of *Chara corallina* cells. J. Expt. Bot. **78**, 1–14 (1973).

LUNDEGÅRDH, H.: An electrochemical theory of salt absorption and respiration. Nature **143**, 203–204 (1939).

LUNDEGÅRDH, H.: Anion respiration: the experimental basis of a theory of absorption, transport and exudation of electrolytes by living cells and tissues. Symp. Soc. Exp. Biol. **8**, 262–296 (1954).

LÜTTGE, U.: Proton and chloride uptake in relation to the development of photosynthetic capacity in greening etiolated barley leaves. In: Ion transport in plants (W.P. ANDERSON, ed.), p. 205–221. London-New York: Academic Press 1973.

LÜTTGE, U., HIGINBOTHAM, N., PALLAGHY, C.K.: Electrochemical evidence of specific action of indole acetic acid on membranes in *Mnium* leaves. Z. Naturforsch. **27B**, 1239–1242 (1972).

LÜTTGE, U., KRAMER, D., BALL, E.: Photosynthesis and apparent proton fluxes in intact cells of greening etiolated barley and maize leaves. Z. Pflanzenphysiol. **71**, 6–21 (1974).

LÜTTGE, U., PALLAGHY, C.K., OSMOND, C.B.: Coupling of ion transport in green cells of *Atriplex spongiosa* leaves to energy sources in the light and in the dark. J. Membrane Biol. **2**, 17–30 (1970).

MACROBBIE, E.A.C.: The active transport of ions in plant cells. Quart. Rev. Biophys. **3**, 251–294 (1970).

MANSFIELD, T.A., JONES, R.T.: Effects of abscisic acid on potassium uptake and starch content of stomatal guard cells. Planta **101**, 147–158 (1971).

MARMÉ, D., BOISARD, J., BRIGGS, W.R.: Binding properties *in vitro* of phytochrome to a membrane fraction. Proc. Natl. Acad. Sci. U.S.A. **70**, 3861–3865 (1973).

MEIDNER, H., MANSFIELD, T.A.: Stomatal responses to illumination. Biol. Rev. Cambridge Phil. Soc. **40**, 483–509 (1965).

MITCHELL, P.: Coupling of phosphorylation to electron and hydrogen transfer by a chemiosmotic type of mechanism. Nature **191**, 144–148 (1961).

MITCHELL, P.: Membranes of cells and organelles: morphology, transport and metabolism. Symp. Soc. Gen. Microbiol. **20**, 121–166 (1970).

MOTT, R.L., STEWARD, F.C.: Solute accumulation in plant cells. I. Reciprocal relations between electrolytes and non-electrolytes. Ann. Bot. (London), N.S., **36**, 621–639 (1972a).

MOTT, R.L., STEWARD, F.C.: Solute accumulation in plant cells. II. The progressive uptake of non-electrolytes and ions in carrot explants as they grow. Ann. Bot. (London), N.S., **36**, 641–653 (1972b).

NEUMANN, J., LEVINE, R.P.: Reversible pH changes in cells of *Chlamydomonas reinhardi* resulting from CO_2 fixation in the light and its evolution in the dark. Plant Physiol. Lancaster **47**, 700–704 (1971).

NOBEL, P.S.: Light-dependent potassium uptake by *Pisum sativum* leaf fragments. Plant Cell Physiol. (Tokyo) **10**, 595–600 (1969).

OSMOND, C.B., GREENWAY, H.: Salt responses of carboxylation enzymes from species differing in salt tolerance. Plant Physiol. Lancaster **49**, 260–263 (1972).
OSTERHOUT, W.J.V.: Physiological studies of single plant cells. Biol. Rev. Cambridge Phil. Soc. **6**, 369–411 (1931).
OSTERHOUT, W.J.V.: The absorption of electrolytes in large plant cells. Bot. Rev. **2**, 283–315 (1936).
OUELLET, C.A., BENSON, A.A.: The path of carbon in photosynthesis. XIII. pH effects on $C^{14}O_2$ fixation by *Scenedesmus*. J. Exp. Botany **3**, 237–245 (1952).
PAILLARD, M.: Direct intracellular pH measurement in rat and crab muscle. J. Physiol. **223**, 297–319 (1972).
PALLAGHY, C.K., LÜTTGE, U.: Light-induced H^+-ion fluxes and bioelectric phenomena in mesophyll cells of *Atriplex spongiosa*. Z. Pflanzenphysiol. **62**, 417–425 (1970).
PITMAN, M.G.: Active H^+ efflux from cells of low-salt barley roots during salt accumulation. Plant Physiol. Lancaster **45**, 787–790 (1970).
PITMAN, M.G., MOWAT, J., NAIR, H.: Interactions of processes for accumulation of salt and sugar in barley plants. Australian J. Biol. Sci. **24**, 619–631 (1971).
POOLE, R.J.: The influence of the intracellular potential on potassium uptake by beetroot tissue. J. Gen. Physiol. **49**, 551–563 (1966).
POOLE, R.J.: The H^+ pump in red beet. In: Ion transport in plants (W.P. ANDERSON, ed.), p. 129–134. London-New York: Academic Press 1973.
POOLE, R.J.: Ion transport and electrogenic pumps in storage tissue cells. Canad. J. Bot. **52**, 1023–1028 (1974).
QUAIL, P., SCHÄFER, E., MARMÉ, D.: Particle-bound phytochrome from maize and pumpkin. Nature **245**, 189–190 (1973).
RACUSEN, R.H.: Membrane potential conformational changes as a mechanism for the phytochrome-induced fixed charge reversal in root cap cells of mung bean. Plant Physiol. Lancaster **51** (Suppl.), 51 (1973).
RANSON, S.L.: The plant acids. In: Plant biochemistry (J. BONNER, J.E. VARNER, eds.), p. 493–525. New York-London: Academic Press 1965.
RASCHKE, K.: Saturation kinetics of the velocity of stomatal closing in response to CO_2. Plant Physiol. Lancaster **49**, 229–234 (1972).
RASCHKE, K., FELLOWS, P.M.: Stomatal movement in *Zea mays*. Shuttle of potassium and chloride between guard cells and subsidiary cells. Planta **101**, 296–316 (1971).
RAVEN, J.A.: Exogenous inorganic carbon sources in plant photosynthesis. Biol. Rev. Cambridge Phil. Soc. **45**, 167–221 (1970).
RAVEN, J.A.: Transport of indoleacetic acid in plant cells in relation to pH and electrical potential gradients, and its significance for polar IAA transport. New Phytologist **74**, 163–172 (1975).
RAVEN, J.A., SMITH, F.A.: The regulation of intracellular pH as a fundamental biological process. In: Ion transport in plants (W.P. ANDERSON, ed.), p. 271–278. London-New York: Academic Press 1973.
RAVEN, J.A., SMITH, F.A.: Significance of hydrogen ion transport in plant cells. Can. J. Botany **52**, 135–148 (1974).
RAYLE, D.L.: Auxin-induced hydrogen-ion secretion in *Avena* coleoptiles and its implications. Planta **114**, 63–73 (1973).
RAYLE, D.L., CLELAND, R.: The *in-vitro* acid growth response: relation to *in-vivo* growth responses and auxin action. Planta **104**, 282–296 (1972).
RAYLE, D.L., JOHNSON, K.D.: Direct evidence that auxin-induced growth is related to hydrogen ion secretion. Plant Physiol. Lancaster **51** (Suppl.), 2 (1973).
RENT, R.K., JOHNSON, R.A., BARR, C.E.: Net H^+ influx in *Nitella clavata*. J. Membrane Biol. **7**, 231–244 (1972).
RICHARDS, J.L., HOPE, A.B.: The role of protons in determining membrane electrical characteristics in *Chara corallina*. J. Membrane Biol. **16**, 121–144 (1974).
ROBERTSON, R.N.: Ion transport and Respiration. Biol. Rev. Cambridge Phil. Soc. **35**, 231–264 (1960).
ROBERTSON, R.N.: Protons, Electrons, Phosphorylation and Active Transport. Cambridge: University Press 1968.
RUBERY, P.H., SHELDRAKE, A.R.: Carrier-mediated auxin transport. Planta **118**, 101–121 (1974).

RUNGIE, J.M., WISKICH, J.T.: Soluble electron-transport activities in fresh and aged turnip tissue. Planta **102**, 190–205 (1972).

RUNGIE, J.M., WISKICH, J.T.: Salt-stimulated adenosine triphosphatase from smooth microsomes of turnip. Plant Physiol. Lancaster **51**, 1064–1068 (1973).

RUTTNER, F.: Zur Frage der Karbonat-Assimilation der Wasserpflanzen. Eine vergleichende Untersuchung. I. Teil: Die beiden Haupttypen der Kohlenstoffaufnahme. Oesterr. Botan. Z. **94**, 265–294 (1947).

SAITO, K., SENDA, M.: The light-dependent effect of external pH on the membrane potential of *Nitella*. Plant Cell Physiol. **14**, 147–156 (1973).

SCHULDINER, S., OHAD, I.: Biogenesis of chloroplast membranes. III. Light-dependent induction of proton pump activity in green cells and its correlation to cytochrome f photooxidation during greening of a *Chlamydomonas reinhardi* mutant (y–l). Biochim. Biophys. Acta **180**, 165–177 (1969).

SCHUTOW, D.A.: Die Assimilation der Wasserpflanzen und die aktuelle Reaktion des Milieus. Planta **2**, 132–151 (1926).

SCOTT, B.I.H.: Electric fields in plants. Ann. Rev. Plant Physiol. **18**, 409–418 (1967).

SHIEH, Y.J., BARBER, J.: Intracellular sodium and potassium concentration and net cation movements in *Chlorella pyrenoidosa*. Biochim. Biophys. Acta **233**, 594–603 (1971).

SINGH, T.N., ASPINALL, D., PALEG, L.G., BOGGESS, S.F.: Stress metabolism. II. Changes in proline concentration in excised plant tissues. Australian J. Biol. Sci. **26**, 57–63 (1973).

SKULACHEV, V.P.: Energy transformations in the respiratory chain. In: Current topics in bioenergetics (D.R. SANADI, ed.), vol. 4, p. 127–190. New York-London: Academic Press 1971.

SLAYMAN, C.L.: Movement of ions and electrogenesis in microorganisms. Am. Zoologist **10**, 377–392 (1970).

SLAYMAN, C.L., LONG, W.S., LU, C.Y.-H.: The relation between ATP and an electrogenic pump in the plasma membrane of *Neurospora crassa*. J. Membrane Biol. **14**, 305–338 (1973).

SLAYMAN, C.L., LU, C.Y.-H., SHANE, L.: Correlated changes in membrane potential and ATP concentration in *Neurospora*. Nature **226**, 274–276 (1970).

SMALL, J.: pH and Plants. London: Baillière, Tindall and Cox 1946.

SMITH, F.A.: Rates of photosynthesis in Characean cells. I. Photosynthetic $^{14}CO_2$ fixation by *Nitella translucens*. J. Exp. Bot. **18**, 509–517 (1967).

SMITH, F.A.: Rates of photosynthesis in Characean cells. II. Photosynthetic $^{14}CO_2$ fixation and ^{14}C-bicarbonate uptake by Characean cells. J. Exp. Bot. **19**, 207–217 (1968).

SMITH, F.A.: The mechanism of chloride transport in Characean cells. New Phytologist **69**, 903–917 (1970).

SMITH, F.A.: Stimulation of chloride transport in *Chara* by external pH changes. New Phytologist **71**, 595–601 (1972).

SMITH, F.A.: The internal control of nitrate uptake into barley roots with differing salt contents. New Phytologist **72**, 769–782 (1973).

SMITH, F.A., LUCAS, W.J.: The role of H^+ and OH^- fluxes in the ionic relations of Characean cells. In: Ion transport in plants (W.P. ANDERSON, ed.), p. 223–231. London-New York: Academic Press 1973.

SMITH, F.A., RAVEN, J.A.: H^+ fluxes, cytoplasmic pH and the control of salt accumulation in plants. In: Membrane transport in plants (U. ZIMMERMANN, J. DAINTY, eds.), p. 380–385. Berlin-Heidelberg-New York: Springer 1974.

SMITH, T.A.: The occurrence, metabolism and functions of amines in plants. Biol. Rev. Cambridge Phil. Soc. **46**, 201–242 (1971).

SPANSWICK, R.M.: The effects of bicarbonate ions and external pH on the membrane potential and resistance of *Nitella translucens*. J. Membrane Biol. **2**, 59–70 (1970).

SPANSWICK, R.M.: Evidence for an electrogenic ion pump in *Nitella translucens*. I. The effects of pH, K^+, Na^+, light and temperature on the membrane potential and resistance. Biochim. Biophys. Acta **288**, 73–89 (1972).

SPANSWICK, R.M.: Electrogenesis in photosynthetic tissues. In: Ion transport in plants (W.P. ANDERSON, ed.), p. 113–128. London-New York: Academic Press 1973.

SPANSWICK, R.M.: Hydrogen ion transport in giant algal cells. Canad. J. Bot. **52**, 1029–1034 (1974).

SPANSWICK, R.M., STOLAREK, J., WILLIAMS, E.J.: The membrane potential of *Nitella translucens*. J. Exp. Bot. **18**, 1–16 (1967).
SPEAR, D.J., BARR, J.K., BARR, C.E.: Localization of hydrogen ion and chloride ion fluxes in *Nitella*. J. Gen. Physiol. **54**, 397–414 (1969).
SPLITTSTOESSER, W.E., BEEVERS, H.: Acids in storage tissues. Effects of salts and ageing. Plant Physiol. Lancaster **39**, 163–169 (1964).
STEEMANN NIELSEN, E.: Photosynthesis of aquatic plants with special reference to the sources. Dansk. Bot. Ark. **12**, 1–71 (1947).
STEEMANN NIELSEN, E.: Uptake of CO_2 by the plant. In: Encyclopedia of plant physiology (W. RUHLAND, ed.), vol. 5, pt. 1, p. 70–84. Berlin-Göttingen-Heidelberg: Springer 1960.
STEVENINCK, R.F.M. VAN: The effects of calcium and tris (hydroxymethyl)aminomethane on potassium uptake during and after the lag phase in red beet tissue. Australian J. Biol. Sci. **18**, 227–233 (1965).
STEVENINCK, R.F.M. VAN: Some metabolic implications of the tris effect in beetroot tissue. Australian J. Biol. Sci. **19**, 271–281 (1966).
STEWARD, F.C., MOTT, R.L.: Cells, solutes and growth: salt accumulation in plants reexamined. Intern. Rev. Cytol. **28**, 275–370 (1970).
STOUTHAMER, A.H., BETTENHAUSEN, C.: Utilization of energy for growth, and maintenance in continuous and batch culture of microorganisms. A reevaluation of the method for the determination of ATP production. Biochim. Biophys. Acta **301**, 53–70 (1973).
THOMAS, M.: Vegetable acids in higher plants. Endeavour **10**, 160–165 (1951).
ULRICH, A.: Metabolism of non-volatile organic acids in excised barley roots as related to cation-anion balance during salt accumulation. Am. J. Bot. **28**, 526–541 (1941).
ULRICH, A.: Metabolism of organic acids in excised barley roots as influenced by temperature, oxygen tension and salt concentration. Am. J. Bot. **29**, 220–227 (1942).
VOLKOV, G.A.: Bioelectrical response of the *Nitella flexilis* cell to illumination. Biochim. Biophys. Acta **314**, 83–92 (1973).
VOTTA, J.J., JAHN, T.L., LEVEDAHL, B.N.: The mechanism of onset of the stationary phase of *Euglena gracilis* growth with 10 mM succinate: intracellular pH values. J. Protozool. **18**, 166–169 (1971).
VREDENBERG, W.J.: Energy control of ion fluxes in *Nitella* as measured by changes in potential, resistance and current-voltage characteristics of the plasmalemma. In: Ion transport in plants (W.P. ANDERSON, ed.), p. 153–170. London-New York: Academic Press 1973.
VREDENBERG, W.J., TONK, W.J.M.: Photosynthetic energy control of an electrogenic ion pump at the plasmalemma of *Nitella translucens*. Biochim. Biophys. Acta **298**, 354–368 (1973).
WALKER, N.A., SMITH, F.A.: Intracellular pH in *Chara corallina* measured by DMO distribution. Plant Sci. Lett. **4**, 125–132 (1975).
WAYGOOD, E.R., MACHE, R., TAN, C.K.: Carbon dioxide, the substrate for phosphoenolpyruvate carboxylase from leaves of maize. Canad. J. Bot. **47**, 1455–1458 (1969).
WEST, I.C., MITCHELL, P.: Stoichiometry of lactose-proton symport across the plasma membrane of *Escherichia coli*. Biochem. J. **132**, 587–592 (1973).
YUNGHANS, H., JAFFE, M.J.: Rapid respiratory changes due to red light or acetylcholine during the early events of phytochrome-mediated photomorphogenesis. Plant. Physiol. Lancaster **49**, 1–7 (1972).
ZELITCH, I.: Stomatal Control. Ann. Rev. Plant Physiol. **20**, 329–350 (1969).
ZELITCH, I., WALKER, D.A.: The role of glycolic acid metabolism in opening of leaf stomata. Plant Physiol. Lancaster **39**, 856–861 (1964).

13. Ion Absorption and Carbon Metabolism in Cells of Higher Plants

C.B. OSMOND

1. Introduction

The absorption of inorganic ions by cells of higher plants involves ionic interaction with the enzymes and intermediates of metabolism at several levels. The primary interaction involves energy metabolism in the membrane-bound enzyme systems responsible for active ion transport across the cell membranes and preceding Chapters have comprehensively reviewed the way in which energy metabolism may fuel ion transport.

This Chapter deals with secondary levels at which ion transport may interact with metabolism. The most clearly defined metabolic responses to ion absorption are those in which metabolism contributes or consumes carboxylate anions as a means of charge balance. In this case, the inorganic cations or anions probably do not have a direct, effector-like action on metabolic processes. Although a range of carboxylates may contribute to charge balance in different tissues (NIERHAUS and KINZEL, 1971), it is clear that only two, malate and oxalate, have been examined in any detail. In this Chapter emphasis is placed on the stoichiometry of changes in carboxylate metabolism in response to ion uptake and on the specific enzyme sytems involved in carboxylate generation. The synthesis of carboxylate anions for charge balance presupposes substantial transport or exchange of H^+ and OH^- across cell and organelle membranes, as discussed in detail by SMITH and RAVEN (Chap. *12*).

Less clearly defined are the areas in which the concentration of particular inorganic ions in the cytoplasm may directly modify a wide range of metabolic events. Attempts to probe the effect of ions on metabolism at this level are at once frustrated by the inadequate specification of cytoplasmic ionic relations. A further serious limitation is the extent to which regulatory phenomena demonstrated *in vitro* may be applied *in vivo*. The regulatory functions of inorganic ions *in vitro,* have many parallels with those of metabolites and there is little hesitation in extrapolating the latter *in vitro* data to problems of metabolic control. Consequently, this Chapter will seek an optimistic, perhaps speculative path towards assessing fine control (allosteric effects, feed back control, etc.) and coarse control (enzyme synthesis) of metabolism in response to inorganic ion concentration in the cytoplasm of higher plant cells.

2. Regulation of Carboxylate Metabolism during Ion Uptake

2.1 Compartmentation and Carboxylate Metabolism

At the outset it is useful to reflect on the strictly inorganic context of many studies of ion transport in cells of higher plants. This emphasis has been due in part to expedient experimental techniques as well as to the influence of elegant studies with giant algal cells in many of which ionic balance is satisfactorily explained in inorganic terms. However, in most tissues of higher plants the concentration of inorganic cations exceeds that of inorganic anions and electrical neutrality is maintained by a substantial pool of carboxylate anions. The extent of carboxylate participation in charge balance in leaf tissues is indicated in Fig. *13.*1 in which 80–90% of the total carboxylate content is balanced by cations. Simple calculations show that the bulk of these carboxylate ions must be outside the small volume of cytoplasm, and at least three compartments for carboxylic acids have been established.

MACLENNAN et al. (1963) estimated the extent of compartmentation in several plant tissues by adopting the following simple premise. If a labeled compound has a higher specific activity than its established precursor during the steady state metabolism of labeled substrates, then only a fraction of the precursor pool is involved in the metabolic sequence. Table *13.*1, derived from this data, suggests that in several tissues most of the succinate is involved in tri-carboxylic acid (TCA) cycle metabolism in the mitochondria whereas a large proportion of malate is not, and that the concentration of carboxylates in equilibrium with the TCA cycle does not often exceed 1 mM. Likewise, in leaves of C_4 plants it has been estimated that only 5–10% of the total malate or aspartate pool is in equilibrium with photosynthetic metabolism, the remainder presumably being involved in ionic balance (JOHNSON and HATCH, 1969; OSMOND, 1971).

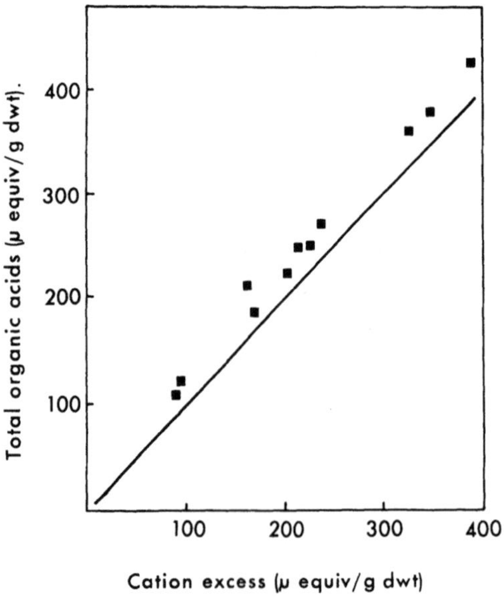

Fig. *13.*1. Relationship between total organic acid content and the excess of inorganic cations over anions in leaves of eleven different plant species. The curve is drawn for a 1:1 relationship. Constructed from data of PIERCE and APPLEMAN (1943)

Table *13*.1. Estimated concentration of carboxylic acids in turnover pools of the TCA cycle and in the vacuole of higher plant cells. (Data of MacLennan et al., 1963)

Acid	TCA cycle turnover pool (mitochondria and cytoplasm) (μmol g_{FW}^{-1})	Storage pool (vacuole) (μmol g_{FW}^{-1})
citrate	0.09–0.96	0.51–8.14
aconitate	0.02–0.67	0.98–4.19
glutamate	0.69–2.48	0.89–2.02
succinate	0.07–0.18	0.02–0.13
malate	0.39–5.50	0.34–15.1
aspartate	0.12–0.74	0.38–3.36

Lips and Beevers (1966a) produced evidence of two malate pools in vacuolated maize root segments. Malate labeled with acetate-^{3}H turned over very rapidly during a chase, whereas that labeled with $^{14}CO_2$ did not (Fig. *13*.2). They postulated that the rapidly metabolized pool was mitochondrial and that the slowly metabolized pool was cytoplasmic. Malate-3-^{14}C fed to the tissue appeared to enter the same pool as that labeled by $^{14}CO_2$ and was only slowly metabolized. In contrast to the behavior of malate, citrate, succinate and aspartate labeled by $^{14}CO_2$, ^{14}C labeled acid or acetate-^{3}H were rapidly metabolized (Lips and Beevers, 1966b; Steer and Beevers, 1967), presumably because all pools of carboxylates other than malate were in rapid equilibrium with the TCA cycle (Fig. *13*.2). These data are also consistent with the hypothesis that the slowly metabolized pool of malate was in the vacuole and that other carboxylates were not readily transported to the vacuole.

Further evidence for the separate cytoplasmic and vacuolar pools of malate in plant tissues is provided by efflux studies (*5*.2.3) which show that these can be identified with the same compartments for inorganic ions (Osmond and Laties, 1969). The exchange of malate ^{14}C from beet discs labeled by treatment with

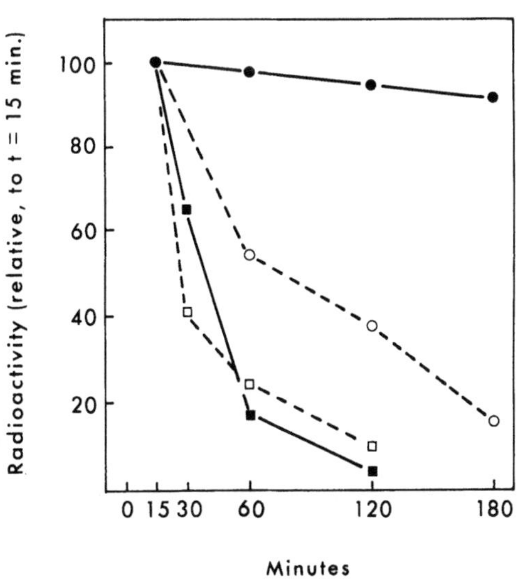

Fig. *13*.2. Changes in radioactivity in carboxylic acids in corn roots in 0.1M phosphate buffer pH 7.5 following 15 min incubation in labeled solutions. Malate labeled by malate-3-^{14}C or $^{14}CO_2$ (●), malate labeled by acetate-^{3}H (○), succinate labeled by succinate-1-^{14}C(■), succinate labeled by acetate-^{3}H (□). Constructed from data of Lips and Beevers (1966b); Steer and Beevers (1967)

$^{14}CO_2$ showed two successive exponential components and discs labeled with malate-3-^{14}C showed three (Fig. *13*.3). This approach has been critically evaluated by CRAM and LATIES (1974) using carrot discs and barley roots to provide more accurate estimates of cytoplasmic content and tonoplast fluxes of malate in these tissues. Influx studies with labeled oxalate in *Atriplex* leaf slices (OSMOND, 1968) and malate in carrot discs and barley roots (CRAM, 1974; CRAM and LATIES, 1974) also show that both acids are accumulated to a slowly exchanging compartment, identified as the vacuole, and a metabolic compartment, the cytoplasm, in which decarboxylation takes place.

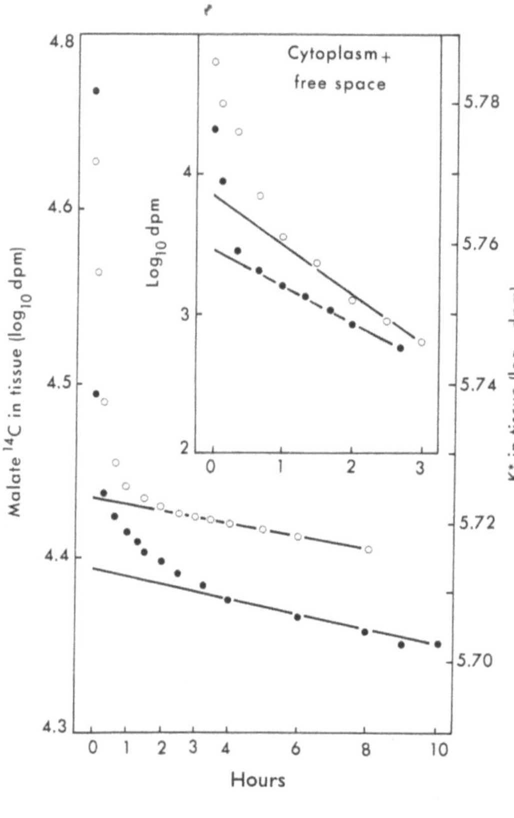

Fig. *13*.3. Exchange of malate-3-^{14}C (●) and K* (^{86}Rb) (○) from beet discs showing the exponential components corresponding to vacuole, cytoplasm and free space. Constructed from data of OSMOND and LATIES (1969) and unpublished data

The separation of two compartments for malate was also indicated in $^{14}CO_2$ efflux studies with maize roots and *Opuntia* roots (TING and DUGGER, 1965, 1966). The release of $^{14}CO_2$ from these tissues following a period of labeling in $^{14}CO_2$ was characterized by two first-order curves with half times of 7 min and 150–180 min respectively. The initial rapid loss of $^{14}CO_2$ may represent the exchange of unfixed $H^{14}CO_3^-$ in the tissue, for the half-time of exchange is similar to that for anion exchange and for HCO_3^- exchange from carrot (CRAM and LATIES, 1974). The slower first-order curve for $^{14}CO_2$ release has a half-time similar to that of the second component of malate and ^{14}C loss from beet and carrot discs (OSMOND and LATIES, 1969; CRAM and LATIES, 1974) but about 5 times longer than that for turnover of acids in the mitochondrial pools of maize roots (LIPS and BEEVERS, 1966a; STEER and BEEVERS, 1967). Presumably then, the slowly released $^{14}CO_2$ measured by TING and DUGGER arises from the decarboxylation of a labeled compound in the cytoplasmic compartment, probably outside the mitochondrion.

In some tissues, the vacuolar pool of malate shows rapid metabolic turnover. Tissues of plants capable of Crassulacean acid metabolism (CAM) accumulate large amounts of free malic acid as a result of dark CO_2 fixation (RANSON and THOMAS, 1960). Depending on environmental conditions, these plants may accumulate 100 µmol g_{FW}^{-1} of acid during a 10–12 h dark period and may metabolize this acid during the first 6 h of the subsequent light period. The flux of acid to the vacuole is about an order of magnitude more rapid than that found during nonautotrophic synthesis of malate in other mature tissues. The accumulation of free acid presumably involves a specialized transport process as indicated by the studies of LÜTTGE and BALL (1974a), differing from that of inorganic anion transport and from the transport of malate salt. One of the principal difficulties in quantitative analysis of malic acid compartmentation in CAM plants is that steady-state conditions rarely prevail for more than a few hours. Nevertheless, KLUGE and HEININGER (1973) described a kinetic analysis of malate distribution in tissue of low and high acid content and were partially successful in estimating the change in cytoplasmic malate concentration.

In summary, the compartmentation data suggest that there are readily characterized pools of malate associated with the cytoplasm and vacuole in higher plant cells. The cytoplasmic pool is probably distinct from that of the mitochondrion. The complex relationship between malate pools in the cytosol and those of the mitochondrion and chloroplast, and the role of malate as a transport metabolite in energy shuttles of the cell, is outside the scope of this Chapter (see Vol. 3).

2.2 Imbalance during Inorganic Ion Absorption

Rarely do tissues of higher plants show exactly equal absorption of cations and anions from simple salt solutions. The changes in the carboxylate content of excised roots and young seedlings during imbalanced ion uptake was established by ULRICH (1941, 1942) and BURSTRÖM (1945). Univalent cation absorption in excess of anions is most pronounced when tissues are bathed in sulfate or phosphate salts, although some barley plants show excess K^+ uptake from KCl (PITMAN et al., 1971) as do freshly cut beet discs in KCl solutions at high pH (HURD, 1958, 1959; VAN STEVENINCK, 1966). The excess K^+ is absorbed in exchange for H^+ released from the tissue (JACKSON and ADAMS, 1963; HIATT and HENDRICKS, 1967; PITMAN, 1970) which becomes slightly more alkaline (ULRICH, 1941, 1942; HIATT and HENDRICKS, 1967; HIATT, 1967b). Fig. *13*.4a taken from JACOBY and LATIES (1971) shows that the K^+/H^+ exchange initiates the stoichiometric synthesis of malate which is maintained by accumulation of K_2-malate to the vacuole. In freshly cut beet discs, *tris* buffer is particularly effective in stimulating K^+ uptake from KCl solutions (VAN STEVENINCK, 1966) possibly because it facilitates K^+/H^+ exchange at the plasmamembrane. OSMOND and LATIES (1968) used pulse chase labeling and compartmental analysis (Fig. *13*.3) to demonstrate that KCl/*tris* treatment of fresh beet discs or K_2SO_4 treatment of aged beet discs resulted in greater synthesis and transport of malate to the vacuolar compartment, than did treatment with KCl alone.

ULRICH (1941, 1942) also observed that anion uptake exceeded cation uptake from solutions such as $CaBr_2$ and $CaCl_2$, and that the absorbed anions replaced

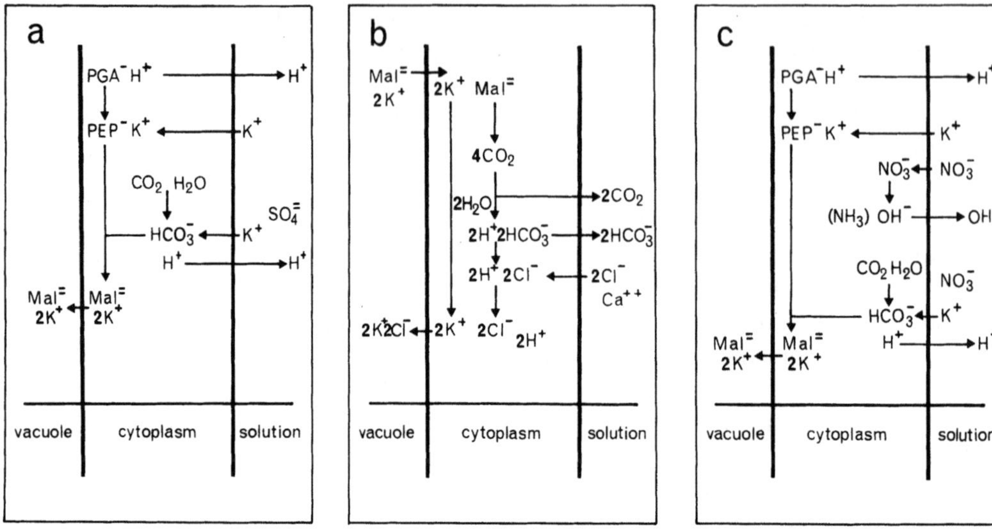

Fig. *13*.4a–c. Probable stoichiometry of ion uptake and malate metabolism in tissues bathed (a) in K_2SO_4, modified after JACOBY and LATIES (1971), (b) in $CaCl_2$ and (c) in KNO_3

organic anions already present in the tissue. Subsequent work has established this response in several tissues (JACOBSON and ORDIN, 1954; SPLITTSTOESSER and BEEVERS, 1964; TORII and LATIES, 1966; HIATT and HENDRICKS, 1967; KHOLDEBARIN and OERTLI, 1970a, b). The anion is absorbed in exchange for HCO_3^- equivalents (JACKSON and ADAMS, 1963) which presumably arise during the decarboxylation of the carboxylic acids. In excised roots in which malate is the principal balancing acid, the stoichiometry of organic acid loss and anion uptake is near 2 (mean value 1.9 from data of ULRICH, 1941; JACOBSON and ORDIN, 1954; HIATT, 1967b). This stoichiometry requires the complete degradation of malate to CO_2, presumably *via* malic enzyme and the TCA cycle, and an increase in tissue H^+ (HIATT, 1967b) as shown in the scheme of Fig. *13*.4b. Although plasma membrane exchange reactions are again responsible for the maintenance of excess anion uptake, the processes involved in the loss of malate from the vacuole are not clear. A simple explanation based on continued decarboxylation of malate and the inhibition of malate synthesis by Cl^- is discussed below (*13*.2.5).

2.3 Metabolic Incorporation of Inorganic Ions

The absorption and metabolic incorporation of anions such as HCO_3^- and NO_3^- results in the synthesis of carboxylate anions which are transferred to the vacuole. Bicarbonate incorporation provides the most simple model, for the absorbed anion is incorporated into the carboxylate molecule (JACOBSON and ORDIN, 1954; HURD, 1958, 1959; JACOBY and LATIES, 1971). The absorption and incorporation of NO_3^-, however, poses a much more complex problem in that the reduction system is itself induced by NO_3^- (BEEVERS and HAGEMAN, 1969) and is controlled independently of carboxylate metabolism.

In a number of studies with excised roots it is apparent that NO_3^- reduction did not take place to any significant extent (LUNDEGÅRDH, 1945; SMITH, 1973). Equimolar quantities of K^+ and NO_3^- were absorbed by the root cells and, in the case of LUNDEGÅRDH's experiments, KNO_3 was translocated to the shoots. Such tissues are ideal for analysis of the absorption of NO_3^-, the stoichiometry with respect to cations, and interactions with other anions. When NO_3^- uptake initiates the induction of nitrate reductase and the reduction of NO_3^-, all of these parameters become much more complex. In the experiments of LUNDEGÅRDH (1945), the inclusion of glucose in the external medium reduced the NO_3^- content of tissues and exudates but did not alter the K^+ content. Presumably glucose provided energy and/or carbon skeletons required for enzyme synthesis and nitrate reduction in LUNDEGÅRDH's experiments. The potential for nitrate reductase induction in detached roots may vary with their sugar content (SMITH and THOMPSON, 1971; MINOTTI and JACKSON, 1970). Further complexities arise in that the absorption and reduction of NO_3^- may vary between tissues. In maize roots, for example, it seems likely that NO_3^- absorption in the region behind the root tip is followed by rapid NO_3^- reduction, whereas further up the root, NO_3^- absorbed is not reduced, but is transported to the shoot (W.A. JACKSON and R.J. VOLK, personal communication).

These complexities no doubt account for the fact that the excised barley roots used by ULRICH (1941, 1942) reduced only part of the NO_3^- absorbed. Malate was synthesized to balance the K^+ remaining after reduction of NO_3^-, thus obviating the accumulation of OH^- or HCO_3^- (*12*.2). Fig. *13*.4c shows the relationship between K^+ and NO_3^- uptake in an excised tissue which is capable of reducing all the NO_3^- absorbed. One of the predictions of this scheme is that K^+ uptake will initially exceed NO_3^- uptake by a factor of 2 and that the external medium will decrease in pH. This stoichiometry was in fact found in ULRICH's (1941, 1942) experiments, in the early stages of KNO_3 uptake by intact wheat seedlings (MINOTTI, et al., 1968) and the decrease in external pH measured by PARR and NORMAN (1964) is consistent with Fig. *13*.4c.

Recent studies show that the anaerobic induction of nitrate reductase depends on the NO_3^- concentration of a small metabolic pool independent of the large storage pool (FERRARI et al., 1973). These data and those of MINOTTI and JACKSON (1970), suggest that NO_3^- which is absorbed to the vacuole is not readily available for reduction. SMITH and THOMPSON (1971) showed that 10 μM NO_3^- in the external solution was sufficient for maximum induction of nitrate reductase in barley roots, an observation consistent with the notion that induction is dependent on a small cytoplasmic pool of NO_3^-. These data suggest that the capacity of roots to reduce NO_3^- saturates at relatively low external concentrations and WALLACE and PATE (1967) showed that the roots of *Pisum* accounted for about half the total nitrate reductase activity of the whole plant at low levels of NO_3^- supply, but that at higher levels, the shoot was the principal site of NO_3^- reduction (Fig. *13*.5). Nitrate absorbed in excess of the reduction capacity in the root and the storage capacity of root vacuoles is translocated to the shoot (WALLACE and PATE, 1965). Nitrate arriving in the shoot is incorporated and the residual cation is balanced by carboxylate synthesis as shown in Fig. *13*.4c. Once again, carboxylate synthesis substitutes for the accumulation of OH^- or HCO_3^-. This process is believed to be the principal contributor to the pattern of ionic balance described in Fig. *13*.1 (DIJKSHOORN, 1962; DEWIT et al., 1963; OSMOND, 1967; BEN-ZIONI et al., 1970).

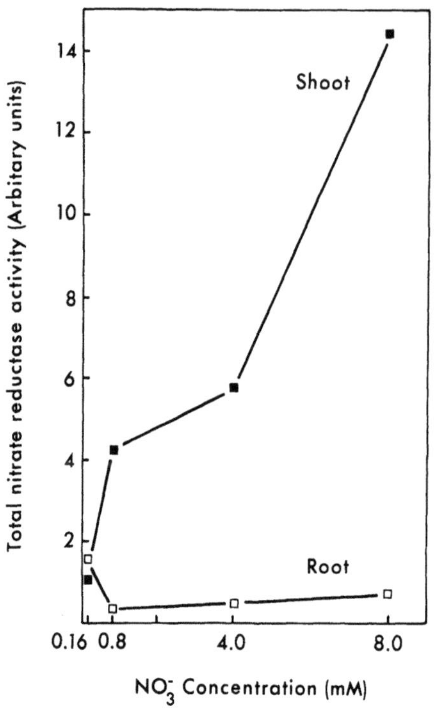

Fig. *13*.5. Influence of external NO_3^- concentration on the distribution and total activity of nitrate reductase in *Pisum*. Constructed from data of WALLACE and PATE (1967)

Overall, these postulates require that the equivalents of reduced nitrogen in the plant should equal the cation excess and the carboxylate content. In the leaves of *Atriplex* and sugar beet, the level of reduced N is closely correlated with the cation excess and the level of oxalate (OSMOND, 1967; VAN EGMOND and HOUBA, 1970). However, in experiments with ryegrass and wheat DIJKSHOORN et al. (1968) and MINOTTI et al. (1968) show that plants grown on NO_3^- cultures contain about twice as much NO_3^- as K^+. The stoichiometry of K^+/NO_3^- uptake changed from 2.0 initially to 0.5 after 24 h (MINOTTI et al., 1968).

DIJKSHOORN (1958) proposed two mechanisms to account for the latter observation. If the root is the primary site of NO_3^- reduction, exchange of OH^- produced during the reduction of NO_3^- for additional NO_3^- from the external medium may take place. In Fig. *13*.4c the NO_3^-/OH^- exchange may establish a closed loop in which NO_3^- uptake and reduction continue independently of K^+ uptake and carboxylate synthesis. The increase in NO_3^- uptake with time in wheat and corn roots is abolished by treatment with inhibitors of protein and nitrate reductase synthesis (JACKSON et al., 1973). This strongly suggests that the increased rate of uptake with time is dependent on concurrent NO_3^- reduction and may be due to the above exchange process, rather than to an inducible NO_3^- transport system.

Alternatively, if the shoot is the prime site of NO_3^- reduction, KNO_3 translocated to the shoot is reduced *in situ* and replaced by carboxylates which balance the accompanying K^+. DIJKSHOORN (1958) and BEN-ZIONI et al. (1971) propose that if the K carboxylate returns to the root and is there metabolized, additional NO_3^- uptake in exchange for HCO_3^- (see Fig. *13*.4b) leads to NO_3^- in excess of K^+.

Although it is well established that the level of carboxylic acids is lower in NH_4^+-grown than in NO_3^--grown plants (CLARK, 1936; COIC et al., 1962; JOY, 1964; HOUBA et al., 1971), the component processes can be defined only in broad outline (KIRKBY, 1969). In plants provided with NH_4NO_3, MINOTTI et al. (1969a, b) and FERGUSON (1969) concluded that NH_4^+ reduced NO_3^- uptake and consequently limited the induction of nitrate reductase. The implications of these changes for carboxylate metabolism await further investigation.

2.4 Pathways of Carboxylate Synthesis

The compartmental analyses described earlier implicate enzymes of the cytosol in the synthesis and degradation of malate. The most favored sequence for malate synthesis is that involving phosphoenolpyruvate carboxylase coupled with malate dehydrogenase. This system is essentially irreversible and has a high affinity for CO_2 or HCO_3^- (WALKER, 1962; TING, 1971, cf: HIATT, 1967a). Roots of maize and other species have a chromatographically distinct form of phosphoenolpyruvate carboxylase (TING and OSMOND, 1973b) which is a cytosol enzyme (DANNER and TING, 1967) and a kinetically distinct, cytosol form of malate dehydrogenase (DANNER and TING, 1967). These authors also demonstrated a cytosol malic enzyme in maize roots which completes the carboxylation/decarboxylation sequence of Fig. *13*.6. Pyruvate formed from cytoplasmic malate presumably enters the TCA-cycle and may be completely oxidised to CO_2. Other carboxylation reactions, such as phosphoenolpyruvate carboxykinase, may result in malate synthesis (HIATT, 1967a) but malic enzyme is unlikely to function as a carboxylase (WALKER, 1962; TING, 1971).

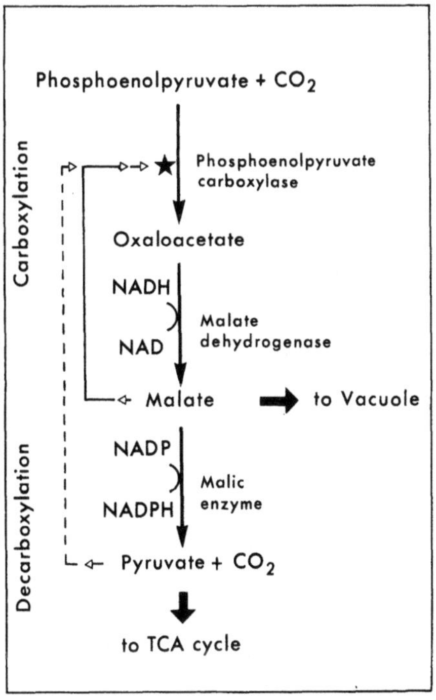

Fig. *13*.6. Probable reaction sequence for malate synthesis and degradation during nonautotrophic CO_2 fixation. Phosphoenolpyruvate carboxylase activity is under feedback control (*) due to the concentration of malate in the cytoplasm

The *in vivo* evidence supporting the non-autotrophic pathway proposed in Fig. *13*.6 is meagre indeed. Malate-^{14}C formed from $^{14}CO_2$ in this reaction sequence should be initially labeled in the 4-C carboxyl carbon. The malate isolated by JACOBSON (1955) from barley roots was about 50% labeled in the 1-C and 4-C carboxyl groups, a distribution which is inconsistent with the above proposals, as well as with the compartmentation of this malate outside the TCA cycle (LIPS and BEEVERS, 1966a). Malate isolated from beet, but not that from carrot discs (SPLITTSTOESSER, 1967) conforms to these predictions, as does malate labeled by dark CO_2 fixation in green tissues (OSMOND and AVADHANI, 1968; SUTTON and OSMOND, 1972; USUDA et al., 1973). A systematic study using reliable degradative techniques could establish the initial pathway and extent of randomization of label in malate during nonautotrophic CO_2 fixation in response to ion uptake.

It is noteworthy that the pathway outlined in Fig. *13*.6 is similar to that responsible for malate synthesis in leaves of C_4 plants and to that involved in malate accumulation in CAM plants. Chromatographically distinct isoenzymes of PEP carboxylase have been associated with this pathway in different metabolic sequences and these isoenzymes have distinct kinetic properties (TING and OSMOND, 1973a, b). As discussed below, this pathway provides a readily regulated means of organic anion synthesis and degradation. In contrast to these events, those leading to the synthesis or degradation of oxalate, another important carboxylate, are much further removed from the "mainstream" of intermediary metabolism. Two enzymes, glycolate oxidase (RICHARDSON and TOLBERT, 1961; MILLERD et al., 1962; OSMOND and AVADHANI, 1968) and oxaloacetase (CHANG and BEEVERS, 1968) may account for oxalate synthesis in green and non-green tissues. Glycolate oxidase may itself decarboxylate oxalate although an oxalate decarboxylase may also be involved.

2.5 Regulation of Malate Synthesis

At least three forms of ionic interaction may be involved in the regulation of malate synthesis in response to the balance of inorganic ion uptake. These are the control of synthesis by means of cytosol pH (which presumably changes during the exchange of K^+ for metabolic H^+ or the exchange of internal Cl^- for external HCO_3^-), by means of the level of malate in the cytosol and by means of the level of Cl^- in the cytosol.

The first of these alternatives has been promoted recently by DAVIES (1973) but was earlier proposed by HIATT (1967a). RAVEN and SMITH (1974 and *12*.2.3) discuss this proposal in detail and describe the problems associated with pH measurements in the cytosol. The "pH stat" proposed by DAVIES (1973) is based on the difference in the pH optima of malic enzyme and phosphoenolpyruvate carboxylase. This is a doubtful distinction, for these optima may be varied by the choice of assay conditions *in vitro*. For example, the pH optimum of malic enzyme becomes substantially more alkaline at saturating malate levels and the pH optimum of phosphoenolpyruvate carboxylase may be modified by inorganic ionic strength (DAVIES, 1973; JOHNSON and HATCH, 1970; OSMOND and GREENWAY, 1972). It is undeniable however that pH shifts of 1 to 2 units between pH 5.5 and 8.5 may favor the synthetic over the degredative system or *vice versa*, if both are functioning *in vivo* at substrate concentrations below the K_m.

At the same time, changes in cytoplasmic pH will alter the proportion of CO_2 and HCO_3^- available to phosphoenolpyruvate carboxylase and in the range pH 7–8, changes in CO_2 are more significant than changes in HCO_3^-. However, this enzyme has an exceedingly high affinity for CO_2 or HCO_3^- (10 μM for either species at pH 7–8, RAVEN, 1970) and although the species involved in this carboxylation is disputed, the regulation of malate synthesis by cytoplasmic HCO_3^- level (JACOBY and LATIES, 1971) is doubtful.

The second form of regulation involves the feedback control of PEP carboxylase due to malate. This feedback system has been identified in all forms of the enzyme examined to date (TING, 1968; MUKERJI and TING, 1971; KLUGE and OSMOND, 1972), and has been specifically implicated in the control of malate synthesis in CAM plants. The K_i value for malate ranges between 1 and 10 mM in different assays. In a tissue capable of accumulating 2 μmol g_{FW}^{-1} h^{-1} malate in response to K_2SO_4 treatment, for example, the cytoplasmic concentration of malate could rise to 10 mM in 10 min when transport to the vacuole ceased (assuming a cytoplasmic volume of 3%). Thus the cytosol concentration of free malic acid in CAM plants, or malate in other tissues may be considered a potential regulator of carboxylate synthesis. Feedback regulation of PEP carboxylase in the sequence of Fig. *13*.6 represents a very sensitive control, for this enzyme is 5 times more active than malic enzyme in extracts of leaves of CAM and C_4 plants and in the roots of spinach, maize and *Atriplex*.

Whether malate synthesis is regulated by K^+/H^+ exchange at the plasmamembrane or by the release of feedback inhibition due to transport of K_2-malate from cytoplasm to vacuole, it is clear that the response to the inorganic ions is indirect. The properties of H^+ transport at the plasma membrane and of K^+ malate transport at the tonoplast are of crucial importance. RAVEN and SMITH (Chap. *12*) discuss H^+ transport in detail but comparatively little is known of carboxylate transport at the tonoplast. Oxalate uptake to the vacuole but not to the metabolic compartment is sensitive to CCCP and uptake of oxalate from 2.5 mM solutions to the vacuole of cells containing about 150 mM oxalate probably involves active processes (OSMOND, 1968). Malate accumulation to the vacuole of barley roots and carrot cells is only slightly sensitive to Cl^- and transport of malate and Cl^- across the tonoplast show quite different kinetics as a function of concentration (CRAM, 1974). Transport of K^+ and Cl^- to the vacuole of leaf slices from the CAM plant *Kalanchoe daigremontiana* is unaffected by content of malic acid or the direction of malic acid fluxes (LÜTTGE and BALL, 1974b). The observation that freshly cut beet discs placed in KCl-*tris* buffer can accumulate malate to the vacuole but only develop a capacity to accumulate Cl^- on ageing (VAN STEVENINCK, 1966) further argues against the probability that malate, Cl^- and malic acid are transported by a common system.

In some instances (e.g., in $CaCl_2$ solutions) the accumulation of inorganic anions in the cytoplasm may directly inhibit CO_2 fixation and oxaloacetate reduction, for both phosphoenolpyruvate carboxylase and malate dehydrogenase are inhibited *in vitro* by Cl^- (OSMOND and GREENWAY, 1972). In this third potentially regulatory process, the inhibition of the carboxylase by Cl^- is competitive with respect to phosphoenolpyruvate and thus Cl^- may function in a manner quite analogous to endogenous malate. A wide range of inorganic ion interactions with enzyme activity have been described *in vitro* and on the following pages (*13*.3) some of the potential sites of interaction with metabolism are discussed.

3. Ion Concentration and Enzyme Activity

3.1 *In vitro* Effects of Ions on Enzymic Activity

The involvement of inorganic ions in metabolism was reviewed by EVANS and SORGER (1966) who distinguished between the role of univalent cations and other mineral elements. This is a useful distinction, for most of the mineral elements other than the univalent cations are implicated as cofactors at relatively low concentrations, and their effects on metabolism are only manifest in deficiency studies. The univalent cations on the other hand are frequently found to modify enzyme reactions at relatively high concentrations *in vitro* and these ions are usually present in plant cells at relatively high concentrations. EVANS and SORGER focused particular attention on pyruvate kinase but recent studies have covered a wide range of enzymes from a variety of higher plants. Table *13*.2 summarizes some of these studies and it must be emphasized that the *in vitro* responses of these enzymes are common to the same enzymes from a wide range of biological materials (DIXON and WEBB, 1964; GREENWAY and OSMOND, 1972; HOCHACHKA and SOMERO, 1973).

The mode of action of the univalent ions on enzyme activity *in vitro* is complex, but a few important features emerge from the studies listed in Table *13*.2:

i) The interference due to univalent ions at high concentration is not due to lower water potential in the medium (GREENWAY and OSMOND, 1972). Thus osmotically equivalent concentrations of mannitol, sucrose or proline are without effect and this may be the basis of the role of compatible solutes, discussed later (*13*.4.4).

ii) Univalent ions frequently result in a shift in the pH optimum of an enzymic reaction (DIXON and WEBB, 1964; WEIMBERG, 1967; GREENWAY and OSMOND, 1972). Ions may thus modify the charge relationships of the protein, perhaps in the region of substrate binding.

iii) In several instances, the univalent ions interfere with the binding of substrate to the enzyme. Thus although the ions may not effect V_{max}, substantial effects on the apparent K_m are observed (TING and OSMOND, 1973a; cf. MUTO and URITANI, 1972). Similar responses observed in the halophilic bacteria indicate that univalent anions may compete with substrate for binding sites or may bind to form an inactive complex with the enzyme (AITKEN and BROWN, 1972). In these same organisms high concentrations of univalent cations are required to maintain the physical integrity (tertiary structure) of the enzyme proteins. It is probable that similar interactions may be involved with proteins from other sources and although enzyme substructure may remain intact, the univalent ions may modify conformation, particularly in the region of the active site.

iv) Although the interference due to univalent ions shows many parallels with the interference due to competing substrates and allosteric effectors, the concentrations of ions involved are usually several orders of magnitude greater than those of the substrates or regulatory metabolic intermediates.

v) The interference due to univalent ions may be principally an anion effect (GREENWAY and OSMOND, 1972) or may involve the ratio of the cations Na^+ and K^+ (MILLER and EVANS, 1957). In either case, as with regulatory metabolic intermediates, the effects are readily reversible on removal of the interfering ions (GREENWAY and SIMS, 1974).

These studies clearly indicate that the response of many enzymes to concentration or ratio of univalent ions is potentially such as to modulate the rate of a wide range of metabolic events. Inorganic ions may thus interact with metabolism at a level of fine control, comparable with that believed to be mediated by the concentration or ratio of intermediary organic substrates and products. Two fundamental questions must be answered. First, is the concentration or ratio of univalent ions in the cytoplasm comparable with the concentration of ions found to be effective *in vitro?* Second, is there evidence *in vivo* for the effects of univalent ions on specific metabolic steps which may be interpreted on the basis of *in vitro* studies?

Table *13*.2. Interactions between inorganic ions and soluble enzymes of higher plants *in vitro*

Enzyme	Source and salt	Ref.
Malate dehydrogenase (oxaloacetate reduction NADH)	Spinach leaf (17 salts)	HIATT and EVANS (1960)
Malate dehydrogenase (oxaloacetate reduction NADH)	10 species, seed, leaf, shoot, particulate and soluble (NaCl)	WEIMBERG (1967, 1968)
Malate dehydrogenase (oxaloacetate reduction NADH)	*Suaeda, Halimone, Salicornia, Beta, Pisum* leaf (NaCl)	FLOWERS (1972a, b)
Malate dehydrogenase (oxaloacetate reduction NADH)	*Atriplex, Phaseolus, Salicornia* leaf, roots (NaCl, KCl, Na_2SO_4)	GREENWAY and OSMOND (1972)
Malate dehydrogenase (oxaloacetate reduction NADH)	*Lemna* ($CaCl_2$)	JEFFERIES et al. (1969)
Malate dehydrogenase (malate oxidation NAD or NADP)	*Pisum* root tips (NaCl)	HANSON-PORATH and POLJAKOFF-MAYBEER (1969)
Glucose-6-phosphate dehydrogenase	*Atriplex, Phaseolus* root and leaf; *Salicornia* leaf (NaCl)	GREENWAY and OSMOND (1972)
Glucose-6-phosphate dehydrogenase	*Suaeda, Beta, Halimone, Pisum* leaf (NaCl)	FLOWERS (1972a, b)
Glucose-6-phosphate dehydrogenase	*Ipomea* root (Several salts)	MUTO and URITANI (1972)
Isocitrate dehydrogenase (NADP)	*Atriplex, Phaseolus* leaf and root; *Salicornia* leaf (NaCl)	GREENWAY and OSMOND (1972)
Aspartate transaminase	*Atriplex, Phaseolus* leaf and root; *Salicornia* leaf (NaCl, KCl, Na_2SO_4)	GREENWAY and OSMOND (1972)
Phosphoenolpyruvate carboxylase	*Atriplex* spp., *Zea* leaf and root (NaCl)	OSMOND and GREENWAY (1972)
Ribulosediphosphate carboxylase	*Atriplex* spp., *Zea* leaf and root (NaCl)	OSMOND and GREENWAY (1972)
Pyruvate kinase	Seed, leaf and embryo of 8 species (KCl, $CaCl_2$, NaCl)	MILLER and EVANS (1957)
Formyltetrahydrofolate synthetase	Spinach and tobacco leaf (Several salts)	HIATT (1965)
Starch synthetase	Spinach and tobacco leaf (Several salts)	NITSOS and EVANS (1969)
Peroxidase	*Suaeda, Pisum, Halimone, Beta* and *Salicornia* leaf (NaCl)	FLOWERS (1972a, b)
Acid phosphatase	*Suaeda, Pisum* leaf (NaCl)	FLOWERS (1972b)

3.2 Ionic Concentration and K^+/Na^+ Ratio in the Cytoplasm of Higher Plants

The complexity of estimating relevant cytoplasmic concentrations in plant cells is shown particularly well by the problem of determining cytoplasmic Cl^- concentration in charophytes (Chap. *5*, Table *5*.8). Despite the large size of the cells and the possibility of separating cytoplasm from vacuole by physical means, there is a 10-fold range in values. For higher plant cells, the only technique to approximate the data obtained with giant algal cells is compartmental analysis (Chap. *5*), though this approach has its limitations. The concentration of ions in the vacuoles can be estimated very closely from analysis of the content of the tissue. Estimation of the concentration in the cytoplasm depends on combining the result of compartmental analysis (an *amount* in the cytoplasm) with the volume of the cytoplasm. This latter quanitity is difficult to determine with the same precision as vacuolar concentration.

In general the results available support the conclusions about cytoplasmic content based on giant algal cells (*6*.5). The ratio of K^+/Na^+ in the cytoplasm of barley roots (PITMAN and SADDLER, 1967) and in *Triglochin maritima* (JEFFERIES, 1973) were generally higher than in the vacuole and higher again than in the external solution (Part B, Chap. *3.3*). This high K^+/Na^+ ratio seems to be characteristic of most organisms (CAMPBELL and PITMAN, 1971; JEFFERIES, 1972; GUTKNECHT and DAINTY, 1969; GINZBURG et al., 1971). It probably accounts for the well known "sparing effect" of Na^+ during K^+ deficiency (Part B, Chap. *3.3*), by allowing the small amount of K^+ available to the whole cells to be retained disproportionately in the cytoplasm.

Despite the ability of the cells to accumulate K^+ preferentially it seems unlikely that K^+/Na^+ ratios in plant cells are as critical for metabolic control as in animals, nor are they so closely regulated. Though many algae show little variation in K^+/Na^+ within a species this is due to the constancy of K^+/Na^+ in the environment. When plants are grown on varied ratios of K^+/Na^+ it is only when concentrations of K^+ in the cytoplasm fall to levels apparently below those required for enzyme operation that growth is retarded.

The concentrations of K^+, Na^+ and Cl^- in the cytoplasm are known with less certainty than the ratios of K^+/Na^+. JEFFERIES' (1973) data suggest that cytoplasmic and vacuolar concentrations are comparable for *Triglochin maritima*. Data for barley give estimates for cytoplasmic concentrations of 80–120 and 40–60 mM for $(K^+ + Na^+)$ and Cl^- respectively (PITMAN and SADDLER, 1967; PITMAN, 1971) when vacuolar K^+ was about 100 mM (assuming a volume of 30–50 µl g_{FW}^{-1}). Similarly for beet slices K^+ and Cl^- concentrations can be estimated at 90 and 30 mM respectively assuming 40 µl g_{FW}^{-1} (PITMAN, 1963, 1964) and for carrot slices CRAM (1968) estimated a Cl^- concentration somewhat less. This is an effective concentration range for many of the interactions summarized in Table *13*.2.

3.3 Effects of Ion Concentration on Products of CO_2 Fixation

Nonautotrophic CO_2 fixation in higher plants has a key role in carboxylate synthesis and ion balance as outlined above (*13*.2). Furthermore, the carboxylase, dehydrogenase, transaminase and malic enzyme associated with this pathway show activation or inhibition due to inorganic anions *in vitro*. It is conceivable that both the rate and products of CO_2 fixation in root tissues may respond to the NaCl level in the cytoplasm. HIATT and HENDRICKS (1967) showed that 6 h pretreatment of barley roots with 100 mM NaCl substantially reduced the subsequent rate of CO_2 fixation of washed roots. However, other explanations such as the consumption of substrate sugar during 6 h of ion uptake, could be entertained. It is also difficult to dissociate the effects of Cl^- concentration *in vivo* from those due to unequal ion uptake (*13*.2.2). For example, unpublished experiments (H. GREENWAY and C.B. OSMOND) show that the rate of CO_2 fixation by low-salt roots was stimulated during the first 30 min of KCl treatment, probably in response to excess K^+ uptake (cf. PITMAN et al., 1971), but was inhibited after 2 h, probably as a result of Cl^- accumulation in the cytoplasm. CRAM (1974) showed that Cl^- inhibited the accumulation of endogenously produced malate in barley roots and that this was not due to Cl^- competition for a common transport process at the tonoplast. These studies thus provide evidence for Cl^- inhibition of malate synthesis in the cytoplasm (cf. excess anion uptake in roots; *13*.2.2, *13*.2.5).

WEBB and BURLEY (1965) reported that Cl^- ions increased the labeling of aspartate relative to malate during nonautotrophic CO_2 fixation in some halophytic species. WEIMBERG (1967) proposed that this observation might reflect a relatively greater sensitivity of malate dehydrogenase to NaCl, compared with aspartate amino transferase. H. GREENWAY and C.B OSMOND have found that treatment of bean roots (*Phaseolus vulgaris*) in 100 mM NaCl and KCl for as little as 15 min increases the labeling of aspartate compared to carboxylic acids during a 1.5 min exposure to $^{14}CO_2$. Similar results were obtained in long-term experiments and in roots and shoots of the halophyte *Atriplex spongiosa.* It would be premature to conclude that such responses were in any way specific to halophytes. Indeed, Cl^- accumulation and the diurnal accumulation of free malic acid may be correlated during the induction of CAM in the halophyte *Mesembryanthemum crystallinum,* (WINTER and VON WILLER, 1972; WINTER, 1973a, b). This response is however, most likely due to water stress, not Cl^-.

3.4 Ion Effects on Carbohydrate Metabolism

Glucose-6-phosphate dehydrogenase may be activated or inhibited by NaCl *in vitro* and it is thus conceivable that cytoplasmic NaCl concentration may regulate the activity of this enzyme at the branch point of the metabolism of glucose by the pentose phosphate and glycolytic pathways. PORATH and POLJAKOFF-MAYBEER (1964, 1968) attempted to assess the pathways of glucose metabolism in the root tips of pea plants grown in NaCl and Na_2SO_4 solutions. They found that addition of NaCl to the incubation mixture did not alter the ratio of $^{14}CO_2$ released from glucose-1-^{14}C and glucose-6-^{14}C. However, plants grown in saline cultures showed a decrease in the C_6/C_1 ratio, which was attributed to an increase in pentose phosphate pathway activity in NaCl treatment. Although this change was correlated with an increase in the specific activity of glucose-6-phosphate dehydrogenase (PORATH and POLJAKOFF-MAYBEER, 1968) the decreased C_6/C_1 ratio was due to a decrease in $^{14}CO_2$ release from glucose-6-^{14}C, rather than to an increase in $^{14}CO_2$ release from glucose-1-^{14}C as might be expected. Although glucose-6-phosphate dehydrogenase is a primary regulator of carbon flux into the pentose phosphate pathway, it would be premature to conclude that NaCl effects on glucose metabolism can be attributed to Cl^- control of this enzyme.

The high K^+/Na^+ ratio of the cytoplasm seems to be important in relation to normal metabolic function in a wide range of plants. EVANS and SORGER (1966) highlighted this phenomenon and discussed the consequences of K^+ deficiency in detail. It is difficult to identify primary and secondary responses in the course of deficiency studies and a wide range of enzymes of carbohydrate metabolism may be involved. The high K^+ requirement of pyruvate kinase *in vitro* suggests that this enzyme may be a primary target for regulation by the K^+/Na^+ ratio in the tissue. H. GREENWAY (personal communication) has demonstrated the inhibitory effect of high Na^+ on this enzyme by measuring the changes in pool sizes of PEP and pyruvate during the glycolytic metabolism of glucose by a pea seed extract *in vitro*. However, evidence for the K^+, Na^+ interaction *in vivo* is meagre.

Thus, although the available estimates suggest cytoplasmic ion concentrations may correspond to the effective concentration of univalent ions *in vitro*, there is little *in vivo* evidence that the enzymes of the cytosol respond as anticipated on the basis of the *in vitro* studies. The most satisfactory explanation of this paradox is that metabolic processes may adjust relatively quickly to the presence of interfering ions. For example, PEP carboxylase may be about 50% inhibited by 50 mM KCl when operating at substrate concentrations in the region of the K_m. This inhibition may be removed by a doubling of substrate concentration. GREENWAY and SIMS (1974) favor this interpretation and show that although KCl reduced the affinity of malic enzyme for its substrate, it also reduced the affinity of the enzyme for allosteric effectors. They further point out that a general increase in the level of metabolites in the cytoplasm to overcome the interfering influence of inorganic ions would provide some degree of osmotic compensation which may be required if regulation of intracellular ion concentration is a feature of tolerance to salinity (*13*.4.3).

4. Modification of Enzymic Activity and Properties in Relation to Ionic Environment

The failure to detect substantial responses of *in vivo* carbon metabolism to large changes in external ionic environment may also be due to the fact that species growing in different habitats contain a metabolic machinery with properties suited to that ionic environment. Two main forms of response may be envisaged;

(i) The level of enzymic activity may be *quantitatively* different in the species which functions effectively in the unusual ionic environment, thus compensating for the interference due to high or low levels of inorganic ions.

(ii) The activity of enzymes from species in different ionic environments may be *qualitatively* different in response to interfering or essential ions. The qualitative difference may be due to the production of alloenzymes which differ in their response to ions or allosteric effectors.

There are indications that lower organisms have adopted both the above strategies (HOCHACHKA and SOMERO, 1973). It is important to establish whether certain higher plants have evolved a basic metabolic system more resistant to the interference of ions or whether plants, such as the halophytes, depend primarily on the physiological regulation of internal ionic environment.

4.1 Micronutrients and Nitrogen Supply

The most clearly identified instances in which the specific activity of soluble enzyme systems responds to ionic environment are those involving micronutrients such as Zn^{2+} (carbonic anhydrase) and the special case of NO_3^- (nitrate reductase). As indicated above (*13*.3.1), the micronutrients are usually implicated as cofactors or are bound to cofactor molecules in very low concentration. Deficiency of micronutrients as central as Fe^{3+} (bound to cytochromes) or of nutrients such as Mg^{2+} (bound to chlorophyll and a cofactor for most reactions involving phosphate transfer) results in a general breakdown of metabolism and it is impossible to separate primary and secondary effects. In other cases however, the level of the micronutrient may limit the extent of synthesis of a particular enzyme with relatively little side effect on general metabolism. The reduced activity of carbonic anhydrase under mild Zn^{2+} deficiency is a good example of specific regulation of enzyme synthesis in response to ionic environment (RANDALL and BOUMA, 1973). Other responses are evidently quite specific, although more complex. BROWNELL and CROSSLAND (1972) propose that symptoms of Na^+ deficiency are confined to those plants with the C_4-pathway of photosynthesis and it is possible that Na^+ is essential for the synthesis of the enzymes of the C_4-cycle peculiar to the mesophyll cells of these plants.

The substrate induction of nitrate reductase is perhaps the best documented case of enzyme induction in higher plants (BEEVERS and HAGEMAN, 1969). As dicussed above (*13*.2.3), many cells have the potential for nitrate reduction and the induction of the enzyme appears to depend on quite low concentrations of NO_3^- in the cytoplasmic compartment. It is interesting to note that other ions such as Cl^- do not interfere with nitrate reductase activation (AUSTENFELD, 1974) and that such changes as do occur are due to water deficits, not to Cl^- ions (PLAUT, 1974).

4.2 Acidic and Calcareous Environments

Clearly defined communities of plants can be identified with acidic and calcareous habitats and attempts have been made to relate the activity and properties of enzymes in the plant to the ionic environment. WOOLHOUSE (1969) examined the Al^{3+}, Pb^{2+} and Ca^{2+} response of root surface acid phosphatases in races of *Agrostis tenuis* from different habitats. JEFFERIES et al. (1969) performed more detailed studies on malate dehydrogenase in response to Ca^{2+} using races of *Lemna*. Culture in 10 mM Ca^{2+} significantly reduced the level of malate dehydrogenase activity (i.e. the amount of active enzyme) compared to 1 mM Ca^{2+} controls, but this reduced activity was compensated by substantial increase in Ca^{2+} activation of the enzyme. Similar differences were found between clonal *Lemna* material collected from low and high Ca^{2+} environments. Ca^{2+} was shown to play an important role in the sub-unit structure and the enzyme from high Ca^{2+} cultures contained a high proportion of high molecular weight units compared with enzyme prepared from low Ca^{2+} cultures. *In vivo* adjustment of the Ca^{2+} sensitivity of malate dehydrogenase in response to sudden changes in external $CaCl_2$ concentrations, without recourse to new enzyme synthesis, was suggested in some experiments. This adjustment appeared to depend on Ca^{2+} effects on the quaternary structure of the enzyme.

4.3 Saline Environments

In general, the specific activity of enzymes extracted from mature tissues of both salt-tolerant and salt-sensitive species does not respond to the level of salinity during growth (WEIMBERG, 1970; GREENWAY and OSMOND, 1972; FLOWERS, 1872b; see however TREICHEL et al., 1974). These studies further show that the *in vitro* response to salts is essentially similar for enzymes isolated from salt tolerant or salt sensitive species. Alloenzymes of PEP carboxylase, for example, show differences in response to NaCl *in vitro* (OSMOND and GREENWAY, 1972; TING and OSMOND, 1973a) but these differences are evidently unrelated to salt tolerance. However, R.L. JEFFERIES (personal communication) has observed a tendency for greater stimulation of malate dehydrogenase by NaCl in extracts of more halophytic species from natural environments, reminiscent of that found for the Ca^{2+} response of calcareous species.

These observations are in marked contrast to those in halophilic bacteria in which extraordinarily high concentrations of salt are required for maximal activity of various enzymes (LARSEN, 1962; BROWN, 1964). Fig. *13*.7 shows that the NaCl concentration required for maximal activity of malate dehydrogenase in *Halobacterium* is about 100 times greater than that required in extracts of the halophyte, *Atriplex*. The salt concentration in the cytoplasm of *Halobacterium* evidently approaches the optimal concentrations required for enzyme activity *in vitro* (BROWN, 1964; GINZBURG et al., 1971). As discussed earlier (*13*.3.1), high levels of salt are required to maintain enzyme integrity in these organisms and may be largely involved in "titration" of the highly acidic proteins of the halophiles (HOCHACHKA and SOMERO, 1973). This requirement is most effectively demonstrated in the stability and aggregation of ribosomal sub-units in the halophiles. Ribosomal activity requires about 4M K^+ in *Halobacterium* whereas in the halophyte *Suaeda*

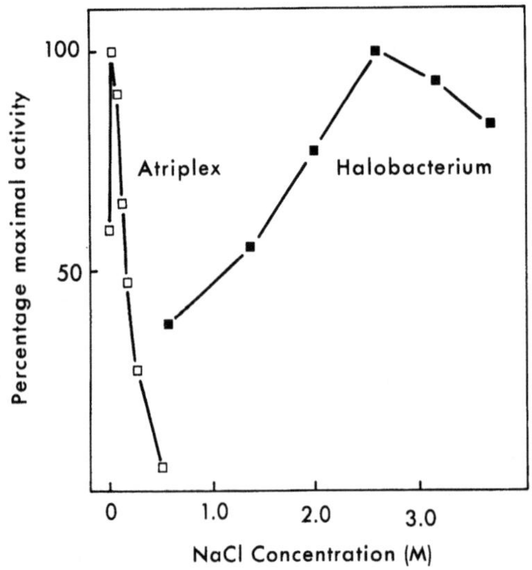

Fig. *13*.7. Concentration of NaCl required for maximum activity of NADH malate dehydrogenase in extracts of the halophyte *Atriplex spongiosa* and the halophilic bacteria *Halobacterium salina*. Constructed from data of GREENWAY and OSMOND (1972) and LARSEN (1962)

optimum K^+ concentration was 50 mM and incorporation of leucine was strongly inhibited by only 100 mM NaCl (HALL and FLOWERS, 1973).

On the other hand, there are reports that the developing tissues of higher plants show some change in enzyme activity in response to salinity treatment. Glycolytic enzyme activity in root tips of salinized pea seedlings was reduced and activity of pentose phosphate pathway enzymes was stimulated (PORATH and POLJAKOFF-MAYBEER, 1968; HANSON-PORATH and POIJAKOFF-MAYBEER, 1969). Recent studies show that increased glucose-6-phosphate dehydrogenase activity following 120 mM NaCl treatment was associated with root injury and was only found in brown, discoloured root tips (A. POLJAKOFF-MAYBEER, H. GREENWAY, personal communication). Evidently this response of enzyme activity to salinity is confounded by problems of senescence. The shoots of 4-day seedlings of *Pennisetum typhoides* show increased activity of glutamate-pyruvate transaminase, aspartate transaminase, malate dehydrogenase; aldolase, sucrose synthetase and sucrose-6-P-synthetase in response to mild salinity (HUBER and SANKHLA, 1973; HUBER et al., 1974). At this stage of development in young seedlings many enzyme systems are undergoing changes in activity. Some of the differences observed by these authors may be due to indirect effects of salt on rate of seedling development as witnessed by the observation that many of the salinity responses may be reproduced by application of phytohormones.

One particularly interesting response is the induction of CAM in some Aizoaceae in response to salinity treatment. Dark CO_2 fixation and the accumulation of malate in the night commence within a few days of salt treatment (WINTER and VON WILLERT, 1972; WINTER, 1973a, b). In these plants TREICHEL et al. (1974) observed an increase in the specific activity of PEP carboxylase required for the commencement of CAM. VON WILLERT (1974) contrasted this response with that in other halophytes. Although it is increasingly evident that this response is due to water stress, not to specific ions, these studies provide an example of the way ion accumulation may indirectly induce a major increase in the activity of a metabolic pathway.

5. Regulation of Internal Ionic Environment; Osmotic and Metabolic Responses

The above discussion suggests that higher plants display a very limited range of metabolic responses to the external ionic environment. There is little evidence for fine control of metabolism by internal ionic activity and regulation of enzyme synthesis in response to ions is restricted. Unlike the bacteria, salt-tolerant higher plants do not appear to contain higher activities or unusually resistant forms of salt-sensitive enzymes. These observations, and the fact that healthy, salt-tolerant plants contain about the same overall concentration of salts as the moribund tissues of salt sensitive species (GREENWAY, 1965) demand that a further alternative be considered. It is possible that plants which tolerate extreme ionic environments do so because they are better able to regulate internal ionic concentration. Superficially, this view is supported by the fact that many of these plants display well-developed salt-secreting systems (Part B, Chap. *5.2*). However, this view also implies that salt-tolerant species are better able to maintain acceptable concentrations of interfering ions in the metabolic compartments, and that tolerant species must have some mechanism for osmotic compensation in the metabolic compartments.

This hypothesis is the basis of the proposal by OERTLI (1968) that extracellular (free space) accumulation of salt may be responsible for salinity damage in leaves, mediated perhaps, by an effect on cell water relations. The cell and organelle membranes are unlikely to be able to withstand a significant pressure differential so that maintenance of low ionic concentration in one compartment relative to another must be accompanied by synthesis of osmotically active compounds in the regulated compartment. BROWN and SIMPSON (1972) proposed the term "compatible solute" to describe organic compounds synthesized for osmoregulatory purposes. These compounds do not interfere with enzymic activity even when present at very high concentrations. In the *in vitro* studies of salt effects on enzymic activity for example, it was pointed out that mannitol did not interfere with activity when present at osmotically equivalent concentrations. Neutral compounds such as glycerol, isofloridoside and proline accumulate in sugar-tolerant yeasts, moderately salt-tolerant bacteria and in salt-tolerant algae during osmotic stress (BROWN and SIMPSON, 1972; TEMPEST et al., 1970; CHRISTIAN and HALL, 1972; KAUSS, 1967; WEGMAN, 1971; BEN-AMOTZ and AVRON, 1973; BOROWITZKA and BROWN, 1974). The location of these substances in the unicellular organisms is not clear but it is likely that they function in osmoregulation in the cytoplasm and small vacuoles.

In higher plants, proline accumulates during water stress and the changes in proline concentration are such as to suggest that this substance acts as a compatible solute in the cytoplasm. SINGH et al. (1973) measured the accumulation of 15 mg g_{FW}^{-1} of proline in barley leaves during a droughting treatment which reduced leaf water potential from -2.5 to -26 bars. It can be calculated that to effect an osmotic adjustment of -25 bars this amount of proline must be restricted to 3% of the tissue volume. That is to say, proline accumulation in the cytoplasm, but not in the vacuole, could account for osmotic adjustment of this magnitude. Presumably the increased concentration of inorganic ions and carboxylates following water loss from the vacuole accounts for osmotic adjustment in this compartment. It should be noted that compounds such as mannitol, glycerol and proline protect membrane-bound enzymes against inactivation due to high local concentrations of electrolytes during freezing and dehydration (SANTARIUS, 1969). Compatible solutes may thus serve dual roles in cytoplasmic osmoregulation and enzyme protection.

Evidence is accumulating that proline concentration also increases in the tissues of halophytes during exposure to saline environments (S.P. TREICHEL, personal communication). W. HUBER and N. SANKHLA (personal communication) have measured increased proline synthesis and increased activity of enzymes of proline synthesis in *Pennisetum typhoides* during mild salinity treatment of germinating sedlings. The particular role of proline as a compatible solute, its localization *in vivo* and comparison between salt-tolerant and salt-sensitive species deserves further attention. Again, as was the case in the induction of CAM in aizoids, salinity and water stress appear to produce very similar metabolic responses and to result in substantial changes in metabolic pathways. In both cases the metabolic response is an indirect consequence of ion uptake.

6. Conclusions

This discussion shows that although many enzyme systems respond directly to inorganic ions *in vitro*, the evidence for direct action *in vivo* is very slim. There are outstanding exceptions in the case of NO_3^- and some micronutrients. These ions regulate the synthesis of specific enzyme systems for which the ions are substrate or cofactors. However, the majority of responses of carbon metabolism to inorganic ions discussed here involve an indirect interaction. That is, both fine control and coarse control of carbon metabolism in response to inorganic ions rarely involve the ions as effectors in the way they appear to function *in vitro*. Fine control of carboxylic acid metabolism, for example, appears to involve the balance of ion uptake, the removal of feedback inhibition by transport of carboxylates from the metabolic compartment, and the effects of coupled changes in cytoplasmic H^+ and OH^- concentration (Chap. *12*). Coarse control of carbon metabolism leading to the induction of enzyme systems in response to the ionic environment also appears to be an indirect response to ions but a direct response to internal osmotic requirement.

The discussion has revealed some important differences, as well as some similarities between the way metabolism in higher plants and lower organisms responds to inorganic ions. When ionic effects on isolated enzymes are considered, it is clear that halophytic higher plants have adopted a strategy very different from that of halophilic bacteria. The halophytes appear to rely on internal regulation of ion concentration in the metabolic compartment rather than the synthesis of enzyme systems which function effectively in the presence of high salt concentrations. Intracellular regulation of ionic activity in different plant species is a largely unknown quantity. It raises problems of osmoregulation in the cytoplasm and in this respect higher plants and the halophytic algae show some evidence for specific metabolic responses involving the synthesis of compatible solutes. Further understanding of the significance of intracellular regulation of ion activity will depend on improved techniques for measuring ionic concentration in different cell compartments. Only then will it be possible to make adequate assessment of the interactions between ion absorption and carbon metabolism, particularly in response to salinity and other forms of ionic stress.

References

AITKEN, D.M., BROWN, A.D.: Properties of halophil nicotinamide-adenine dinucleotide phosphate—specific isocitrate dehydrogenase. True Michaelis constants, reaction mechanisms and molecular weights. Biochem. J. **130**, 645–662 (1972).

AUSTENFELD, F.-A.: Der Einfluß des NaCl und anderer Alkalisalze auf die Nitratreduktaseaktivität von *Salicornia europaea*. L. Z. Pflanzenphysiol. **71**, 288–296 (1974).

BEEVERS, L., HAGEMAN, R.H.: Nitrate reduction in higher plants. Ann. Rev. Plant Physiol. **20**, 495–522 (1969).

BEN-AMOTZ, A., AVRON, M.: The role of glycerol in the osmotic regulation of the halophilic alga *Dunaliella parva*. Plant Physiol. **51**, 875–878 (1973).

BEN-ZIONI, A., LIPS, S.H., VAADIA, Y.: Correlations between nitrate reduction, protein synthesis and malate accumulation. Physiol. Plant. **23**, 1039–1047 (1970).

BEN-ZIONI, A., VAADIA, Y., LIPS, S.H.: Nitrate uptake by roots as regulated by nitrate reduction products of the shoots. Physiol. Plant. **24**, 288–290 (1971).

BOROWITZKA, L.J., BROWN, A.D.: The salt relations of marine and halophilic species of the unicellular green alga *Dunaliella*. The role of glycerol as a compatible solute. Arch. Microbiol. **96**, 37–52 (1974).

BROWN, A.D.: Aspects of bacterial response to the ionic environment. Bacteriol. Rev. **28**, 269–329 (1964).

BROWN, A.D., SIMPSON, J.R.: Water relations of sugar tolerant yeasts: the role of intracellular polyols. J. Gen. Microbiol. **72**, 589–591 (1972).

BROWNELL, P.F., CROSSLAND, C.J.: The requirement for sodium as a micronutrient by species having the C_4 dicarboxylic photosynthetic pathway. Plant Physiol. **49**, 794–797 (1972).

BURSTRÖM, H.: Studies on the buffer systems of cells. Arch. Botanik **32** (7), 1–18 (1945).

CAMPBELL, L.C., PITMAN, M.G.: Salinity and plant cells. In: Salinity and water use (T. TALSMA, J.R. PHILIP, eds.), p. 207–223. London: MacMillan 1971.

CHANG, C.C., BEEVERS, H.: Biogenesis of oxalate in plant tissues. Plant Physiol. **43**, 1821–1828 (1968).

CHRISTIAN, J.H.B., HALL, J.M.: Water relations of *Salmonella oranieburg*: accumulation of potassium and amino acids during respiration. J. Gen. Microbiol. **70**, 497–506 (1972).

CLARK, H.E.: Effect of NH_4^+ and NO_3^- nitrogen on the composition of the tomato plant. Plant Physiol. **11**, 5–24 (1936).

COIC, Y., LESAINT, C., LE ROUX, F.: Effets de la nature ammoniacale ou nitrique de l'alimentation azotée et du changement de la nature de cette alimentation sur le metabolisme des anions et cations chez la tomate. Ann. Physiol. Vegetale **4**, 117–125 (1962).

CRAM, W.J.: Compartmentation and exchange of chloride in carrot root tissue. Biochim. Biophys. Acta. **163**, 339–353 (1968).

CRAM, W.J.: Chloride fluxes in cells of isolated root cortex of *Zea mays*. Australian J. Biol. Sci. **26**, 757–779 (1973).

CRAM, W.J.: Effects of Cl^- on HCO_3^- and malate fluxes and CO_2 fixation in carrot and barley root cells. J. Exptl. Bot. **25**, 11–27 (1974).

CRAM, W.J., LATIES, G.G.: The kinetics of bicarbonate and malate exchange in carrot and barley root cells. J. Exptl. Bot. **25**, 253–268 (1974).

DANNER, J., TING, I.P.: CO_2 metabolism of corn roots II. Intracellular distribution of enzymes. Plant Physiol. **42**, 719–724 (1967).

DAVIES, D.D.: Control of and by pH. In: Symp. Soc. Exp. Biol. **27**, 513–529 (1973).

DIJKSHOORN, W.: Nitrate accumulation, nitrogen balance and cation-anion ratio during the regrowth of perennial rye grass. Ned. J. Agric. Sci. **6**, 211–221 (1958).

DIJKSHOORN, W.: Metabolic regulation of the alkaline effect of nitrate utilisation in plants. Nature **194**, 165–167 (1962).

DIJKSHOORN, W., LATHWELL, D.J., DE WIT, C.T.: Temporal changes in carboxylate content of rye-grass with stepwise change in nutrition. Plant Soil **29**, 369–390 (1968).

DIXON, M., WEBB, E.C.: Enzymes. pp. 950 2nd ed. London: Longmans 1964.

EGMOND, F. VAN, HOUBA, V.J.G.: Production of carboxylates (C–A) by young sugar beet plants grown in nutrient solution. Ned. J. Agric. Sci. **18**, 182–187 (1970).

EVANS, H.J., SORGER, G.T.: Role of mineral elements with emphasis on the univalent cations. Ann. Rev. Plant Physiol. **17**, 47–76 (1966).

FERGUSON, A.R.: The nitrogen metabolism of *Spirodela oligorrhiza* II. Control of the enzymes of nitrate assimilation. Planta **88**, 353–363 (1969).

FERRARI, T.E., YODDER, O.C., FILNER, P.: Anaerobic nitrate production by plant cells and tissues: evidence for two nitrate pools. Plant Physiol. **51**, 423–431 (1973).

FLOWERS, T.J.: Salt tolerance in *Suaeda maritima* L. (Dum). The effect of sodium chloride on growth, respiration and soluble enzymes in a comparative study with *Pisum*. J. Exptl. Bot. **23**, 310–321 (1972a).

FLOWERS, T.J.: The effect of sodium chloride on enzyme activities from four halophyte species of Chenopodiaceae. Phytochemistry **11**, 1881–1886 (1972b).

GINZBURG, M., SACHS, L., GINZBURG, B.Z.: Ion metabolism in *Halobacterium* II. Ion concentrations in cells at different levels of metabolism. J. Membrane Biol. **5**, 78–101 (1971).

GREENWAY, H.: Plant response to saline substrates VII. Growth and ion uptake throughout plant development in two varieties of *Hordeum vulgare*. Australian J. Biol. Sci. **18**, 763–779 (1965).

GREENWAY, H., OSMOND, C.B.: Salt responses of enzymes from species differing in salt tolerance. Plant Physiol. **49**, 256–259 (1972).

GREENWAY, H., SIMS, A.P.: Ionic regulation of malic enzyme from the halophyte *Triglochin maritima*. Australian J. Plant Physiol. **1**, 15–29 (1974).

GUTKNECHT, J., DAINTY, J.: Ionic relationships of marine algae. Oceanogr. Marine Biol. Ann. Rev. **6**, 163–200 (1969).

HALL, J.L., FLOWERS, T.J.: The effect of salt on protein synthesis in the halophyte *Suaeda maritima*. Planta **110**, 361–368 (1973).

HANSON-PORATH, E., POLJAKOFF-MAYBEER, A.: The effect of salinity on the malic dehydrogenase of pea roots. Plant Physiol. **44**, 1031–1034 (1969).

HIATT, A.J.: Formic acid activation in plants II. Activation of formyltetrahydrofolate synthetase by magnesium, potassium and other univalent cations. Plant Physiol. **40**, 189–193 (1965).

HIATT, A.J.: Reactions *in vitro* of enzymes involved in CO_2 fixation accompanying salt uptake by barley roots. Z. Pflanzenphysiol. **56**, 233–245 (1967a).

HIATT, A.J.: Relationship of cell sap pH to organic acid change during ion uptake. Plant Physiol. **42**, 294–298 (1967b).

HIATT, A.J., EVANS, H.J.: Influence of salts on activity of malic dehydrogenase from spinach leaves. Plant Physiol. **35**, 662–672 (1960).

HIATT, A.J., HENDRICKS, S.B.: The role of CO_2 fixation in accumulation of ions by barley roots. Z. Pflanzenphysiol. **56**, 220–232 (1967).

HOCHACHKA, P.W., SOMERO, G.N.: Strategies of Biochemical Adaptation, 358 pp. Philadelphia: W.B. Saunders 1973.

HOUBA, V.J.G., EGMOND, F., VAN, WITTICH, E.M.: Changes in production of organic nitrogen and carboxylates (C–A) in young sugar-beet plants grown in nutrient solutions of different nitrogen composition. Ned. J. Agric. Sci. **19**, 39–47 (1971).

HUBER, W., RUSTAGI, P.N., SANKHLA, N.: Ecophysiological studies on Indian arid zone plants III. Effect of sodium chloride and gibberellin on the activity of the enzyme of carbohydrate metabolism in leaves of *Pennisetum typhoides*. Oecologia **15**, 77–86 (1974).

HUBER, W., SANKHLA, N.: Ecophysiological studies on Indian arid zone plants II. Effect of salinity and gibberellin on the activity of the enzymes of amino acid metabolism in leaves of *Pennisetum typhoides*. Oecologia **13**, 271–277 (1973).

HURD, R.G.: The effect of pH and bicarbonate ions on the uptake of salts by discs of reed beet. J. Exptl. Bot. **9**, 159–174 (1958).

HURD, R.G.: An effect of pH and bicarbonate on salt accumulation by discs of storage tissues. J. Exptl. Bot. **10**, 345–358 (1959).

JACKSON, P.C., ADAMS, H.R.: Cation-anion balance during potassium and sodium absorption by barley roots. J. Gen. Physiol. **46**, 369–386 (1963).

JACKSON, W.A., FLESHER, D., HAGEMAN, R.H.: Nitrate uptake by dark grown corn seedlings: some characteristics of apparent induction. Plant Physiol. **51**, 120–127 (1973).

JACOBSON, L.: CO_2 fixation and ion absorption in barley roots. Plant Physiol. **30**, 264–269 (1955).

JACOBSON, L., ORDIN, L.: Organic acid metabolism and ion absorption in roots. Plant Physiol. **29**, 70–74 (1954).

JACOBY, B., LATIES, G.G.: Bicarbonate fixation and malate synthesis in relation to salt-induced stoichiometric synthesis of organic acid. Plant Physiol. **47**, 525–531 (1971).

JEFFERIES, R.L.: Aspects of salt-marsh ecology with particular reference to inorganic nutrition. In: The estuarine environment (R.S.K. BARNES, J. GREEN, eds.), p. 61–85. London: Applied Science Publishers 1972.

JEFFERIES, R.L.: The ionic relations of seedlings of the halophyte *Triglochin maritima* L. In: Mineral absorption by plants (W.P. ANDERSON, ed.), p. 297–321. London: Academic Press 1973.

JEFFERIES, R.L., LAYCOCK, D., STEWART, G.R., SIMS, A.P.: The phenotypic modifications of the properties of enzymes in response to changing ionic environments. In: Ecological aspects of the mineral nutrition of plants (I.H. RORISON, ed.), p. 281–307. Oxford: Blackwells 1969.

JOHNSON, H.S., HATCH, M.D.: The C_4 dicarboxylic acid pathway of photosynthesis. Identification of intermediates and products and quantitative evidence for the route of carbon flow. Biochem. J. **114**, 127–134 (1969).

JOHNSON, H.S., HATCH, M.D.: Properties and regulation of leaf nicotinamideadenine dinucleotide phosphate-malate dehydrogenase and "malic" enzyme in plants with the C_4 dicarboxylic acid pathway of photosynthesis. Biochem. J. **119**, 273–280 (1970).

JOY, K.W.: Accumulation of oxalate in tissues of sugar-beet and the effect of nitrogen supply. Ann. Bot. (London), N.S., **28**, 689–701 (1964)

KAUSS, H.: Isofloridosid und Osmoregulation bei *Ochromonas malhamensis*. Z. Pflanzenphysiol. **56**, 453–465 (1967).

KHOLDEBARIN, B., OERTLI, J.J.: The effect of tris-buffer on salt uptake and organic acid synthesis by leaf tissues under light and dark conditions. Z. Pflanzenphysiol. **62**, 231–236 (1970a).

KHOLDEBARIN, B., OERTLI, J.J.: Changes of organic acids during salt uptake by leaf tissues under light and dark conditions. Z. Pflanzenphysiol. **62**, 237–244 (1970b).

KIRKBY, E.A.: Ion uptake and ionic balance in plants in relation to the form of nitrogen nutrition. In: Ecological aspects of the mineral nutrition of plants (I.H. RORISON, ed.), p. 215–235. Oxford: Blackwells 1969.

KLUGE, M., HEININGER, B.: Untersuchungen über den Efflux von Malat aus den Vacuolen der assimilierenden Zellen von *Bryophyllum* und mögliche Einflüsse dieses Vorganges auf den CAM. Planta **113**, 333–343 (1973).

KLUGE, M., OSMOND, C.B.: Studies on phosphoenolpyruvate carboxylase and other enzymes of Crassulacean acid metabolism in *Bryophyllum tubiflorum* and *Sedum praealtum*. Z. Pflanzenphysiol. **66**, 97–105 (1972).

LARSEN, H.: Halophilism. In: The bacteria, a treatise on structure and function IV. The physiology of growth (I.C. GUNSALUS and R.Y. STAINER, eds.), p. 297–342. New York: Academic Press 1962.

LIPS, S.H., BEEVERS, H.: Compartmentation of organic acids in corn roots I. Differential labeling of 2 malate pools. Plant Physiol. **41**, 709–712 (1966a).

LIPS, S.H., BEEVERS, H.: Compartmentation of organic acids in corn roots II. The cytoplasmic pool of malic acid. Plant Physiol. **41**, 713–717 (1966b).

LUNDEGÅRDH, H.: Absorption, transport and exudation of inorganic ions by the roots. Arch. Botanik **32**, (12) 1–139 (1945).

LÜTTGE, U., BALL, E.: Proton and malate fluxes in cells of *Bryophyllum daigremontianum* leaf slices in relation to potential osmotic pressure of the medium. Z. Pflanzenphysiol. **73**, 326–338 (1974a).

LÜTTGE, U., BALL, E.: Mineral ion fluxes in slices of acidified and deacidified leaves of the CAM plant *Bryophyllum daigremontianum*. Z. Pflanzenphysiol. **73**, 339–348 (1974b).

MACLENNAN, D.H., BEEVERS, H., HARLEY, J.L.: Compartmentation of acids in plant tissues. Biochem. J. **89**, 316–327 (1963).

MILLER, G., EVANS, H.J.: The influence of salts on pyruvate kinase from tissues of higher plants. Plant Physiol. **32**, 346–354 (1957).

MILLERD, A., MORTON, R.K., WELLS, J.R.E.: Role of isocitrate lyase in synthesis of oxalic acid in plants. Nature **196**, 955–956 (1962).

MINOTTI, P.L., JACKSON, W.A.: Nitrate reduction in roots and shoots of wheat seedlings. Planta **95**, 36–44 (1970).

MINOTTI, P.L., WILLIAMS, D.G., JACKSON, W.A.: Nitrate uptake and reduction as affected by calcium and potassium. Soil. Sci. Soc. Amer. Proc. **32**, 692–698 (1968).

MINOTTI, P.L., WILLIAMS, D.G., JACKSON, W.A.: Nitrate uptake by wheat as influenced by ammonium and other cations. Crop Sci. **9**, 9–14 (1969a).

MINOTTI, P.I., WILLIAMS, D.G., JACKSON, W.A.: The influence of ammonium on nitrate reduction in wheat seedlings. Planta **86**, 267–271 (1969b).

MUKERJI, S.K., TING, I.P.: Phosphoenolpyruvate carboxylase isoenzymes: separation and properties of three forms from cotton leaf tissue. Arch. Biochem. Biophys. **143**, 297–317 (1971).

MUTO, S., URITANI, I.: Glucose-6-phosphate dehydrogenase from sweet potato: effect of various ions and their ionic strength on enzyme activity. Plant Cell Physiol. **13**, 111–118 (1972).

NIERHAUS, D., KINZEL, H.: Vergleichende Untersuchungen über die organischen Säuren in Blättern höherer Pflanzen. Z. Pflanzenphysiol. **64**, 107–123 (1971).

NITSOS, R.E., EVANS, H.J.: Effects of univalent cations on the activity of particulate starch synthetase. Plant Physiol. **44**, 1260–1266 (1969).

OERTLI, J.J.: Extra cellular salt accumulation, a possible mechanism of salt injury in plants. Agrochimica **12**, 461–469 (1968).

OSMOND, C.B.: Acid metabolism in *Atriplex*. I. Regulation of oxalate synthesis by the apparent excess cation absorption in leaf tissue. Australian J. Biol. Sci. **20**, 575–587 (1967).

OSMOND, C.B.: Ion absorption in *Atriplex* leaf tissue. I. Absorption by mesophyll cells. Australian J. Biol. Sci. **21**, 1119–1130 (1968).

OSMOND, C.B.: Metabolite transport in C_4 photosynthesis. Australian J. Biol. Sci. **24**, 159–163 (1971).

OSMOND, C.B., AVADHANI, P.N.: Acid metabolism in *Atriplex*. II. Oxalate synthesis during acid metabolism in the dark. Australian J. Biol. Sci. **21**, 917–927 (1968).

OSMOND, C.B., GREENWAY, H.: Salt responses of carboxylation enzymes from species differing in salt tolerance. Plant Physiol. **49**, 260–263 (1972).

OSMOND, C.B., LATIES, G.G.: Compartmentation of malate in relation to ion absorption in beet. Plant Physiol. **44**, 7–14 (1969).

PARR, J.F., NORMAN, A.G.: pH control in nitrate uptake studies with excised roots. Plant Soil **21**, 185–190 (1964).

PIERCE, E.C., APPLEMAN, C.O.: Role of ether soluble organic acids in the cation-anion balance in plants. Plant Physiol. **18**, 224–238 (1943).

PITMAN, M.G.: The determination of the salt relations of the cytoplasmic phase in cells of beetroot tissue. Australian J. Biol. Sci. **16**, 647–660 (1963).

PITMAN, M.G.: The effect of divalent cations on the uptake of salt by beetroot tissue. J. Exptl. Bot. **15**, 444–456 (1964).

PITMAN, M.G.: Active H^+ efflux from cells of low-salt barley roots during salt accumulation. Plant Physiol. **45**, 787–790 (1970).

PITMAN, M.G.: Uptake and transport of ions in barley seedlings I. Estimation of chloride fluxes in cells of excised roots. Australian J. Biol. Sci. **24**, 407–421 (1971).

PITMAN, M.G., MOWAT, J., NAIR, H.: Interactions of processes for accumulation of salt and sugar in barley plants. Australian J. Biol. Sci. **24**, 619–631 (1971).

PITMAN, M.G., SADDLER, H.D.W.: Active sodium and potassium transport in cells of barley roots. Proc. Natl. Acad. Sci. U.S.A. **57**, 44–49 (1967).

PLAUT, Z.: Nitrate reductase activity in wheat seedlings during exposure to and recovery from water stress and salinity. Physiol. Plant. **30**, 212–217 (1974).

PORATH, E., POLJAKOFF-MAYBER, A.: Effect of salinity on metabolic pathways in pea root tips. Israel J. Bot. **13**, 115–121 (1964).

PORATH, E., POLJAKOFF-MAYBER, A.: The effect of salinity in the growth medium on carbohydrate metabolism in pea roots. Plant Cell Physiol. **9**, 195–203 (1968).

RANDALL, P.J., BOUMA, D.: Zinc deficiency, carbonic anhydrase and photosynthesis in leaves of spinach. Plant Physiol. **52**, 229–232 (1973).

RANSON, S.L., THOMAS, M.: Crassulacean acid metabolism. Ann. Rev. Plant Physiol. **11**, 81–110 (1960).

RAVEN, J.A.: Exogenous inorganic carbon sources in plant photosynthesis. Biol. Rev. Cambridge Phil. Soc. **45**, 167–221 (1970).

RAVEN, J.A., SMITH, F.A.: Significance of hydrogen ion transport in plant cells. Canad. J. Bot. **52**, 1035–1048 (1974).

RICHARDSON, K.E., TOLBERT, N.E.: Oxidation of glycolic acid to oxalic acid by glycolic acid oxidase. J. Biol. Chem. **236**, 1280–1284 (1961).

SANTARIUS, K.A.: Der Einfluß von Elektrolyten auf Chloroplasten beim Gefrieren und Trocknen. Planta **89**, 23–46 (1969).

SINGH, T.N., ASPINALL, D., PALEG, L.G., BOGGESS, S.F.: Stress metabolism I. Nitrogen metabolism and growth in the barley plant during water stress. Australian J. Biol. **26**, 45–56 (1973).

SMITH, F.A.: The internal control of nitrate uptake into excised barley roots with differing salt contents. New Phytologist **72**, 769–782 (1973).

SMITH, F.W., THOMPSON, J.F.: Regulation of nitrate reductase in excised barley roots. Plant Physiol. **48**, 219–223 (1971).

SPLITTSTOESSER, W.E.: Synthesis of malate by carrot and beet root tissues incubated with bicarbonate-^{14}C and distribution of ^{14}C within the malate molecule. Proc. Amer. Soc. Hort. Sci. **90**, 235–238 (1967).

SPLITTSTOESSER, W.E., BEEVERS, H.: Acids in storage tissues. Effects of salts and aging. Plant Physiol. **39**, 163–169 (1964).

STEER, B.T., BEEVERS, H.: Compartmentation of organic acids in corn roots. III. Utilisation of exogenously supplied acids. Plant Physiol. **42**, 1197–1201 (1967).

STEVENINCK, R.F.M. VAN: Some metabolic implications of the tris effect in beetroot tissue. Australian J. Biol. Sci. **19**, 271–281 (1966).

SUTTON, B.G., OSMOND, C.B.: Dark fixation of CO_2 by Crassulacean plants. Evidence for a single carboxylation step. Plant Physiol. **50**, 360–365 (1972).

TEMPEST, D.W., MEERS, J.L., BROWN, C.M.: Influence of environment on the content and composition of microbial free amino acids pools. J. Gen. Microbiol. **64**, 171–185 (1970).
TING, I.P.: CO_2 metabolism in corn roots III. Inhibition of P-enolpyruvate carboxylase by L-malate. Plant Physiol. **43**, 1919–1924 (1968).
TING, I.P.: Nonautotrophic CO_2 fixation. In: Photosynthesis and photorespiration (M.D. HATCH, C.B. OSMOND, R.O. SLAYTER, eds.), p. 169–185. New York: Wiley Interscience 1971.
TING, I.P., DUGGER, W.M.: Transhydrogenation in root tissue: mediation by carbon dioxide. Science **150**, 1727–1728 (1965).
TING, I.P., DUGGER, W.M.: CO_2 fixation in *Opuntia* roots. Plant Physiol. **41**, 500–505 (1966).
TING, I.P., OSMOND, C.B.: Photosynthetic phosphoenolpyruvate carboxylases characteristics of alloenzymes from leaves of C_3 and C_4 plants. Plant Physiol. **51**, 439–447 (1973a).
TING, I.P., OSMOND, C.B.: Multiple forms of plant phosphoenolpyruvate carboxylase associated with different metabolic pathways. Plant Physiol. **51**, 448–453 (1973b).
TORII, K., LATIES, G.G.: Organic acid synthesis in response to excess cation absorption in vacuolate and non-vacuolate sections of corn and barley roots. Plant Cell Physiol. **7**, 395–403 (1966).
TREICHEL, S.P., KIRST, G.O., WILLERT, D.J., VON: Veränderung der Aktivität der Phosphoenol-pyruvat-Carboxylase durch NaCl bei Halophyten verschiedener Biotope. Z. Pflanzenphysiol. **71**, 437–449 (1974).
ULRICH, A.: Metabolism of non-volatile organic acids in excised barley roots as relation to cation-anion balance during salt accumulation. Amer. J. Bot. **28**, 526–537 (1941).
ULRICH, A.: Metabolism of organic acids in excised barley roots as influenced by temperature oxygen tension and salt concentration. Amer. J. Bot. **29**, 220–227 (1942).
USUDA, H., SAMEJIMA, M., MIYACHI, S.: Distribution of radioactivity in carbon atoms of malic acid formed during light-enhanced dark CO_2 fixation in maize leaves. Plant Cell Physiol. **14**, 423–426 (1973).
WALKER, D.A.: Pyruvate carboxylation and plant metabolism. Biol. Rev. Cambridge Phil. Soc. **37**, 215–256 (1962).
WALLACE, W., PATE, J.S.: Nitrate reductase in the field pea (*Pisum arvense* L.). Ann. Bot. (London), N.S., **29**, 655–671 (1965).
WALLACE, W., PATE, J.S.: Nitrate assimilation in higher plants with special reference to the cocklebur (*Xanthium pennsylvanicum* Wallr.). Ann. Bot. (London), N.S., **31**, 213–228 (1967).
WEBB, K.L., BURLEY, J.W.A.: Dark fixation of $^{14}CO_2$ by obligate and faculative salt marsh halophytes. Canad. J. Bot. **43**, 281–285 (1965).
WEGMANN, K.: Osmotic regulation of photosynthetic glycerol production in *Dunaliella*. Biochim. Biophys. Acta. **234**, 317–323 (1971).
WEIMBERG, R.: Effect of sodium chloride on the activity of a soluble malate dehydrogenase from pea seeds. J. Biol. Chem. **242**, 3000–3006 (1967).
WEIMBERG, R.: An electrophoretic analysis of the isozymes of malate dehydrogenase in several plants. Plant Physiol. **43**, 622–628 (1968).
WEIMBERG, R.: Enzyme levels in pea seedlings grown on highly saline media. Plant Physiol. **46**, 466–470 (1970).
WILLERT, D.J. VON: Einfluß von NaCl und Blattalter auf die CO_2 Dunkelfixierung bei Halophyten — vergleichende Untersuchungen bei *Mesembryanthemum crystallinum* und *Atriplex spongiosa*. Z. Pflanzenphysiol. **72**, 62–74 (1974).
WINTER, K.: Zum Problem der Ausbildung des Crassulaceensäurestoffwechsels bei *Mesembryanthemum crystallinum* unter NaCl-Einfluß. Planta **109**, 135–145 (1973a).
WINTER, K.: CO_2-Fixierungsreaktionen bei der Salzpflanze *Mesembryanthemum crystallinum* unter variierten Außenbedingungen. Planta **114**, 75–85 (1973b).
WINTER, K., WILLERT, D.J. VON: NaCl induzierter Crassulaceensäurestoffwechsel bei *Mesembryanthemum crystallinum*. Z. Pflanzenphysiol. **66**, 166–170 (1972).
WIT, C.T. DE, DIJKSHOORN, W., NOGGLE, J.C.: Ionic balance and growth of plants. Verslag. Landbouwk. Onderzoek. **69**, 15, p. 1–68 (1963).
WOOLHOUSE, H.: Differences in the properties of the acid phosphatases of plant roots and their significance in the evolution of edaphic ecotypes. In: Ecological aspects of the mineral nutrition of plants (I.H. RORISON, ed.), p. 357–380. London: Blackwells 1969.

Author Index

Page numbers *in italics* refer to the bibliography

Symbols, Units, and Abbreviations

Symbols

Some symbols are defined, and used, only in parts of particular Chapters. If no definition is given then the symbols have the following meanings.

Symbol	Description	Unit
A	area	m^2
a	activity	mM
b	partition coefficient	
C	capacity	F
c	concentrations	mM ($=$mol m^{-3})
$\bar{c}$	mean concentration	mM ($=$mol m^{-3})
D	diffusion coefficient	$m^2\ s^{-1}$
d	diameter	m
G	Gibbs free energy	J
g	gravitational force (centrifuge). Note 1,000 g often written as 1 K (Part A, Ch 10)	
g	membrane conductance	$S\ m^{-2}$
H	enthalpy	J
I	current	A
J	generalized flow ($L \cdot X$)	
J_j	net flux, $=\phi_{oi,j}-\phi_{io,j}$	flux
J_s	solute flux	e.g. mol $m^{-2}\ s^{-1}$
J_v	volume flow	$m\ s^{-1}$
K_i	inhibitor constant	mM
K_m	Michaelis-Menten constant	mM
k	rate constant	s^{-1}
L	generalized conductance coefficient, i.e. (J/X)	
L_p	hydraulic conductivity	$m\ s^{-1}\ Pa^{-1}$
l	length	m
ln	logarithm to base e	
log	logarithm to base 10	
n	number of moles	
P	hydrostatic pressure	Pa
P	permeability	$m\ s^{-1}$
P_K	permeability to K^+ (etc.)	
P_w	diffusion permeability to water	

Q	quantity of substance, location or ion shown by subscript	
Q^*	quantity of isotope	
q	electrical charge	C
R	electrical resistance	Ω
R	relative water content (Part A, Ch. *2*) (% of content at full turgor)	
r	electrical resistance	Ω
	or resistivity of membrane	$\Omega\,m^2$
r	radius	m
s	specific radio-activity	(counts) $mol^{-1}\,s^{-1}$
T	temperature	K or °C
t	time	s
u	electric mobility	$m^2\,V^{-1}\,s^{-1}$
$\bar{V}$	partial molar volume	$m^3\,mol^{-1}$
V	volume	
V_{max}	maximum reaction velocity; Michaelis-Menten enzyme kinetics	$mol\,s^{-1}$
v	velocity of reaction	$mol\,s^{-1}$
W	weight	
x	mole fraction	
X	generalized force	
z	valency	
α, β, γ	permeability coefficient ratios	
γ	activity coefficient	
ε	elastic coefficient (Part A, Chap. *2*)	
ε	electric permittivity	
μ	chemical potential	$J\,mol^{-1}$
μ_w^o	standard state of μ_w	$J\,mol^{-1}$
$\bar{\mu}$	electrochemical potential	$J\,mol^{-1}$
$\bar{\mu}_{vo}$	difference in $\bar{\mu}$ between v and o	$J\,mol^{-1}$
π	osmotic pressure, or potential	Pa or $J\,m^{-3}$
σ	reflection coefficient	
τ	transport number	
τ	matric potential (Part A, Chap. *2*)	Pa or $J\,m^{-3}$
Φ	net flux based on radioactive measurements[1] eg:—	$mol\,m^{-2}\,s^{-1}$ or $mol\,g_{FW}^{-1}\,s^{-1}$ etc.
Φ_{ov}, Φ_{in}	net tracer uptake to cell or tissue relative to s_o	
Φ_{vo}, Φ_{out}	net tracer efflux from cell or tissue relative to s_v	
Φ_{ox}	net tracer transport from solution to xylem, relative to s_o	

[1] The net flux Φ can be expressed in terms of the fluxes (ϕ) at each membrane (see Part A, Chap. *5*), and Φ may thus be described as a "complex" flux in terms of the "simple" fluxes (ϕ). The usual result of experiments using tracers to measure uptake is Φ which should not be taken as equivalent to a ϕ except in certain well defined conditions.

Φ^*	net tracer uptake, or efflux ($\Phi \cdot s$)	
ϕ	unidirectional flux	mol m^{-2} s^{-1}
	or	mol g_{FW}^{-1} s^{-1} etc.
ϕ_{oc}	subscripts show direction of fluxes	
$\phi_{oc,K}$	specifies the ion, and direction	
Ψ	water potential	Pa or J m^{-3}
ψ	Electrical potential difference	V
ψ_{vo}	ψ of vacuole relative to outside	
ψ_K	equilibrium potential for K^+ (etc.) between two phases	
ω	permeability coefficient	
[ ]	denotes concentration	
Δ	difference in	
∇	gradient operator $\left(=\frac{d}{dx}+\frac{d}{dy}+\frac{d}{dz}\right)$	

Subscript Labels

Subscript labels used with symbols have the following general meanings. Chemical symbols (K, Cl) when used as subscripts specify the particular ion. The order of labels shows direction, as with unidirectional fluxes.

Label	Description
a, b	two arbitrary phases in contact
c	cytoplasmic
ch	refers to a chemical reaction
d	diffusion (as in P_d)
i	inside
j	j^{th} species
M	membrane
max	maximum level reached (eg. V_{max})
o	outside, solution
p	4th component of isotope exchange model (Part A, Chap. 5)
s,S	solute (e.g. c_s)
T	total (Part A, Chap. 5)
v	vacuole
w	water
x	xylem

Constants

Values of certain constants are:

Constant	Value
F	Faraday constant $=96490$ C mol^{-1}

R	gas constant 8.314 J mol^{-1} K^{-1}
RT/F	has the value 25.7 mV at 25° C.

Units

The International System of Units (SI) is used in this book, but with retention of the terms mM and μM for concentrations, as these have direct equivalents in the SI system (mM=mol m^{-3}; μM=mmol m^{-3}). The unit mol m^{-3} is used where it is more appropriate to express content relative to volume (e.g. stomata; Part A, Chap. *10*). Where possible fluxes are expressed in meter/second units or where referred to amount of tissue, relative to gram fresh weight or dry weight. Note that 1 pmol cm^{-2} s^{-1}=10 nmol m^{-2} s^{-1}; 1 mmol kg^{-1} h^{-1}=1 μmol g^{-1} h^{-1} =0.278 nmol g^{-1} s^{-1}. 1 μl=1 mm^3=10^{-9} m^3.

Prefixes:

Standard prefixes are used to all units, e.g.

M	mega	10^3
c	centi	10^{-2}
m	milli	10^{-3}
μ	micro	10^{-6}
n	nano	10^{-9}
p	pico	10^{-12}

Units:

Unit	**Description**
A	ampere; electrical current
bar	unit of pressure; 100 kPa or 0.987 atmospheres
C	coulomb; electrical charge
Ci	curie; radioactivity ($3.7 \cdot 10^{10}$ s^{-1})
d	day
°C	degrees Celsius; temperature
eq	equivalent; mol × z
F	farad; electrical capacitance
g	gram
g_{FW}	gram fresh weight
g_{DW}	gram dry weight
Hz	hertz; frequency
h	hour
ha	hectare (10^4 m^2)
J	joule (N m)
K	kelvin (degrees)
lx	lux; illumination (lm m^{-2})
M	molar; used mainly as (mM) as this is equivalent to (mol m^{-3}), the S.I. unit

m	meter
min	minute; time
mol	mole; amount of substance, ion or compound, whose mass in grams equals the molecular weight
N	newton; force
osmol	sum of mole contribution to osmotic pressure irrespective of ionic species. Note osmol m^{-3} = mOsmolar = mOsm
Pa	pascal; pressure ($J\,m^{-3}$; $N\,m^{-2}$)
rad	absorbed dose of ionizing radiation ($10^{-2}\,J\,kg^{-1}$)
S	siemens; electrical conductance
s	second
V	volt; electrical potential difference
W	watt; power ($J\,s^{-1}$)

Abbreviations

ABA	abscisic acid
AC	alternating current
Acetyl CoA	S-acetyl coenzyme A
ACT-D	actinomycin D
ADP	adenosine 5′-diphosphate
AFS	apparent free space
AMP	adenosine 5′-monophosphate
c-AMP	cyclic AMP, adenosine 2′:3′-cyclic monophosphate
AP	action potential
APW	artificial pond water
ATP	adenosine 5′-triphosphate
ATPase	adenosine triphosphatase (Cl-ATPase, K-ATPase show ion required to stimulate activity)
BA	benzyl adenine
pBQ	*p*-benzoquinone
C_3	refers to photosynthetic carbon reduction cycle
C_4	refers to photosynthetic dicarboxylic acid pathway
CAM	refers to crassulacean acid metabolism
CCCP	carbonyl cyanide, *m*-chlorophenyl hydrazone
CDP	cytidine 5′-diphosphate
CMP	cytidine 5′-monophosphate
CHM	cycloheximide
CMU	3′-(4-chlorophenyl)-1′, 1-dimethyl urea
CTP	cytidine 5′-triphosphate
DCCD	N,N′-dicyclohexyl carbodiimide
DCMU	3′-(3,4 dichlorophenyl)-1′, 1-dimethyl urea
2,4-D	2,4-dichlorophenoxyacetic acid
3,5-D	3,5-dichlorophenoxyacetic acid
DFS	Donnan free space
DHAP	dihydroxyacetone phosphate
Dio-9	antibiotic of unknown structure
DMO	5,5-dimethyl-2,4-oxazolidine dione
DNA	desoxyribonucleic acid
DNP	2,4-dinitrophenol
DSPD	disalicylidenepropanediamine
DW	dry weight
EDTA	ethylenediaminetetraacetic acid
EDDHA	ethylenediamine-di(*o*-hydroxy phenylacetate)
EMF	electromotive force
ER	endoplasmic reticulum
r ER	rough ER
s ER	smooth ER
FAD	flavin adenine dinucleotide

FC	fusicoccin
FCCP	carbonyl cyanide, *p*-tri-fluoro-methoxy phenyl hydrazone
FPA	DL-*p*-fluorophenylalanine
FW	fresh weight
GA, GA_3	gibberellic acid
GTP	guanosine 5′-triphosphate
IAA	indole-3-acetic acid
IDP	inosine 5′-diphosphate
IMP	inosine 5′-monophosphate
ITP	inosine-5′-triphosphate
LS	low-salt (roots)
3-0-MG	3-0-methyl-glucose
α-NAA	naphthalene-1-acetic acid
β-NAA	naphthalene-2-acetic acid
NAD	nicotinamide adenine dinucleotide;
$NADH_2$	—, reduced form
NADP	nicotinamide adenine dinucleotide phosphate;
$NADPH_2$	—, reduced form
−ive	negative
OAA	oxaloacetic acid
~P	"high energy" phosphate
P_i	inorganic phosphate
P_r	phytochrome (red light absorbing form = inactive form)
P_{fr}	phytochrome (far-red light absorbing form = active form)
PCMB	*p*-chloromercuribenzoic acid
PCMBS	*p*-chloromercuribenzene sulfonic acid (= *p*-chloromercuriphenyl sulfonic acid)
PD	potential difference
PEP	phosphoenolpyruvate
PEP carboxylase	phosphoenolpyruvate carboxylase
PGA	phosphoglyceric acid
Phe	phenyl alanine
PMS	phenazine methosulfate
ppm	parts per million
PS I	photosystem I
PS II	photosystem II
+ive	positive
Q_{10}	temperature coefficient, ratio of rates at temperatures differing by 10° C
RNA	ribonucleic acid
m-RNA	messenger RNA
t-RNA	transfer RNA
RUDP	D-ribulose-1,5-diphosphate
SCC	short circuit current
SEM	standard error of the mean
SF	senescence factor
sp., spp.	species
TCA	tricarboxylic acid cycle
α-TEG	thioethyl-D-glucopyranoside
tris	tris (hydroxy methyl) aminomethane
poly-U	poly (uridine) nucleotide
UDP	uridine 5′-diphosphate
UMP	uridine 5′-monophosphate
UTP	uridine 5′-triphosphate
UV	ultraviolet light
WFS	water free space

Subject Index

Italic page numbers refer to Part A; roman page numbers refer to Part B.

Encyclopedia of Plant Physiology, New Series

Editors: A. Pirson, M.H. Zimmermann

Volume 1

Transport in Plants I

Phloem Transport

Editors: M.H. Zimmermann, J.A. Milburn

93 figures. XIX, 535 pages. 1975

With contributions by Canny, M.J.P.; Dainty, J.; Dixon, A.F.G.; Eschrich, W.; Fensom, D.S.; Geiger, D.R.; Heyser, W.; Höll, W.; Milburn, J.A.; Nonweiler, T.R.F.; Parthasarathy, M.V.; Pate, J.S.; Peel, A.J.; Sovonick, S.A.; Spanner, D.C.; Tammes, P.M.L.; Tyree, M.T.; Van Die, J.; Ziegler, H.; Zimmermann, M.H.

Contents: Structural Considerations in Phloem Transport. – Nature of Substances in Phloem. – Phloem Transport: Assessment of Evidence. – Possible Mechanisms of Phloem Transport. – Phloem Loading: Storage and Circulation. – Author Index. – Subject Index.

Volume 2 (In two parts)

Transport in Plants II

Part A: Cells

Editors: U. Lüttge, M.G. Pitman

With a foreword by R.N. Robertson.

97 figures. 64 tables. Approx. 440 pages. 1976

With contributions by Cram, W.J.; Dainty, J.; Findlay, G.P.; Hodges, T.K.; Hope, A.B.; Jennings, D.H.; Lüttge, U.; Osmond, C.B.; Pitman, M.G.; Poole, R.J.; Raven, J.A.; Smith, F.A.; Walker, N.A.

Contents: Foreword. – Theoretical and Biophysical Approaches. – Particular Cell Systems. – Regulation, Metabolism and Transport. – Author Index. – Subject Index.

Part B: Tissues and Organs

Editors: U. Lüttge, M.G. Pitman

129 figures. 47 tables. Approx. 480 pages. 1976

With contributions by Anderson, W.P.; Epstein, E.; Hill, A.E.; Hill, B.S.; Hsiao, T.C.; Jeschke, W.D.; Läuchli, A.; Lüttge, U.; Pate, J.; Pitman, M.G.; Schnepf, E.; Spanswick, R.M.; Steveninck, R.F.M. van; Sutcliffe, J.F.

Contents: Pathways of Transport in Tissues. – Particular Tissue Systems. – Control and Regulation of Transport in Tissues and Integration in Whole Plants. – Epilog: Integration. – Author Index. – Subject Index.

Springer-Verlag Berlin Heidelberg New York

MIX
Papier aus verantwortungsvollen Quellen
Paper from responsible sources
FSC® C105338

If you have any concerns about our products,
you can contact us on
ProductSafety@springernature.com

In case Publisher is established outside the EU,
the EU authorized representative is:
Springer Nature Customer Service Center GmbH
Europaplatz 3, 69115 Heidelberg, Germany

Printed by Libri Plureos GmbH
in Hamburg, Germany